POST-MODERN ALGEBRA

PURE AND APPLIED MATHEMATICS

A Wiley-Interscience Series of Texts, Monographs, and Tracts

Founded by RICHARD COURANT
Editor Emeritus: PETER HILTON and HARRY HOCHSTADT
Editors: MYRON B. ALLEN III, DAVID A. COX, PETER LAX,
JOHN TOLAND

A complete list of the titles in this series appears at the end of this volume.

POST-MODERN ALGEBRA

JONATHAN D. H. SMITH

ANNA B. ROMANOWSKA

A Wiley-Interscience Publication

JOHN WILEY & SONS, INC.

New York • Chichester • Weinheim • Brisbane • Singapore • Toronto

This book is printed on acid-free paper. ♾

Published simultaneously in Canada.

For ordering and customer information call, 1-800-CALL-WILEY.

Library of Congress Cataloging in Publication Data:

Smith, J. D. H. (Jonathan D. H.)
Post-modern algebra / Jonathan D. H. Smith and Anna B. Romanowska.
p. cm. -- (Pure and applied mathematics)
"A Wiley-Interscience publication."
Includes index.
ISBN 0-471-12738-8 (cloth : alk. paper)
1. Algebra. 2. I. Romanowska, A. B. (Anna B.) II. Title.
III. Series: Pure and applied mathematics (John Wiley & Sons : Unnumbered)
QA155.S62 1999
512 dc21 9823909

Printed in the United States of America

10 9 8 7 6 5 4 3 2 1

CONTENTS

PREFACE

This book is intended as a graduate-level introduction to algebra. Traditionally, such introductions cover groups, rings, fields, and modules from an abstract, axiomatic standpoint going back to van der Waerden's *Modern Algebra*. Although this tradition has served well for several decades, a number of factors are now rendering it obsolete. On the one hand, its axiomatic approach and its limited choice of topics have misled many into regarding algebra as irrelevant to most applications of mathematics. On the other hand, its preoccupation with historical issues (such as "ruler and compass constructions") has left little time for rich, subtle, and powerful contemporary developments such as universal algebra and category theory. *Post-Modern Algebra* is designed to address these disadvantages. The traditional topics of groups, rings, fields, and modules are accompanied by ordered sets, monoids, monoid actions, quasigroups, loops, lattices, Boolean algebras, categories, and Heyting algebras. Rather than being introduced through abstract axioms, these concepts emerge from generic applications or from natural mathematical considerations. The various structures are unified by the techniques of universal algebra and category theory.

Readers are assumed to have a solid grounding in undergraduate mathematics. The introductory Chapter O is written in an open style that leaves instructors room to tailor a presentation to their students' requirements. In particular, instructors will need to help less well-prepared students with various elementary proofs that are relegated to exercises. Even well-prepared students may need to expend some effort on acquiring fluency in algebraic notation, especially in disabusing themselves of the unfortunate myth that the only proper place for a function is on the left of its argument.

Chapter O presents the basic set-theoretic constructions such as products, disjoint unions, binary relations, and sets of functions. Its final section introduces monoids and semigroups. Chapter I deals with groups and quasigroups, on the basis of monoid and group actions. Groups themselves are introduced formally in the key section I 1.3 on group actions. The elements of group theory are presented in parallel with the elements of quasigroup theory, especially in the many instances where associativity is irrelevant. The final section of Chapter I deals with loops, isotopy, and loop transversals.

Isotopy is an important example showing that the isomorphism concept does not enjoy complete primacy in algebra. The conclusion of the section provides an introduction to coding theory using the loop transversal concept.

Chapter II presents rings, fields, and modules under the title of "Linear Algebra." The key theme is that linear algebra works with abelian groups and homomorphisms where general algebra works with sets and functions. The monoid actions and loop transversals of Chapter I are used in Chapter II. In particular, readers accustomed to seeing coding theory as an application of factorization in polynomial rings over fields should note that Section II.3 takes the opposite approach, using coding theory terminology to consider Euclidean domains.

Chapter III gives an introduction to categories and lattices. These topics are usually treated in a more advanced and abstract fashion. The presentation in Chapter III is governed by the introductory level of the book. The foundational approach adopted is the usual one for mathematics at this level: naive set theory recognizing the set/class distinction (but avoiding explicit treatment of "conglomerates," "Grothendieck universes," etc.). Categories are not presented axiomatically as a new foundation for mathematics, but just as a common generalization of partially ordered sets and monoids. Category theoretical concepts are introduced in parallel with their algebraic applications, for example: products and coproducts with lattices, slice categories with semidirect products of groups, coequalizers with group and module presentations, and adjoint functors together with Galois theory, Heyting algebras, topology and tensor products. The guiding theme, first broached in Section III 1.3, is that algebraic constructions specify initial objects in appropriate categories.

Chapter IV covers the rudiments of universal algebra. At its most basic level, universal algebra generalizes dynamical systems as introduced in Section O 4.2. A dynamical system is a set (the state space) equipped with a single unary operation (the evolution operator), while universal algebra studies sets equipped with many operations of arbitrary (finite) arity. This viewpoint (in Section IV 1.4) then leads to "universal geometry," the polarity between operations and invariant relations that generalizes the polarity between group actions and invariant relations identified by Klein's *Erlanger Programm* as the basis of (classical) geometry. Delving deeper into universal algebra, three equivalent approaches are adopted. The first and most concrete, via classes of algebras satsifying identities, is presented in Section IV.2. The second, via algebraic theories, is presented in Section IV.3. This approach culminates in the major Theorem 3.4.4 that accounts for most of the adjunctions arising from algebraic constructions. The final approach, via the monad concept that extends closure operators from posets to arbitrary categories, is presented in Section IV.4. The chapter concludes with recognition of the equivalence between these three approaches to algebra.

A wide range of exercises is offered throughout the book. At first, the exercises are located immediately following the corresponding block of text.

Later, they are collected at the ends of sections, and their order there does not necessarily reflect their level of difficulty or the order of the relevant material within the section. Apart from simple exercises designed to familiarize readers with the notation and concepts appearing, there are other exercises, of various grades of hardness, designed to give a foretaste of mathematical research. An essential step in the solution of a research problem is to determine which techniques are likely to lead to its solution. The order of the exercises is not intended to preempt such choices. Readers should be warned that there are one or two exercises which lead up blind alleys: Dealing with the frustration that follows apparently wasted effort is also an unavoidable concomitant of research. Some exercises require knowledge of mathematical topics not covered in the text. For example, Exercise I 3.1G (which appears before the formal introduction of the complex numbers as matrices in Exercise II 2.4Q) requires some complex analysis. Several exercises are designed as brief introductions to mathematical topics not otherwise covered in the text. For example, Exercise III 1.1J leads toward domain theory, while Exercises III 3.4L and III 3.4M lead toward algebraic geometry. Instructors with particular interests might wish to elaborate on such exercises.

Much of the material in the text has been used in (American) graduate-level courses at Iowa State University and Warsaw University of Technology. At Iowa State University, Chapters O and I have been covered in the Fall semester, followed by Chapters II and III in the Spring. We are grateful to our students, particularly Robert Haber, Jeehyun Lee, Krzysztof Parzyszek, Clyde Ruby, and Piotr Syrzycki for catching many mistakes in earlier versions of the material. The second author's work was greatly facilitated by the Mathematical Institute, Warsaw University of Technology, in granting several leaves of absence, and by the Mathematics Department of Iowa State University with its continual hospitality. Our special thanks go to Ruth DeBoer for her valiant efforts in typing and retyping several hundred pages of intricate text, and to Lisa Van Horn for guiding the transformation from typed manuscript to printed volume.

JONATHAN D.H. SMITH
ANNA B. ROMANOWSKA

Ames, Iowa
October 1998

O
INTRODUCTION

1. MODERN AND POST-MODERN ALGEBRA

The standard twentieth-century approach to algebra followed the pattern set by van der Waerden's extremely influential 1931 German Springer text *Modern Algebra*. It comprised an abstract, axiomatic coverage of groups, rings, fields, modules, and linear algebra. The approach was "modern" in 1931, being developed especially in Göttingen by Hilbert, Nöther, and others to supersede the late nineteenth-century approach which regarded algebra as (manual) computation with matrices and polynomial equations. There is an interesting parallel with the contemporary development of "modern" architecture by schools such as the Bauhaus, also in Germany. "Modern" architecture set out to replace the elaborate details and ornaments of nineteenth-century architecture with simple, functional forms. Now, although modern architecture became very successful and prevailed for decades, problems eventually emerged to discredit it. (See e.g. Tom Wolfe's *From Bauhaus to Our House*.) The public came to view it as brutal and sterile. Parallel dissatisfaction with modern algebra is becoming apparent: The axiomatic method is boring, and abstraction (which still has its place) is no longer regarded as an end in itself. Architecture has responded to the withering of modernism by moving on to "post-modernism," restoring earlier idioms of detail and ornament, and exploring new forms of construction. Algebra, too, is now moving into a "post-modern" phase. There is renewed interest in explicit computation, reviving nineteenth-century techniques such as invariant theory. At the same time it is becoming clear that the range of algebraic structures covered by "modern algebra" (groups, rings, and modules) is too narrow, and that increasing emphasis will have to be placed on other structures such as ordered sets, monoids, quasigroups, etc.

2. ALGEBRA: THE CENTRAL DISCIPLINE OF MATHEMATICS

Mathematics is widely regarded as the study of the underlying abstract structure that is common to various sciences. One measure of the level of sophistication reached by a particular science is the degree to which it has reduced its problems to mathematical questions. The role of algebra within mathematics is comparable to this role played by mathematics amongst the sciences. Algebra abstracts and studies the structures appearing in various branches of mathematics, and one way in which a branch of mathematics may develop is by reducing its problems to algebraic questions. Perhaps the most highly developed branch of mathematics is number theory, and it is worth noting that the algebraic topics covered by "modern algebra" all emerged from the algebraization of number theory. Part of the driving force behind the movement towards "post-modern algebra" is the extent to which other branches of mathematics have now developed sufficiently to demand study of their own algebraic structures, which do not always coincide with those produced from number theory.

As an elementary example of an abstract algebraic structure common to various branches of mathematics, consider the non-abelian group of order 6. It appears in geometry as the group of symmetries of an equilateral triangle. It appears in combinatorics as the group of permutations of a three-element set. In trigonometry it underlies the relationships between the basic trigonometric functions:

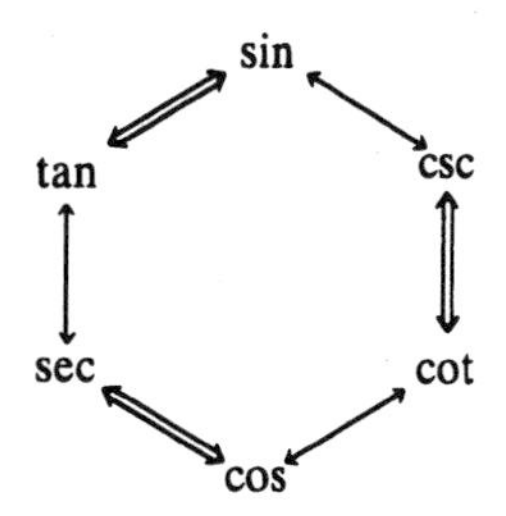

(Here $\leftrightarrow$ denotes an interchange of "opposite" and "hypotenuse," while $\Leftrightarrow$ denotes the interchange of "adjacent" and "hypotenuse." Interchange of "adjacent" and "opposite" applies or removes the prefix "co-.") In matrix theory or linear algebra the group appears as the set of invertible 2×2 matrices, with entries in the two-element unital ring $\mathbb{Z}_2$, under matrix multiplication. It even appears in chemistry as the symmetry group of e.g. the chloroform molecule.

EXERCISE

2. Can you find any other places where the non-abelian group of order 6 appears? In the examples above or elsewhere that you know, can you attach any significance to the various subgroups? (In the trigonometric example, the two-element subgroup interchanging "adjacent" and "opposite" applied or removed the prefix "co-." In the group of symmetries of

the equilateral triangle, what is special about the symmetries forming the three-element subgroup?)

Mastery of algebra makes it easier to study other mathematical disciplines. Familiar structures reappear, while unfamiliar ones are often best approached by the extent to which they differ from similar known ones. A good knowledge of algebraic concepts enables one to avoid excessive computations, and to reduce unavoidable computations to a minimum.

3. SETS WITH STRUCTURE AND SETS WITHOUT STRUCTURE

The previous section characterized algebra externally by its relationship to other branches of mathematics. Algebra can also be characterized internally as the "study of sets with structure." For example, groups are sets endowed with multiplication and inversion operations satisfying certain axioms. From this point of view, the most basic objects of algebraic study are sets themselves. The algebraic properties of sets are not completely trivial, and they often serve as a framework for studying more elaborate algebraic structures. This is particularly striking in linear algebra. A large part of the theory of vector spaces just consists of using the "basis" concept to transfer set constructions to vector spaces. Here are some examples, not all of which may be familiar. Let V be a (say real) vector space with finite basis B, and W a vector space with finite basis C:

(3.1) Any linear transformation $f: V \to W$ is specified uniquely as the extension of a set map $f: B \to W$.

(3.2) A linear transformation $f: V \to W$ injects (is one to one) iff there is a basis D of W extending f(B) such that the set map $f: B \to D$ injects.

(3.3) A linear transformation $f: V \to W$ surjects (is onto) iff a subset of $f(B)$ is a basis for W.

(3.4) A linear transformation $f: V \to W$ is an isomorphism iff $f(B)$ is a basis for W.

(3.5) The vector space W is a subspace of V iff a basis for W is a subset of a basis for V.

(3.6) The direct sum $V \oplus W$ has the disjoint union $B \cup C$ as a basis.

(3.7) The tensor product $V \otimes W$ has the direct product $B \times C$ as a basis.

(3.8) The exterior algebra ΛV has the power set 2^B as a basis.

EXERCISE

3. See which of (3.1)–(3.8) you understand. If you understand one of the statements, then try to prove it. If you don't know what the "disjoint union" in (3.6) means, then take (3.6) as a definition of the disjoint union $B \uplus C$ of two sets B, C. What is the disjoint union of the sets $B = \{x_1, x_2, x_3\}$ and $C = \{x_3, x_4\}$? In particular, how many elements does it have? You may not know what tensor products $V \otimes W$ and exterior algebras ΛV are. But if $\dim V = n$ and $\dim W = m$, what are $\dim V \otimes W$ and $\dim \Lambda V$? (In advanced mathematics the ability to extract partial information from something that you don't completely understand is extremely useful.)

3.1. Set Mappings

When algebra studies sets with structure (e.g. groups), it also studies structure-preserving maps between those sets (e.g. homomorphisms of groups). Thus, in studying sets, it studies maps or functions $f : A \to B$ from a set A to a set B. (There is no structure to be preserved, so there are no restrictions on the functions studied.) There are several possible ways of writing the image of an element x of A under the map f, e.g. with the map on the left as $f(x)$, on the right as xf, as a superfix x^f, etc. Examples from elementary mathematics are the notations $\sin x$, $x!$, x^2. Since text is read from left to right, the most natural notation is usually xf, since then the composite with a further function $g : B \to C$ sends x to xfg and the functions f, g are read from left to right in the correct order of application: first f, then g. However, the notation $f(x)$ is used by many writers and is natural in some situations, such as those of Exercise II 2.1E and (IV 3.1.2). The composite is sometimes then written as $g \circ f$, so that $(g \circ f)(x) = g(f(x))$. One should be accustomed to both conventions, and should be able to work with each. In this book, $g \circ f$ will usually denote the composite when writing the functions on the left of their arguments, and fg will denote the composite writing the functions on the right (either on the line as xf or above as x^f). Another valuable notational convention is to use an ordinary arrow $f : A \to B$ to connect the domain $A = \operatorname{Dom} f$ and codomain $B = \operatorname{Cod} f$ of a function f, while using a barred arrow $f : x \mapsto f(x)$ to describe the effect $f(x)$ of the function f on an element x of the domain of f. These notations may be combined to give a compact description of a function, even without naming the function, e.g.

$$\mathbb{N} \to \mathbb{R};\ n \mapsto \sqrt{n}\,.$$

The barred arrow is especially effective at enhancing the clarity when the elements of the domain set of a function are themselves sets, e.g. cosets of a normal subgroup or vector subspace.

The identity function on a set A will be written as 1_A or id_A, thus

$$1_A : A \to A; a \mapsto a.$$

A function $f : A \to B$ *is injective, injects,* or *is one to one* iff

$$\forall x, y \in A, \quad xf = yf \Rightarrow x = y. \tag{3.1.1}$$

(Read: for all x, y in A, $xf = yf$ implies $x = y$.) It is a *monomorphism* iff

$$\forall g : C \to A, h : C \to A, \quad gf = hf \Rightarrow g = h. \tag{3.1.2}$$

It is *right invertible* iff

$$\exists r : B \to A. \quad fr = 1_A \tag{3.1.3}$$

(read: there exists $r : B \to A$ such that $fr = 1_A$); the map r here is called a *right inverse* or a *retraction* for f.

Proposition 3.1.1. *For a function $f : A \to B$, the following conditions are equivalent*:

(a) *f injects*;
(b) *f is a monomorphism.*

Furthermore, if $A \neq \varnothing$, they are equivalent to:

(c) *f is right invertible.* □

EXERCISES

3.1A. Prove Proposition 3.1.1.

3.1B. Restate (3.1.1)–(3.1.3) as they would appear using the left-handed $f(x)$ notation rather than the right-handed xf notation for images.

There is an important underlying principle in algebra called *duality.* Basically, one obtains duals of notions by reversing the directions of arrows and the order of composition of mappings. The duals of (3.1.1)–(3.1.3) are as follows. A function $f : B \to A$ *is surjective, surjects,* or *is onto* iff

$$\forall x \in A, \exists y \in B. \quad yf = x. \tag{3.1.4}$$

(Read: For all x in A, there is a y in B such that $yf = x$.) It is an *epimorphism* iff

$$\forall g : A \to C, h : A \to C, fg = fh \Rightarrow g = h. \tag{3.1.5}$$

It is *left invertible* iff

$$\exists s : A \to B. \quad sf = 1_A; \tag{3.1.6}$$

the map s here is called a *left inverse* or a *section* for f.

Mnemonic. A Right inverse is a Retraction. The Latin for "left" is *sinister*. A left inverse, or Sinistral inverse, is a Section. This mnemonic goes awry under the "$f(x)$" convention.

EXERCISES

3.1C. Formulate and prove the dual of Proposition 3.1.1.

3.1D. Prove that if A and B are finite sets of the same cardinality, then a function $f : A \to B$ injects iff it surjects. Give an example of a map $f : C \to C$ which is injective but not surjective. Dually, give an example of a map $f : C \to C$ which is surjective but not injective.

3.1E. Prove that each retraction is surjective. Formulate and prove the dual statement.

3.1F. For a subset S of a set A, a function $f : A \to B$ is an *extension* of a function $g : S \to B$ iff g is the restriction $f|_S$ of f to S.

(a) Show that an extension of a surjective function is surjective.

(b) Show that a restriction of an injective function is injective.

Combining (3.1.1)–(3.1.3) with (3.1.4)–(3.1.6) gives the following. A function $f : A \to B$ *is bijective*, or *bijects*, iff it is both injective and surjective. It is *invertible*, or an *isomorphism*, iff there is a map $g : B \to A$ which is simultaneously a left and right inverse for f, i.e. for which

$$fg = 1_A \quad \text{and} \quad gf = 1_B. \tag{3.1.7}$$

The map g is called an *inverse* of f. Moreover, in this case the sets A and B are said to be *equipollent* or *isomorphic*, written $A \cong B$.

EXERCISES

3.1G. Prove that if f has an inverse, then that inverse is unique.

3.1H. For a function $f : A \to B$, prove that the following conditions are equivalent:

(a) f bijects;

(b) f is both a monomorphism and an epimorphism;

(c) f is invertible.

Warning. For sets with non-trivial structure and functions preserving that structure, the analogues of Proposition 3.1.1 or its dual need no longer hold. (Cf. Example III 1.3.1.) Beware of texts which confuse the distinction between, say, surjections and epimorphisms.

3.2. Cartesian Products and Disjoint Unions

Given two sets A and B, their *Cartesian product* may be defined as the set

$$A \times B := \{(x, y) | x \in A,\ y \in B\} \tag{3.2.1}$$

of ordered pairs of elements from A and B respectively. There are *projections*

$$\pi_A : A \times B \to A; (x, y) \mapsto x \tag{3.2.2}$$

and

$$\pi_B : A \times B \to B; (x, y) \mapsto y \tag{3.2.3}$$

onto the factors A and B. Note that if one of the factors is empty, then so is the product. The most important property of the Cartesian product construction is given by the following proposition.

Proposition 3.2.1. *Given any set C and maps $f: C \to A$, $g: C \to B$, there is a unique map $(f, g): C \to A \times B$ such that $(f, g)\pi_A = f$ and $(f, g)\pi_B = g$.*

Proof. $(f, g): C \to A \times B$; $c \mapsto (cf, cg)$ satisfies $(f, g)\pi_A = f$ and $(f, g)\pi_B = g$. Conversely, if $h: C \to A \times B$ satisfies $h\pi_A = f$ and $h\pi_B = g$, then $ch\pi_A = cf$ and $ch\pi_B = cg$ for all c in C, so that $ch = (cf, cg)$ and $h = (f, g)$. □

Diagramatically, Proposition 3.2.1 may be expressed as

$$\begin{array}{ccccc} A & \xleftarrow{\pi_A} & A \times B & \xrightarrow{\pi_B} & B \\ {\scriptstyle f}\uparrow & & \uparrow{\scriptstyle (f,g)} & & \uparrow{\scriptstyle g} \\ C & \xleftarrow[1_C]{} & C & \xrightarrow[1_C]{} & C \end{array} . \tag{3.2.4}$$

In fact, Proposition 3.2.1 captures such a characteristic property of the Cartesian product that it may be used as a definition.

Definition 3.2.2. The *direct product* $A \times B$ of two sets A and B is a set $A \times B$ equipped with maps $\pi_A : A \times B \to A$ and $\pi_B : A \times B \to B$ such that, for any set C and set maps $f: C \to A$, $g: C \to B$, there is a unique map $(f, g): C \to A \times B$ such that $(f, g)\pi_A = f$ and $(f, g)\pi_B = g$. □

Proposition 3.2.1 shows that the Cartesian product construction (3.2.1) gives one way of realizing the direct product of A and B. Another way would be to take the Cartesian product $B \times A$ with $\pi_A : B \times A \to A$; $(x, y) \mapsto y$ and $\pi_B : B \times A \to B$; $(x, y) \mapsto x$ (note the difference from (3.2.2) and (3.2.3)). In fact, all the "different" direct products of A and B are isomorphic (e.g. $A \times B \to B \times A$; $(x, y) \mapsto (y, x)$ is an isomorphism between two Cartesian products).

EXERCISES

3.2A. Prove that any two direct products of a pair of sets A and B are isomorphic.

3.2B. Let A, B, and C be sets. Prove that the direct products $A \times B$ and $B \times A$ are isomorphic. (The observation above showed that the Cartesian products $A \times B$ and $B \times A$ are isomorphic.) Prove that there is an isomorphism of direct products $(A \times B) \times C \cong A \times (B \times C)$.

3.2C. Let A be a set. Prove $\varnothing \times A \cong \varnothing \cong A \times \varnothing$. Is there a set I such that, for any set A, one has $I \times A \cong A \cong A \times I$? If such a set exists, is it unique? If there are two distinct such sets, say I and J, what is the relationship between I and J?

3.2D. Let A be a set. For each positive integer n, define the *direct power* A^n inductively by $A^1 = A$ and $A^{n+1} = A^n \times A$. Prove $A^{n+m} \cong A^n \times A^m$ for positive integers n and m. Can A^0 be defined so that $A^{n+m} \cong A^n \times A^m$ holds for all natural numbers n and m? Can A^{-1} be defined so that $A^{-1} \times A^2 \cong A$?

3.2E. Prove that the map π_A of Definition 3.2.2 is a surjection if B is non-empty.

Dual to the notion of the direct product $A \times B$ of two sets A and B is their disjoint union $A \uplus B$ defined by the dual picture

$$
\begin{array}{ccccc}
A & \xrightarrow{\iota_A} & A \uplus B & \xleftarrow{\iota_B} & B \\
{\scriptstyle f}\downarrow & & \downarrow{\scriptstyle f \uplus g} & & \downarrow{\scriptstyle g} \\
C & \xrightarrow[1_C]{} & C & \xleftarrow[1_C]{} & C
\end{array} \quad . \tag{3.2.5}
$$

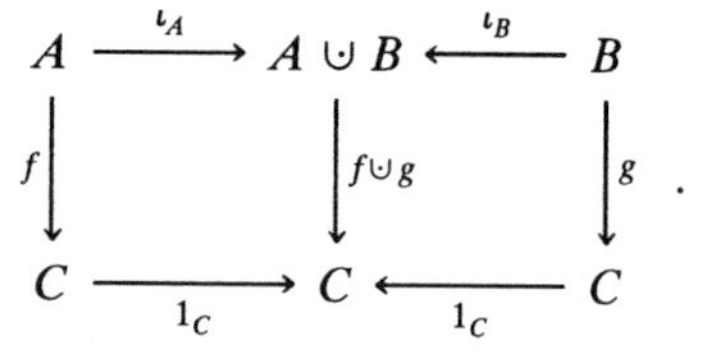

Definition 3.2.3. The *disjoint union* $A \uplus B$ of two sets A and B is a set $A \uplus B$ equipped with maps $\iota_A : A \to A \uplus B$ and $\iota_B : B \to A \uplus B$ (called *insertions*) such that, for any set C and set maps $f : A \to C$, $g : B \to C$, there is a unique map $f \uplus g : A \uplus B \to C$ such that $\iota_A(f \uplus g) = f$ and $\iota_B(f \uplus g) = g$. $\square$

Just as the Cartesian product ordered pair construction gave one (very natural) way of realizing the direct product, so there is a standard (although somewhat less natural) way of realizing the disjoint union. If A and B are disjoint sets ($A \cap B = \varnothing$), then the union $A \cup B$ works as a disjoint union $A \uplus B$, with $\iota_A : A \to A \cup B;\ a \mapsto a$ and $\iota_B : B \to A \cup B;\ b \mapsto b$. Trouble arises here if A and B have a common element x. How should $x(f \uplus g)$ be defined if $xf \neq xg$? The standard solution is to take $A \uplus B = (A \times \{0\}) \cup (B \times \{1\})$, with $\iota_A : A \to A \uplus B;\ a \mapsto (a, 0)$ and $\iota_B : B \to A \uplus B;\ b \mapsto (b, 1)$. Thus if $x \in A \cap B$, $x\iota_A = (x, 0) \neq (x, 1) = x\iota_B$. The copies $A\iota_A$ and $B\iota_B$ of A and B in $A \uplus B$ have been made disjoint.

For a whole family $\langle A_i | i \in I \rangle$ of sets, the *product* of $\langle A_i | i \in I \rangle$ is written as $\prod_{i \in I} A_i$ or $\prod\langle A_i | i \in I \rangle$, with projections $\pi_i : \prod_{i \in I} A_i \to A_i$. Analogously to Definition 3.2.2, it satisfies the property that, given a family $\langle f_i : C \to A_i | i \in I \rangle$ of maps, there is a unique map $f : C \to \prod_{i \in I} A_i$ such that $f\pi_i = f_i$ for each i in I. The *disjoint union* of $\langle A_i | i \in I \rangle$ is written as $\sum_{i \in I} A_i$ or $\biguplus_{i \in I} A_i$ or $\sum\langle A_i | i \in I \rangle$, with insertions $\iota_k : A_k \to \sum_{i \in I} A_i$. Analogously to Definition 3.2.3, it satisfies the property that, given a family $\langle f_i : A_i \to C | i \in I \rangle$ of maps, there is a unique map $f : \sum_{i \in I} A_i \to C$ such that $\iota_i f = f_i$ for each i in I. A map $f : \sum_{i \in I} A_i \to \prod_{j \in J} B_j$ is then specified uniquely by the $I \times J$-*matrix* $[f_{ij} | i \in I,\ j \in J]$ of maps $f_{ij} : A_i \to B_j$ given by

$$
\begin{array}{ccc}
\Sigma A_i & \xrightarrow{f} & \Pi B_j \\
\uparrow{\scriptstyle \iota_i} & & \downarrow{\scriptstyle \pi_j} \\
A_i & \xrightarrow[f_{ij}]{} & B_j
\end{array} \tag{3.2.6}
$$

EXERCISES

3.2F. Prove that the standard construction of the disjoint union $A \uplus B$ satisfies the requirements of Definition 3.2.3.

3.2G. Prove that any two disjoint unions of a pair of sets A and B are isomorphic.

3.2H. Prove that the "insertions" ι_A and ι_B of Definition 3.2.3 are monomorphisms.

3.2I. What are the relationships amongst the cardinalities $|A|$, $|B|$, $|A \cup B|$, $|A \cap B|$, $|A \times B|$, $|A \uplus B|$?

3.2J. Let A be a set. Prove $\varnothing \uplus A \cong A \cong A \uplus \varnothing$.

3.2K. Let A, B, and C be sets. Prove $A \uplus B \cong B \uplus A$ and $(A \uplus B) \uplus C \cong A \uplus (B \uplus C)$.

3.2L. Prove (3.6).

3.2M. Characterize the direct sum $V \oplus W$ of two (real) vector spaces V and W by a diagram similar to (3.2.5), with A, B, and C replaced by vector spaces V, W, U, and with linear transformations in place of arbitrary set maps.

3.2N. Show that $A \times B$ may be realized as the disjoint union of $|B|$ copies of A. [Hint: The standard construction of $A \cup\!\!\!\cdot\, A$ yields the Cartesian product $A \times \{0, 1\}$.]

3.2O. Describe the function $f \cup\!\!\!\cdot\, g$ of (3.2.5) in terms of the "**if then else**" statement used in various programming languages.

3.2P. Given the family $\langle A_i | i \in I \rangle$ of sets, show that the disjoint union $\Sigma\{A_i \in I\}$ may be realized as the subset $\bigcup\{\{i\} \times A_i | i \in I\}$ of $I \times \bigcup\{A_i | i \in I\}$, with $\iota_i : A_i \to \bigcup\{\{i\} \times A_k | i \in I\};\ a \mapsto (i, a)$.

3.2Q. Given the family $\langle A_i | i \in I \rangle$ of sets, define the map $p : \Sigma_{i \in I} A_i \to I$ by $p_i : A_i \to \{i\}$ and $\iota_i p_i = p$ for each i in I. Show that the product $\Pi\{A_i | i \in I\}$ may be realized as the set of sections $s : i \mapsto a_i$ for p [in the sense of (3.1.6)], with $\pi_i : \Pi_{j \in I} A_j \to A_i;\ s \mapsto is$.

3.3. Relations, Good Definition, and the First Isomorphism Theorem

A *(binary) relation* α on a set A is a subset of the direct square A^2 (cf. Exercise 3.2D). A membership $(x, y) \in \alpha$ may also be written $x \alpha y$. Typical properties of a relation are listed below. The relation α is said to be:

(3.3.1) *reflexive* if $\forall x \in A,\ (x, x) \in \alpha$;

(3.3.2) *symmetric* if $x \alpha y \Rightarrow y \alpha x$;

(3.3.3) *transitive* if $(x \alpha y$ and $y \alpha z) \Rightarrow x \alpha z$;

(3.3.4) *antisymmetric* if $(x \alpha y$ and $y \alpha x) \Rightarrow x = y$.

A relation α is a *(partial) order* on A if it is reflexive, transitive, and antisymmetric. The set A together with such an order α is often called an *ordered set*, a *partially ordered set*, or (by an acronym) a *poset*. For example, for a given set B, consider the *power set* 2^B or $\mathscr{P}(B)$ consisting of all subsets of B. Then the containment relations $\subseteq$ and $\supseteq$ on the power set 2^B are partial orders. The "not greater than" and "not less than" relations $\leq$ and $\geq$ on the set $\mathbb{R}$ of real numbers are partial orders. The study of partial orders is a major branch of post-modern algebra, with significant extra-mathematical applications in computer science, economics, and operations research.

A relation α on a set A is an *equivalence relation* if it is reflexive, symmetric, and transitive. For an equivalence relation α on a set A, the *equivalence class* of an element x of A under α is the subset

$$x^\alpha := \{y \in A | x \alpha y\} \tag{3.3.5}$$

of A. (Many writers use the clumsier notation x/α in place of x^α.) Note that the reflexivity of α gives at least one element x of x^α. The set of equivalence classes

$$A^\alpha := \{x^\alpha | x \in A\} \tag{3.3.6}$$

is called the *quotient* of the set A by the equivalence relation α (so many writers denote it by A/α). The map

$$\text{nat}\,\alpha : A \to A^\alpha;\ x \mapsto x^\alpha \tag{3.3.7}$$

is called the (*natural*) *projection* of A onto the quotient A^α. [Note how the notations (3.3.5) and (3.3.6) then become examples of the superfix notation for functions.] The smallest equivalence relation on A is the *diagonal*

$$\hat{A} := \{(x,x)|x \in A\}, \tag{3.3.8}$$

which represents the equality relation. The largest is the *universal* relation $A^2 : \forall x, y \in A,\ xA^2y$.

EXERCISES

3.3A. The motivating examples of equivalence relations were congruences in number theory. For a natural number n, define a relation $\langle n \rangle$ on $\mathbb{Z}$ by

$$(x,y) \in \langle n \rangle \Leftrightarrow \exists q \in \mathbb{Z}.\quad x - y = qn.$$

Prove that $\langle n \rangle$ is an equivalence relation on $\mathbb{Z}$. The relationship $(x,y) \in \langle n \rangle$ is often written as $x \equiv y \bmod n$, read "x congruent to y mod(ulo) n." Identify $\langle 0 \rangle$, $\langle 1 \rangle$, $\mathbb{Z}^{\langle 0 \rangle}$, and $\mathbb{Z}^{\langle 1 \rangle}$. For $n > 1$, the quotient $\mathbb{Z}^{\langle n \rangle}$ is usually denoted by $\mathbb{Z}_n$, the "set of integers mod n." What is $|\mathbb{Z}_n|$? The map (3.3.7) in these cases is called "reduction mod n."

3.3B. If α and β are equivalence relations on A, show that $\alpha \cap \beta$ is also. If $\{\alpha_i | i \in I\}$ is a set of equivalence relations on A, show that $\bigcap_{i \in I} \alpha_i$ is also.

3.3C. If α and β are equivalence relations on A, decide (by proof or counterexample) whether $\alpha \cup \beta$ necessarily is also.

3.3D. Let σ be a reflexive and symmetric relation on A. For each natural number n, define σ^n inductively by $\sigma^0 := \hat{A}$ and $\sigma^{n+1} := \{(x,y) \in A^2 | \exists t \in A.\ x\sigma^n t \sigma y\}$. Prove that $\sigma^T := \bigcup_{n \in \mathbb{N}} \sigma^n$ is an equivalence relation on A. The relation σ^T is called the *transitive closure* of σ. Prove that σ^T is a subset of each equivalence relation on A containing σ.

Many of the sets studied in algebra (and other branches of mathematics) are quotients (3.3.6) or subsets of such quotients, i.e. sets of equivalence classes (e.g. integers mod n as in Exercise 3.3A). A function taking its domain as such a set assigns values to each equivalence class. These values are often best described in terms of particular members of the equivalence class. For example, the negation function on the integers mod n may be described putatively as

$$\mathbb{Z}_n \to \mathbb{Z}_n;\ m^{\langle n \rangle} \mapsto (-m)^{\langle n \rangle}.$$

A problem arises: Is such a definition sound? If a different member of the equivalence class had been used to describe the function value, could a different function value emerge? Such a function definition is called *good*, or the function is called *well-defined*, if the putative definition is independent of the arbitrary choice of a representative member for the typical equivalence class being taken as the function argument. The negation function is well-defined above if $(\ell, m) \in \langle n \rangle$ implies $(-\ell, -m) \in \langle n \rangle$.

Equivalence classes and the question of good definition feature in the most basic theorem in algebra, the First Isomorphism Theorem. This theorem appears in "modern algebra" for groups, rings, and modules, but it is really a theorem about sets and functions.

First Isomorphism Theorem 3.3.1. *Let $f: A \to B$ be a function. Then there is an equivalence relation* ker f *on A, called the* kernel *of the function f, such that f factorizes*

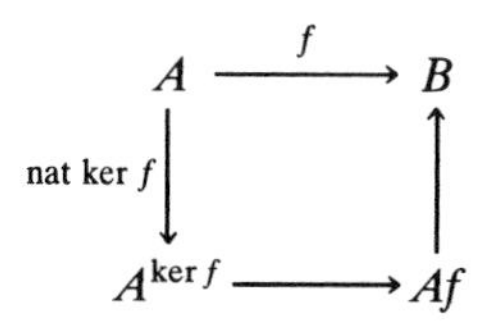

as the composition of the natural projection $A \to A^{\ker f}$, the isomorphism $A^{\ker f} \to Af$; $x^{\ker f} \mapsto xf$, and the subset embedding $Af \to B$; $y \mapsto y$. □

EXERCISE

3.3E. Prove the First Isomorphism Theorem by verifying that ker f is an equivalence relation, that $A^{\ker f} \to Af$; $x^{\ker f} \mapsto xf$ is a well-defined isomorphism, and that the function f agrees with the composite of the projection, the isomorphism, and the embedding.

3.4. Exponentiation, Quantifiers, and Negation

For sets A and B, the construction of the set $\underline{\underline{\mathrm{Set}}}(B, A)$ or A^B of all set functions from B to A may be viewed as an exponentiation ("raising A to the exponent B"). For example, if A and B are finite, then $|A^B| = |A|^{|B|}$. Now for natural numbers l, m, n, one of the laws of exponentiation is $l^{nm} = (l^m)^n$. The analogue for sets X, Y, Z would be an isomorphism

$$X^{Z \times Y} \cong (X^Y)^Z. \tag{3.4.1}$$

There is indeed such an isomorphism, given by a technique often described as "Currying" in computer science or "parametrization of a family of functions" in analysis. Given an element f of $X^{Z \times Y}$, i.e. a function $f : Z \times Y \to X$; $(z, y) \mapsto f(z, y)$, each element z of Z determines ("parametrizes") a function

(3.4.2) $$f_z : Y \to X;\ y \mapsto f(z, y).$$

Conversely, an element $z \mapsto f_z$ of $(X^Y)^Z$, regarded as a parametrized set $\{f_z : Y \to X | z \in Z\}$, determines a function $(y, z) \mapsto f_z(y)$ in $X^{Y \times Z}$. Thus the bijection (3.4.1) is given by

$$X^{Z \times Y} \to (X^Y)^Z;\ (f : (z, y) \mapsto x) \mapsto (z \mapsto (f_z : y \mapsto x)).$$

Such a bijection is sometimes displayed diagrammatically as

(3.4.3) $$\frac{(z, y) \mapsto x}{z \mapsto (y \mapsto x)},$$

or as

(3.4.4) $$\underline{\underline{\mathrm{Set}}}(Z \times Y, X) \cong \underline{\underline{\mathrm{Set}}}(Z, X^Y).$$

Later, many more examples of such bijections will be encountered.

One use of set exponentiation is in classifying subsets of a set A. Let 2 denote the two-element set $\{0, 1\}$; in logic, its elements are often interpreted as $0 =$ "false" and $1 =$ "true". Each subset B of A determines an element χ_B of 2^A, the *characteristic function* of the subset B with $\chi_B(a) =$ **if** $a \in B$ **then** 1 **else** 0. One sometimes writes $\chi_B(a) = [a \in B]$, the *truth value* of the proposition $a \in B$. Note that $B = \chi_B^{-1}\{1\}$. Conversely, a function f in 2^A determines a subset $f^{-1}\{1\}$ of A whose characteristic function is f. The bijection $\mathscr{P}(A) \to 2^A$; $B \mapsto \chi_B$ is often used to identify $\mathscr{P}(A)$ and 2^A; thus one speaks of 2^A as the power set of A. Now a proposition $P(a)$ which is either true or false for each element a of A yields a function $A \to 2$; $a \mapsto [P(a)]$ mapping a to the truth value of the proposition $P(a)$. Under the identification of elements of 2^A with subsets of A, the proposition $P(a)$, via the function $a \mapsto [P(a)]$, determines the subset $\{a \in A | P(a) \text{ holds}\}$ or $\{a \in A | P(a)\}$ of A. For example, the negation $\neg(a \in B)$ determines the complement $A - B$, with characteristic function $\chi_{A-B} = 1 - \chi_B$. More generally, the negation $\neg P(a)$ [read as "not $P(a)$" or "it is not true that $P(a)$ holds"] determines $\{a \in A | \neg P(a)\} = A - \{a \in A | P(a)\}$.

Now consider two sets X, Y and a proposition $P(x, y)$ involving elements x of X and y of Y. The universal quantifier yields a new proposition

(3.4.5) $$A(y) = (\forall x \in X, P(x, y)).$$

Corresponding to the subset $\{(x, y) | P(x, y)\}$ of $X \times Y$, one obtains a subset $\{y | A(y)\} = \{y | \forall x \in X, P(x, y)\}$ of Y. Similarly, the existential quantifier yields a new proposition

(3.4.6) $$E(y) = (\exists x \in X.\, P(x, y)).$$

This time, the subset $\{(x,y)|P(x,y)\}$ of $X \times Y$ yields the subset $\{y|E(y)\} = \{y|\exists x \in X.\ P(x,y)\}$ of Y. If X is non-empty, then $\{y|\forall x \in X, P(x,y)\} \subseteq \{y|\exists x \in X.\ P(x,y)\}$. Moreover, negation interchanges the roles of the quantifiers. If X is non-empty, the negation of (3.4.5) is

$$\neg A(y) = (\exists x \in X.\ \neg P(x,y)), \tag{3.4.7}$$

while the negation of (3.4.6) is

$$\neg E(y) = (\forall x \in X, \neg P(x,y)). \tag{3.4.8}$$

For example, the set of non-negative real numbers may be described as $\{y \in \mathbb{R}|\exists x \in \mathbb{R}.\ x^2 = y\}$, while its complement, the set of negative real numbers, may be described as $\{y \in \mathbb{R}|\forall x \in \mathbb{R}, x^2 \neq y\}$. This algebra of quantifiers is particularly useful in the analysis of concepts such as the uniform convergence of a set $\{f_n : \mathbb{R} \to \mathbb{R}|n \in \mathbb{N}\}$ of functions. The (Cauchy version of) uniform convergence is the property $\forall \varepsilon > 0, \exists N \in \mathbb{N}.\ \forall n \geq N, \forall m \geq N, \forall x \in \mathbb{R}, |f_n(x) - f_n(x)| < \varepsilon$. If one wished to show that the set was not uniformly convergent, then one could aim to demonstrate the negation of the property, namely $\exists \varepsilon > 0.\ \forall N \in \mathbb{N}, \exists n \geq N.\ \exists m \geq N.\ \exists x \in \mathbb{R}.\ |f_n(x) - f_m(x)| \geq \varepsilon$. There is an interchange of quantifier phrases:

$$\text{“}\forall a \in A,\text{”} \leftrightarrow \text{“}\exists a \in A.\text{”} \tag{3.4.9}$$

followed by the negation of the punchline.

EXERCISES

3.4A. For sets X, Y, Z, prove:
(a) $(X \times Y)^Z \cong X^Z \times Y^Z$;
(b) $X^{Y \cup Z} \cong X^Y \times X^Z$;
(c) $(X^Y)^Z \cong (X^Z)^Y$.

3.4B. Show that $|\emptyset^X| =$ **if** $X = \emptyset$ **then** 1 **else** 0.

3.4C. Determine binary operations $+ : 2^2 \to 2$ and $\cdot : 2^2 \to 2$ such that, for each set A, each element a of A, and each pair B, C of subsets of A, one has $\chi_{B \cup C}(a) = \chi_B(a) + \chi_C(a)$ and $\chi_{B \cap C}(a) = \chi_B(a) \cdot \chi_C(a)$.

3.4D. Formulate and prove (by induction) the extension

$$\frac{(y_1, y_2, \ldots, y_n) \mapsto x}{y_1 \mapsto (y_2 \mapsto \cdots (y_n \mapsto x) \cdots)}$$

of (3.4.3).

3.4E. (a) Show that a function $f : X \to Y$ induces a function

$$f^{-1} : 2^Y \to 2^X;\ B \mapsto f^{-1}(B).$$

(b) For $f : X \to Y$ and $g : Y \to Z$, show that $(fg)^{-1} = g^{-1}f^{-1}$.
(c) For $f : X \to Y$ and $B \subseteq Y$, show that $f\chi_B = \chi_{f^{-1}(B)}$.

3.4F. Which of the following are equivalent:

(a) $\forall \varepsilon > 0, \exists N \in \mathbb{N}. \forall n \geq N, \forall m \geq N, \forall x \in \mathbb{R}, |f_n(x) - f_m(x)| < \varepsilon;$

(b) $\forall x \in \mathbb{R}, \forall \varepsilon > 0, \quad \exists N \in \mathbb{N}. \forall n \geq N, \forall m \geq N, |f_n(x) - f_m(x)| < \varepsilon;$

(c) $\forall \varepsilon > 0, \exists N \in \mathbb{N}. \forall x \in \mathbb{R}, \forall n \geq N, \forall m \geq N, |f_n(x) - f_m(x)| < \varepsilon;$

(d) $\forall x \in \mathbb{R}, \exists N \in \mathbb{N}. \forall \varepsilon > 0, \quad \forall n \geq N, \forall m \geq N, |f_n(x) - f_m(x)| < \varepsilon?$

3.4G. Write out the negations of the assertions (b) and (c) in Exercise 3.4F.

3.5. Ordered Sets and Induction

Consider the set $(\mathbb{N}, \leq)$ of natural numbers, ordered by the "not greater than" relation $\leq$. This order relation underlies inductive proofs of a proposition $P(n)$ involving a general natural number n.

Proposition 3.5.1 (Principle of Mathematical Induction). *Consider a proposition $P(n)$ involving a general natural number n. Then $P(n)$ is true for all n if the following two conditions hold:*

(Induction Basis) $P(0)$ *is true;*

(Induction Step) $\forall n \in \mathbb{N}, P(n) \Rightarrow P(n + 1)$.

Proof. Consider the subset $S = \{n \in \mathbb{N} | [P(n)] = 1\}$ of the set $\mathbb{N}$ of natural numbers. Since $0 \in S$ by the induction basis and $n \in S \Rightarrow n + 1 \in S$ by the induction step, the set S comprises $\mathbb{N}$. □

In proofs by induction, the proposition $P(n)$ is known as the *Induction Hypothesis*. The art of proof by induction is to find a hypothesis $P(n)$ that is sufficiently strong to imply $P(n + 1)$ and sufficiently weak to be a consequence of $P(n - 1)$.

Corollary 3.5.2 (Principle of Complete Induction). *The proposition $P(n)$ is true for all n if*

$$(\forall m < n, [P(m)] = 1) \Rightarrow [P(n)] = 1. \tag{3.5.1}$$

Proof. Define a new proposition $Q(n) = P(0) \& P(1) \& \ldots \& P(n)$, i.e. $[Q(n)] = \prod_{m \leq n}[P(m)]$. Assuming (3.5.1), $Q(n)$ will be proved by induction. For $n = 0$, the condition of (3.5.1) is vacuous, so (3.5.1) asserts that $P(0)$, i.e. $Q(0)$, is true. This is the induction basis. For the induction step, suppose $Q(n)$ is true. This means $\forall m < n + 1, [P(m)] = 1$. By (3.5.1), $P(n + 1)$ follows. Thus $Q(n + 1) = Q(n) \& P(n + 1)$ also holds. Finally, note $Q(n) \Rightarrow P(n)$. □

A partial ordering α on a set A induces a partial ordering $\alpha \cap B^2$ on each subset B of A. Now a partial ordering γ on a set C is said to be *total* or *linear* if

$$\forall x, y \in C, \ (x, y) \in \gamma \text{ or } (y, x) \in \gamma. \tag{3.5.2}$$

One also says that the poset (C, γ) is a *chain* or a *flag*. For example, $(\mathbb{R}, \leq)$ and the induced $(\mathbb{N}, \leq)$ form chains. An element u of a poset $(A, \leq)$ is an *upper bound* for a subset S of A if $s \leq u$ for all s in S. Note that the subset $\mathbb{N}$ of $(\mathbb{R}, \leq)$ has no upper bound. An element m of a poset $(A, \leq)$ is said to be *maximal* if $m \leq a \in A \Rightarrow m = a$. For example, $\{1, 2\}$ and $\{1, 3\}$ are maximal elements in the poset of proper subsets of $\{1, 2, 3\}$ ordered by inclusion. A poset $(A, \leq)$ is *inductive* if it contains an upper bound for each subset C whose induced order $(C, \leq)$ is a chain. Despite its popular name, the following statement is an axiom due to Kuratowski.

Zorn's Lemma 3.5.3. *Each inductive poset has a maximal element.* □

A partial order (A, β) is said to *extend* a partial order (A, α) if α is a subset of β. The following result provides a useful way to represent posets in computers. Its proof illustrates the application of Zorn's Lemma.

Proposition 3.5.4. (*a*) *Each partial order extends to a chain.*
(*b*) *Each partial order is an intersection of total orders.*

Proof. Let (A, α) be a partial order. Consider the set E of order relations on A extending α, ordered by inclusion [i.e. with order induced from $(\mathscr{P}(A^2), \subseteq)$]. A total order extending α is maximal in $(E, \subseteq)$: No new order relations can be introduced, because of antisymmetry. Conversely, maximal elements of E are total. Indeed, if x and y are incomparable in an extension β of α, i.e. $(x, y) \notin \beta$ and $(y, x) \notin \beta$, consider $X = \{a \in A | (x, a) \in \beta\}$ and $Y = \{a \in A | (a, y) \in \beta\}$. Note $X \cap Y = \varnothing$, since $z \in X \cap Y \Rightarrow x \beta z \beta y \Rightarrow x \beta y$. Then $\beta \cup (Y \times X)$ is a proper extension of β.

Now the poset $(E, \subseteq)$ is inductive: An upper bound of a chain is provided by the union of $\{\alpha\}$ and the elements of the chain. By Zorn's Lemma, the poset has a maximal element. This maximal element is a total order on A extending (A, α). Thus (a) is proved. For (b), one obtains α as the intersection of the maximal orderings that contain it. These maximal orderings are total. □

EXERCISES

3.5A. The *Pigeonhole Principle* asserts that if there is an injection $f: A \to B$ between finite sets, then $|A| \leq |B|$. Prove the Pigeonhole Principle by induction on the cardinality n of B.

3.5B. Prove $\Sigma_{r=1}^{n} r^3 = (\Sigma_{r=1}^{n} r)^2$.

3.5C. Is the empty poset inductive?

3.5D. Express the inclusion order on each of $\mathscr{P}(\{1, 2\})$ and $\mathscr{P}(\{1, 2, 3\})$ as an intersection of total orders.

3.5E. Determine the partial order given as the intersection of the chains $0 < 1 < 2 < 3 < 4 < 5$ and $0 < 2 < 4 < 1 < 3 < 5$.

4. SEMIGROUPS AND MONOIDS

Let X be a set. Let S be a set of functions $f: X \to X$ with the *closure* property

$$\forall f,\ g \in S,\ fg \in S. \tag{4.1}$$

For example, one may take the set $S = X^X$ of all functions from X to X. Or, for a fixed element x of X, one may take $S = O_x := \{f : X \to X | Xf = \{x\}\}$. The composition of functions is associative:

$$\forall f, g, h \in S,\quad (fg)h = f(gh). \tag{4.2}$$

If $1 = 1_S$ lies in S, then

$$\forall f \in S,\quad 1f = f = f1. \tag{4.3}$$

Abstracting these properties of sets of functions on a set that are closed under composition, one obtains the following definitions.

Definition 4.1. A *semigroup* $(S, \cdot)$ is a set S equipped with a binary operation $S^2 \to S$; $(f, g) \mapsto fg$ of *multiplication*, denoted by $f \cdot g$ or juxtaposition fg, such that the associative law (4.2) is satisfied. □

Definition 4.2. A *monoid* $(S, \cdot, 1)$ is a semigroup $(S, \cdot)$ equipped with a chosen element 1 or 1_S, called the *identity* (*element*) of S, such that (4.3) holds. □

In referring to a set-with-structure, such as a semigroup $(S, \cdot)$ or a monoid $(S, \cdot, 1)$, it is often convenient to suppress explicit mention of the structure, merely speaking of "the semigroup S" or "the monoid S."

Warning. Some texts (especially older French ones) interchange these usages of the words "semigroup" and "monoid."

EXERCISES

4A. Let X be a set. Prove that the set of right invertible maps $f: X \to X$ forms a monoid under composition.

4B. Let Y be a subset of a set X. Prove that the set $O_Y = \{f : X \to X | Xf \subseteq Y\}$ forms a semigroup under composition. Is it a monoid?

4C. Let M be a monoid. Prove that 2^M is a monoid under the "complex multiplication" $XY := \{xy | x \in X, y \in Y\}$ with identity $\{1\}$.

4D. Prove that the identity element of a monoid is unique.

4E. Is $\varnothing^{\varnothing}$ a monoid?

The products $(f_1 f_2)f_3$ and $f_1(f_2 f_3)$ in a semigroup may both be written unambiguously as $f_1 f_2 f_3$. More generally, it may be shown (by induction on n) that longer products may be written unambiguously (without explicit parentheses) as $f_1 f_2 \cdots f_n$.

Given two semigroups $(A, \cdot)$, $(B, \cdot)$, a *semigroup homomorphism* $f:(A, \cdot) \to (B, \cdot)$ is a set function $f: A \to B$ "preserving the semigroup structure," in the sense that

$$\forall x, y \in A, \quad xf \cdot yf = (xy)f. \tag{4.4}$$

Given two monoids $(A, \cdot, 1_A)$ and $(B, \cdot, 1_B)$, a *monoid homomorphism* $f:(A, \cdot, 1_A) \to (B, \cdot, 1_B)$ is a semigroup homomorphism $f:(A, \cdot) \to (B, \cdot)$ such that

$$1_A f = 1_B. \tag{4.5}$$

EXERCISE

4F. Given monoids $(A, \cdot, 1_A)$ and $(B, \cdot, 1_B)$, decide by proof or counterexample whether a semigroup homomorphism $f:(A, \cdot) \to (B, \cdot)$ is necessarily a monoid homomorphism.

A *monoid isomorphism* $\theta:(A, \cdot, 1_A) \to (B, \cdot, 1_B)$ is a bijective monoid homomorphism. If such exists, then the monoids $(A, \cdot, 1_A)$ and $(B, \cdot, 1_B)$ are said to be *isomorphic*, written $A \cong B$ or $(A, \cdot, 1_A) \cong (B, \cdot, 1_B)$. Each element x of a semigroup $(A, \cdot)$ determines a *right multiplication*

$$R_x : A \to A;\ a \mapsto ax \tag{4.6}$$

and a *left multiplication*

$$L_x : A \to A;\ a \mapsto xa. \tag{4.7}$$

The following theorem shows that Definition 4.2 has not abstracted too far from monoids of functions on a set, in the sense that every abstract monoid $(M, \cdot, 1)$ is isomorphic to a monoid of functions $f: M \to M$ under composition.

Theorem 4.3. *Let $(M, \cdot, 1)$ be a monoid. Then $R: M \to M^M;\ x \mapsto R_x$ gives a monoid isomorphism $M \to M\,R$ of $(M, \cdot, 1)$ with a monoid of functions on the set M under composition.*

Proof. Certainly R injects, since $R_x = R_y$ implies $x = 1x = 1R_x = 1R_y = 1y = y$. Thus $R: M \to MR \subseteq M^M$ bijects. Moreover, R is a monoid homomorphism, since $\forall m \in M, mR_1 = m1 = m = m\ \mathrm{id}_M$ implies $R_1 = \mathrm{id}_M$ and $\forall m \in M, mR_x R_y = (mx)R_y = (mx)y = m(xy) = mR_{xy}$ implies $R_x R_y = R_{xy}$. □

EXERCISES

4G. Does $L: M \to M^M; x \mapsto L_x$ give a monoid homomorphism from $(M, \cdot, 1)$ to the monoid of functions on M under composition?

4H. Let X be a set. Show that the projection $X^2 \to X; (x, y) \mapsto x$ gives a semigroup multiplication on X.

4I. Is every semigroup isomorphic to a semigroup of functions under composition?

A subset R of a semigroup S is called a *subsemigroup* if $x, y \in R \Rightarrow xy \in R$, i.e. if R is a semigroup under the multiplication of S.

EXERCISES

4J. If R_1 and R_2 are subsemigroups of S, show that $R_1 \cap R_2$ is also. If $\{R_i | i \in I\}$ is a set of subsemigroups of S, show that $\bigcap_{i \in I} R_i$ is also.

4K. If R_1 and R_2 are subsemigroups S, decide (by proof or counterexample) whether $R_1 \cup R_2$ necessarily is also.

A subset R of a monoid $(M, \cdot, 1)$ is a *submonoid* if it is a subsemigroup of $(M, \cdot)$ and if $1 \in R$. Note that the inclusion $R \to M; r \mapsto r$ then becomes a monoid homomorphism. If $(A, \cdot, 1)$ and $(B, \cdot, 1)$ are monoids, then their direct product $(A \times B, \cdot, 1)$ is the Cartesian product $A \times B$ equipped with the *componentwise product*

$$(a, b)(a', b') = (aa', bb') \tag{4.8}$$

and with (the "componentwise identity") $(1, 1)$ as identity element. An equivalence relation α on a monoid M is a *(monoid) congruence* if it is a submonoid of the direct square $(M^2, \cdot, 1)$. There is then a well-defined product $m^\alpha \cdot n^\alpha = (mn)^\alpha$ on the quotient M^α, making it a monoid with identity element 1^α. Suppose that the function $f: A \to B$ of the First Isomorphism Theorem 3.3.1 is actually a monoid homomorphism $f: (A, \cdot, 1) \to (B, \cdot, 1)$. Then the projection, isomorphism, and subset embedding of the theorem all become monoid homomorphisms.

EXERCISES

4L. Repeat Exercises 4J and 4K (or 3.3B and 3.3C) for submonoids.

4M. Let X be a subset of a monoid $(M, \cdot, 1)$. For positive integers n, define X^n inductively by $X^1 := X$ and $X^{n+1} := \{xy \in M | x \in X^n, y \in X\}$. Also set $X^0 := \{1\}$. Define $\langle X \rangle := \bigcap_{n \in \mathbb{N}} X^n$. Show that $\langle X \rangle$ is a submonoid of $(M, \cdot, 1)$ contained in each submonoid of $(M, \cdot, 1)$ containing X. The monoid $\langle X \rangle$ is called the *submonoid* of $(M, \cdot, 1)$ *generated* by X.

4N. Let S denote the set of submonoids of a monoid $(M, \cdot, 1)$. Does $(A, B) \mapsto \langle A \cup B \rangle$ define a monoid multiplication on S?

4O. Give a formal statement and proof of the First Isomorphism Theorem for monoids.

4P. Let α be a congruence on a monoid $(M, \cdot, 1)$. Show that the equivalence class 1^α is a submonoid of $(M, \cdot, 1)$.

4Q. Show that a monoid congruence α on a monoid $(M, \cdot, 1)$ is not uniquely determined by the "normal" submonoid 1^α of Exercise 4P.

4R. Consider the monoid $(\mathbb{Z}, +, 0)$ of integers under addition. Show that the monoid congruences on $\mathbb{Z}$ are precisely the congruences $\langle n \rangle$ of number theory (cf. Exercise 3.3A).

4S. Let $f:(M, \cdot, 1) \to (N, \cdot, 1)$ be a monoid homomorphism. If S is a submonoid of N, show that $f^{-1}(S)$ is a submonoid of M.

4T. Let $f:(M, \cdot, 1) \to (N, \cdot, 1)$ be an invertible monoid homomorphism. Show that $f^{-1}:(N, \cdot, 1) \to (M, \cdot, 1)$ is a monoid homomorphism.

4U. Let $f:(A, \cdot, 1) \to (B, \cdot, 1)$ and $g:(B, \cdot, 1) \to (C, \cdot, 1)$ be monoid homomorphisms. Show that $fg:(A, \cdot, 1) \to (C, \cdot, 1)$ is also a monoid homomorphism.

4.1. Free Monoids and Codes

Let A be a set, referred to as an *alphabet*. (Typical examples to bear in mind are the *binary alphabet* $\{0, 1\}$, the *Morse alphabet* $\{\cdot, -, \square\}$, the *RNA alphabet* $\{U, C, A, G\}$, the *alphabet* $\{A, B, C, \dots, X, Y, Z\}$ or the *English alphabet* $\{A, B, C, \dots, X, Y, Z, \square\}$.) A (*non-empty*) *word in an alphabet A* is a concatenation $a_1 a_2 \dots a_n$ of (not necessarily distinct) elements of A. For example, 01101000 is a word in the binary alphabet, WORD is a word in the alphabet, and EAST□OF□EDEN is a word in the English alphabet (which includes the space or blank written here as □). Let A^+ denote the set of all non-empty words in A. Concatenation gives a semigroup multiplication on A^+, e.g. $(a_1 \dots a_m, b_1 \dots b_n) \mapsto a_1 \dots a_m b_1 \dots b_n$. Under this multiplication, A^+ becomes the so-called *free semigroup* over A. Adjoining an identity element 1 to A^+, called the *empty word*, one obtains the *free monoid* A^* over A. The *length* of a non-empty word $a_1 \dots a_n$ in A is the number n of "letters" appearing in it. The length of the empty word is 0. Length gives a monoid homomorphism $A^* \to (\mathbb{N}, +, 0)$. In the notation of Exercise 4M, A^n is the set of words of length n in A.

Definition 4.1.1. A subset C of A^* is a *code* over the alphabet A if

$$(4.1.1)\quad \forall m, n \in \mathbb{Z}^+, \forall c_1, \ldots, c_m, d_1, \ldots, d_n \in C,$$
$$c_1 \ldots c_m = d_1 \ldots d_n \Rightarrow m = n, c_1 = d_1, \ldots, c_m = d_m. \quad \square$$

This means that any word in C^+ can be decoded uniquely as a concatenation $c_1 \cdots c_n$ of codewords in C.

EXERCISES

4.1A. Show that $\{0, 01, 10\}$ is not a code over $\{0, 1\}$.

4.1B. For each positive integer n, show that A^n is a code over the alphabet A (called the *uniform code* of length n over A). In molecular biology, the uniform code of length 3 over the RNA alphabet plays a significant role (comprising the so-called "codons" that determine protein synthesis).

4.1C. Fix one particular element of the alphabet A (e.g. the blank $\square$ in the Morse or English alphabet) and call it the "comma." A *comma code* C is a subset of A^* consisting of words, in each of which the comma appears just once, namely at the end. Show that C is a code over A.

4.1D. A word v in A^* is said to appear as a *proper suffix* in a word w if there is a word u in A^+ such that $w = uv$. A subset C of A^* is a *suffix code* if no element of C appears as a proper suffix of another. Show that a suffix code is a code over A.

4.1E. Show that A^* is the submonoid $\langle A \rangle$ of A^* generated by A (cf. Exercise 4M).

4.1F. If $|A| = 1$, show that a subset C of A^* is a code iff $|C| = 1$.

Proposition 4.1.2. *A subset C of A^* is a code if there is an alphabet B and an injective monoid homomorphism $k : B^* \to A^*$ such that $Bk = C$.*

Proof. Suppose $c_1 \ldots c_m = d_1 \ldots d_n$, say $c_i = e_i^k$ and $d_j = f_j^k$ for $e_1, \ldots, e_m, f_1, \ldots, f_n \in B$. Then $(e_1 \ldots e_m)^k = e_1^k \ldots e_m^k = c_1 \ldots c_m = d_1 \ldots d_n = f_1^k \ldots f_n^k = (f_1 \ldots f_n)^k$, since k is a monoid homomorphism. Since k injects, it follows that $e_1 \ldots e_m = f_1 \ldots f_n$ in B^*, whence $m = n$, $e_1 = f_1, \ldots, e_m = f_m$. Applying k, one obtains $c_1 = e_1^k = f_1^k = d_1, \ldots, c_m = e_m^k = f_m^k = d_m$, so that C is a code. $\square$

The homomorphism k of Proposition 4.1.2 is called the *coding homomorphism* from B^* to A^*. It shows how the letters of the alphabet B are encoded as codewords, from C, in the alphabet A. For example, the standard coding homomorphism k from words in the English alphabet to words in the Morse alphabet has $Ak = \cdot - \square$, $Bk = - \cdots \square$, $\ldots, Zk = - \cdot\cdot \square$, $\square k = \square$. This is the "Morse Code" which opened up the whole field of telecommunications in the nineteenth century.

The free monoid B^* over the alphabet B satisfies the following "universality property":

(4.1.2) For each monoid $(M, \cdot, 1)$ and set map $f : B \to M$, there is a unique monoid homomorphism $\bar{f} : (B^*, \cdot, 1) \to (M, \cdot, 1)$ such that $\bar{f}|_B = f$.

EXERCISE

4.1G. Prove (4.1.2).

The property (4.1.2) may be expressed diagrammatically as

$$\begin{array}{ccc} B & \longrightarrow & B^* \\ {\scriptstyle f}\downarrow & & \downarrow{\scriptstyle \bar{f}} \\ M & = & M \end{array} \qquad (4.1.3)$$

[compare (3.2.4) and (3.2.5)]. The property shows that the coding homomorphism $k : B^* \to A^*$ of Proposition 4.1.2 is specified uniquely by $k|_B : B \to C$; $b \mapsto bk$.

EXERCISES

4.1H. Show that a subset C of A^* is a code iff the embedding $e : C \to A^*$; $c \mapsto c$ extends under (4.1.2) to an isomorphism $\bar{e} : C^* \to \langle C \rangle$ of the free monoid C^* over C with the submonoid $\langle C \rangle$ of A^* generated by C.

4.1I. Compare (3.1) and (4.1.2). Are all vector spaces "free"?

4.1J. Let a subset C of the free monoid A^* be a code over the alphabet A. For each positive integer n, show that the set C^n (as defined in Exercise 4M) is isomorphic with the direct power C^n (as defined in Exercise 3.2D).

4.2. Dynamical Systems and Cyclic Monoids

A *dynamical system* is a pair (X, T) consisting of a set X called the *state space* and a map $T : X \to X$ called the *evolution operator*. The interpretation is that elements x of X, called *states*, describe the state of a system (e.g. the positions and momenta of a collection of particles). The map T describes the evolution or development of the system over one unit of time, so that a system in state x at one point in time finds itself in state xT one time unit later. The implicit assumption is that the state x gives a complete description of the system at the first time point, and that the system's evolution to the uniquely specified subsequent state xT is completely deterministic.

The map T generates a submonoid $T^{\mathbb{N}} = \{T^n | n \in \mathbb{N}\}$ of the monoid $(X^X, \cdot, 1_X)$ of maps from X to X under composition. A monoid M generated by a single element T, i.e. with $M = \langle\{T\}\rangle$, is called a *cyclic monoid*. The first task in analyzing dynamical systems is to classify all the possible cyclic monoids $T^{\mathbb{N}}$. If the cyclic monoid $T^{\mathbb{N}}$ is infinite, then there is a monoid isomorphism $(\mathbb{N}, +, 0) \to (T^{\mathbb{N}}, \cdot, 1); n \mapsto T^n$. Otherwise, the monoid is finite, and there is a coincidence $T^q = T^{q+r}$ (with $r > 0$) between powers of T. The *index* i of $T^{\mathbb{N}}$ is then defined to be the least natural number q for which $\exists r > 0.\ T^q = T^{q+r}$. Note $T^i = T^{i+r}$ implies $\forall j \geq i,\ T^j = T^{j+r}$. The *period* p of $T^{\mathbb{N}}$ is then defined to be the least positive integer r for which $T^i = T^{i+r}$. Note $T^i = T^{i+p}$ implies $T^i = T^{i+sp}$ for all natural numbers s.

Theorem 4.2.1. *Let $(M, \cdot, 1)$ be a non-trivial cyclic monoid. If M is infinite, the monoid is isomorphic to the monoid of natural numbers under addition. If M is finite, then it is specified up to isomorphism by its index i and period p. For each natural number i and positive integer p, there is a finite cyclic monoid of index i and period p.* □

EXERCISES

4.2A. Complete the proof of Theorem 4.2.1.

4.2B. An *idempotent* of a monoid M is an element e such that $ee = e$. Prove that a cyclic monoid M has a unique idempotent iff $i = 0$ or $|M| = \infty$, and that otherwise M has a unique non-identity idempotent, namely T^j for $M = \langle T \rangle$ and $i \leq j \equiv 0 \bmod p$.

4.2C. Show that an infinite cyclic monoid $T^{\mathbb{N}}$ is free over $\{T\}$.

The *orbit* of an element x of a dynamical system (X, T) is the set

$$xT^{\mathbb{N}} := \{xT^n | n \in \mathbb{N}\}. \tag{4.2.1}$$

Each orbit gives a small dynamical system $(xT^{\mathbb{N}}, T|_{xT^{\mathbb{N}}})$. The structure of the dynamical system (X, T) may be analyzed by specifying the cyclic monoids $(T|_B)^{\mathbb{N}}$ (via Theorem 4.2.1) for maximal orbits B, together with the way these maximal orbits fit together.

EXERCISE

4.2D. Analyze the dynamical systems $(\mathbb{N}, n \mapsto 2n)$, $(\mathbb{Z}_4, n^{\langle 4 \rangle} \mapsto (2n)^{\langle 4 \rangle})$, $(\mathbb{N}, n \mapsto n^2)$.

4.3. Semilattices and Ordered Sets

A *semilattice* is a semigroup $(S, \cdot)$ that is *commutative*–

$$\forall x, y \in S, \quad xy = yx \tag{4.3.1}$$

and *idempotent*–

$$\forall x \in S, \quad xx = x \tag{4.3.2}$$

(cf. Exercise 4.2B). As examples of semilattices, consider the power set $(2^B, \cap)$ of a set B under the operation of intersection, or the real numbers $(\mathbb{R}, \wedge)$ under the operation $\wedge : \mathbb{R}^2 \to \mathbb{R}; (r, s) \mapsto \min\{r, s\}$. In these examples the semilattice multiplications involve taking "greatest lower bounds" in a partial order. If $\leq$ is a partial order on a set A, then the *greatest lower bound* $\text{glb}\{a, b\}$ of a subset $\{a, b\}$ of A is an element c of A such that $c \leq a$, $c \leq b$, and $(d \leq a$ and $d \leq b) \Rightarrow d \leq c$. A pair of elements in a partially ordered set may or may not have a greatest lower bound.

EXERCISE

4.3A. Show that $\hat{A}$ is a partial order on A, and that no two-element subset of A has a greatest lower bound under the partial order $\hat{A}$ on A. A poset of the form $(A, \hat{A})$ is called an *antichain*.

Let $(A, \leq)$ be a partially ordered set in which each subset $\{a, b\}$ has a greatest lower bound. Then $A^2 \to A; (a, b) \mapsto \text{glb}\{a, b\}$ gives a semilattice multiplication on A. In fact, such partially ordered sets essentially determine all semilattices. Let $(S, \cdot)$ be a semilattice. Define a relation $\leq_{\cdot}$ or $\leq$ on S by

$$x \leq y \Leftrightarrow x \cdot y = x. \tag{4.3.3}$$

EXERCISES

4.3B. Show that (4.3.3) defines a partial order $\leq$ on S.

4.3C. Show that each subset $\{a, b\}$ of S has a greatest lower bound under the partial order $\leq$ on S defined by (4.3.3).

4.3D. Show that $\text{glb}\{a, b\} = a \cdot b$ for all $\{a, b\} \subseteq S$.

4.3E. Let $\leq$ be partial order on a set A such that each subset $\{a, b\}$ has a greatest lower bound. Prove that $a \leq b$ iff $\text{glb}\{a, b\} = a$.

4.3F. If a semilattice $(S, \cdot)$ is a monoid $(S, \cdot, 1)$, what role does the identity element play with respect to the partial order $\leq$ defined by (4.3.3)?

The reduction of semilattices to partially ordered sets with greatest lower bounds is an elementary but typical example of the connections that arise between different branches of algebra–in this case between semigroups and ordered sets. Some of the deepest and most useful parts of algebra are founded on these connections, e.g. the representation theory of groups which studies how groups appear as groups of invertible elements of rings. The most serious disadvantage of abstract algebra, of the axiomatic approach, is that it tends to isolate the study of one particular set of axioms, breaking algebra up into a collection of disjoint specialities. In working through algebra, one should be careful not to view the various topics in isolation, either from themselves or from the rest of mathematics and the world beyond.

4.4. Monoids of Relations

Let A be a set. Let 2^{A^2} be the set of binary relations α on A, i.e. the set of subsets of the direct square A^2. A binary operation $\circ$ of *relation product* is defined on 2^{A^2} by

$$x\ \alpha \circ \beta\ y \Leftrightarrow \exists t \in A.\ x\ \alpha\ t\ \beta\ y. \tag{4.4.1}$$

The operation $\circ$ is associative. Moreover, the equality or diagonal relation $\hat{A}$ is an identity element. The set 2^{A^2} becomes a monoid $(2^{A^2}, \circ, \hat{A})$ known as the *relation monoid* on the set A. Of course, the relation monoid also carries a lot of additional structure: for example, the semilattice structure $(2^{A^2}, \cap)$ that it shares with any power set.

In general, the relation product (4.4.1) is not commutative, i.e. $\alpha \circ \beta$ need not coincide with $\beta \circ \alpha$. A set of relations $\{\alpha_i | i \in I\}$ is said to be *permutable* if $\alpha_i \circ \alpha_j = \alpha_j \circ \alpha_i$ for all i, j in I. For example, for any reflexive and symmetric relation α, the set $\{\alpha^n | n \in \mathbb{N}\}$ of Exercise 3.3D is permutable. If two equivalence relations α_1 and α_2 are permutable, then their common product $\alpha_1 \circ \alpha_2 = \alpha_2 \circ \alpha_1$ is again an equivalence relation.

Permutable relations play a key role in the structure of direct products. To begin, suppose that the set A is the Cartesian product $A_1 \times A_2$. Let α_i be the kernel of the projection $\pi_i : A \to A_i$. Then $\alpha_1 \circ \alpha_2 = A^2 = \alpha_2 \circ \alpha_1$. Indeed, given any pairs $a = (a_1, a_2)$ and $a' = (a_1', a_2')$ of elements of A, one has the relationships

$$\begin{array}{ccc} (a_1, a_2') & \overset{\alpha_2}{\text{———}} & (a_1', a_2') \\ \alpha_1 \Big| & & \Big| \alpha_1 \\ (a_1, a_2) & \underset{\alpha_2}{\text{———}} & (a_1', a_2) \end{array} \tag{4.4.2}$$

Moreover, $\alpha_1 \cap \alpha_2 = \hat{A}$. Conversely, suppose that a set A carries a permutable pair α_1, α_2 of equivalence relations whose intersection is $\hat{A}$ and whose common relation product is A^2. Consider the map θ given by the following instance of (3.2.4):

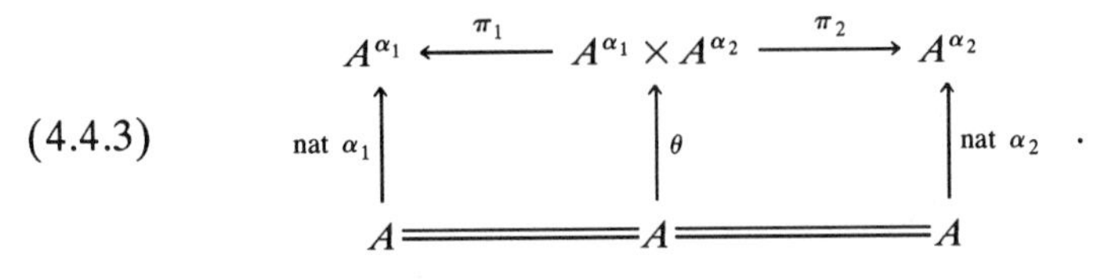

Certainly θ injects, since $a\theta = a'\theta \Rightarrow a\ \alpha_1 \cap \alpha_2\ a' \Rightarrow a = a'$. It also surjects. Consider $(b^{\alpha_1}, c^{\alpha_2})$ in $A^{\alpha_1} \times A^{\alpha_2}$. Now $(b, c) \in A^2 = \alpha_1 \circ \alpha_2 \Rightarrow \exists a \in A.$ $b\alpha_1 a \alpha_2 c$. Then $a\theta = (a^{\alpha_1}, a^{\alpha_2}) = (b^{\alpha_1}, c^{\alpha_2})$. The set A becomes the direct product of the sets A^{α_1} and A^{α_2} by means of the projections $\theta\pi_1 : A \to A^{\alpha_1}$ and $\theta\pi_2 : A \to A^{\alpha_2}$.

EXERCISES

4.4A. Show that, if two equivalence relations α and β are permutable, then their common product $\alpha \circ \beta = \beta \circ \alpha$ is again an equivalence relation.

4.4B. Give an example of two relations on a set A that are not permutable.

4.4C. For a positive integer n, define a relation α_n on the Cartesian plane $\mathbb{R}^2$ by $(x, y)\alpha_n(x', y') \Leftrightarrow x - x' = n(y - y')$. Show that the set $\{\alpha_n | n \in \mathbb{Z}^+\}$ is permutable, and that $\alpha_m \cap \alpha_n = \widehat{\mathbb{R}^2}$ if $m \neq n$.

4.4D. Display the multiplication table of the monoid of relations on the two-element set $\{0, 1\}$.

4.4E. Give an example to show that, for $A \cong \prod_{i \in I} A^{\alpha_i}$ to hold, the following conditions on a set $\{\alpha_i | i \in I\}$ of equivalence relations on A are insufficient: $\forall i \neq j,\ \alpha_i \circ \alpha_j = A^2 = \alpha_j \circ \alpha_i$ and $\alpha_i \cap \alpha_j = \hat{A}$.

4.4F. Let $\alpha_1, \alpha_2, \alpha_3$ be equivalence relations on a set A. Suppose that $\beta_1 = \alpha_2 \circ \alpha_3 = \alpha_3 \circ \alpha_2$, $\beta_2 = \alpha_3 \circ \alpha_1 = \alpha_1 \circ \alpha_3$, and $\beta_3 = \alpha_1 \circ \alpha_2 = \alpha_2 \circ \alpha_1$ are equivalence relations on A, with $\alpha_1 \cap \beta_1 = \alpha_2 \cap \beta_2 = \alpha_3 \cap \beta_3 = \hat{A}$ and $\alpha_1 \circ \beta_1 = \beta_1 \circ \alpha_1 = \alpha_2 \circ \beta_2 = \beta_2 \circ \alpha_2 = \alpha_3 \circ \beta_3 = \beta_3 \circ \alpha_3 = A^2$. Show that A is the direct product $A^{\alpha_1} \times A^{\alpha_2} \times A^{\alpha_3}$.

4.4G. Let α_1 and α_2 be congruence relations on a monoid M, with $\alpha_1 \circ \alpha_2 = \alpha_2 \circ \alpha_1 = M^2$ and $\alpha_1 \cap \alpha_2 = \hat{M}$. Show that the monoid M is monoid isomorphic with the Cartesian product monoid $M^{\alpha_1} \times M^{\alpha_2}$ [cf. (4.8)].

4.4H. For relations a, β, γ on a set A, show that $(\alpha \cap \beta) \circ \gamma \subseteq (\alpha \circ \gamma) \cap (\beta \circ \gamma)$. Give an example where the inclusion is proper.

4.4I. For relations α, β, γ on a set A, show that $(\alpha \cup \beta) \circ \gamma = (\alpha \circ \gamma) \cup (\beta \circ \gamma)$.

4.4J. Identify $\mathbb{Z}_{60}$ with $\prod_{3 \leq n \leq 5} \mathbb{Z}_n$ via $m^{\langle 60 \rangle} \mapsto (m^{\langle 3 \rangle}, m^{\langle 4 \rangle}, m^{\langle 5 \rangle})$. Interpret $\prod_{3 \leq n \leq 5} \mathbb{Z}_n$ (cf. Exercise 3.2Q) as the set of sections $s : \{3, 4, 5\} \to \sum_{3 \leq n \leq 5} \mathbb{Z}_n$ of the disjoint union $p : \sum_{3 \leq n \leq 5} \mathbb{Z}_n \to \{3, 4, 5\}$ of the maps $p_n : \mathbb{Z}_n \to \{n\}$, as illustrated opposite:

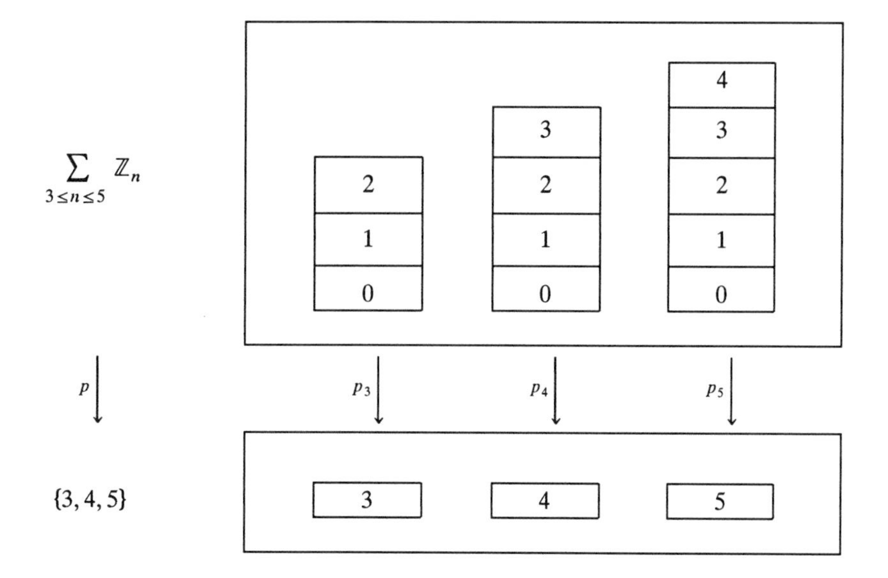

For example, $26 \equiv 2 \bmod 3$, $26 \equiv 2 \bmod 4$, and $26 \equiv 1 \bmod 5$, so that $26^{\langle 60 \rangle}$ is represented by the section

.

Determine the congruence classes modulo 60 represented by the following sections:

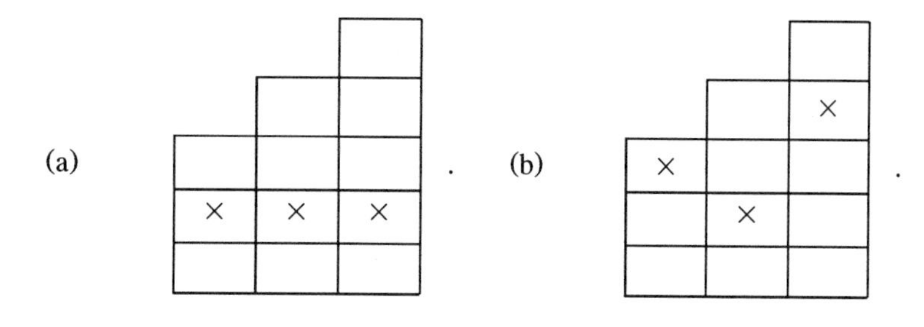

I

GROUPS AND QUASIGROUPS

1. MONOID ACTIONS

Let $(M, \cdot, 1)$ be a monoid. A *(right) M-set* or *(right) M-action* is a set X together with a monoid homomorphism *(representation)*

$$(1.1) \qquad R: M \to X^X;\ m \mapsto (x \mapsto xm)$$

from M to the monoid of all functions from X to X. For example, Theorem O 4.3 shows that the monoid M may be considered as an M-set, known as the *right regular representation* of M. Or, if (X, T) is a dynamical system as in Section O 4.2, then X is a right $T^{\mathbb{N}}$-set. Shifting the focus from the monoid to the set, one could define an M-set (X, M) as a set X equipped with a set M of maps or operations $m_X: X \to X;\ x \mapsto xm$, such that the properties

$$(1.2) \qquad \forall m, n \in M, \forall x \in X, (xm)n = x(mn)$$

and

$$(1.3) \qquad \forall x \in X, x1 = x$$

are satisfied. Semigroup actions may be defined in a similar way, either via a semigroup homomorphism in place of the monoid homomorphism (1.1), or by dropping the requirement (1.3). Now for a monoid $(M, \cdot, 1)$, the *opposite* of M is the monoid $M^{\mathrm{op}} = (M, \check{\circ}, 1)$ with the multiplication

$$(1.4) \qquad M^2 \to M;\ (m, n) \mapsto m \check{\circ} n = nm.$$

Then a *left M-set* or *M-action* is a right M^{op}-set or M^{op}-action. It is natural to write the operations of a left M-set as

$$(1.5) \qquad m_X: X \to X;\ x \mapsto mx,$$

so that (1.2) may take the form

$$\forall m, n \in M, \forall x \in X, n(mx) = (nm)x, \tag{1.6}$$

with the multiplication on the right-hand side of the equation in M rather than in M^{op}.

For a right M-set X, the action may also be described by the map

$$X \times M \to X; (x, m) \mapsto xm. \tag{1.7}$$

Mimicking (O 4.6) and (O 4.7), one may then define

$$R_m : X \to X; x \mapsto xm \tag{1.8}$$

for m in M and

$$L_x : M \to X; m \mapsto xm \tag{1.9}$$

for x in X. The M-set X is said to be *trivial* if each operation (1.8) is just the identity map on X.

EXERCISES

1A. Let B be a subset of a set A. Show that $Y : 2^A \to 2^A; X \mapsto X \cap Y$ for Y in 2^B yields a semigroup action of the semilattice $(2^B, \cap)$ (as in Section O 4.3) on the power set 2^A.

1B. In the context of Exercise 1A, the semigroup $(2^B, \cap)$ is actually a monoid $(2^B, \cap, B)$. Is the semigroup action of Exercise 1.1A also a monoid action?

1C. In the context of Exercise 1A, does $Y : 2^A \to 2^A; X \mapsto X \cup Y$ yield a semigroup action of $(2^B, \cap)$ on 2^A?

1D. Show that the map L of Exercise O 4G yields a left M-action.

1E. Let A be an alphabet. Show that the free monoid A^* is isomorphic to its opposite.

1F. For which alphabets A does A^* differ from $A^{*\mathrm{op}}$?

1G. Can you find an example of a monoid that is not isomorphic to its opposite?

1H. Let $(M, \cdot, 1)$ be a monoid. Show that the power set 2^M becomes a right M-set under the action

$$2^M \times M \to 2^M; (Y, m) \mapsto L_m^{-1}(Y),$$

using the notation of (O 4.7).

Fix a monoid $(M, \cdot, 1)$. Given two (right) M-sets $(A, M), (B, M)$, an *M-homomorphism* is a function $f: A \to B$ such that $afm_B = am_A f$ for all a in A and m in M. An *isomorphism* of M-sets is a bijective M-homomorphism. Two M-sets (A, M) and (B, M) are *isomorphic* if there is an isomorphism $f: (A, M) \to (B, M)$. One writes $(A, M) \cong (B, M)$ in this case. The *direct product* $(A \times B, M)$ of two M-sets (A, M) and (B, M) is the Cartesian product (O 3.2.1) equipped with the operations

$$m_{A \times B} : A \times B \to A \times B; \; (a, b) \mapsto (am_A, bm_B).$$

The *disjoint union* $(A \uplus B, M)$ of (A, M) and (B, M) is the disjoint union $A \uplus B$ of Definition O 3.2.3 equipped with the operations $m_{a \uplus B} = m_A \iota_A \uplus m_B \iota_B$ specified by (O 3.2.5) as follows:

$$\begin{array}{ccccc}
A & \xrightarrow{\iota_A} & A \uplus B & \xleftarrow{\iota_B} & B \\
{\scriptstyle m_A \iota_A} \downarrow & & \downarrow {\scriptstyle m_A \iota_A \uplus m_B \iota_B} & & \downarrow {\scriptstyle m_B \iota_B.} \\
A \uplus B & \xrightarrow[1_{A \uplus B}]{} & A \uplus B & \xleftarrow[1_{A \uplus B}]{} & A \uplus B
\end{array}$$

Write $\underline{\underline{M}}$ for the class of all (right) M-sets. For M-sets (A, M) and (B, M), write $\underline{\underline{M}}(A, B)$ for the set of all M-homomorphisms from A to B. Note how (O 3.2.4) and (O 3.2.5) yield bijections $\underline{\underline{M}}(C, A \times B) \cong \underline{\underline{M}}(C, A) \times \underline{\underline{M}}(C, B)$ and $\underline{\underline{M}}(A \uplus B, C) \cong \underline{\underline{M}}(A, C) \times \underline{\underline{M}}(B, C)$ for any M-set (C, M).

Now suppose that there is a whole family $\langle (A_i, M) \mid i \in I \rangle$ of M-sets. Let $m_i : A_i \to A_i$ denote the operation of the monoid element m on the M-set A_i. The *product* $\prod_{i \in I}(A_i, M)$ is the M-set $\prod_{i \in I} A_i$ on which the action of the monoid element m is given by $m\pi_i = \pi_i m_i$. The *disjoint union* $\sum_{i \in I}(A_i, M)$ is the M-set $\sum_{i \in I} A_i$ on which the action of the monoid element m is given by $\iota_i m = m_i \iota_i$. Note that $(A_1 \times A_2, M) \cong \prod_{i \le i \le 2}(A_i, M)$ and $(A_1 \uplus A_2, M) \cong \sum_{1 \le i \le 2}(A_i, M)$.

Let (A, M) be an M-set. The *orbit* xM of an element x of A is the set $\{xm \mid m \in M\}$ [cf. (O 4.2.1)]. A subset S of an M-set (A, M) is *invariant*, or an *M-subset*, if it contains the orbit sM of each of its elements s. Thus an M-subset S of (A, M) is itself an M-set (S, M), with the operation m_S as the restriction of m_A to S for each m in M. The insertion $S \to A$; $s \mapsto s$ yields an M-homomorphism if S is an M-subset. It is convenient to write $S \subseteq A$ to denote that S is a subset of A, and $S \le A$ to denote that S is an M-subset of A. An M-set (A, M) always has the *improper* M-subset (A, M) and the *trivial* M-subset $(\varnothing, M)$. It is *irreducible* if it has no others. It is *indecomposable* if it is not expressible as the disjoint union of proper, non-trivial M-subsets. Finally, an element x of an M-set (A, M) is a *fixed point* of the action if it forms a singleton M-subset $(\{x\}, M)$. The set $\mathrm{Fix}(A, M)$ or $\mathrm{Fix}\, M$ of all fixed points of (A, M) itself forms an M-subset of (A, M), of course. The action

(A, M) is said to be *fixed-point free* if $\text{Fix}(A, M)$ is trivial. In particular, non-singleton irreducible actions are fixed-point free.

If $f:(A, M) \to (B, M)$ is an M-homomorphism, then the First Isomorphism Theorem O 3.3.1 yields $(\ker f, M)$ as an M-subset of $(A \times A, M)$ and (Af, M) as an M-subset of (B, M). Moreover, $m: x^{\ker f} \mapsto xm^{\ker f}$ yields a well-defined action of M on $A^{\ker f}$. Thus the *First Isomorphism Theorem for M-actions* factorizes the M-homomorphism $f: A \to B$ as a product of M-homomorphisms $A \to A^{\ker f}$, $A^{\ker f} \to Af$, and $Af \to B$. In particular, the M-sets $A^{\ker f}$ and Af are isomorphic. Conversely, an equivalence relation α on an M-set (A, M) is called a *congruence* (*relation*) or an *M-congruence* if it is an M-subset of $(A \times A, M)$. In this case, $m: x^{\alpha} \mapsto xm^{\alpha}$ yields a well-defined action of M on A^{α}, making the natural projection (O 3.3.7) an M-homomorphism. The M-set (A, M) is said to be *simple* or *primitive* if its only congruences are the *improper* one, namely the universal relation A^2, and the *trivial* one, namely the diagonal or equality relation $\hat{A}$. Congruence relations on the right regular representation (M, M) of a monoid M have a special form, as shown by the following:

Proposition 1.1. *Let α be an M-congruence on the right regular representation of a monoid M. Then the equivalence class 1^{α} of the identity element of M is a submonoid of M.*

Proof. Since α is reflexive, the identity element 1 lies in 1^{α}. Now suppose that m and m' lie in 1^{α}. Since α is an M-subset of $M \times M, (1, m) \in \alpha \Rightarrow (m', mm') \in \alpha$. Then $1 \alpha m' \alpha mm'$ implies $mm' \in 1^{\alpha}$ by the transitivity of α. □

EXERCISES

1I. Show that the inverse of an isomorphism of M-sets is itself an isomorphism. Deduce that isomorphism of M-sets is an equivalence relation on any set of M-sets.

1J. Use the uniqueness of $f \uplus g$ in Definition O 3.2.3 to show that

$$(m_A \iota_A \uplus m_B \iota_B)(n_A \iota_A \uplus n_B \iota_B) = (mn)_A \iota_A \uplus (mn)_B \iota_B$$

for m, n in M. Deduce that $(A \uplus B, M)$ really is an M-set.

1K. Consider $\mathbb{Z}_4$ as a monoid $(\mathbb{Z}_4, \cdot, 1^{\langle 4 \rangle})$ under the multiplication of congruence classes. Consider the right regular representation of $\mathbb{Z}_4$ as a $\mathbb{Z}_4$-set. Show that it is indecomposable, but not irreducible.

1L. Show that $\{1^{\langle 4 \rangle}, 3^{\langle 4 \rangle}\}$ forms a submonoid M of $(\mathbb{Z}_4, \cdot, 1^{\langle 4 \rangle})$. Realize $\mathbb{Z}_4$ as an M-set (A, M) by omitting the operations of even congruence classes in the right regular representation of $(\mathbb{Z}_4, \cdot, 1^{\langle 4 \rangle})$. Decompose (A, M) as a disjoint union of irreducible M-sets. Determine its fixed points.

1M. Show that an irreducible M-set is indecomposable.

1N. Given M-sets (A, M) and (B, M), show that $\underline{\underline{M}}(A \times M, B)$ is an M-set under the action $m : f \mapsto (f^m : (x, n) \mapsto (x, \overline{nm})f)$.

1O. Let $\{B_i | i \in I\}$ be a set of M-subsets of an M-set A. Show that $\bigcup_{i \in I} B_i$ and $\bigcap_{i \in I} B_i$ are M-subsets of A. How are the fixed-point sets of B_i, $\bigcup_{i \in I} B_i$ and $\bigcap_{i \in I} B_i$ related?

1P. Let $\mathbf{M}$ be the set of all M-subsets of the right regular representation of a monoid M. Show that $\mathbf{M}$ is an M-subset of the M-set $(2^M, M)$ of Exercise 1H.

1Q. If B is an M-subset of an M-set A, show that there is an M-homomorphism

$$\chi_B : A \to \mathbf{M};\ x \mapsto L_x^{-1}(B)$$

[using the notation (1.9)], with $\chi_B^{-1}\{M\} = B$.

1R. In the context of Exercise 1Q, show that each M-homomorphism $f : A \to \mathbf{M}$ yields an M-subset $f^{-1}\{M\}$ of A.

1S. If the monoid M has only one element, show that any set is an M-set and any function is an M-homomorphism. Describe the map χ_B of Exercise 1Q in this case.

1T. Let B be an M-subset of an M-set A. Show that the complement of $\bigcup_{m \in M} R_m^{-1}(B)$ [using the notation of (1.8)] is an M-subset of A.

1U. Let M be a monoid and let X be a set. Show that the action $X \times M \times M \to X \times M$; $(x, m, n) \mapsto (x, mn)$ makes $X \times M$ an M-set. Define $\eta : X \to X \times M$; $x \mapsto (x, 1)$. Show that, for any M-set A and map $f : X \to A$, there is a unique M-homomorphism $\bar{f} : X \times M \to A$ with $f = \eta\bar{f}$. [By analogy with (O 4.1.2), the M-set $X \times M$ is called the *free M-set over the set X*.]

1V. In the context of Exercise 1U, show that $\{x\} \times M$ is an M-subset of $X \times M$ for each x in X. Show that $X \times M$ is the disjoint union $\Sigma_{x \in X}\{x\} \times M$ of these M-subsets.

1W. Let (A, M) be an M-set. Show that the diagonal $\hat{A}$ is an M-subset of $(A \times A, M)$. Is the complement $A^2 - \hat{A}$ an M-subset of $(A \times A, M)$?

1X. Let (A, M) be an M-set. For any subset S of A, show that $SM = \{sm | s \in S, m \in M\}$ is an M-subset. In particular, for x in A, $xM = \{x\}M \leq (A, M)$.

1Y. Consider the right regular representation $(\mathbb{Z}, \mathbb{Z})$ of the monoid $(\mathbb{Z}, +, 0)$ of integers under addition. Show that the $\mathbb{Z}$-congruences on $\mathbb{Z}$ are precisely the congruences $\langle n \rangle$ of number theory (cf. Exercises O 3.3A and 4R). Exhibit a submonoid of $(\mathbb{Z}, +, 0)$ which is not of the form 0^α for a $\mathbb{Z}$-congruence α.

1.1. Automata

Dynamical systems (X, T) as introduced in Section O 4.2 consisted of a state space X and a single evolution operator T. If $\{T\}^*$ is the free monoid over the singleton alphabet $\{T\}$, then the dynamical system (X, T) becomes a $\{T\}^*$-set $(X, \{T\}^*)$. Automata are generalizations of dynamical systems in this sense. Instead of the single evolution operator T, any element of a whole set or alphabet A of operators or *elementary events* may act on the state space X by an *elementary action* $A \to X^X$. Thus an *automaton* is defined as a right A^*-set (X, A^*) for the free monoid A^* over an alphabet A of elementary events. The monoid homomorphism $R: A^* \to X^X$ is the extension $\bar{f}$ [as in (O 4.1.2)] of the elementary action $f: A \to X^X$. The elements of A^*, i.e. the (possibly empty) words in the alphabet A, are then considered as *compound events*, represented by *compound actions* on the state space. Automata are excellent models of monoid actions: They are often very helpful in providing intuitions about general properties of monoid actions. Here are two examples of automata, one static (the remote control) and one dynamic, or "in real time" (the stopwatch).

Example 1.1.1 (The On / Off / Volume Control). Consider the following simplified version of a remote control for a television monitor. The remote control has to switch the television on and off, and to adjust the volume over one of four levels from 0 (completely muted) through 1 (quiet) and 2 (moderate) to 3 (loud). The remote control has three buttons. The first, labeled "O" for "on/off," switches the television on and off. The second button is labeled "U" for "up," and the third is labeled "D" for "down." If the television is switched off, then pressing the buttons U and D has no effect. If the television is turned on, then pressing and releasing button U raises the volume by one level. However, if the television is already set for the maximum volume, then pressing U has no effect. Similarly, pressing and releasing D reduces the volume by one level, unless the television is already muted or switched off. While it is switched off, the television "remembers" the volume level to which it was set when it was turned off, and returns to that level when it is turned on again.

The television set becomes an A^*-set X for the free monoid A^* over the alphabet $A = \{O, U, D\}$. The state space X has eight elements or states, denoted by n_i or f_i for $i = 0, 1, 2, 3$. If the television is running at volume level i, then it is in state n_i. If it is turned off, remembering that it was set to volume level i, then it is in state f_i. The effect of the remote control buttons may be summarized by the following *transition diagram*:

$$\begin{array}{ccccccccc}
D \circlearrowleft n_0 & \underset{D}{\overset{U}{\rightleftarrows}} & n_1 & \underset{D}{\overset{U}{\rightleftarrows}} & n_2 & \underset{D}{\overset{U}{\rightleftarrows}} & n_3 \circlearrowright U \\
O \uparrow \downarrow O & & O \uparrow \downarrow O & & O \uparrow \downarrow O & & O \uparrow \downarrow O \\
f_0 & & f_1 & & f_2 & & f_3 \\
\circlearrowleft & & \circlearrowleft & & \circlearrowleft & & \circlearrowleft \\
U, D & & U, D & & U, D & & U, D
\end{array}$$

The action of compound events may be read off the transition diagram. For example, $f_1OU^5D = n_2$ and $n_2OU^5D = f_2$. Of course, the action of the empty word is inaction, e.g. $n_3 1 = n_3$. □

Example 1.1.2 (The Digital Stopwatch). A digital stopwatch records time intervals in increments of one-hundredth of a second up to 99 minutes and $59\frac{99}{100}$ seconds. It has a start/stop button and a reset button. Let $N = 6 \times 10^5$. Then for $0 \le i < N$, the stopwatch has N states c_i in which it is running ("counting" hundredths) and displaying an elapsed time of i hundredths, as well as N states h_i in which it is stopped ("holding") while displaying the elapsed time of i hundredths. Denote the elementary event of pressing and releasing the start/stop button by S, and by R for the reset button. If the watch is running, then the reset button has no effect. If it is stopped, then the reset button leaves the watch stopped, but returns the display to zero. Pressing S then starts the watch. Pressing S while the watch is running will stop the watch, holding the displayed time. Pressing S again will restart the watch, the displayed time increasing from the value that was held while the watch was stopped. This part of the "logic" of the watch is comparable to the logic of the remote control. The new, dynamic element is provided by the elementary event T, denoted the elapse of one-hundredth of a second. Thus the full alphabet of elementary events is $A = \{R, S, T\}$. The state space X is the (disjoint) union of $\{c_i | 0 \le i < N\}$ and $\{h_i | 0 \le i < N\}$. Part of the transition diagram is as follows:

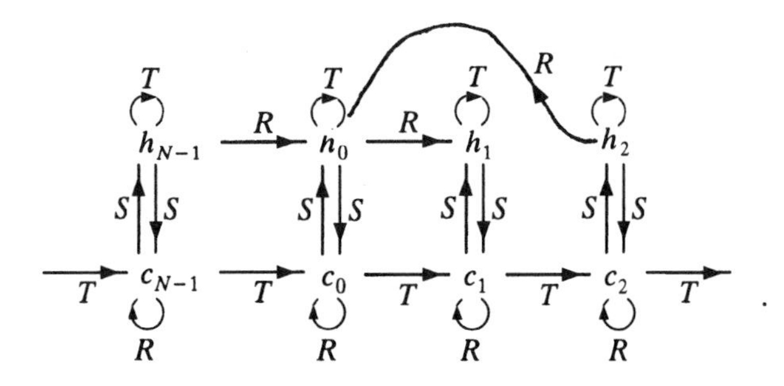

(To save space, the remaining 1,199,992 states have been omitted.) □

EXERCISES

1.1A. During the course of a typical day, the television of Example 1.1.1 is to be controlled as follows. Having been switched off late the previous night at the quiet volume level (so as not to disturb the neighbors), it is turned on and set to the moderate volume level. Later, it has to be muted for a telephone conversation, during which friends agree to come for a visit. When the conversation is over, the volume is reset to "moderate." After a while, the doorbell rings, at which point the television is turned off. The friends appear at the door, wishing to watch an interesting program. Since one of the friends is hard of hearing, the television has to be turned on and set at the loud level. After the friends leave, by which time it is already late, the television has to be brought down to the quiet level, before finally being turned off for the night. Give a compound event, as an element of the free monoid over the three-element set $\{O, U, D\}$, that will achieve the desired control. Then give a second compound event, comprising exactly five more elementary events than the first, that achieves the same effect.

1.1B. A VHF television has 13 channels, numbered from 01 to 13. Its remote control has an on/off button and 10 buttons labeled with the digits from 0 to 9. To select a new channel (while the television is on), the two digits denoting the channel have to be selected in sequence, the button for the second digit being pressed no later than five seconds after the button for the first digit has been pressed. If more than five seconds are allowed to elapse, the television forgets that the first digit was pressed, and simply holds the current channel. Construe the television with its remote control as an automaton, giving the state space, the alphabet of elementary events, and the transition diagram. [Hints: (a) Admit "elapse of five seconds" as one of the elementary events. (b) Disregard control of volume, brightness, contrast, tint, saturation, etc.]

1.1C. A certain country does not recognize polygamy. Its citizens (past and present) may have just one of the following civil states: single, married, divorced, widowed, and deceased. Using the elementary events of marriage, divorce, decease of (current) spouse, and decease of subject, set up an automaton to describe the possible changes in civil status. [Hint: Set up inapplicable elementary events as having trivial action, e.g. (married, marriage) $\mapsto$ married or (divorced, decease of (current) spouse) $\mapsto$ divorced.]

1.1D. Discuss the modeling of a simple pocket calculator as an automaton.

1.1E. Let X be a subset of a monoid M. Let $R: M \to M^M$ be the right regular representation of the monoid M. Construe each element x of X as an elementary event acting on the state space M by the right multiplication $R_x: M \to M;\ m \mapsto mx$. Thus the monoid M becomes an automaton (M, X^*), with representation $R^X: X^* \to M^M$. If $\langle X \rangle$ is the submonoid of $(M, \cdot, 1)$ generated by X (cf. Exercise O 4M), show that $\langle X \rangle = R^{-1}(X^* R^X)$.

1.2. The Class of All Actions

Up to this point, attention has been restricted to various sets upon which a single monoid acts. The time has now come to zoom out and consider the actions of various monoids together. The concept of automaton as introduced in Section 1.1 leads to a useful intuition of an M-set (X, M) as a "system" with its own internal workings, namely the actions of the monoid elements as transformations of the state space X.

Let $(M,\cdot,1)$ and $(N,\cdot,1)$ be monoids. Let (A, M) be an M-set and let (B, N) be an N-set. There are then two basic ways to construct sets acted on by the direct product monoid $(M \times N,\cdot,1)$ [cf. (O 4.8)]. The disjoint union $A \uplus B$ of Definition O 3.2.3 becomes an $(M \times N)$-set $(A \uplus B, M \times N)$ under the actions $a\iota_A(m, n) = am\iota_A$ and $b\iota_B(m, n) = bn\iota_B$ for a in A, b in B, m in M, and n in N. The direct product $A \times B$ of Definition O 3.2.2 becomes an $(M \times N)$-set $(A \times B, M \times N)$ under the "componentwise" action $(a, b)(m, n) = (am, bn)$. If N happens to coincide with M, one should be careful to note the difference between the direct product $(A \times B, M)$ of Section 1, which is an M-set, and the direct product $(A \times B, M \times M)$, which is an $(M \times M)$-set.

Example 1.2.1. Consider the automata $(A, \langle T\rangle)$ and $(B, \langle T\rangle)$ with transition diagrams as follows:

$$(A,\langle T\rangle): \quad a \overset{T}{\longleftrightarrow} a'$$

$$(B,\langle T\rangle): \quad b \overset{T}{\longleftrightarrow} b' \quad .$$

Then the transition diagram for the disjoint union $(A \uplus B, \langle T\rangle \times \langle T\rangle)$ is

$$(1,T) \circlearrowleft a \overset{(T,1)}{\longleftrightarrow} a' \circlearrowright (1,T)$$

$$(T,1) \circlearrowleft b \underset{(1,T)}{\longleftrightarrow} b' \circlearrowright (T,1) \quad .$$

while the transition diagram for the direct product $(A \times B, \langle T\rangle \times \langle T\rangle)$ is

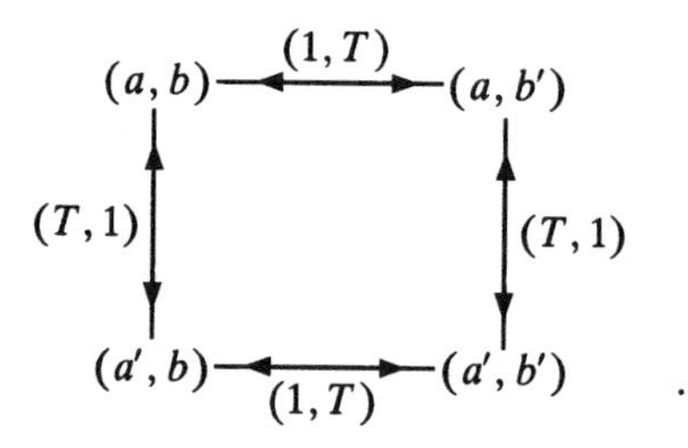

By contrast, the transition diagram for the disjoint union $(A \cup B, \langle T \rangle)$ in the class of $\langle T \rangle$-sets is simply

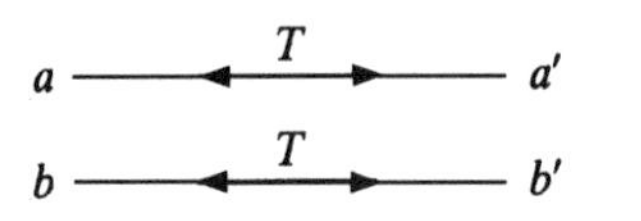

while the transition diagram for $(A \times B, \langle T \rangle)$ is

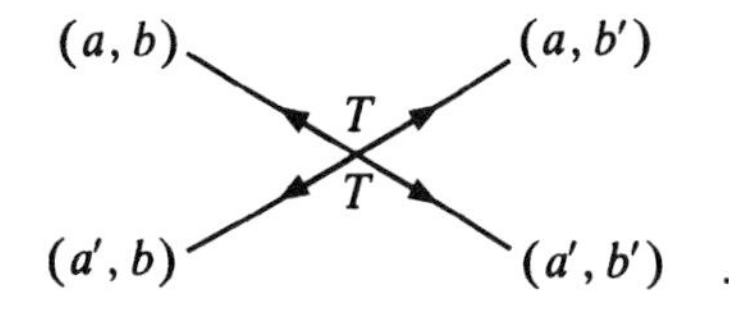

Note how the action of the diagonal element (T, T) in $(A \cup B, \langle T \rangle \times \langle T \rangle)$ and $(A \times B, \langle T \rangle \times \langle T \rangle)$ reflects the action of the element T in $(A \cup B, \langle T \rangle)$, and $(A \times B, \langle T \rangle)$, respectively. □

For a fixed monoid M, one may compare different M-sets by means of M-homomorphisms. On the other hand, one may compare different monoids by means of monoid homomorphisms. The actions of different monoids on different sets may then be compared by a suitable combination of both kinds of homomorphism. To begin with, consider an N-set (B, N) specified by the representation $R' : N \to B^B$ as in (1.1). It is occasionally convenient to denote the N-set as (B, N, R'), emphasizing the normally tacit role of the representation. Given a monoid homomorphism $f : M \to N$ (as in Section O 4), one may then form the composite $fR' : M \to N \to B^B$. This is a monoid homomorphism representing M by action on B, making B into the M-set (B, M, fR').

Definition 1.2.2. Let (A, M, R) be an M-set and let (B, N, R') be an N-set. Then an *action morphism* $(s, f) : (A, M, R) \to (B, N, R')$ is a pair consisting of a monoid homomorphism $f : M \to N$ and an M-homomorphism $s : (A, M, R) \to (B, M, fR')$. The action morphism (s, f) is called a *similarity* if f is a monoid isomorphism and if s is an isomorphism of M-sets. In this case the M-set A and the N-set B are said to be *similar*. □

Example 1.2.3. In Example 1.2.1, consider the diagonal $\Delta : \langle T\rangle \to \langle T\rangle \times \lfloor T\rangle;\ m \mapsto (m, m)$. Then $(1_{A \uplus B}, \Delta) : (A \uplus B, \langle T\rangle) \to (A \uplus B, \langle T\rangle \times \langle T\rangle)$ and $(1_{A \times B}, \Delta) : (A \times B, \langle T\rangle) \to (A \times B, \langle T\rangle \times \langle T\rangle)$ are action morphisms. Define $\theta : A \uplus B \to A \times B$ by the matrix [cf. (O 3.2.6)] with $\theta_{AA} = 1_A$, $(\theta_{AB} : a \mapsto b,\ a' \mapsto b')$, $(\theta_{BA} : b \mapsto a',\ b' \mapsto a)$, and $\theta_{BB} = 1_B$. Then $(\theta, 1_{\langle T\rangle}) : (A \uplus B, \langle T\rangle) \to (A \times B, \langle T\rangle)$ is a similarity. □

EXERCISES

1.2A. In Example 1.2.1, are the actions $(A \uplus B, \langle T\rangle \times \langle T\rangle)$ and $(A \times B, \langle T\rangle \times \langle T\rangle)$ similar?

1.2B. Discuss the similarity of the actions of the non-abelian group of order 6 given in Section O 2.

1.2C. Show that action morphisms $(s, f) : (A, M, R) \to (B, N, R')$ and $(t, g) : (B, N, R') \to (C, P, R'')$ compose to give an action morphism

$$(s, f)(t, g) = (st, fg) : (A, M, R) \to (C, P, R'').$$

(Hint: Consider the diagrams

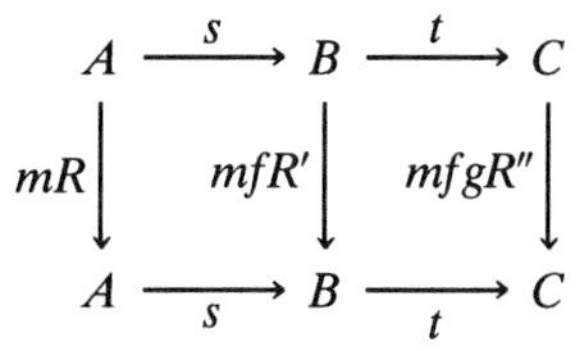

$$\begin{array}{ccccc} A & \xrightarrow{\ s\ } & B & \xrightarrow{\ t\ } & C \\ {\scriptstyle mR}\downarrow & & \downarrow{\scriptstyle mfR'} & & \downarrow{\scriptstyle mfgR''} \\ A & \xrightarrow[s]{} & B & \xrightarrow[t]{} & C \end{array}$$

for m in M.)

1.2D. Given actions (A, M) and (B, N), exhibit action morphisms $\pi_A : (A \times B, M \times N) \to (A, M)$ and $\pi_B : (A \times B, M \times N) \to (B, N)$ such that, for any action (C, P) and action morphisms $(s, f) : (C, P) \to (A, M), (t, g) : (C, P) \to (B, N)$, there is a unique action morphism $((s, f), (t, g)) : (C, P) \to (A \times B, M \times N)$ such that

$$((s, f), (t, g))\pi_A = (s, f) \text{ and } ((s, f), (t, g))\pi_B = (t, g).$$

(Hint: Compare Definition O 3.2.2.)

1.2E. Present the dual of Exercise 1.2D. (Hint: Compare Definition O 3.2.3.)

The class of all monoid actions is denoted by $(\underline{\underline{\text{Mon}}}; \underline{\underline{\text{Set}}})_0$. For given actions (A, M) and (B, N), the set of all action morphisms $(s, f) : (A, M) \to (B, N)$ is denoted by $(\underline{\underline{\text{Mon}}}; \underline{\underline{\text{Set}}})\,(A, B)$. For a given action (C, P), the direct product construction gives a bijection $(\underline{\underline{\text{Mon}}}; \underline{\underline{\text{Set}}})\,((C, P), (A \times B, M \times N)) \cong (\underline{\underline{\text{Mon}}}; \underline{\underline{\text{Set}}})\,((C, P), (A, M)) \times (\underline{\underline{\text{Mon}}}; \underline{\underline{\text{Set}}})\,((C, P), (B, N))$—see Exercise 1.2D. Dually, the disjoint union gives a bijection $(\underline{\underline{\text{Mon}}}; \underline{\underline{\text{Set}}})\,((A \uplus B,$

$M \times N), (C, P)) \cong \underline{\underline{(\text{Mon}; \text{Set})}}((A, M), (C, P)) \times \underline{\underline{(\text{Mon}; \text{Set})}}((B, N), (C, P))$ —see Exercise 1.2E. Besides the direct product and the disjoint union, there is another fundamental construction available in the class $\underline{\underline{(\text{Mon}; \text{Set})}}_0$. Consider the action (A, M) as being a *microscopic* system and (B, N) as being a *macroscopic* system. Each state b of (B, N) is replaced by a microscopic system $(A \times \{b\}, M)$ that is M-isomorphic to the prototype microscopic system (A, M) via the M-isomorphism

$$\kappa_b : (A, M) \to (A \times \{b\}, M); a \mapsto (a, b). \tag{1.2.1}$$

The direct product set $A \times B$, construed as the disjoint union $\Sigma_{b \in B} A \times \{b\}$ (cf. Exercise O 3.2N), becomes the state space for a composite system acted on by a new monoid $M \wr (B, N)$ known as the *wreath product of the monoid M by the action* (B, N). The underlying set of the wreath product $M \wr (B, N)$ is the direct product $(\Pi_{b \in B} M) \times N$. As in Exercise O 3.2Q the product $\Pi_{b \in B} M$ may be realized as the set of sections $s : B \to \Sigma_{b \in B} M$ of the map $p: \Sigma_{b \in B} M \to B$ given by $p_b : M \to \{b\}$ and $\iota_b p_b = p$ for each b in B. The product in $M \wr (B, N)$ is defined by

$$(s, n)(t, n') = (b \mapsto bs \cdot bnt, nn'). \tag{1.2.2}$$

The action of (s, n) on an element (a, b) of $A \times B$ is then given by

$$(a, b)(s, n) = (a \cdot bs, bn). \tag{1.2.3}$$

In words, the element (s, n) first acts on each microsystem $A \times \{b\}$ by applying bs, and then applies the action n at the macroscopic level. The action $(A \times B, M \wr (B, N))$ is denoted by $(A, M) \wr (B, N)$, known as the *wreath product action.*

EXERCISES

1.2F. For the actions $(A, \langle T \rangle)$ and $(B, \langle T \rangle)$ of Example 1.2.1, construct the wreath product $(A, \langle T \rangle) \wr (B, \langle T \rangle)$.

1.2G. Show that the multiplication (1.2.2) in $M \wr (B, N)$ is associative. What is its identity element?

1.2H. Show that (1.2.3) really does define an action of $M \wr (B, N)$ on the set $\Sigma_{b \in B} A \times \{b\}$.

1.2I. Show that there is an action morphism $(g, f) : (A \times B, M \wr (B, N)) \to (B, N)$ with $(a, b)g = b$ and $(s, n)f = n$.

1.2J. Show that there is an action morphism $(1_{A \times B}, j) : (\Sigma_{b \in B} A, \Pi_{b \in B} M) \to (A \times B, M \wr (B, N))$ with $j : s \mapsto (s, 1_N)$.

1.2K. Show that the wreath product is associative, in the sense that $((A, M) \wr (B, N)) \wr (C, P)$ and $(A \wr M) \wr ((B \wr N) \wr (C, P))$ are similar.

1.3. Group Actions

Let S be a subset of the set X^X of all mappings from a set X to itself. Define a relation σ on X by

$$(x, y) \in \sigma \Leftrightarrow \exists s \in S. \quad xs = y.$$

If the set S has the property

$$\forall s, t \in S, \, st \in S$$

of *closure* under composition of mappings, so that S is a semigroup, then the relation σ is transitive. If S has the property

$$\exists 1 \in S. \quad \forall s \in S, \, 1s = s = s1$$

of containing the identity element of X^X, so that S is a monoid, then σ is a reflexive and transitive relation, i.e. a *pre-ordering* or "quasi-ordering." Finally, if

$$\forall s \in S, \exists s^{-1} \in S. \quad ss^{-1} = 1 = s^{-1}s, \tag{1.3.1}$$

so that each element of S is invertible and S contains the (unique: Exercise O 3.1G) inverse of each of its elements, then the relation σ is symmetric, and thus becomes an equivalence relation. An abstract monoid S satisfying (1.3.1) is a *group*. Thus the series

transitive, pre-ordering, equivalence

of increasingly stronger properties of the relation σ corresponds to the series

semigroup, monoid, group

of increasingly stronger properties of the set S.

If G is a group, then G-sets A have a number of special properties not shared by more general monoid actions. Let B be a G-subset of the G-set A. Then $xg \in B$ for $x \in A$, $g \in G$ implies $x = x1 = xgg^{-1} \in Bg^{-1} \subseteq B$, so the complement $\overline{B} = \{x \in A | x \notin B\}$ of the G-subset B is itself a G-subset of A. In particular, indecomposable G-sets are irreducible (contrast with Exercise 1K). Such G-sets are also described as being *transitive*. A non-empty irreducible G-subset of a G-set (A, G) is called an *orbit* of (A, G). The set of orbits of the (right) G-set (A, G) is denoted by A/G. Note that a non-empty subset B of A is an orbit of (A, G) iff $B = xG$ for each element x of B.

Proposition 1.3.1. *Let G be a group and let (A, G) be a (right) G-set. Then (A, G) is the disjoint union $\Sigma A/G$.*

Proof. If two irreducible G-subsets of (A, G) have a non-trivial intersection, then they coincide, since their intersection is itself a G-subset. Thus the union $\bigcup A/G$ of the orbits is the disjoint union $\Sigma A/G$. If $\Sigma A/G < A$, pick x in $A - \Sigma A/G$. Then $x \in xG \subseteq \Sigma A/G$, a contradiction. □

A second special property of groups not shared by all monoids is that each group G is isomorphic to its opposite, via the inversion map

$$J : G \to G; g \mapsto g^{-1} \tag{1.3.2}$$

(cf. Exercises 1.3A and 1.3B). Thus each left G-set A, considered as a right G^{op}-set (A, G^{op}), is similar to the right G-set (A, G) given by declaring $(1, J) : (A, G^{\mathrm{op}}) \to (A, G)$ to be a similarity. Nevertheless, it is still convenient to consider left G-sets and to use the notation (1.6). The set of orbits of a left G-set (A, G) is then denoted by $G \backslash A$.

Irreducible (right) G-sets may be described in terms of subgroups of G. Recall that a *subgroup* H of G is a submonoid containing the inverse of each of its elements. The group G becomes a left H-set or right H^{op}-set (G, H^{op}) under the action

$$L : H \to G^G; h \mapsto (L_h : g \mapsto hg). \tag{1.3.3}$$

The orbits of this action are the *right cosets* $Hx = \{hx | h \in H\}$ of the various elements x of G with respect to the subgroup H. The set of orbits is denoted by $H \backslash G$. A set T of representatives for the orbits, so that $G = \bigcup_{t \in T} Ht = \Sigma_{t \in T} Ht$, is called a *right transversal* to H in G. The set of orbits becomes a G-set, known as a *homogeneous space*, under the action

$$g : H \backslash G \to H \backslash G; Hx \mapsto Hxg \tag{1.3.4}$$

for g in G. Note that $(H \backslash G, G)$ is irreducible, since $x^{-1}y : Hx \mapsto Hy$ for any Hx, Hy in $H \backslash G$.

If t is an element of a right G-set X, then the *stabilizer* (or "inertial subgroup" or "isotropy subgroup") of t is the set

$$G_t = \{g \in G | tg = t\}.$$

Note that G_t is a subgroup of G. Indeed, there is a G-homomorphism

$$L_t : (G, G) \to (X, G); g \mapsto tg. \tag{1.3.5}$$

The stabilizer G_t, clearly closed under inversion, is the submonoid $1^{\ker L_t}$ of G given by Proposition 1.1. Moreover, for each element g of G, the congruence class $g^{\ker L_t}$ is the coset $G_t g$, so that the right G-sets $(G_t \backslash G, G)$ and $(G^{\ker L_t}, G)$ coincide. The First Isomorphism Theorem for G-sets applied to the homomorphism (1.3.5) then yields the following:

Proposition 1.3.2. *Let G be a group and let (X, G) be a non-empty irreducible (right) G-set. Then for an element t of X, there is a G-isomorphism $(G_t \backslash G, G) \cong (X, G)$. In particular, if u is also an element of X, then $(G_t \backslash G, G) \cong (G_u \backslash G, G)$.* □

Together, Propositions 1.3.1 and 1.3.2 show that any G-set is a disjoint union of homogeneous spaces:

Theorem 1.3.3 (Structure Theorem for G-Sets). *Let G be a group and let (A, G) be a right G-set. Let T be a set of representatives for the orbits A/G. Then*

$$(A, G) \cong \sum_{t \in T} (G_t \backslash G, G). \quad \square \tag{1.3.6}$$

Example 1.3.4. Let G be the direct product $\{1, T\} \times \{1, T\}$ of the two-element group $\{1, T\}$ with itself. Set $X = \{a, a', b, b'\}$. Consider the G-set (X, G) given by the disjoint union $(A \cup B, \langle T \rangle \times \langle T \rangle)$ of Example 1.2.1. The element a of the G-subset $\{a, a'\}$ has stabilizer $G_a = \{(1,1), (1,T)\}$. The element b of the G-subset $\{b, b'\}$ has stabilizer $G_b = \{(1,1), (T,1)\}$. Theorem 1.3.3 thus yields $(X, G) \cong \Sigma_{t \in \{a, b\}}(G_t \backslash G)$. □

As a final point of distinction between group actions and general monoid actions on a set X, one may narrow down the codomain of the representation (1.1). The codomain of a monoid representation is the monoid of all functions from X to X, denoted by X^X in accord with the observation that $|X^X| = |X|^{|X|}$. By a similar convention, the group of all bijections from X to X is denoted by $X!$ [so that $|X!| = |X|!$, at least for finite sets X: Exercise 1.3K. A potential comparable notation $\Gamma(X+1)$ is not to be recommended]. The representation yielding a group action (X, G) may then be written as

$$R: G \to X! \tag{1.3.7}$$

in place of (1.1). Theorem O 4.3 specializes to *Cayley's Theorem* yielding the *right regular representation*

$$R: G \to G!;\ g \mapsto \left(R_g : x \mapsto xg\right) \tag{1.3.8}$$

of a group G. One may also consider the *left regular representation*

$$JL: G \to G!;\ g \mapsto \left(L_{g^J} : x \mapsto g^{-1}x\right) \tag{1.3.9}$$

of G as a right G-set.

EXERCISES

1.3A. Show that each element of a group has a unique inverse (cf. Exercise O 3.1G).

1.3B. Show that inversion (1.3.2) furnishes an isomorphism of each group with its opposite.

1.3C. Let $f: G \to H$ be a monoid homomorphism between the groups G and H. Show that $(xf)^{-1} = x^{-1}f$ for all x in G.

1.3D. Let G be the group of all rotations of the Cartesian plane $\mathbb{R}^2$ about the origin. Show that $(\mathbb{R}^2, G) \cong (\{(0,0)\}, G) \cup \Sigma_{d>0}(G,G)$. [Hint: The summand (G,G) corresponding to a positive real d is the circle of radius d about the origin.]

1.3E. Let G be a group. Show that G is a right $(G \times G)$-set under the *biregular representation* $T: G \times G \to G!; (g,h) \mapsto (x \mapsto g^{-1}xh)$.

1.3F. Let G be the non-abelian group of order 6. Apply Theorem 1.3.3 to the direct product $(G \times G, G \times G)$ in $\underline{\underline{G \times G}}$ of the biregular representation of G with itself.

1.3G. Let G be a group of even order. Consider G as a right $\langle J \rangle$-set for the map J of (1.3.2). Apply Theorem 1.3.3 to show that G contains a subgroup of order 2.

1.3H. Let σ be a pre-order on a set S. Define a relation α on S by $x \alpha y$ iff $x \sigma y$ and $y \sigma x$.

(a) Show that α is an equivalence relation on S.

(b) Show that "$x^\alpha \le y^\alpha$ iff $x \sigma y$" yields a well-defined order relation on S^α.

1.3I. Formulate and prove the First Isomorphism Theorem for Groups.

1.3J. Let H be a subgroup of a group G.

(a) Show that for each g in G, the map R_g of (1.3.8) yields an H^{op}-isomorphism between the right cosets H and Hg [defined as below (1.3.3)].

(b) Prove $(G, H^{\text{op}}) = \Sigma_{H \backslash G}(H, H^{\text{op}})$.

(c) Exhibit a set isomorphism $G \cong H \times (H \backslash G)$ (Hint: cf. Exercise O 3.2N).

(d) Prove *Lagrange's Theorem*: If G is finite, then $|H|$ divides $|G|$.

1.3K. (a) For finite sets X, Y, let $\text{Iso}(X,Y)$ denote the set of bijections from X to Y. Prove, by induction, that $|\text{Iso}(X,Y)| =$ **if** $|X| = |Y|$ **then** $|X|!$ **else** 0.

(b) Conclude $|X!| = |X|!$ for finite X.

1.3L. For subsets A, B of a group G, write $AB = \{xy \in G \mid x \in A,\ y \in B\}$.

(a) If A and B are subgroups of G, show that AB is a subgroup of G if and only if $AB = BA$.

(b) Give an example of subgroups A, B of a group G such that AB is not a subgroup of G.

(c) If A and B are finite subgroups of G, show that $|AB| \cdot |A \cap B| = |A| \cdot |B|$.

(d) Let A, B and C be subgroups of G, with $A \subseteq C \subseteq AB$. Show that $C = (AB) \cap (AC) = A(B \cap C)$.

1.3M. An element x of a monoid $(M, \cdot, 1)$ is *invertible* or a *unit* if and only if its image R_x under the right regular representation is invertible. Show that the set M^* of invertible elements of M forms a submonoid of M that is a group, the *group of units* of M.

1.4. Free Groups

Let X be a set. The aim is to construct a group XG and map $\iota : X \to XG$ such that, given any set map $f : X \to M$ from X to a group M, there is a unique homomorphism (of monoids or groups: cf. Exercise 1.3C) $f^G : XG \to M$ such that $\iota f^G = f$. By analogy with (O 4.1.2), the group XG is called the *free group over* X. The construction of XG will be achieved using an automaton. (In this sense at least, the construction of free groups is automatic.)

Form the disjoint union $A = X \cup XJ$ of two copies of the set X. The insertions are

$$i : X \to A;\ x \mapsto x \tag{1.4.1}$$

and $j : X \to A$; $x \mapsto x^J$. Define the map $J : A \to A$ with $J^2 = 1_A$ by $J : x \mapsto x^J$. [This map models the inversion (1.3.2).] Let P denote the subset of the free monoid A^* consisting of those words in the alphabet A in which no letter a is followed immediately by a^J. For example, the words $xhhx^Jx^Jyy$, 1, and x^Jy^Jxy lie in P. On the other hand, the words xyy^Jy^J and y^Jx^Jx do not. [In the latter, the letter x^J is followed by $x = (x^J)^J$.] The set P becomes the state space of an automaton (P, A^*). The elementary action of a letter a^J of the alphabet is defined as follows. Using the notation of (O 4.6), observe that P is the disjoint union $(P \cap A^*R_a) \cup (P - (P \cap A^*R_a))$ of two subsets, namely the set of words pa in P ending in a, and its complement, the set of words q that do not end in a. The action of a^J on P is then given as the disjoint union $(pa \mapsto p) \cup (q \mapsto qa^J)$. This action is invertible, and its two-sided inverse is the action of a. Since the composition of bijections is bijective, the compound actions also biject. Moreover, the inverse of the compound action of a word $a_1 \dots a_n$ in A^* is the compound action of the word $a_n^J \dots a_1^J$. Thus the image of A^* under the representation $R : A^* \to P^P$ defining the automaton (P, A^*) is actually a subgroup of $P!$. This subgroup is defined to be the free group XG on the set X. Thus the representation yields a monoid homomorphism

$$R : A^* \to XG. \tag{1.4.2}$$

There is then a map

$$\iota : X \to XG;\ x \mapsto xiR \tag{1.4.3}$$

constructed with the help of (1.4.1).

Example 1.4.1. Consider the case where X is the two-element set $\{x, y\}$. A fragment of the transition diagram of the automaton (P, A^*) is as follows:

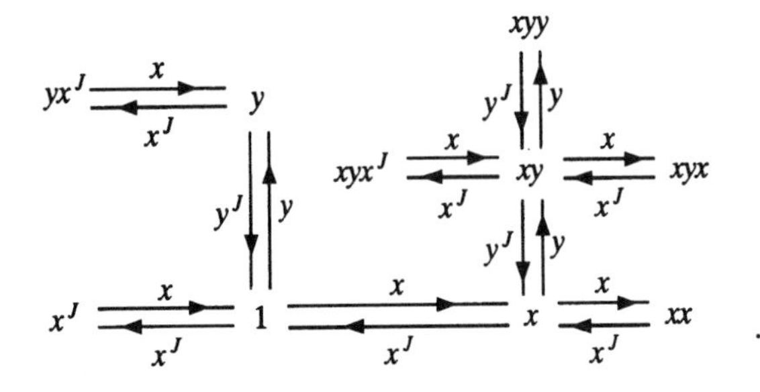

Note how the elementary action of x shifts one step to the right, and x^J shifts one step to the left. Similarly, y and y^J shift one step up and down, respectively. Bearing this in mind, the whole transition diagram may be summarized by the unlabeled tree of Figure 1.1, in which the central vertex denotes the point 1 of P. □

Suppose that $u = a_1 \ldots a_m$ and v are words in A, with letters $a_1, \ldots, a_m$. The word v is said to be (obtained from u by) an *elementary reduction* of u if $a_i a_{i+1} = aa^J$ for some $1 \leq i < m$ and a in A, and then $v = a_1 \ldots a_{i-1} a_{i+2} \ldots a_m$. Successive application of at most $\lfloor m/2 \rfloor$ elementary reductions to u reduces it to a word in P that is the result $1R_u$ of acting on 1 in P by the element u of A^*. Two words u, v in A are said to be *group equivalent* if and only if $1R_u = 1R_v$. Note that group equivalence is an equivalence relation, the kernel of the set map $L_1 : A^* \to P;\ u \mapsto 1u$. Moreover, P is a full set of representatives for the equivalence classes. The element $1R_u$ of P is called the (*completely*) *reduced form* or *normal form* of u. Given two elements $u = a_1 \ldots a_m$ and $v = b_1 \ldots b_n$ in A^*, it takes at most $\lfloor m/2 \rfloor + \lfloor n/2 \rfloor$ steps to reduce each to their respective normal forms $1R_u, 1R_v$. The procedure of completely reducing each of the two words and comparing their normal forms is called a *decision procedure* for group-equivalence. In general, given an equivalence relation α on a set S, there may or may not exist a decision procedure for deciding, in a finite or otherwise bounded number of steps, whether two general elements u, v of S are related by α.

A set may $f : X \to M$ from X to a group M extends to a set map $f : A \to M$ given by the following diagram [which is an instance of (O 3.2.5)]:

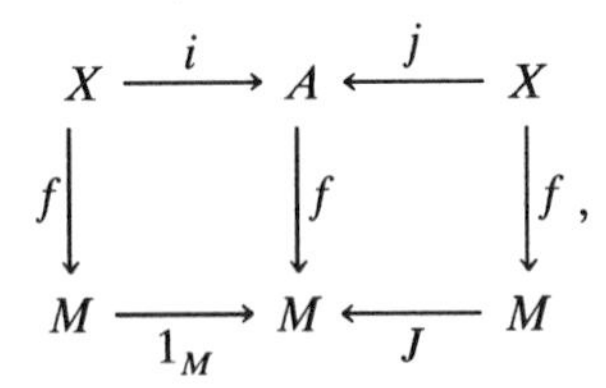

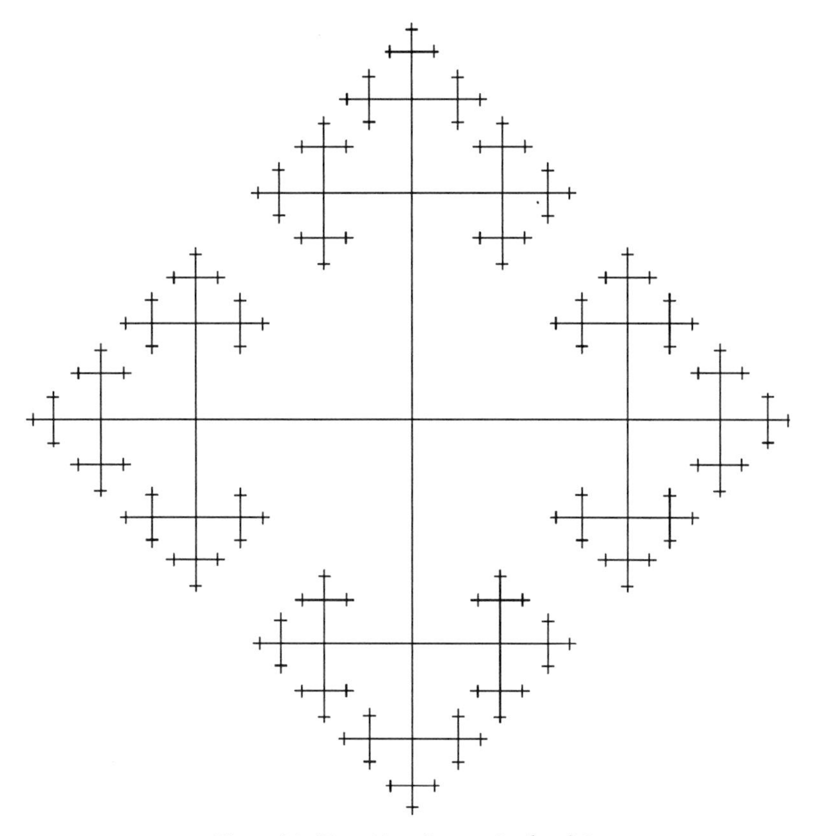

Figure 1.1. Transition diagram for $\{x, y\}G$.

so that $a^J f = (af)^{-1}$ in the group M. By the universality property (O 4.1.2) for the free monoid A^* on A, there is then a unique monoid homomorphism $\bar{f}: A^* \to M$ extending $f: A \to M$. If $u = saa^J t$ in A^* and $v = st$, so that v is an elementary reduction of u, then $u\bar{f} = saa^J t\bar{f} = s\bar{f}af a^J f t\bar{f} = s\bar{f}af(af)^{-1}t\bar{f} = s\bar{f}t\bar{f} = st\bar{f} = v\bar{f}$. It follows that group-equivalent words u, v in A^* have the same images $u\bar{f}, v\bar{f}$ in the group M. Consideration of the particular instance (1.4.3) of a set map $f: X \to M$ from X to a group yields

(1.4.4) $$\forall p \in P, \forall u, v \in A^*, 1R_u = 1R_v \Rightarrow pR_u = pR_v.$$

In other words, the map

(1.4.5) $$L_1: XG \to P;\ R_u \mapsto 1R_u$$

injects. Since $p = 1R_p$ for p in P, (1.4.5) also surjects. Define a multiplication on P by

(1.4.6) $$p \cdot q = pR_q.$$

Then the bijection L_1 becomes an isomorphism of groups, so that one may also consider the set P as the free group on X. (Indeed, $L_1 : R_p R_q \mapsto 1R_p R_q = pR_q = p \cdot q$. Here each element R_u of XG is represented by R_p using the normal form p of u.) Now for elements p, q of P, the product (1.4.6) in the free group P is a normal form for the concatenation pq in A^*. Thus the restriction

$$\bar{f} : P \to M \tag{1.4.7}$$

to P of the monoid homomorphism $\bar{f} : A^* \to M$ becomes a monoid homomorphism, and hence a group homomorphism, from the group P to the group M. Finally, define

$$f^G = \left(L_1 \bar{f} : XG \to P \to M\right). \tag{1.4.8}$$

Then f^G is the desired group homomorphism with $\iota f^G = f$. This completes the construction of the free group XG over X.

It remains to note that the set X is a subset of the free group P. One thus refers to the set X as a set of *generators* of P and to P as the *free group generated by the set* X. (Of course, P may have other sets of generators, e.g. XJ.) Finally, (1.4.3) injects, as the composition of the embedding $X \to P$ with the inverse of the bijection (1.4.5).

EXERCISES

1.4A. Copy a portion of the tree of Figure 1.1 and label the vertices on your copy. The copy should include the vertices 1, yxy^J, and $x^J y^J xy$.

1.4B. Consider the right multiplications (or "translations") $R_{(1,0)}$ and $R_{(0,1)}$ in the group $(\mathbb{R}^2, +, (0,0))$ of translations of the Cartesian plane $\mathbb{R}^2$. Define a set map $f : X \to \mathbb{R}^2$ from the set $X = \{x, y\}$ of Example 1.4.1 to the group $\mathbb{R}^2$ by $f : x \mapsto R_{(1,0)}, y \mapsto R_{(0,1)}$. Determine the corresponding map $\bar{f}$ of (1.4.7) from the tree of Figure 1.1 to the plane $\mathbb{R}^2$. What is the kernel of $\bar{f}$?

1.4C. Show that the right regular representation (1.3.8) of the free group P as in (1.4.6) is the concatenation of the restriction to P of (1.4.2) with the embedding of XG in $P!$.

1.4D. Let X be a subset of a group M. Let $e : X \to M$; $x \mapsto x$ be the set map embedding X in M, and let $e^G : XG \to M$ be the group homomorphism with $\iota e^G = e$. The image XGe^G of this group homomorphism is called the *subgroup* $\langle X \rangle$ or $\langle X \rangle_M$ of *M generated by X*.

(a) Show that $\langle X \rangle$ is a subgroup of M.

(b) Show that $\langle X \rangle$ is the intersection of the set of subgroups of M that contain X.

(c) For P as in (1.4.6), show that $P = \langle X \rangle$.

(d) Considering the group M as a monoid, what is the difference between $\langle X \rangle$ as in this exercise and the concept $\langle X \rangle$ of Exercises O 4M and I 1.1E?

1.4E. Work through the constructions of this section for the sets $X = \varnothing$ and $X = \{a\}$.

1.4F. Consider the free monoid $(A^*, \cdot, J)$ equipped with the "pseudo-inversion" $J : A^* \to A^*;\ a_1 \ldots a_n \mapsto a_n^J \ldots a_1^J$. Is this structure a group?

1.5. Free Commutative Monoids and Partitions

Groups are obtained from monoids by adding an inversion operation (1.3.2). Free groups may then be constructed using free monoids, as in Section I 1.4. In the other direction, one may obtain new types of structure from monoids by imposing extra restrictions, such as commutativity (O 4.3.1). This raises the question of constructing the free commutative monoid on a set X. For each x in X, consider the *Kronecker delta function* $(\delta_x : X \to \mathbb{N}) = (\{x\} \to \{1\}) \cup ((X - \{x\}) \to \{0\})$. Since $(\mathbb{N}, +, 0)$ is a monoid, the freeness (O 4.1.2) of the free monoid X^* yields a monoid homomoprhism $\bar{\delta}_x : X^* \to \mathbb{N}$. The image $u\bar{\delta}_x$ of a word u in X is called the *exponent* or *multiplicity of occurrence* of x in u. Then the *free commutative monoid* $X^{*\kappa}$ on X is the quotient of X^* by the congruence $\kappa = \bigcap_{x \in X} \ker \bar{\delta}_x$, equipped with the injection

$$\iota : X \to X^{*\kappa};\ x \mapsto x^\kappa = \{x\}. \tag{1.5.1}$$

Given a set map $f : X \to M$ from X to a commutative monoid M, the monoid homomorphism $\bar{f} : X^* \to M$ given by (O 4.1.2) has the property $u \kappa v \Rightarrow u\bar{f} = v\bar{f}$. There is thus a well-defined homomorphism of commutative monoids

$$\bar{f}^\kappa : X^{*\kappa} \to M : u^\kappa \mapsto u\bar{f} \tag{1.5.2}$$

with $\iota \bar{f}^\kappa = f$, exhibiting the freeness of $X^{*\kappa}$.

The free commutative monoid $X^{*\kappa}$ on a set X is sometimes interpreted as the set of *finite multisubsets* of the set X. In the usual subset notation, elements are only counted once. Thus $\{2, 1, 3\}$ and $\{1, 1, 2, 1, 3, 3\}$ both represent the same subset $\{1, 2, 3\}$ of $\mathbb{Z}$. On the other hand, in the element 112133^κ of $\mathbb{Z}^{*\kappa}$, the element 1 of $\mathbb{Z}$ occurs $112133\bar{\delta}_1 = 3$ times, 2 occurs $112133\bar{\delta}_2 = 1$ time, and 3 occurs $112133\bar{\delta}_3 = 2$ times. Thus 112133^κ represents the *multiset* $\langle 1, 1, 2, 1, 3, 3 \rangle$. This multiset is distinct from $\langle 2, 1, 3 \rangle$ or 213^κ, in which each of the elements 1, 2, 3 occurs with multiplicity 1. The identity element of $X^{*\kappa}$ represents the empty multiset $\langle\ \rangle$. Note how the multiplication in $X^{*\kappa}$ yields a *disjoint union* of multisets, e.g.

$$\langle 1,1,2,1,3,3\rangle\langle 2,1,3\rangle = \langle 1,1,2,1,3,3,2,1,3\rangle = \langle 1,1,1,1,2,2,3,3,3\rangle,$$

comprising four 1's, two 2's, and three 3's. The *size* or *cardinality* of a multisubset u^κ of X is the finite natural number $\Sigma_{x \in X} u\bar{\delta}_x$. For example, the size of $\langle 1,1,2,1,3,3,2,1,3\rangle$ in $\mathbb{Z}^{*\kappa}$ is $\cdots 0 + 0 + 4 + 2 + 3 + 0 + 0 + \cdots = 9$. The *elements* of the multiset u^κ are the letters comprising the word u in X. Thus the elements of $\langle 1,2,1,1\rangle = \langle 1,1,1,2\rangle$ are $1, 1, 1$, and 2. By convention, a set may be considered as a multiset in which each element has multiplicity 1.

If the set X is a singleton $\{x\}$, then the map $\bar{\delta}_x^\kappa : X^{*\kappa} \to (\mathbb{N}, +, 0)$ is an isomorphism of commutative monoids. The natural number 1 is the image of $x^\kappa = \{x\}$ [cf. (1.5.1)], so the natural number n is the image of $\{x\}\{x\}\cdots\{x\}$ – n copies of $\{x\}$ multiplied in $X^{*\kappa}$. It is convenient to identify $\mathbb{N}$ with $X^{*\kappa}$ under this isomorphism, for each singleton X.

For any set X, the set of non-identity elements of the free commutative monoid $(X^{*\kappa}, \cdot, 1)$ on X is a subsemigroup of $(X^{*\kappa}, \cdot\,)$ which forms the *free commutative semigroup* on the set X (cf. Exercise 1.5E). For any semigroup S or $(S, \cdot\,)$, let S^c denote the free commutative semigroup on the underlying set S. Denote the singleton semigroup $(\{1\}, \cdot\,)$ by 1. As discussed in the preceding paragraph, one may identify 1^c with the semigroup $(\mathbb{Z}^+, +\,)$ of positive integers under addition. The unique semigroup homomorphism $1^c \to 1$ is then realized by removing braces and carrying out the multiplication in $(\{1\}, \cdot\,)$. For example, $3 = \{1\}\{1\}\{1\} \mapsto 1.1.1 = 1$. Now form $(\mathbb{Z}^+, +\,)^c = 1^{cc}$. Elements of 1^{cc} are called (*integer*) *partitions*. For example, consider

$$\{\{1\}\}\{\{1\}\{1\}\}\{\{1\}\{1\}\}\{\{1\}\{1\}\{1\}\} \tag{1.5.3}$$

or $\{1\}\{2\}\{2\}\{3\}$. Such an element may also be written in the *sum form* $1 + 2 + 2 + 3$ (as a "sum waiting to happen"), or in the *multiset form* $\langle 1,2,2,3\rangle$, or in the *product form* $1^1 2^2 3^1 = 1^1 2^2 3^1 4^0 5^0 \ldots$. The elements of the multiset are called the *parts* of the partition. The size of the multiset, i.e. the number of parts, is called the *length* of the partition. Note that the length of a partition is its image under the semigroup homomorphism $\lambda : 1^{cc} \to 1^c$ induced by the set mapping $1^c \to 1$. When the partition is written in the basic form as in (1.5.3), then λ acts by removing the inner braces: $\{\{1\}\}\{\{1\}\{1\}\}\{\{1\}\{1\}\}\{\{1\}\{1\}\{1\}\}\lambda = \{1\}\{1.1\}\{1.1\}\{1.1.1\} = \{1\}\{1\}\{1\}\{1\} = 4$. The *sum* of a partition is its image under the semigroup homomorphism $\sigma : 1^{cc} \to 1^c$ induced by the identity set mapping on 1^c. The sum acts on basic forms by removing the outer braces: $\{\{1\}\}\{\{1\}\{1\}\}\{\{1\}\{1\}\}\{\{1\}\{1\}\{1\}\}\sigma = \{1\}\{1\}\{1\}\{1\}\{1\}\{1\}\{1\}\{1\} = 8$. If the partition is in its sum form as a "sum waiting to happen," then the homomorphism σ "makes the sum happen": $(1 + 2 + 2 + 3)\sigma = 1 + 2 + 2 + 3 = 8$. A partition with sum n is said *to partition* n or to be a *partition of* n.

Elements of 1^c have been identified as positive integers, and elements of 1^{cc} as integer partitions. One more level is also used: Elements of 1^{ccc} are known as *Segre characteristics*. They have been written classically as non-empty multisets of non-empty multisets of positive integers, with (square) brackets outermost and parentheses or round brackets innermost. Thus the classically written Segre characteristic [(21)(11)(11)] denotes the multiset $\langle\langle 2, 1\rangle, \langle 1, 1\rangle, \langle 1, 1\rangle\rangle$ of multisets. The sum of the sums of the partitions in a Segre characteristic is called the *sum* of the characteristic. It is the image of the characteristic under the semigroup homomorphism $1^{ccc} \to 1^c$ induced by the set map $\sigma : 1^{cc} \to 1^c$. If the Segre characteristic were to be written in a basic form with three levels of nested braces, then the sum would be realized by removing the outer two levels.

EXERCISES

1.5A. For each of the following words u in $\{1, 2, 3\}^*$, list all the elements of the congruence class u^κ:

(a) 132; (b) 111; (c) 121.

1.5B. Show that the product of the monoid homomorphisms $\overline{\delta}_x^\kappa : X^{*\kappa} \to \mathbb{N}$ yields an injective monoid homomorphism $X^{*\kappa} \to \mathbb{N}^X$. When is this homomorphism surjective?

1.5C. Show that each multisubset of a set X is a disjoint union of singleton multisubsets.

1.5D. Show that the map $X \to 2^X;\ x \mapsto \{x\}$ induces a monoid homomorphism $\varepsilon : X^{*\kappa} \to (2^X, \cup, \emptyset)$. The image of a multisubset u^κ of X under ε is called the *set of elements* of the multisubset.

1.5E. State and prove the defining universality property for the free commutative semigroup $X^{*\kappa} - \{1\}$ on a set X.

1.5F. Show that the inclusion $P \hookrightarrow \mathbb{Z}^+$ of the set P of prime numbers in the set of positive integers extends to a monoid isomorphism $P^{*\kappa} \to (\mathbb{Z}^+, \cdot\,)$. (Hint: This is the Fundamental Theorem of Arithmetic.)

1.5G. List all 7 partitions of 5 in each of the four forms: basic, sum, multiset, and product.

1.5H. List all 14 Segre characteristics that sum to 4.

2. GROUPS AND QUASIGROUPS

The first section of this chapter approached a group Q as a special kind of monoid, namely one in which the right and left multiplications biject. More generally, one may consider a set Q with a binary operation of *multiplication*, conventionally denoted by $\cdot$ or juxtaposition, having the same property that the right and left multiplications biject. Such a set-with-structure is called a *quasigroup.* The multiplication is not required to be associative.

Example 2.1. Consider the integers $(\mathbb{Z}, -)$ equipped with the binary operation of subtraction. Note that this operation is not associative. [Pick three integers m, n, p at random, and compare $(m - n) - p$ with $m - (n - p)$. If they agree, try again.] Nevertheless, the "multiplications" $R_m : \mathbb{Z} \to \mathbb{Z};\ x \mapsto x - m$ and $L_m : \mathbb{Z} \to \mathbb{Z};\ x \mapsto m - x$ biject. Geometrically, considering the integers as a discrete series of points along the real axis, R_m is a translation, while L_m is a combination of a reflexion in the origin with a translation. □

It is sometimes advantageous to write the right and left multiplications by an element x in the form $R(x)$ and $L(x)$, rather than in the form R_x and L_x used originally in (O 4.6) and (O 4.7). The image ax of an element a under right multiplication by x may then be written as $aR(x)$. Similarly, $aL(x) = xa$. This notation is already helpful in the next example.

Example 2.2. Any group $(Q, \cdot, 1)$ forms a quasigroup $(Q, \cdot)$, one which happens to be associative. Note that the right multiplication $R(x)$ has the right multiplication $R(x^{-1})$ as its right and left inverse, while $L(x^{-1})$ is the inverse of $L(x)$. Now the empty set forms a quasigroup $(\varnothing, \cdot)$, in which the multiplication is the unique map $\varnothing \times \varnothing \to \varnothing$ [namely the one whose graph is the empty subset of $(\varnothing \times \varnothing) \times \varnothing$]. For any three elements x, y, z of $\varnothing$, the associative law $(xy)z = x(yz)$ holds. However, the associative quasigroup $(\varnothing, \cdot)$ is not a group, since there is no identity element. □

Example 2.3. Consider the real numbers $(\mathbb{R}, \circ)$ equipped with the binary operation of arithmetic mean, i.e. $x \circ y = (x + y)/2$. Then $(\mathbb{R}, \circ)$ is a quasigroup that is not associative. However, it is idempotent [in the sense of (O 4.3.2)]. □

Note that describing a product as "non-associative" means that it may or may not be associative. Thus the products of Examples 2.2 and 2.3 are both non-associative.

Example 2.4. Houses in the Great Plains are almost always built so that their four walls face north, south, east, and west. A research organization wishes to compare four types of exterior wall covering: aluminum, brick, cedar, and vinyl. Walls facing different directions are subject to different weathering influences: the low winter sun from the south, the prevailing wind from the west, the shade on the north side that encourages fungal growth, etc. The research organization owns four houses, with addresses 1, 2, 3, and 4. It decides to equip each of these houses with each of the four wall coverings, so that the influences due to the differing locations of the houses are spread out evenly. To distribute the effects of the directional weathering evenly, each wall covering should be used exactly once facing each of the four possible directions. The wall coverings are assigned to the various walls according to the following plan:

(2.1)

	N	S	E	W
1	A	V	B	C
2	C	B	A	V
3	V	A	C	B
4	B	C	V	A

For example, house 2 has cedar to the north, brick to the south, aluminum to the east and vinyl to the west. The critical feature of the plan is apparent in the 4×4 square in the bottom right-hand corner. Each of the four symbols A, B, C, V appears exactly once in each row and in each column. For an n-element set Q, an arrangement of the elements into an $n \times n$ square such that each element appears exactly once in each column and each row is called a *Latin square* on Q. Latin squares are very useful in the design of experiments like the one above, in which one wishes to distribute the influence of various factors as evenly as possible. Now the construction of a Latin square on a finite set Q is essentially equivalent to the specification of a quasigroup multiplication on Q. The multiplication table of such a quasigroup may be obtained from the Latin square by labeling the rows and columns with the elements of Q in some order. Thus the Latin square of (2.1) leads to the quasigroup structure on the set of types of wall covering with the following multiplication table:

(2.2)

$\cdot$	A	B	C	V
A	A	V	B	C
B	C	B	A	V
C	V	A	C	B
V	B	C	V	A

Note that the unique occurrence of each of the four elements in the row labeled B means that the left multiplication $L(B)$ bijects. Similarly, the unique occurrence of each of the elements in the column labeled B means that $R(B)$ bijects. Conversely, given the multiplication table of a finite quasigroup $(Q,\cdot)$, one may obtain a Latin square by removing the left-hand and upper borders. □

Example 2.5. A large organization owns a lakeside retreat and training center, including a small yacht that is berthed there. The organization sends groups of seven management trainees to the center for week-long seminars, starting on Monday and finishing on Sunday. Each day of such a week, a group of three of the trainees is selected to crew the yacht, under the supervision of a permanent captain. The selection is such that each pair of trainees appears together exactly once in a crew. Thus each trainee spends a total of three days on the yacht. The trainees are assigned numbers from 1 to 7, and the crew schedule is then as follows:

(2.3)

Day	*Crew*
Monday	2, 4, 6
Tuesday	1, 4, 5
Wednesday	3, 4, 7
Thursday	1, 2, 3
Friday	2, 5, 7
Saturday	1, 6, 7
Sunday	3, 5, 6

.

This schedule corresponds to a quasigroup $(Q, \cdot)$ on the set Q of trainees defined by $x \cdot y = z$ if and only if $x = y = z$ or $\{x, y, z\}$ forms a crew. The multiplication table of the quasigroup becomes

(2.4)

$\cdot$	1	2	3	4	5	6	7
1	1	3	2	5	4	7	6
2	3	2	1	6	7	4	5
3	2	1	3	7	6	5	4
4	5	6	7	4	1	2	3
5	4	7	6	1	5	3	2
6	7	4	5	2	3	6	1
7	6	5	4	3	2	1	7

.

Note that $(Q, \cdot)$ is commutative, so that $R(x) = L(x)$ for each x. Then the bijectivity of $R(x)$ on $Q - \{x\}$ corresponds to the fact that x appears exactly once together with each other crew member y.

EXERCISES

2A. Show that a non-empty, associative quasigroup has a unique identity element.

2B. Verify that $(\mathbb{R}, \circ)$ in Example 2.3 really is a quasigroup that is not associative.

2C. Show that the positive reals form a quasigroup under the operation of the geometric mean.

2D. Construct a 5×5 Latin square without using the addition or subtraction table of the integers modulo 5. (Hint: Try filling in the table one location at a time, maintaining the Latin square property. If you get stuck, backtrack just enough to get unstuck, and try again.)

2E. Can you construct a three-person crew schedule analogous to (2.3), but assigning nine trainees over a 12 day time period?

2F. Which of the following conditions on a non-empty set S of integers is necessary and sufficient for S to be a subgroup of the group $(\mathbb{Z},+,0)$ of integers:

(i) S is closed under the associative operation of addition;

(ii) S is closed under the non-associative operation of subtraction?

2.1. Multiplication Groups of Quasigroups

Let $(Q,\cdot)$ be a quasigroup. There are maps $R: Q \to Q!;\ x \mapsto R(x)$ and $L: Q \to Q!;\ x \mapsto L(x)$. The map R injects because each left multiplication bijects:

$$qR(x) = qR(y) \Rightarrow qx = qy \Rightarrow xL(q) = yL(q) \Rightarrow x = y.$$

Similarly L injects, since each right multiplication bijects. The disjoint union $R(Q) \dot{\cup} L(Q)$ of the images of R and L is the set of generators for a free group $\tilde{G}$ or $\mathrm{UMlt}(Q,\cdot)$ known as the *universal multiplication group* of the quasigroup $(Q,\cdot)$. The disjoint union $(R(Q) \hookrightarrow Q!) \dot{\cup} (L(Q) \hookrightarrow Q!)$ of the respective embeddings of $R(Q)$ and $L(Q)$ in $Q!$ extends, by the freeness of $\tilde{G}$, to a group homomorphism $\tilde{G} \to Q!$. The image of this group homomorphism is called the *(combinatorial) multiplication group* G or $\mathrm{Mlt}(Q,\cdot)$ of the quasigroup $(Q,\cdot)$. In the terminology of Exercise 1.4D, it is the subgroup of $Q!$ generated by the (not necessarily disjoint) union $R(Q) \cup L(Q)$. The group homomorphisms $\tilde{G} \to Q!$ and $G \hookrightarrow Q!$ are representations making the quasigroup $(Q,\cdot)$ a $\tilde{G}$-set $(Q,\tilde{G})$ and a G-set (Q,G). Note that these actions are irreducible.

Example 2.1.1. Let $(Q,\cdot,1)$ be a commutative group. Since $R(x) = L(x)$ for all x in Q, the multiplication group $\mathrm{Mlt}\,Q$ of Q is the subgroup of $Q!$ generated by $R(Q)$. However, the subset $R(Q)$ of $Q!$ is already a subgroup of $Q!$, namely the image of the right regular representation (1.3.8). By Theorem O 4.3, it follows that the map

$$R: Q \to \mathrm{Mlt}\,Q;\ x \mapsto R(x) \tag{2.1.1}$$

is an isomorphism of groups. If $(Q,\cdot,1)$ is the additive group $(\mathbb{Z}_n,+,0)$ of integers modulo a positive integer n, then the group action $(\mathbb{Z}_n, \mathrm{Mlt}(\mathbb{Z}_n,+))$ is called the *cyclic action* or *cycle* or *n-cycle* C_n. These names are also applied to similar actions. By abuse of language, the abstract group $\mathrm{Mlt}(\mathbb{Z}_n,+)$ or $(\mathbb{Z}_n,+,0)$ is also called the *cyclic group* C_n. In analogous fashion, group actions similar to $(\mathbb{Z}, \mathrm{Mlt}(\mathbb{Z},+))$ are described as *countable cycles* C_∞. □

Example 2.1.2. Let $(Q,+,0)$ be a commutative group written additively, and let $(Q,-)$ be the corresponding quasigroup with subtraction as multiplication. The inversion map (1.3.2) in the group $(Q,+,0)$ is the negation map

$J: Q \to Q;\ x \mapsto -x$. Note that negation gives a group isomorphism $J:(Q,+,0) \to (Q,+,0)$. For q in Q, let $R_+(q): Q \to Q;\ x \mapsto x + q$ denote right "multiplication" (i.e. addition) for the group $(Q,+,0)$. Then for the quasigroup $(Q,-)$, one has

$$R(q) = R_+(qJ) = JR_+(q)J \tag{2.1.2}$$

and

$$L(q) = JR_+(q). \tag{2.1.3}$$

Now $R(q)^{-1} = JR_+(q)^{-1}J = JR_+(qJ)J = R(qJ)$ and $L(q)^{-1} = R_+(q)^{-1}J = R_+(qJ)J = JJR_+(qJ)J = JR_+(q) = L(q)$. One may also verify $R(q)R(r) = R(q+r)$, $R(q)L(r) = L(q+r)$, $L(q)R(r) = L(q-r)$, and $L(q)L(r) = R(q-r)$, either directly by application of each side of the equations to an element x of Q, or by use of the relations (2.1.2) and (2.1.3). Thus the union $R(Q) \cup L(Q)$, which is disjoint for $|Q| > 2$, is already a subgroup of $Q!$, since any word in the multiplications or their inverses reduces to an element of $R(Q) \cup L(Q)$. This means that $\mathrm{Mlt}(Q,-) = R(Q) \cup L(Q)$, with $\mathrm{Mlt}(Q,-) = R(Q) \,\dot\cup\, L(Q)$for $|Q| > 2$. If $(Q,+,0)$ is the additive group $(\mathbb{Z}_n,+,0)$ of integers modulo a positive integer n, then group actions similar to $(\mathbb{Z}_n, \mathrm{Mlt}(\mathbb{Z}_n,-))$ are described as *dihedral actions* D_n. By abuse of language, the abstract group $\mathrm{Mlt}(\mathbb{Z}_n,-)$ is also called the *dihedral group* D_n. In analogous fashion, group actions similar to $(\mathbb{Z}, \mathrm{Mlt}(\mathbb{Z},-))$ are described as *countable dihedral actions* D_∞, while groups isomorphic to $\mathrm{Mlt}(\mathbb{Z},-)$ are called *countable dihedral groups* D_∞. □

Example 2.1.3. Let Q be a group. The subset $R(Q)$ of $Q!$ is a subgroup of $Q!$, namely the image of the right regular representation (1.3.8). The subset $L(Q)$ is also a subgroup, the image of the left regular representation (1.3.9). The multiplication group Mlt Q of Q is the image of the biregular representation T of Exercise 1.3E. By the First Isomorphism Theorem for Groups (cf. Theorem O 3.3.1, Exercise 1.3I), the multiplication group of Q is the quotient of Q^2 by the kernel of T. □

EXERCISES

2.1A. Verify the formulas for the products of left and right multiplications given in Example 2.1.2, by each of the two suggested methods.

2.1B. Show that the action C_6 is the direct product of the actions C_2 and C_3. (Warning: it does not immediately suffice to exhibit a monoid or group isomorphism $\mathbb{Z}_6 \cong \mathbb{Z}_2 \times \mathbb{Z}_3$.)

2.1C. Given quasigroups $(Q_1,\cdot)$ and $(Q_2,\cdot)$ with respective multiplication groups G_1 and G_2, the *direct product* $(Q_1 \times Q_2,\cdot)$ is the Cartesian product $Q_1 \times Q_2$ equipped with the componentwise multiplication $(x_1, x_2)(y_1, y_2) = (x_1y_1, x_2y_2)$.

(a) Show that the direct product is a quasigroup.

(b) Is the action $(Q_1 \times Q_2, \mathrm{Mlt}(Q_1 \times Q_2))$ the product $(Q_1, G_1) \times (Q_2, G_2)$ in $(\underline{\underline{\mathrm{Mon}}}; \underline{\underline{\mathrm{Set}}})$?

2.1D. Let a Latin square be given on the finite set Q. The multiplication table of a quasigroup $(Q,\cdot)$ is obtained (as in Example 2.4) by labeling the rows and columns of the Latin square with the elements of Q in some order. Which of the following depend on the particular choice of labeling:

(i) the abstract group $\mathrm{Mtl}(Q,\cdot)$;

(ii) the action $(Q, \mathrm{Mlt}(Q,\cdot))$;

(iii) the similarity type of the action $(Q, \mathrm{Mlt}(Q,\cdot))$?

2.1E. Let a quasigroup $(Q,\cdot)$ be given by the following multiplication table on the set $\mathbb{Z}_3$:

$\cdot$	0	1	2
0	0	2	1
1	2	1	0
2	1	0	2

.

(i) Show that the action $(Q, \mathrm{Mlt}(Q,\cdot))$ is a dihedral action D_3.

(ii) Is there a map $f: Q \to \mathbb{Z}_3$ such that $xf - yf = (x \cdot y)f$ for all x, y in Q?

2.1F. Repeat Exercise 2.1E for the quasigroup $(Q, \circ)$ with multiplication table

$\circ$	0	1	2
0	1	0	2
1	0	2	1
2	2	1	0

.

2.1G. Two subsets X, Y of a group G are said to *centralize mutually* or to be *permutable* if $xy = yx$ for all $x \in X$, $y \in Y$. Let Q be a non-empty quasigroup such that the subsets $L(Q)$ and $R(Q)$ of $Q!$ are permutable. Show that Q is a group. Conclude that a non-empty, associative quasigroup is a group.

2.1H. Show that D_4 is similar to $C_2 \wr C_2$.

The element $R(1)$ of $\mathrm{Mlt}(\mathbb{Z}_n, +)$ has action

$$R(1): 0 \mapsto 1 \mapsto 2 \mapsto \cdots \mapsto (n-2) \mapsto (n-1) \mapsto 0 \tag{2.1.4}$$

on the set $\mathbb{Z}_n = \{0, 1, \dots, n-1\}$. This action may be denoted more succinctly by the *cycle notations* $(012\dots(n-1))$ or $(12\dots(n-1)0)$ or $\dots$ or

$((n-1)0\ldots(n-2))$. The notation carries over to any cyclic action C_n by similarity. For example, $L(V)$ in (2.2) may be written as $(ABCV)$ or $(BCVA)$ or $(CVAB)$ or $(VABC)$. Occasionally, one inserts commas as separators, writing, e.g. (A, B, C, V) instead of $(ABCV)$.

Now consider a dynamical system (X, T) (as in Section O 4.2) in which X has finite order n and T bijects. The operator T of such a system is called a *permutation* (*of the set* X). (Warning: If X is infinite, many writers restrict the extension of the use of the term "permutation" to the case of bijective operators T that only move finitely many elements of X. Cf. Section 3.2.) Let $\langle T\rangle$ be the subgroup of $X!$ generated by T. Forgetting inversion, $\langle T\rangle$ is a cyclic monoid of index 0 and period p. Recalling inversion and Lagrange's Theorem (Exercise 1.3J), p divides $n!$. Now the dynamical system (X, T) yields the group action $(X, \langle T\rangle)$. Moreover, the orbits of (X, T) in the sense of (O 4.2.1) are the orbits of the $\langle T\rangle$-set X. Each orbit $(\{x_i, x_iT, \ldots, x_iT^{n_i-1}\}, \langle T\rangle)$ is a cyclic action C_{n_i}. The analysis of the dynamical system (X, T), in the sense discussed at the end of Section O 4.2, is then achieved by Proposition 1.3.1. It yields $(X, \langle T\rangle) = \Sigma_i C_{n_i}$. Note here that the *cycle type* $T\tau = n_1 + n_2 + \cdots + n_i + \cdots$ of T is (the sum form of) an integer partition of n. One may then extend the cycle notation by writing

$$T = \prod_i \left(x_i, x_iT, \ldots, x_iT^{n_i-1}\right), \tag{2.1.5}$$

the *cycle decomposition* of T. If T is not the identity on X, one usually suppresses mention of the cycles in (2.1.5) with $n_i = 1$. For example, $L(7)$ in (2.4) may be written as (16)(25)(34), suppressing the cycle (7). Note that the elements of cycles of length 1 form the set $\mathrm{Fix}\langle T\rangle$ or $\mathrm{Fix}\, T$ of fixed points of $(X, \langle T\rangle)$. Thus

$$(\mathrm{Fix}\, T, \langle T\rangle) = \sum \left\{C_{n_i} | n_i = 1\right\}, \tag{2.1.6}$$

and in particular $|\mathrm{Fix}\, T| = |\{i | n_i = 1\}|$.

The cycle notation (2.1.5) provides an efficient calculus in the group of bijections or *permutation group* $X!$ on the finite set X. For example, in the multiplication group of the quasiqroup (2.2), one may compute

$$\begin{aligned} L(C)L(V) &= (AVB)(ABCV) \\ &= \left(A \overset{L(C)}{\mapsto} V \overset{L(V)}{\mapsto} A\right)\left(B \overset{L(C)}{\mapsto} A \overset{L(V)}{\mapsto} B\right)\left(C \overset{L(C)}{\mapsto} C \overset{L(V)}{\mapsto} V \overset{L(C)}{\mapsto} B \overset{L(V)}{\mapsto} C\right) \\ &= (A)(B)(CV) = (CV). \end{aligned}$$

Moreover, with T as in (2.1.5),

$$T^{-1} = \prod_i \left(x_iT^{n_i-1}, \ldots, x_iT, x_i\right). \tag{2.1.7}$$

Finally, if $X = Y \cup Z$ with $T|_Z = 1_Z$ and $U|_Y = 1_Y$, so that the elements appearing in non-trivial cycles of T are disjoint from the elements appearing in non-trivial cycles of U, then the elements T and U of $X!$ centralize each other (i.e. $\{T\}$ and $\{U\}$ centralize in the sense of Exercise 2.1G). This fact is often summarized in the form "disjoint cycles permute."

EXERCISES

2.1I. Determine the multiplication group of the quasigroup with multiplication table (2.2).

2.1J. Show that two dynamical systems (X, T) and (Y, U) with finite state spaces and bijective operators yield similar actions if and only if the cycle types $T\tau$ and $U\tau$ coincide.

2.1K. Show that the *order* of the element (2.1.5) of $X!$, i.e. its period p, is the lowest common multiple of the parts n_i of the partition $n_1 + n_2 + \cdots + n_i + \cdots$. (Hint: Disjoint cycles permute.)

2.2. Divisions and Quasigroup Homomorphisms

A quasigroup has been defined as a set $(Q, \cdot)$ with a binary multiplication operation such that all the right and left multiplications biject. Recalling that homomorphic images of monoids are monoids, and that homomorphic images of groups are groups, the definition of a quasigroup suffers from a major disadvantage: Homomorphic images of quasigroups need not be quasigroups. More specifically, there is a surjective map $f:(Q, \cdot) \to (P, \cdot)$ from a quasigroup Q to a set P equipped with a binary multiplication, such that the map is a homomorphism of the multiplication in the sense that $xf \cdot yf = (xy)f$ for all x, y in Q, but nevertheless $(P, \cdot)$ is not a quasigroup.

Example 2.2.1. Let P be the set of polynomials $p(x)$ with real coefficients. Define a binary "multiplication" operation on P by $p(x) \cdot q(x) = p(x) + q'(x)$, i.e. as the sum of the first factor and the derivative of the second. Let Q be the set of sequences $q = (q_0(x), q_1(x), \dots)$ of polynomials such that $q_n(x) = q'_{n+1}(x)$ for each natural number n. Define the multiplication on Q by componentwise multiplication in P, i.e. $p \cdot q = (p_0(x), p_1(x), \dots)(q_0(x), q_1(x), \dots) = (p_0(x) + q'_0(x), p_1(x) + q'_1(x), \dots)$. Note that $R(q)$ has the two-sided inverse $R(-q)$ for $-(q_0(x), q_1(x), \dots) = (-q_0(x), -q_1(x), \dots)$, while $L(p)$ has the two-sided inverse $(q_0(x), q_1(x), \dots) \mapsto (q_1(x) - p_1(x), q_2(x) - p_2(x), \dots)$. Thus $(Q, \cdot)$ is a quasigroup. On the other hand, $(P, \cdot)$ is not a quasigroup; e.g. left multiplication by zero in P is the non-injective map that differentiates its argument. Finally, the map $f: Q \to P$; $(q_0(x), q_1(x), \dots) \mapsto q_0(x)$ is a surjective homomorphism, having left inverse $q_0(x) \mapsto (q_0(x), q_1(x), \dots)$ with $q_{n+1}(x) = \int_0^x q_n(t)\, dt$ for natural numbers n. □

If one wishes to study quasigroups together with their homomorphisms, i.e. if one wishes to apply algebraic methods to the study of quasigroups, then the basic concept has to be given a different but equivalent definition. To begin, consider a quasigroup $(Q,\cdot)$, defined as having bijective right and left multiplications. One may then introduce two new operations of division, namely the *right division*

$$/ : Q^2 \to Q; (x, y) \mapsto x/y = xR(y)^{-1} \tag{2.2.1}$$

and the *left division*

$$\backslash : Q^2 \to Q; (y, x) \mapsto y\backslash x = xL(y)^{-1}. \tag{2.2.2}$$

Just as the product $x \cdot y$ is read as "x times y," so x/y may be read as "x divided by y (from the right)" and $x\backslash y$ may be read as "x dividing y (from the left)." From the definitions (2.2.1) and (2.2.2), it is apparent that the set $(Q,\cdot,/,\backslash)$ equipped with the multiplication and divisions satisfies

$$\begin{aligned} &IL : y\backslash(y \cdot x) = x;\ IR : x = (x \cdot y)/y;\\ &SL : y \cdot (y\backslash x) = x;\ SR : x = (x/y) \cdot y \end{aligned} \tag{2.2.3}$$

for all x, y in Q. For example, IR expresses the equation $R(y)R(y)^{-1} = 1$ in the multiplication group of $(Q,\cdot)$, while SL expresses $L(y)^{-1}L(y) = 1$. Now consider a set $(Q,\cdot/,\backslash)$ equipped with a binary multiplication and divisions such that (2.2.3) is satisfied. By IR, the right multiplications inject. By SR, they surject. By IL and SL, the left multiplications biject. Thus $(Q,\cdot)$ is a quasigroup. Summarizing:

Proposition 2.2.2. *A set with multiplication is a quasigroup if and only if it carries right and left divisions satisfying* (2.2.3). □

From now on, in speaking of "a quasigroup Q," it will be most helpful to think of the set-with-structure $(Q,\cdot,/,\backslash)$, i.e. to recall the right and left divisions as well as the multiplication. In expressions involving the multiplication and divisions, multiplication denoted by juxtaposition binds more strongly than the divisions or multiplication denoted by the $\cdot$ symbol. Thus $x \cdot yz$ denotes $x(yz)$, and $(x \cdot y)/(z \cdot t)$ may be written as xy/zt. Now a *quasigroup homomorphism* $f : Q \to P$ is a set map between the underlying sets of quasigroups $(Q,\cdot,/,\backslash)$ and $(P,\cdot,/,\backslash)$ such that $xf \cdot yf = (xy)f$, $xf/yf = (x/y)f$, and $xf\backslash yf = (x\backslash y)f$ for all x, y in Q. A subset S of Q is a *subquasigroup* of Q, written $S \leq (Q,\cdot,/,\backslash)$ or $S \leq Q$, if it is closed under all three operations multiplication, right division, and left division of Q. A *congruence* on a quasigroup Q is an equivalence relation on Q that is a subquasigroup of the direct square quasigroup Q^2 (using the direct product of quasigroups as in Exercise 2.1C). The quotient Q^α of the quasigroup Q by the congruence α then forms a quasigroup $(Q^\alpha, \cdot, /, \backslash)$ with well-defined operations $x^\alpha \cdot y^\alpha = (x \cdot y)^\alpha$, $x^\alpha / y^\alpha = (x/y)^\alpha$ and $x^\alpha \backslash y^\alpha = (x \backslash y)^\alpha$. A quasigroup Q is said to be *simple* if its only congruences are the *improper* congruence Q^2 and the *trivial* congruence $\hat{Q}$.

For elements y, z of a quasigroup Q, define the elements

$$\rho(y, z) = R(y \backslash y)^{-1} R(y \backslash z) \tag{2.2.4}$$

of the universal and combinatorial multiplication groups of Q. Note that $y\rho(y, z) = yR(y \backslash y)^{-1} R(y \backslash z) = yR(y \backslash z) = z$, since $yR(y \backslash y) = y \cdot (y \backslash y) = y$ by (2.2.3) *SL*. Moreover $\rho(y, y) = 1$. Indeed, for fixed y in Q, $\{\rho(y, q) | q \in Q\}$ is a right transversal to the stabilizers of y in G and $\tilde{G}$. Now setting

$$(x, y, z)P = x\rho(y, z), \tag{2.2.5}$$

it follows that $(y, y, z)P = z$ and $(z, y, y)P = z$ for all y, z in Q. Since $(x, y, z)P = (x/(y \backslash y)) \cdot (y \backslash z)$, the operation P preserves a quasigroup congruence α in the sense that $x_i \alpha x_i'$ for $1 \leq i \leq 3$ implies $(x_1, x_2, x_3)P \; \alpha \; (x_1', x_2', x_3')P$. An important consequence, which ultimately guarantees the good algebraic behavior of quasigroups, is given by the following:

Proposition 2.2.3. *The congruence relations on a quasigroup are permutable.*

Proof. Suppose that α and β are congruence relations on a quasigroup Q, with $x \alpha y \beta z$. Consider the following "book-keeping," in which the bottom line is obtained by applying the ternary operation P columnwise:

$$\begin{array}{ccccc}
x & \alpha & x & \beta & x \\
x & \alpha & y & \beta & z \\
z & \alpha & z & \beta & z \\
\hline
z = (x, x, z)P & \alpha & (x, y, z)P & \beta & (x, z, z)P = x
\end{array}$$

Then $x \; \beta \; (x, y, z)P \; \alpha \; z$, so that $\beta \circ \alpha \supseteq \alpha \circ \beta$. Similarly, $\alpha \circ \beta \supseteq \beta \circ \alpha$, whence $\alpha \circ \beta = \beta \circ \alpha$: The congruence relations α and β are permutable (in the sense of Section O 4.4). □

Another effect of the operation (2.2.5) is a rather curious characterization of quasigroup congruences.

Proposition 2.2.4. *Let Q be a quasigroup. Then a subquasigroup α of Q^2 is a congruence of Q if and only if it contains the diagonal subquasigroup $\hat{Q}$.*

Proof. A congruence, as a reflexive relation, contains the diagonal. Conversely, suppose that $\hat{Q} \leq \alpha \leq Q^2$. Symmetry and transitivity of α are to be demonstrated. Suppose $x\,\alpha\,y\,\alpha\,z$. Then:

$$\begin{array}{ccc} x & \alpha & x \\ x & \alpha & y \\ y & \alpha & y \\ \hline y = (x, x, y)P & \alpha & (x, y, y)P = x \end{array}$$

so that α is symmetric. Moreover:

$$\begin{array}{ccc} x & \alpha & y \\ y & \alpha & y \\ y & \alpha & z \\ \hline x = (x, y, y)P & \alpha & (y, y, z)P = z \end{array}$$

so that α is also transitive, as required. □

EXERCISES

2.2A. Let $(Q, \cdot, 1)$ be a group and let $(S, \cdot)$ be a semigroup. Let $f : (Q, \cdot) \to (S, \cdot)$ be a surjective semigroup homomorphism. Show that S is a group.

2.2B. Consider the quasigroup $(\mathbb{R}^{\times}, \cdot)$ of non-zero real numbers under multiplication. Describe the corresponding right and left division operations (2.2.1), (2.2.2) in terms of the usual division operation $\div$.

2.2C. Determine the right and left division operations of the quasigroup $(\mathbb{Z}, -)$ of Example 2.1.

2.2D. Demonstrate why the kernel of the function $f : Q \to P$ of Example 2.2.1 fails to be a congruence on the quasigroup Q.

2.2.E. Let $(Q, \cdot, /, \backslash)$ and $(P, \cdot, /, \backslash)$ be quasigroups. Suppose that a set map $f : Q \to P$ satisfies $xf \cdot yf = (xy)f$ for all x, y in Q. Show that f is a quasigroup homomorphism.

2.2F. If $f_i : (Q_{i-1}, \cdot, /, \backslash) \to (Q_i, \cdot, /, \backslash)$ is a quasigroup homomorphism for each $i = 1, \ldots, n$, verify that the composite map $f_1 f_2 \cdots f_n : (Q_0, \cdot, /, \backslash) \to (Q_n, \cdot, /, \backslash)$ is also a quasigroup homomorphism.

2.2G. Formulate and prove the First Isomorphism Theorem for Quasigroups.

2.2H. For a group, write the operation P of (2.2.5) in terms of multiplication and inversion. Do any other ternary group operations P have the property that $(y, y, z)P = z = (z, y, y)P$? Can you find infinitely many such operations (in the sense that they yield distinct elements of the free group on $\{x, y, z\}$)?

2.2I. Determine the right and left divisions in the quasigroups (2.2) and (2.4).

2.2J. Show that the order of a subquasigroup of a finite quasigroup need not divide the order of the quasigroup. (Cf. Exercise 1.3J.)

2.2K. Let α be a congruence on a quasigroup Q.

(a) For y, z in Q, show that the map $\rho(y, z)$ of (2.2.4) is a bijection from y^{α} to z^{α}.

(b) Show that α is uniquely determined by any one of its classes.

(c) If Q is finite, with element x, show that $|x^{\alpha}|$ divides $|Q|$.

(d) Show that quasigroups of prime order are simple.

2.2L. An *idempotent* of a quasigroup Q is an element e with $e \cdot e = e$.

(a) Show that an element x of a quasigroup Q is an idempotent iff $\{x\} \leq Q$.

(b) If e is an idempotent and α is a congruence, show that e^{α} is a subquasigroup of Q.

(c) Give an example of an element x and a congruence α of a quasigroup Q such that x^{α} is a subquasigroup of Q, but x is not an idempotent.

2.2M. A quasigroup Q is *entropic* iff $(xy)(zt) = (xz)(yt)$ for all x, y, z, t in Q. (The name means "inner turning," referring to the swapping of y and z.)

(a) Show that a quasigroup is entropic iff each of the operations $\cdot : Q^2 \to Q, / : Q^2 \to Q, \backslash : Q^2 \to Q$ is a homomorphism from Q^2 to Q.

(b) Show that commutative groups are entropic.

(c) Give an example of a non-empty entropic quasigroup that is not a commutative group.

2.3. Restriction and Induction

Let H be a subgroup of the group G. Each G-set (X, G) yields an H-set $(X, G)\downarrow^G_H = (X, H)$ obtained "by forgetting the G-operations outside H," i.e. furnished by the representation $H \hookrightarrow G \to X!$ of H. Sometimes, $(X, G)\downarrow^G_H$ is called the "H-reduct of the G-set X." Conversely, given an H-set (X, H), a G-set $(X, H)\uparrow^G_H$ will be constructed, together with an H-homomorphism $\eta : (X, H) \to (X, H)\uparrow^G_H\downarrow^G_H$, such that each H-homomorphism $f : (X, H) \to (A, G)\downarrow^G_H$ to the H-reduct of a G-set (A, G) may be extended to a unique G-homomorphism $\bar{f} : (X, H)\uparrow^G_H \to (A, G)$ with $\eta\bar{f} = f$. The G-set $(X, H)\uparrow^G_H$ is described as the G-set *induced* by the H-set (X, H) or [by analogy with (O 4.1.2) or Section 1.4] as the *free G-set over the H-set* (X, H).

The construction of the induced G-set $(X, H)\uparrow^G_H$ is based on two different structures carried by the set $X \times G$. Firstly, $X \times G$ is the free G-set

$(X \times G, G)$ over the set X, as in Exercise 1U. Secondly, $X \times G$ is the H-set $(X \times G, H)$ that is the direct product (in the class $\underline{\underline{H}}$ of H-sets) of (X, H) with the H-reduct (G, H) of the left regular representation (1.3.9) of G. Now consider the kernel of the H-homomorphism projecting $(X \times G, H)$ onto the trivial H-set $((X \times G)/H, H)$ of orbits of $(X \times G, H)$. This kernel α is a G-congruence on the free G-set $(X \times G, G)$, since $(x, g)g' = (x, gg')\alpha(xh, h^{-1}gg') = (xh, h^{-1}g)g'$ for x in X, h in H, and g, g' in G. Thus the projection $X \times G \to (X \times G)/H;\ (x, g) \mapsto (x, g)H$ may be used to put a G-action on $(X \times G)/H$, well-defined by

$$g' : (x, g)H \mapsto (x, gg')H \tag{2.3.1}$$

for g' in G. This yields the G-set $((X \times G)/H, G) = (X, H)\uparrow^G_H$. The map

$$\eta : X \to (X \times G)/H;\ x \mapsto (x, 1)H \tag{2.3.2}$$

is an H-homomorphism from (X, H) to $((X \times G)/H, G)\downarrow^G_H$, since $x\eta h = (x, 1)Hh = (x, h)H = (xh, h^{-1}h)H = (xh, 1)H = xh\eta$ for x in X and h in H. Finally, given an H-homomorphism $f : (X, H) \to (A, G)\downarrow^G_H$, the map

$$\bar{f} : (X \times G)/H \to A;\ (x, g)H \mapsto xfg \tag{2.3.3}$$

becomes a G-homomorphism uniquely specified by its property $\eta\bar{f} = f$.

EXERCISES

2.3A. If H is a subgroup of a finite group G, and if (X, H) is a finite H-set, show that $|(X \times G)/H| = |X| \times |G|/|H|$.

2.3B. Consider the right regular representation $(\langle J \rangle, \langle J \rangle)$ of the subgroup $\langle J \rangle$ of the dihedral group D_n. Determine the similarity type of the free D_n-set on the $\langle J \rangle$-set $(\langle J \rangle, \langle J \rangle)$.

2.3C. Let H be a subgroup of a subgroup K of a group G. For an H-set X, show that the G-sets $(X, H)\uparrow^K_H\uparrow^G_K$ and $(X, H)\uparrow^G_H$ are isomorphic.

2.3D. For $H \leq G$ and H-sets $(X, H), (Y, H)$, show that the G-sets $((X, H) \cup (Y, H))\uparrow^G_H$ and $(X, H)\uparrow^G_H \cup (Y, H)\downarrow^G_H$ are isomorphic.

2.3E. A submonoid H of a monoid M is said to be a *subgroup* of M if $\forall h \in H, \exists h^{-1} \in H.\ hh^{-1} = 1 = h^{-1}h$.

(a) Show that a subgroup of M is a subgroup of the group of units of M (cf. Exercise 1.3M).

(b) Describe the construction of the free M-set on an H-set (X, H).

2.3F. Show that $xH \mapsto x\eta G$ furnishes a bijection from the orbit set of (X, H) to the orbit set of $(X, H)\uparrow^G_H$.

2.4. Quasigroup and Group Conjugacy Classes

Let $(Q, \cdot, /, \backslash)$ be a quasigroup, with multiplication group G. The G-action (Q, G) is irreducible. On the other hand, the direct square action (Q^2, G) is always reducible, having the G-subset $(\hat{Q}, G)$ given by the diagonal $\hat{Q}$. More generally, each quasigroup congruence corresponds to a G-subset of Q^2.

Proposition 2.4.1. *A pre-ordering α on the quasigroup Q is a quasigroup congruence if and only if it is a G-subset of Q^2.*

Proof. First, suppose that α is a quasigroup congruence. For q in Q and (x, y) in α, $(x, y)R(q) = (xR(q), yR(q)) = (xq, yq) = (x, y)(q, q) \in \alpha$ and $(x, y)L(q) = (q, q)(x, y) \in \alpha$, so that α is a G-subset of Q^2. Conversely, suppose that a pre-ordering α is a G-subset of Q^2. By Proposition 2.2.4, it must be shown to be a subquasigroup of Q^2. Suppose $x \alpha y$ and $z \alpha t$. Then $(xz, yz) \in \alpha R(z) \subseteq \alpha$ and $(yz, yt) \in \alpha L(y) \subseteq \alpha$, whence $(xz, yt) \in \alpha$ by transitivity. Thus α is closed under multiplication. Further, α contains $((y, x)((z, t)L(y)^{-1}))L(x)^{-1} = (zL(x)^{-1}, tL(y)^{-1}) = (x, y)\backslash(z, t)$, and so is closed under left division. Closure under right division follows similarly. □

Corollary 2.4.2. *The quasigroup Q is simple if and only if the action of the multiplication group is primitive.* □

Let G be a group, and let (Q, G) be a non-empty transitive or irreducible G-set. The orbits of the direct square action (Q^2, G), often construed as relations on Q, are known as the *orbitals* of the G-set (Q, G). Now fix an element e in Q, and consider the stabilizer G_e, with action $(Q, G_e) = (Q, G)\downarrow^G_{G_e}$.

Proposition 2.4.3. *The direct square action (Q^2, G) is similar to the free G-set over (Q, G_e).*

Proof. There is a G_e-homomorphism

$$f:(Q, G_e) \to (Q^2, G_e);\ q \mapsto (e, q), \tag{2.4.1}$$

inducing the G-homomorphism $\bar{f}:(Q \times G)/G_e \to Q^2;\ (q, g)G_e \mapsto (eg, qg)$. Let $\{\rho(e, q) | q \in Q\}$ with $e\rho(e, q) = q$ be a right transversal from G to G_e [e.g. as in (2.2.4) if G is the multiplication group of a quasigroup Q]. Then $(p, q) \mapsto (q\rho(e, p)^{-1}, \rho(e, p))G_e$ furnishes a two-sided inverse to $\bar{f}$. □

Corollary 2.4.4. (Cf. Exercise 2.3F.) *The map*

$$Q/G_e \to Q^2/G;\ qG_e \mapsto (e, q)G \tag{2.4.2}$$

furnishes a bijection between the set of orbits of (Q, G_e) and the set of orbitals of (Q, G). □

If G is the multiplication group of a quasigroup Q, then the orbitals of (Q, G) are known as the *quasigroup conjugacy classes* of the quasigroup Q. If 1 is the identity element of a group Q, then the orbits of the stabilizer G_1 on Q are known as the (*group*) *conjugacy classes* of the group Q. Thus the map (2.4.2) of Corollary 2.4.4 provides a one-to-one correspondence between the quasigroup conjugacy classes and the group conjugacy classes of a group.

Let Q be a group. Then the composite of the diagonal embedding $\Delta : Q \to Q^2; g \mapsto (g, g)$ with the biregular representation $T : Q^2 \to G; (g, h) \mapsto (x \mapsto g^{-1}xh)$ is a group homomorphism

$$T : Q \to G_1; g \mapsto (T(g) : x \mapsto g^{-1}xg) \tag{2.4.3}$$

to the stabilizer G_1. This group homomorphism surjects, since $1 = g^{-1} \cdot 1 \cdot h \Rightarrow h = g$. Bijective quasigroup homomorphisms from a quasigroup Q to itself are called *automorphisms* of the quasigroup Q. They form a subgroup $\operatorname{Aut} Q$ of $Q!$, the *automorphism group* of Q. If Q is a group, each $T(g): Q \to Q; x \mapsto x^g$ is an automorphism of Q, a so-called *inner automorphism* or *conjugation by* g (whence the name "conjugacy class" for the orbits of $G_1 = \{T(g) | g \in Q\}$). The stabilizer G_1 is then called the *inner automorphism group* $\operatorname{Inn} Q$ of the group Q. Elements of $\operatorname{Aut} Q - \operatorname{Inn} Q$ are called *outer*. Note that for an arbitrary element x of a quasigroup Q (e.g. a non-identity element of a group Q), the elements of the stabilizer G_x are not necessarily automorphisms of Q.

Proposition 2.4.5. *Let G be a group, and let (X, G) be a G-set. Then for x in X and g in G, one has $G_x^g = G_{xg}$.*

Proof. For $s \in G_x$, one has $xgs^g = xgg^{-1}sg = xsg = xg$, so $G_x^g \subseteq G_{xg} = G_{xg}^{gg^{-1}} \subseteq G_{xgg^{-1}}^g = G_x^g$. □

If a subquasigroup P of a quasigroup Q is a class of a congruence α on Q, then P is said to be a *normal subquasigroup* of Q, written $P \triangleleft Q$. As in Exercise 2.2K(b), P then determines α uniquely. In this case, the quotient Q^α is often written as Q/P. If Q is a group, then Proposition 1.1 shows that the α-class 1^α is a submonoid of Q. Moreover, 1^α is a normal subgroup of Q. If $f: Q \to Q_1$ is a group homomorphism, then the normal subgroup $1^{\ker f}$ is called the (*group*) *kernel* $\operatorname{Ker} f$ of f. Normal subquasigroups may be recognized as follows.

Proposition 2.4.6. *Let P be a non-empty subquasigroup of a quasigroup Q. Then the following conditions are equivalent*:

(*a*) $P \triangleleft Q$;
(*b*) $\forall e \in P,\ PG_e = P$;
(*c*) $\exists e \in P.\ PG_e \subseteq P$.

Proof. (b) $\Rightarrow$ (c): P is non-empty.

(c) $\Rightarrow$ (a): Consider the relation

$$\alpha = \{(e,p)g | p \in P,\ g \in G\}. \tag{2.4.4}$$

It is reflexive, containing $(q,q) = (e,e)\rho(e,q)$ for each q in Q. Suppose $x\alpha y\alpha z$, say $x = eg_1$, $y = p_1g_1 = eg_2$, and $z = p_2g_2$ for p_i in P, g_i in G, $1 \le i \le 2$. Since $e\rho(e,p_1)g_1g_2^{-1} = p_1g_1g_2^{-1} = eg_2g_2^{-1} = e$, one has $p := p_2g_2g_1^{-1}\rho(e,p_1)^{-1} \in PG_e \subseteq P$. Then $p' := p_2g_2g_1^{-1} = p\rho(e,p_1) = (p/(e\backslash e))(e\backslash p_1) \in P$, since P is a subquasigroup of Q. Finally, $(x,z) = (eg_1, p_2g_2) = (eg_1, p'g_1) \in \alpha$, so that α is also transitive, and is thus a pre-ordering. Now $\alpha = (\{e\} \times P)G$ is a G-subset of Q^2, and so a quasigroup congruence on Q, by Proposition 2.4.1. Note $p \in P \Rightarrow (e,p) \in \alpha$, so $P \subseteq e^\alpha$. Conversely, $q \in e^\alpha \Rightarrow \exists p \in P,\ g \in G.\ (e,q) = (e,p)g \Rightarrow e = eg \Rightarrow g \in G_e \Rightarrow q = pg \in PG_e \subseteq P$. Thus $e^\alpha = P \triangleleft Q$.

(a) $\Rightarrow$ (b): Suppose $P = e^\alpha$ for a congruence α on Q. Consider an element p of P and an element g of G_e. Then $(e,pg) = (e,p)g \in \alpha g = \alpha$, the latter equality holding by Proposition 2.4.1. Thus $pg \in e^\alpha = P$, and so $PG_e \subseteq P \subseteq PG_e$. □

Corollary 2.4.7. *In a group G, a subgroup H is normal if and only if it coincides with each of its conjugates H^g.* □

For a subgroup H of a group G, the *core* $K_G(H)$ is the intersection $\bigcap_{g\in G} H^g$ of all the conjugates of H. In the other direction, the *normalizer* $N_G(H)$ of the subgroup H in the group G is $\{g \in G | H^g = H\}$. It is a subgroup of G that contains H as a normal subgroup (Exercise 2.4H). The quotient $N_G(H)/H$ is called the *Weyl group* $W_G(H)$ of H in G. If G is a commutative group, there are no non-identity conjugations (Exercise 2.4A), so that all subgroups of G are normal. [Alternatively, cf. Exercise 2.2M(b) and 2.4L.]

EXERCISES

2.4A. Show that each inner automorphism of an abelian group is trivial. Exhibit an automorphism of an abelian group that is not inner (i.e. a so-called *outer* automorphism).

2.4B. Give an example of an element x of a quasigroup Q with multiplication group G such that the stabilizer G_x is not a subgroup of the automorphism group Aut Q.

2.4C. Describe the quasigroup conjugacy classes of the quasigroups $(\mathbb{Z}_n, +)$ and $(\mathbb{Z}_n, -)$ of Examples 2.1.1 and 2.1.2. (Hint: Use Corollary 2.4.4.)

2.4D. Describe the group conjugacy classes of the dihedral group D_3 (cf. Exercise 1.3F), and determine all its normal subgroups.

2.4E. Give an example of a quasigroup with no proper normal subquasigroups, not even singletons.

2.4F. Let H be a subgroup of a group G. Prove that the core of H is a normal subgroup of G, namely the kernel of the representation (1.3.7) afforded by the homogeneous space (1.3.4) of right cosets of H.

2.4G. Show that the core of a subgroup H of a group G is the largest normal subgroup of G contained in H.

2.4H. Show that the normalizer of a subgroup H of a group G is the largest subgroup of G containing H as a normal subgroup.

2.4I. Let A and B be subgroups of a group G, with $B \leq N_G(A)$.

(a) Show that AB is a subgroup of $N_G(A)$.

(b) Show that $A \cap B$ is a normal subgroup of B.

(c) Show that $AB/A \cong B/(A \cap B)$.

2.4J. Let X be a non-empty set. Consider the free group XG on X and the "parity" group homomorphism $p = f^G : XG \to (\mathbb{Z}_2, +)$ extending the set map $f : X \to \{1\}$.

(a) Show that a word $u = a_1 \ldots a_m$ in $(X \cup X^J)^*$ maps to 0 under $(X \cup X^J)^* \xrightarrow{R} XG \xrightarrow{R} \mathbb{Z}_2$ [with R as in (1.4.2)] if and only if m is even.

(b) Show that the group kernel $\operatorname{Ker} p$ is a normal subgroup of XG, with $XG/\operatorname{Ker} p \cong \mathbb{Z}_2$. Elements of $\operatorname{Ker} p$ are called *even* elements of XG, while the remaining elements are called *odd*.

2.4K. Let H be a subgroup of a group G, with $|H \setminus G| = 2$. Show that H is a normal subgroup of G.

2.4L. Show that each subquasigroup P of an entropic quasigroup Q is normal, the quotient $Q/P = \{Px | x \in Q\}$ having multiplication $Px \cdot Py = P \cdot xy$ (cf. Exercise 2.2M).

2.4M. Is it true that a non-empty subset P of a quasigroup Q is a congruence class if and only if $PG_e = P$ for all e in P?

2.5. Stability and the Class Equation

Let Q be a quasigroup with multiplication group G. Consider the relation

$$\sigma = \sigma(Q) = \{(x, y) \in Q^2 | G_x = G_y\} \tag{2.5.1}$$

on Q. For (x, y) in σ and g in G, one has $G_x = G_y \Rightarrow G_{xg} = G_x^g = G_y^g = G_{yg}$ by Proposition 2.4.5, whence $(x, y)g$ also lies in σ and σ becomes a G-subset of Q^2. Since σ is clearly an equivalence relation, Proposition 2.4.1 shows that σ is a congruence on Q, the so-called *stability congruence* of Q. If Q is a group, then the normal subgroup 1^σ is the *center* $Z(Q) = \{z | G_z = G_1\} = \{z | \forall q \in Q, q^{-1}zq = z\} = \{z | \forall q \in Q, zq = qz\}$ of Q, i.e. the set of elements commuting with each element of the group.

The stability congruence is intimately connected with the map $\rho : Q^2 \to G$ of (2.2.4). For an element x of Q, recall that $\{\rho(x, y) | y \in Q\}$ is a right transversal to G_x in G, so that $G = \Sigma_{y \in Q} G_x \rho(x, y)$. Now consider the subtransversal $\{\rho(x, y) | y \in S\}$ to G_x in the normalizer $N_G(G_x)$. By Proposition 2.4.5, one has $y \in S \Leftrightarrow G_x^{\rho(x,y)} = G_y = G_x \Leftrightarrow yG_x = y$, so that $N_G(G_x) = \Sigma_{y \in \mathrm{Fix}\, G_x} G_x \rho(x, y)$.

Proposition 2.5.1. *(a) For y in Q, $yG_x = y \Leftrightarrow \rho(x, y) \in Z(G)$.*
(b) $N_G(G_x) = G_x Z(G)$.

Proof. (a) Firstly, suppose $\rho(x, y) \in Z(G)$. Then for $g \in G_x$, one has $yg = x\rho(x, y)g = xg\rho(x, y) = x\rho(x, y) = y$. Conversely, suppose $yG_x = y$. Let $z \in Q$, $g \in G$. Now $xL(x)^{-1}L(z/(x \backslash x))g = zg = xL(x)^{-1}L((zg)/(x \backslash x))$, implying that $yL(x)^{-1}L(z/(x \backslash x))g = yL(x)^{-1}L((zg)/(x \backslash x))$, i.e. $zR(x \backslash x)^{-1}R(x \backslash y)g = zgR(x \backslash x)^{-1}R(x \backslash y)$. Thus $z\rho(x, y)g = zg\rho(x, y)$, whence $\rho(x, y) \in Z(G)$.
(b) $G_x Z(G) \subseteq N_G(G_x) = \Sigma_{y \in \mathrm{Fix}\, G_x} G_x \rho(x, y) \subseteq G_x Z(G)$. □

Since the stability congruence $\sigma(Q)$ is a reflexive relation, it contains the diagonal $\hat{Q}$ as a subquasigroup. The next result shows that $\hat{Q}$ is a normal subquasigroup of the quasigroup $\sigma(Q)$, and locates the quotient $\sigma(Q)/\hat{Q}$.

Proposition 2.5.2. *The restriction of the map $\rho : Q^2 \to G$ to the stability congruence $\sigma(Q)$ yields a quasigroup homomorphism $\rho : \sigma(Q) \to Z(G)$ into the center of the multiplication group, inducing an embedding of the quotient $\sigma(Q)/\hat{Q}$ into $Z(G)$.*

Proof. Let $q \in Q$ and $(x_i, y_i) \in \sigma(Q)$ for $i = 1, 2$. Then

$$\begin{aligned}
q &= q\rho(x_1x_2, x_1x_2) = q\rho(x_2, x_2) \\
&\Rightarrow x_2L(x_1)L(x_1x_2)^{-1}L(q/(x_1x_2 \backslash x_1x_2)) = x_2L(x_2)^{-1}L(q/(x_2 \backslash x_2)) \\
&\Rightarrow y_2L(x_1)L(x_1x_2)^{-1}L(q/(x_1x_2 \backslash x_1x_2)) = y_2L(x_2)^{-1}L(q/(x_2 \backslash x_2)) \\
&\Rightarrow (q/(x_1x_2 \backslash x_1x_2))(x_1x_2 \backslash x_1y_2) = (q\rho(x_1, x_1)/(x_2 \backslash x_2))(x_2 \backslash y_2) \\
&\Rightarrow x_1R(y_2)L(x_1x_2)^{-1}L(q/(x_1x_2 \backslash x_1x_2)) \\
&\qquad = x_1L(x_1)^{-1}L(q/(x_1 \backslash x_1))R(x_2 \backslash x_2)^{-1}R(x_2 \backslash y_2) \\
&\Rightarrow y_1R(y_2)L(x_1x_2)^{-1}L(q/(x_1x_2 \backslash x_1x_2)) \\
&\qquad = y_1L(x)^{-1}L(q/(x_1 \backslash x_1))R(x_2 \backslash x_2)^{-1}R(x_2 \backslash y_2) \\
&\Rightarrow q\rho(x_1x_2, y_1y_2) = q\rho(x_1, y_1)\rho(x_2, y_2).
\end{aligned}$$

As in Exercise 2.2E, it follows that $\rho : \sigma(Q) \to G$ is a quasigroup homomorphism. By Proposition 2.5.1(a), the image of this homomorphism lies in $Z(G)$.

The kernel of ρ is a congruence having the diagonal $\hat{Q}$ as an equivalence class, so $\hat{Q} \triangleleft \sigma(Q)$. The First Isomorphism Theorem for quasigroups then shows that ρ induces an embedding of $\sigma(Q)/\hat{Q}$ into $Z(G)$. □

A quasigroup Q with $\sigma(Q) = Q^2$ is said to be *abelian*. Note that the empty quasigroup is abelian. Otherwise, an abelian quasigroup is an abelian group:

Proposition 2.5.3. *Let Q be a non-empty quasigroup, with multiplication group G. Then the following conditions are equivalent*:

(*a*) $\sigma(Q) = Q^2$;
(*b*) $\forall x \in Q, G_x \triangleleft G$;
(*c*) $\exists x \in Q. G_x \triangleleft G$;
(*d*) *G is abelian*;
(*e*) *Q is an abelian group.*

Proof. (a) ⇒ (b): For x in Q, g in G, Proposition 2.4.5 yields $G_x^g = G_{xg} = G_x$, the latter equality holding since $(xg, x) \in \sigma(Q)$. Corollary 2.4.7 then shows $G_x \triangleleft G$.

(b) ⇒ (c): Q is non-empty.

(c) ⇒ (d): $G_x = \bigcap_{g \in G} G_x^g = \bigcap_{g \in G} G_{xg} = \{1\}$ since $G \leq Q!$. By Proposition 2.5.1(b), $G = N_G(G_x) = G_x Z(G) = \{1\}Z(G) = Z(G)$ is abelian.

(d) ⇒ (e): Consider q_i in Q, for $1 \leq i \leq 3$. Then $(q_1 q_2) q_3 = q_2 L(q_1) R(q_3) = q_2 R(q_3) L(q_1) = q_1(q_2 q_3)$, so that Q is associative, say with identity element 1. Then $q_1 q_2 = 1R(q_1)R(q_2) = 1R(q_2)R(q_1) = q_2 q_1$, so that Q is an abelian group.

(e) ⇒ (a): $G_1 = \{1\} = G_x$ for all x in Q. □

Proposition 2.5.3 has an interesting consequence. There are non-abelian groups, known as *Hamiltonian* groups, in which each subgroup is normal (Exercise 2.5E). Proposition 2.5.3 then shows that a Hamiltonian group G cannot be the (combinatorial) multiplication group of a quasigroup. If it were, then a stabilizer G_x would be normal, and G would be abelian.

Now let e be an element of a finite quasigroup Q with multiplication group G, for example the identity element 1 of a group Q. Let T be a set of representatives for the orbits of G_e on Q, say $T = \{e = t_1, t_2, \ldots, t_s\}$. Order the representatives so that $i \leq j \Rightarrow |t_i G_e| \leq |t_j G_e|$. By Proposition 1.3.1, one has

$$|Q| = \sum_{i=1}^{s} |t_i G_e|. \tag{2.5.2}$$

This equation is known as the *class equation*. If G_{e,t_i} is the stabilizer of t_i in G_e, then Proposition 1.3.2 gives a G_e-isomorphism $(G_{e,t_i} \backslash G_e, G_e) \cong$

$(t_i G_e, G_e)$. In particular, Lagrange's Theorem (Exercise 1.3J) yields $|t_i G_e| = |G_{e,t_i} \setminus G_e| = |G_e| / |G_{e,t_i}||G_e|$.

Proposition 2.5.4. *Let Q be a quasigroup with multiplication group G of prime-power order. Then Q has a non-trivial stability congruence.*

Proof. Suppose $|G| = p^n$ for a prime p. Now $|Q| = |G|/|G_e| = p^m$ for $0 < m \leq n$. The class equation becomes $p^m = 1 + \Sigma_{i=2}^{s} |t_i G_i|$, with $|t_i G_e||G_e| = p^{n-m}$. If $|t_i G_e| > 1$ for $i > 1$, then the class equation modulo p becomes $0 \equiv 1 \pmod{p}$, a contradiction. Thus $|t_i G_e| = 1$ for $1 \leq i \leq p$, and $\{e, t_2, \ldots, t_p\}$ is contained in a non-trivial $\sigma(Q)$-class. □

Corollary 2.5.5. *A non-trivial group Q of prime-power order has a non-trivial center.*

Proof. The multiplication group G, as a quotient of Q^2, has prime-power order. □

EXERCISES

2.5A. Determine the stability congruence of the quasigroup $(\mathbb{Z}_n, -)$ of Example 2.1.2 in each of the cases (a) n even and (b) n odd.

2.5B. Show that a non-trivial quasigroup of prime-power order may have a trivial stability congruence. [Hint: Consider (2.2).]

2.5C. Determine the center of the dihedral group D_4.

2.5D. By considering the class equation, show that a quasigroup of order 8 cannot have a multiplication group of order 40. [Hint: Recall Exercise 2.2K(c).]

2.5E. Let Q be the group of 4×4 invertible matrices generated by

$$\begin{bmatrix} 0 & 0 & 1 & 0 \\ 0 & 0 & 0 & 1 \\ -1 & 0 & 0 & 0 \\ 0 & -1 & 0 & 0 \end{bmatrix} \text{ and } \begin{bmatrix} 0 & 0 & 0 & 1 \\ 0 & 0 & -1 & 0 \\ 0 & 1 & 0 & 0 \\ -1 & 0 & 0 & 0 \end{bmatrix}.$$

Show that Q is Hamiltonian, and that Q is not isomorphic to the group D_4.

2.5F. Let Q be a quasigroup in which each element is an idempotent (cf. Exercise 2.2L). Show that the stability congruence of Q is trivial.

2.5G. Let Q be a group. Show that the center $Z(Q)$ is invariant under the automorphism group $\text{Aut}\, Q$ of Q.

2.5H. Let Q be a Hamiltonian group and let A be an abelian group.

(a) Show that $Q \times A$ is Hamiltonian.

(b) Show that $Q \times Q$ is not Hamiltonian.

3. SYMMETRY

Let G be a group, and let (X, G) be a (right) G-set, determined by the representation

$$R : G \to X! . \tag{3.1}$$

The G-set (X, G) and the representation are said to be *effective*, or *faithful*, if (3.1) injects. The group G is said to be a *symmetry group on the set* X, or *a permutation group* of *degree* $|X|$ in the case of finite X, if (3.1) embeds G as a subgroup of $X!$. For example, the combinatorial multiplication group of a quasigroup is a permutation group on the quasigroup. If a G-set X is faithful, then one may use the First Isomorphism Theorem for groups to identify G with its image under (3.1), thereby construing G as a symmetry group on X. In particular, suppose that a symmetry group G on the set X is given. In the class $\underline{\underline{G}}$ of G-sets, one may then form the direct power $(X, G)^n = (X^n, G)$ for a positive integer n. The corresponding representation of G is still faithful, since the diagonal $\Delta : X \to X^n;\ x \mapsto (x,x, \dots, x)$ yields an injective G-homomorphism $\Delta : (X, G) \to (X^n, G)$. One may thus construe G, which was originally a symmetry group on X, as a symmetry group on the direct power X^n for each positive integer n.

In Section O 3.3, a (binary) relation on a set X was defined as a subset α of the direct square X^2. More generally, an n-ary *relation* on the set X is defined as a subset of the direct power X^n for a positive integer n. For $n = 1, 2, 3, 4$, "n-ary" becomes unary, binary, ternary, and quaternary, respectively. A *relational structure* (X, R) is a set X together with a multiset R of relations on X. For example, a monoid $(M, \cdot, 1)$ may be considered as a relational structure (M, R), in which R consists of the unary relation $\{1\}$ and the ternary relation $\{(x,y,x \cdot y) | x, y \in M\}$, the *multiplication table* of $(M, \cdot)$. Similarly, a quasigroup $(Q, \cdot, /, \backslash)$ is a relational structure (Q, R) in which R consists of the single ternary relation $\{(x, y, x \cdot y) | x, y \in Q\}$, the *multiplication table* of $(Q, \cdot)$. [Note that the multiplication tables of $(Q, /)$ and $(Q, \backslash)$ are obtained from the table of $(Q, \cdot)$ by permuting components, e.g. (x, y, z) has $z = x \cdot y$ iff (z, y, x) has $x = z/y$.] An *automorphism* of a relational structure (X, R) is a bijection $T : X \to X$ such that each relation α in R is invariant under T. In other words, if $\alpha \subseteq X^n$, then α is a $\langle T \rangle$-subset of $(X^n, \langle T \rangle)$. The set of all automorphisms of a relational structure (X, R) forms a subgroup of $X!$, the *automorphism group* $\mathrm{Aut}(X, R)$ of (X, R). Note that, if Q is a quasigroup with multiplication table μ, then the automorphism group Aut Q of the quasigroup Q, as defined in Section 2.4, coincides with the automorphism group Aut $(Q, \langle \mu \rangle)$ of the relational structure $(Q, \langle \mu \rangle)$.

The binary diagonal relation $\hat{X} = X(\Delta : X \to X^2) = \{(x, x) | x \in X\}$ is contained in each reflexive binary relation α, by the very definition of reflexivity. The complement $X^2 - \hat{X} = \{(x, y) | x \neq y\}$ is called the (*binary*) *diversity* relation on X. Subsets of the binary diversity relation are called binary *antireflexive* relations on X or *simple directed graphs on the vertex set* X. Each element (x, y) of a simple directed graph α on X may be represented graphically by an arrow $x \to y$, a so-called *directed edge* of the graph α. One sometimes draws these arrows in the form $x \xrightarrow{\alpha} y$, describing the directed edge as being "labeled" or " colored" by α. If an antireflexive relation α on X is symmetric, then it is called a *simple undirected graph on the vertex set* X. The pair (x, y), (y, x) of elements of α, or the equivalent doubleton $\{x, y\}$, may be represented graphically by an *(undirected) edge* $x - y$, possibly in the form $x \xrightarrow{\alpha} y$ "labeled" or "colored" by α. Directed or undirected graphs α on a vertex set X yield relational structures $(X, \langle \alpha \rangle)$, usually written just as (X, α).

Now consider an index $k > 1$. A k-ary relation α on X is said to be *antireflexive* if each element $(x_1, \ldots, x_k)$ of α contains no repeated components. The largest k-ary antireflexive relation $\{(x_1, \ldots, x_k) \in X^k \mid k = |\{x_1, \ldots, x_k\}|\}$ on X is called the *(k-ary) diversity* relation on X. A G-set (X, G) is *k-transitive* if the k-ary diversity relation is an irreducible G-subset of (X^k, G). For example, if X is finite, then $(X, X!)$ is $|X|$-transitive. By Corollary 2.4.4, an irreducible G-set X with element e is 2-transitive if and only if the G_e-set $(X - \{e\}, G_e)$ is irreducible. A G-set is *1-transitive* if and only if it is transitive, i.e. irreducible.

EXERCISES

3A. Draw the non-diagonal quasigroup conjugacy classes of the dihedral group D_3 as directed graphs (cf. Exercises 1.3F, 2.4D).

3B. Determine the automorphism group of the stability congruence of the quasigroup $(\mathbb{Z}_4, -)$.

3C. Determine the automorphism groups of the undirected graph

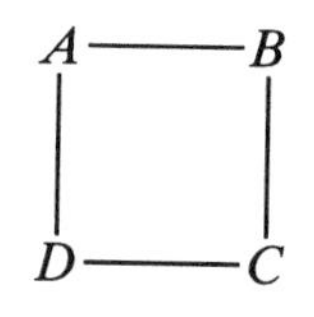

and the following directed graphs:

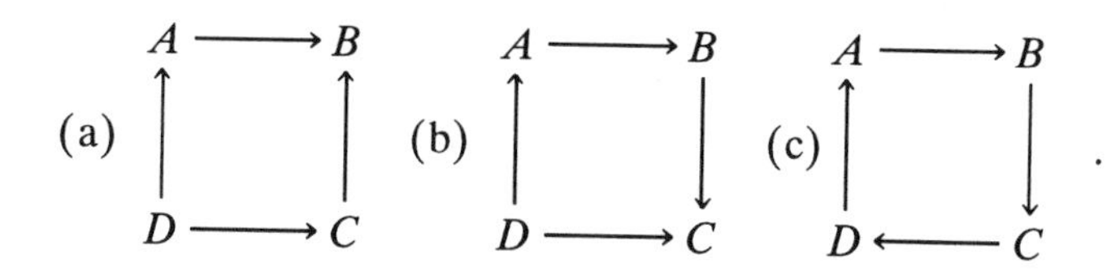

3D. Draw the graph $J(r,2)$ of the relation $\{(A,B)\mid 1 = |A \cap B|\}$ on the set of two-element subsets of an r-element set, for the cases $r = 4$ and $r = 5$. [The graphs $J(r,2)$ are examples of the so-called *Johnson graphs*: More generally one considers the relation $\{(A,B)\mid k-1 = |A \cap B|\}$ on the set of k-element subsets of an r-element set, for $k \le r/2$, yielding the graph $J(r,k)$. The graph $J(4,2)$ is also known as the *octahedral graph*, and $J(5,2)$ as the complement of the *Petersen graph*.]

3E. Consider the uniform code of length r over a two-element alphabet A (cf. Exercise O 4.1B). Say that two words are related if they differ by a single letter. The corresponding graph $H(r)$ is known as the *Hamming graph* or the *r-hypercube*. Draw $H(r)$ for $1 \le r \le 4$.

3F. Let $\{R_i \mid i \in I\}$ be a set of relation sets on X. Show that

$$\operatorname{Aut}\left(X, \bigcup\{R_i|i \in I\}\right) = \bigcap\{\operatorname{Aut}(X, R_i)|i \in I\}.$$

3G. Let $\{R_i \mid i \in I\}$ be a set of relation sets on X. Show that

$$\operatorname{Aut}\left(X, \bigcap\{R_i|i \in I\}\right) = \left\langle \bigcup\{\operatorname{Aut}(X, R_i)|i \in I\}\right\rangle$$

the subgroup of $X!$ generated by the set $\bigcup\{\operatorname{Aut}(X, R_i) \mid i \in I\}$.

3H. A graph is *rigid* if its only automorphism is the identity. Determine the minimal size of the vertex set of a rigid, simple, undirected graph.

3I. For $k \ge 1$, show that a k-transitive G-set is l-transitive for $1 \le l \le k$.

3J. Show that a 2-transitive G-set is simple.

3K. Let (X,G) be an irreducible G-set with element e. For $k > 1$, show that (X,G) is k-transitive if and only if $(X - \{e\}, G_e)$ is $(k-1)$-transitive. (Hint: Corollary 2.4.4 covers the case $k = 2$.)

3L. (a) Let G be a finite k-transitive permutation group of degree n. Prove that the *falling factorial*

$$[n]_k = n(n-1)(n-2)\cdots(n-k+1)$$

divides $|G|$.

(b) Show that the only n-transitive permutation group on a set X of order n is the permutation group $X!$.

3.1. Permutation Groups

As observed after Proposition 2.5.3, there are finite groups that cannot be realized as multiplication groups of quasigroups. On the other hand, each finite group can be realized as a permutation group, for example by the right regular representation (1.3.8). The question then arises as to whether each permutation group can be realized as the automorphism group of a relational structure. The following result, known as the *Krasner-Wielandt Theorem*, gives an affirmative answer.

Theorem 3.1.1. *Let G be a permutation group on a set X of (finite) cardinality n. For $1 \leq k \leq n$, consider the set X^k/G of orbits of G on X^k.*

(a) $\forall 1 \leq k < n$, $\mathrm{Aut}(X, X^{k+1}/G) \leq \mathrm{Aut}(X, X^k/G)$.

(b) $\exists 1 \leq k \leq n$. $G = \mathrm{Aut}(X, X^k/G)$.

Proof. (a) For each subgroup H of $X!$, there is an injective H-homomorphism $D: X^k \to X^{k+1}$; $(x_1, x_2, \ldots, x_k) \mapsto (x_1, x_1, x_2, \ldots, x_k)$. Now D induces a map $X^k/G \to X^{k+1}/G$; $\alpha \mapsto \alpha D$. For α in X^k/G and T in Aut $(X, X^{k+1}/G)$, αD is a $\langle T \rangle$-subset of $(X^k D, \langle T \rangle)$. Then α is a $\langle T \rangle$-subset of $(X^k, \langle T \rangle)$. It follows that T lies in $\mathrm{Aut}(X, X^k/G)$.

(b) Let $b: \{1,2, \ldots, n\} \to X$ biject. Consider $(1b, \ldots, nb)G$ in X^n/G. For T in $\mathrm{Aut}(X, X^n/G)$, one has $(1b, \ldots, nb)GT = (1b, \ldots, nb)G$. In particular, $\exists g \in G$. $(1b, \ldots, nb)T = (1b, \ldots, nb)g$. Then $T = g \in G$. Thus $\mathrm{Aut}(X, X^n/G) \leq G$. Conversely, $G \leq \mathrm{Aut}(X, X^n/G)$. □

If $G = \mathrm{Aut}(X, X^k/G)$, then G is said to be *k-closed*. For example, $X!$ is 1-closed.

The Krasner-Wielandt Theorem emphasizes the importance of determining the orbit sets of permutation group actions (X, G) and their direct powers (X^k, G). The decisive step is to evaluate the cardinality $|X/G|$. This is achieved by the following result (due to Cauchy and Frobenius) known as *Burnside's Lemma*.

Theorem 3.1.2. *The number of orbits in a permutation group action is the average number of fixed points. Thus in the permutation group action (X, G),*

$$|X/G| = \frac{1}{|G|} \sum_{T \in G} |\mathrm{Fix}\, T|. \tag{3.1.1}$$

Proof. In $X \times G$, consider the subset

$$\{(x, T) | xT = x\} = \sum_{x \in X} \{x\} \times G_x = \sum_{T \in G} (\mathrm{Fix}\, T) \times \{T\}.$$

It has cardinality

$$\begin{aligned}\sum_{T \in G} |\mathrm{Fix}\, T| &= \sum_{x \in X} |G_x| = \sum_{x \in X} |G|/|xG| = |G| \sum_{x \in X} 1/|xG| \\ &= |G| \sum_{Y \in X/G} \sum_{x \in Y} 1/|xG| = |G| \sum_{Y \in X/G} \sum_{x \in Y} 1/|Y| \\ &= |G| \sum_{Y \in X/G} 1 = |G| \cdot |X/G|.\end{aligned}$$

The second equality in the chain of equalities follows by Proposition 1.3.2. Dividing through by $|G|$ yields (3.1.1). □

The *permutation character* $\pi_{(X,G)}$ or π_G or π_X or π of a permutation group action (X, G) is the function

$$\pi : G \to \mathbb{N}; T \mapsto |\text{Fix } T|. \tag{3.1.2}$$

Noting that $\text{Fix}(A \times B, M) = \text{Fix}(A, M) \times \text{Fix}(B, M)$ for any monoid actions (A, M) and (B, M), Burnside's Lemma yields the formula

$$|X^k/G| = \frac{1}{|G|} \sum_{g \in G} \pi(g)^k \tag{3.1.3}$$

for each natural number k. These formulae may be summarized using the *Poincaré Series*

$$p(z) = \sum_{k=0}^{\infty} |X^k/G| z^k \tag{3.1.4}$$

of the action (X, G). The series may be considered purely formally, or as the series expansion of an analytic function $p : D \to \mathbb{C}$ whose domain is the open disc $D = \{z \in \mathbb{C} | z\bar{z} < |X|^{-2}\}$. Then

$$\begin{aligned}\sum_{k=0}^{\infty} |X^k/G| z^k &= \frac{1}{|G|} \sum_{k=0}^{\infty} \sum_{g \in G} (z\pi(g))^k \\ &= \frac{1}{|G|} \sum_{g \in G} \sum_{k=0}^{\infty} (z\pi(g))^k \\ &= \frac{1}{|G|} \sum_{g \in G} [1 - z\pi(g)]^{-1},\end{aligned}$$

so that

$$p(z) = \frac{1}{|G|} \sum_{g \in G} [1 - z\pi(g)]^{-1}. \tag{3.1.5}$$

Example 3.1.3. Consider the dihedral action $D_{2n} = \text{Mlt}(\mathbb{Z}_{2n}, -)$ of Example 2.1.2. Recall $D_{2n} = R(\mathbb{Z}_{2n}) \cup L(\mathbb{Z}_{2n})$. Note $x \in \text{Fix } R(q) \Leftrightarrow xJ \in \text{Fix } R_+(q)$. Thus $\text{Fix } R(0) = \mathbb{Z}_{2n}$, while $\text{Fix } R(q) = \varnothing$ for $q \neq 0$. Also $x \in \text{Fix } L(q) \Leftrightarrow q = 2x$. Thus $|\text{Fix } L(q)| = 2$ for $q \in 2\mathbb{Z}_{2n}$, while $|\text{Fix } L(q)| = 0$ for $q \in 2\mathbb{Z}_{2n} + 1$. By (3.1.5), $p(z) = \{3n - 1 + n(1 - 2z)^{-1} + (1 - 2nz)^{-1}\}/4n = 1 + z + \sum_{k=2}^{\infty} 2^{k-2}(1 + n^{k-1})z^k$. For example, there are $1 + n$ orbits in $(\mathbb{Z}_{2n}^2, D_{2n})$, namely the quasigroup conjugacy classes of $(\mathbb{Z}_{2n}, -)$. The two $2n$-element orbits $\hat{\mathbb{Z}}_{2n} = (0,0)D_{2n}$ and $(0,n)D_{2n}$ constitute the stability congruence $\sigma(\mathbb{Z}_{2n}, -)$. Since the maximum size of an orbit is $|D_{2n}| = 4n$, the remaining $n-1$ orbits all have to have this maximum size $4n$, since $4n(n-1) + 2n + 2n = (2n)^2 = |\mathbb{Z}_{2n}^2|$. Under (2.4.1), these $n-1$ orbits of D_{2n} correspond to the $n-1$ doubleton orbits $\{\pm i\}, 0 < i < n$, of the stabilizer $\langle J \rangle$ of 0 in D_{2n}. In particular, $(0,1)D_{2n} = (0,1)C_{2n} \cup (0,-1)C_{2n} = (0,1)C_{2n} \cup (1,0)C_{2n}$ is a symmetric, antireflexive relation determining the simple, undirected graph with edge set $\{\{x, x+1\} | x \in \mathbb{Z}_{2n}\}$. Note that $D_{2n} = \mathrm{Aut}\,(\mathbb{Z}_{2n}, \{(0,1)D_{2n}\})$—indeed this property is often used to define the action D_{2n}. Moreover, $D_{2n} = \mathrm{Aut}(\mathbb{Z}_{2n}, \mathbb{Z}_{2n}^{2n}/D_{2n}) \leq \mathrm{Aut}(\mathbb{Z}_{2n}, \mathbb{Z}_{2n}^{2}/D_{2n}) \leq \mathrm{Aut}(\mathbb{Z}_{2n}, \{(0,1)D_{2n}\}) = D_{2n}$, so that D_{2n} is 2-closed. [The first equality here is by Theorem 3.1.1(b), the first inequality by Theorem 3.1.1(a), and the second inequality by Exercise 3F. □

EXERCISES

3.1A. (a) What is the cardinality of $\mathbb{Z}_3^2/C_3$?

(b) What is the cardinality of $\mathbb{Z}_3^3/C_3$?

(c) Determine a complete set of representatives for the orbits of C_3 on $\mathbb{Z}_3^3$.

3.1B. Determine the Poincaré series of the cyclic action C_3, and use it to verify your answers to Exercise 3.1A (a), (b).

3.1C. Determine the Poincaré series of the dihedral action D_3. How many orbits does D_3 have on $\mathbb{Z}_3^3$?

3.1D. Determine the automorphism group of the simple undirected graph on $\mathbb{Z}_5$ with edge set $\{\{x, x+1\} | x \in \mathbb{Z}_5\}$.

3.1E. Show that the action of the automorphism group of the simple, undirected graph $(\mathbb{Z}_n, \{\{x, x+1\} | x \in \mathbb{Z}_n\})$ is similar to the action of $\mathrm{Mlt}(\mathbb{Z}_n, -)$ on $\mathbb{Z}_n$.

3.1F. Repeat the work of Example 3.1.3 for the dihedral action D_{2n+1}.

3.1G. A complex function $p(z)$ is the analytic continuation of the Poincaré series of a permutation action (X, G).

(a) Locate the poles of $p(z)$, and interpret their meaning in terms of properties of the permutation action (X, G).

(b) Show how to obtain $|X|$ and $|G|$ from $p(z)$. [Hint: Consider the smallest real pole of $p(z)$, and the corresponding residue.]

3.1H. Show that the dihedral action D_{2n} is not 1-closed for $n > 1$.

3.1I. (a) Is D_4 the automorphism group of the unary relational structure $(\mathbb{Z}_4, \{\{0,2\}, \{1,3\}\})$?

(b) Is D_4 the automorphism group of the simple undirected graph $(\mathbb{Z}_4, \{\{0,2\}, \{1,3\}\})$?

3.1J. Show that D_n is not the automorphism group of a unary relation on $\mathbb{Z}_n$ for any $n > 2$.

3.1K. Give an example of two dissimilar permutation group actions with the same permutation character. [Hint: Cf. Exercise 2.5E.] Prove that the two actions really are dissimilar.

3.1L. The *Bell numbers* B_n are defined by the recursion $B_0 = 1, B_{n+1} = \sum_{k=0}^{n} \binom{n}{k} B_k$. Thus $B_1 = 1, B_2 = 2, B_3 = 5, B_4 = 15, B_5 = 52, \ldots$. Show that a permutation action (X, G) is k-transitive if and only if its Poincaré series is $p(z) = \sum_{j=0}^{k} B_j z^j + O(z^{k+1})$.

3.2. Symmetric and Alternating Groups

For a finite set X, the group $X!$ of all bijections from X to X [aliter the group of permutations of X, aliter the automorphism group of the relational structures $(X, \varnothing)$ and $(X, \{\varnothing\})$] is known as the *symmetric group* on X or *the* permutation group of the set X (cf. Section 2.1). The action $(\mathbb{Z}_n, \mathbb{Z}_n!)$ is called the *symmetric group action* S_n. (Here one usually uses $\{1, 2, \ldots, n\}$ as the set of representatives for the equivalence classes constituting $\mathbb{Z}_n = \mathbb{Z}^{\langle n \rangle}$.) These names are also applied to similar actions. By abuse of language, the abstract group $\{1, 2, \ldots, n\}!$ is also called the *symmetric group* S_n. Now for $m \leq n$, the inclusions $\{1, \ldots, m\} \hookrightarrow \{1, \ldots, n\}$ and $\{1, \ldots, m\}! \hookrightarrow \{1, \ldots, n\}!$ yield an action morphism $S_m \to S_n$. Thus the group $S_\infty = \bigcup_{n=1}^{\infty} S_n$ acts on the set $\mathbb{Z}^+$ of positive integers, yielding an action $(\mathbb{Z}^+, \bigcup_{n=1}^{\infty} S_n)$. Actions similar to $(\mathbb{Z}^+, \bigcup_{n=1}^{\infty} S_n)$ are described as *countable permutation actions* S_∞, while a group isomorphic to $\bigcup_{n=1}^{\infty} S_n$ is called *the countable permutation group* S_∞. Note that S_∞ is the proper subgroup of $\mathbb{Z}^+!$ consisting of those bijections whose fixed-point set has a finite complement. Such bijections are called *permutations*.

For a set X of finite cardinality n, let T and U be elements of $X!$. The conjugate of the cycle $C = (y_1, \ldots, y_m)$ by the permutation U is

$$(y_1, \ldots, y_m)^U = (y_1 U, \ldots, y_m U), \tag{3.2.1}$$

since

$$(y_1, \ldots, y_m)^U = U^{-1} C U : y_i U \overset{U^{-1}}{\mapsto} y_i \overset{C}{\mapsto} y_{i+1} \overset{U}{\mapsto} y_{i+1} U;\ yU \overset{U^{-1}}{\mapsto} y \overset{C}{\mapsto} y \overset{U}{\mapsto} yU$$

for $y \notin \{y_1, \ldots, y_m\}$ and $(\{1, \ldots, m\}, +) = (\mathbb{Z}_m, +)$. If (2.1.5) is the cycle decomposition of T, the fact that conjugation by U is an automorphism of $X!$ yields

$$T^U = \prod_i (x_i U, x_i T U, \ldots, x_i T^{n_i - 1} U). \tag{3.2.2}$$

Note that the cycle types $T\tau$ and $T^U\tau$ coincide. Conversely, suppose that S and T are elements of $X!$ with $S\tau = T\tau$, say $S = \prod_i (y_i, y_i S, \ldots, y_i S^{n_i - 1})$ and $T = \prod_i (x_i, x_i T, \ldots, x_i T^{n_i - 1})$. Then for $U : x_i T^j \mapsto y_i S^j$, one has $T^U = S$. Summarizing

Proposition 3.2.1. *Elements of the symmetric group are conjugate if and only if they have the same cycle type.* □

Proposition 3.2.1 helps determine the size of each conjugacy class and the overall number of conjugacy classes in the symmetric group. Consider a partition μ of n written in product form $\mu = 1^{e_1} 2^{e_2} \ldots n^{e_n}$. Define $\varepsilon : 1^{cc} \to \mathbb{Z}^+;\ \mu \mapsto 1^{e_1} 2^{e_2} \ldots n^{e_n}$, the unique semigroup homomorphism $1^{cc} \to (\mathbb{Z}^+, \cdot)$ extending the identity map $\mathbb{Z}^+ = 1^c \to \mathbb{Z}^+$. Define $\gamma : 1^{cc} \to \mathbb{Z}^+;$ $\mu \mapsto e_1! e_2! \ldots e_n!$.

Proposition 3.2.2. *Let X be a set of finite cardinality n.*

(a) *The image of the cycle type map $\tau : X! \to 1^{cc}$ is the set $\sigma^{-1}(n)$ of partitions of n.*

(b) *The conjugacy class $\tau^{-1}(\mu)$ in $X!$, consisting of the permutations having a given partition μ of n as cycle type, has cardinality $n!/\mu\gamma \cdot \mu\varepsilon$.*

Proof. (a) For T in S_n, one has $T\tau\sigma = n$. Conversely, consider $\mu = n_1 + n_2 + \cdots$ in $\sigma^{-1}(n)$. For $X = \sum_{i=1}^{\mu\lambda} \{x_{i1}, \ldots, x_{in_i}\}$, the cycle type of $\prod_{i=1}^{\mu\lambda} (x_{i1}, \ldots, x_{in_i})$ is μ.

(b) Let $\mu = 1^{e_1} 2^{e_2} \ldots n^{e_n}$ in product form. There are $n!$ words of length n in X^* containing each element of X exactly once. Each such word $x_{111} \ldots x_{1e_1 1} x_{211} x_{212} \ldots x_{2e_2 1} x_{2e_2 2} \ldots x_{ne_n 1} \ldots x_{ne_n n}$ yields a permutation $\prod_{l=1}^{n} \prod_{j=1}^{e_l} (x_{lj1}, \ldots, x_{ljl})$. For each cycle length l, each of $e_l!$ permutations of the e_l subwords $x_{lj1} \ldots x_{ljl} (1 \le j \le e_l)$ yields the same element of $X!$. Moreover, for each of the e_l cycles of length l, each of the l cyclic permutations of the subword $x_{lj1} \ldots x_{ljl}$ yields the same cycle. Thus the number of distinct permutations of the given cycle type is $n!/e_1! \ldots e_n! \cdot 1^{e_1} \cdot 1^{e_2} \cdots 1^{e_n}$, as claimed. □

Note that the number $|\sigma^{-1}(n)|$ of conjugacy classes of S_n may be obtained from the generating function

$$\sum_{n=1}^{\infty} |\sigma^{-1}(n)| z^n = \prod_{l=1}^{\infty} (1 - z^l)^{-1}. \tag{3.2.3}$$

For example, $(1 + z + z^2 + z^3 + z^4 + \cdots)(1 + z^2 + z^4 + \cdots)$ $(1 + z^3 + \cdots)(1 + z^4 + \cdots) \cdots = 1 + \cdots + (z^4 \cdot 1 \cdot 1 \cdot 1 + z^2 \cdot z^2 \cdot 1 \cdot 1 + 1 \cdot z^4 \cdot 1 \cdot 1$ $+ z \cdot 1 \cdot z^3 \cdot 1 + 1 \cdot 1 \cdot 1 \cdot z^4) + \cdots = 1 + \cdots + 5z^4 + \cdots$, so that there are 5 conjugacy classes in S_4, corresponding to the respective cycle types $1^4 2^0 3^0 4^0$,

$1^2 2^1 3^0 4^0$, $1^0 2^2 3^0 4^0$, $1^1 2^0 3^1 4^0$, and $1^0 2^0 3^0 4^1$. By Proposition 3.2.2(b), their respective orders are $1, 6, 3, 8$, and 6. In the action S_4, their respective numbers of fixed points are $4, 2, 0, 1, 0$. Thus the Poincaré series of the action S_4 is $[9 + 8(1 - z)^{-1} + 6(1 - 2z)^{-1} + 1(1 - 4z)^{-1}]/24 = 1 + z + 2z^2 + 5z^3 + 15z^4 + 51z^5 + \cdots$.

Permutations of cycle type $1^{n-2}2^1$ in S_n are known as *transpositions*. By Proposition 3.2.2(b), there are $n(n - 1)/2$ of them. They are useful for generating other permutations. Since

$$(123\dots r) = (12)(13)\dots(1r), \tag{3.2.4}$$

each cycle is a product of transpositions. The cycle decomposition (2.1.5) then shows that every permutation is a product of transpositions. A subset G of a group M is a *set of generators* for M if $M = \langle G \rangle$ (in the sense of Exercise 1.4D).

Proposition 3.2.3.

(a) The set $\{(r, r + 1) | 1 \leq r < n\}$ generates S_n.
(b) The set $\{(12), (12\dots n)\}$ generates S_n.

Proof. (a) It will be shown that each transposition (rs), with $r < s$, is generated by the given set. By (3.2.1), one has $(1s) = (12)T((23)(34)\dots(s - 1, s))$. Then again by (3.2.1), one has $(rs) = (1s)T((12)(23)\dots(r - 1, r))$.

(b) Since $(12)T((12\dots n))^{r-1} = (r, r + 1)$ for $1 \leq r < n$, each element of the generating set of (a) is generated by the generating set of (b). □

The transpositions forming the generating set of Proposition 3.2.3(a) are often displayed in the graphical form

$$(12) - (23) - (34) - \cdots - (n - 1, n). \tag{3.2.5}$$

Two transpositions x and y are connected by an edge if and only if their product xy has order 3. If there is no edge between x and y, then they permute, or in other words xy has order 2.

Now let D be the full set of $n(n - 1)/2$ transpositions in S_n, considered as a set of generators of S_n. The embedding $D \hookrightarrow S_n$ extends to a surjective group homormorphism $DG \to S_n$. The image of the normal subgroup of even elements of the free group DG (cf. Exercise 2.4J) is a normal subgroup A_n of S_n, called the *alternating group* A_n. Thus the elements of A_n are precisely the *even permutations*, those permutations expressible as the product of an even number of transpositions. The remaining permutations are described as *odd*. Note $|A_n| = n!/2$. As a subgroup of S_n, the group A_n acts on $\mathbb{Z}_n$. Actions similar to $(\mathbb{Z}_n, A_n)$ are called *alternating group actions* A_n. Moreover, since the inclusion $\{1, \dots, m\}! \hookrightarrow \{1, \dots, n\}!$ maps A_m to A_n for

$m \leq n$, one obtains the *countable alternating group* $A_\infty = \bigcup_{n=1}^{\infty} A_n$ as a subgroup of S_∞. Then actions similar to $(\mathbb{Z}^+, A_\infty)$ are called *countable alternating actions*.

By (3.2.4), a cycle $(12\ldots r)$ is even if and only if r is odd. Thus the permutation T of (2.1.5) is even if and only if $\Sigma_i(n_i - 1)$ is even. In other words, a permutation T is even if and only if the sum $T\tau\sigma$ of its cycle type $T\tau$ is congruent modulo 2 to the length $T\tau\lambda$ of its cycle type, i.e. iff its *sign* sgn $T := (-1)^{T\tau\sigma - T\tau\lambda}$ is 1. For example, the 12-element subset A_4 of S_4 is the union of the S_4-conjugacy classes of cycle type $1^4 2^0 3^0 4^0, 1^0 2^2 3^0 4^0$, and $1^1 2^0 3^1 4^0$. Thus the Poincaré series of the action A_4 is $[3 + 8(1-z)^{-1} + 1(1-4z)^{-1}]/12 = 1 + z + 2z^2 + 6z^3 + 22z^4 + \cdots$. Within A_n, having the same cycle type is still necessary for two (even) permutations to be conjugate, but no longer sufficient. For example, (123) and (132) are not conjugate in A_3 or A_4 (Exercise 3.2J).

Proposition 3.2.4. *The action A_n is $(n-2)$-transitive.*

Proof. For elements $x = (x_3, \ldots, x_n)$ and $y = (y_3, \ldots, y_n)$ of the $(n-2)$-ary diversity relation on $\mathbb{Z}_n$, there is an element T of $\mathbb{Z}_n!$ with $xT = y$. If T is odd, one then has $xT(y_1 y_2) = y$ for $\mathbb{Z}_n = \{y_1, y_2, y_3, \ldots, y_n\}$. □

EXERCISES

3.2A. Verify that $\bigcup_{n=1}^{\infty} S_n$ is a group.

3.2B. (a) Show that C_n and D_n are subgroups of S_n for all positive integers n.

(b) Are C_∞ and D_∞ subgroups of S_∞?

3.2C. Show that the S_∞-actions S_∞ and $S_\infty \uplus S_\infty$ are not S_∞-isomorphic.

3.2D. Show that S_∞ is a proper normal subgroup of $\mathbb{Z}^+!$.

3.2E. Show that the permutations in S_4 of cycle types 1^4 and 2^2 form a normal subgroup V of order 4, the so-called *Vierergruppe*.

3.2F. Let T be a permutation in S_n of cycle type μ. How many permutations in S_n centralize T (cf. Section 2.1)?

3.2G. (a) For an integer partition μ, show that $\mu\sigma!/\mu\gamma \cdot \mu\varepsilon > 1$ unless $\mu\sigma \leq 2$ or $\mu\sigma = \mu\lambda$.

(b) Determine the center $Z(S_n)$ of the symmetric group S_n.

3.2H. Express each element of S_4 as a product of the generators (3.2.5), namely for $n = 4$.

3.2I. Given a group Q, consider a map $f: \{(12), (23)\} \to Q$; $(12) \mapsto s$, $(23) \mapsto t$. Show that f extends to a homomorphism $f: S_3 \to Q$ if and only if $s^2 = 1$, $t^2 = 1$, and $(st)^3 = 1$ in Q.

3.2J. Show that (123) and (132) are not conjugate in A_3 or A_4. Are they conjugate in A_5?

3.2K. Show that conjugation by (12) yields an outer automorphism of A_3 and A_4.

3.2L. Show that A_n is generated by the set of cycles of length 3 in S_n.

3.2M. Prove that

$$A_n = \left\{T \in S_n < S_\infty \middle| \prod_{1 \le i < j \le n} (j - i) = \prod_{1 \le i < j \le n} (jT - iT)\right\}.$$

(The subtractions and products take place in $\mathbb{Z}$.)

3.2N. Consider the map $x \mapsto (12)$, $y \mapsto (23)$ from the set X of Example 1.4.1 to the symmetric group S_3. This map extends to a unique group homomorphism from the free group XG to S_3. Copy a portion of the tree of Figure 1.1, and label the vertices by their corresponding images in S_3. Make sure that you have included enough of the tree to ensure that each element of S_3 appears as a label.

3.2O. The *trace* μt of an integer partition $\mu = 1^{e_1}2^{e_2}\dots$ is the multiplicity e_1 with which 1 appears as a part. Use this concept to determine the Poincaré series of the symmetric group action S_n.

3.3. Sylov's Theorem, p-Groups, and Simplicity

Fix a prime number p. A (not necessarily finite) group P is said to be a *p-group* if the order of each element of P is a power of the prime p. Subgroups, homomorphic images, and products of finite sets of p-groups are themselves p-groups. However, since the group $(\mathbb{Z}, +)$ embeds into $\prod_{r=1}^{\infty} \mathbb{Z}^{\langle p^r \rangle}$ via the product of the maps $n \mapsto n^{\langle p^r \rangle}$, infinite products of p-groups are not necessarily p-groups. For a group G, the set of p-subgroups is ordered by inclusion. Maximal elements are called *Sylov p-subgroups* of G. Denote the set of Sylov p-subgroups by $\mathrm{Syl}_p(G)$. For a positive integer h, denote the set of subgroups of G having order h by $\mathrm{Sb}_h(G)$. If h is a prime power dividing finite $|G|$, this set is non-empty:

Proposition 3.3.1. *Let G be a group of finite order $p^e m$. Then $|\mathrm{Sb}_{p^e}(G)| \langle p \rangle 1$ in $\mathbb{Z}$.*

Proof. Let A be the set of subsets of G of order p^e. Note that A is a G-subset of the G-set 2^G of Exercise 1H. Let T be a set of representatives for the orbits A/G. By the Structure Theorem for G-sets, $(A, G) \cong \sum_{S \in T}(G_S \backslash G, G)$. For S in T, one has $G_S S^J = S^J$, so that S^J is a G_S^{op}-subset of the right G_S^{op}-set (G, G_S^{op}) of (1.3.3). As such, S^J is a disjoint union of orbits, namely right cosets of G_S. Thus $|G_S|$ divides $|S^J| = p^e$. In particular, $|G_S \backslash G| = m \Leftrightarrow G_S \in A \Leftrightarrow (G_S \backslash G, G) \le (A, G)$.

Indeed, $H \mapsto (Hy/G, G)$ yields a bijection from $\mathrm{Sb}_{p^e}(G)$ to the set of irreducible m-element G-subsets of (A, G). Thus $|A| = \sum_{S \in T} |G_S y/G| \langle pm \rangle \sum_{H \in \mathrm{Sb}_{p^e}(G)} |H \backslash G| = |\mathrm{Sb}_{p^e}(G)| \cdot m$. Suppose $m = p^f l$ with p not dividing l. Then $|A| = p^{e+f} l \cdot (p^{e+f} l - 1) \dots (p^{e+f} l - (p^e - 1))/p^e(p^e - 1) \dots 1 = m \prod_{k=1}^{p^e - 1} (p^{e+f} l - k)/k \langle pm \rangle m$, since $1 \langle p \rangle \prod_{k=1}^{p^e - 1} (p^{e+f} l - k)/k$. Thus $|\mathrm{Sb}_{p^e}(G)| \cdot m \langle pm \rangle m$, whence the result. □

Corollary 3.3.2. *Let G be a finite group.*

(a) **(Cauchy's Theorem.)** *If a prime p divides $|G|$, then G contains an element of order p.*
(b) *For a prime p, the group G is a p-group if and only if $|G|$ is a power of p.*
(c) *For a prime p, if $|G| = p^e m$ with p not dividing m, then* $\mathrm{Syl}_p(G) = \mathrm{Sb}_{p^e}(G)$.

Proof. (a) Take a non-identity element of an element of $\mathrm{Sb}_p(G)$.

(b) If $|G|$ is not a power of p, so some other prime q divides $|G|$, then G contains an element of order q, by (a). Conversely, if G contains an element x whose order r is not a power of p, then Lagrange's Theorem [Exercise 1.3J(d)] shows that $|\langle\{x\}\rangle| = r$ divides $|G|$, which cannot then be a power of p.

(c) The proof follows from (b). □

Theorem 3.3.3 (Sylov's Theorem). *Let G be a finite group, and let p be a prime.*

(a) *In $\mathbb{Z}$, one has $|\mathrm{Syl}_p(G)| \langle p \rangle 1$. In particular, $\mathrm{Syl}_p(G)$ is non-empty.*
(b) *The inner automorphism group of G acts transitively on $\mathrm{Syl}_p(G)$. In particular, $|\mathrm{Syl}_p(G)|$ divides $|G|$.*

Proof. (a) By Corollary 3.3.2 (c) and Proposition 3.3.1.

(b) Let $|G| = p^e m$ with p not dividing m. Let P and Q be Sylov p-subgroups of G. Let K be the subgroup of Mlt G generated by $L(P)$ and $R(Q)$. Consider the action of K on G. For g in G, the size of the orbit of g is $|PgQ| = |g^{-1}PgQ| = |P^g Q| = |P^g| \cdot |Q|/|P^g \cap Q|$ [cf. Exercise 1.3L(c)]. If P^g were never to agree with Q, then the size of all the orbits, and hence their sum $|G|$, would be divisible by p^{e+1}. □

Let (X, G) be a permutation group with permutation character π. A non-trivial subgroup H of G is *regular* if $\pi(h) = 0$ for all non-identity elements of H. If N is a regular normal subgroup of G (e.g. the Vierergruppe in S_4: Exercise 3.2E) and e is an element of X, then $N - \{1\}$ becomes a G_e-subset of the G_e-set (G, G_e) whose representation is the restriction $T: G_e \to \mathrm{Inn}\, G$ of (2.4.3). One then obtains a G_e-isomorphism

$$(N - \{1\}, G_e) \to (X - \{e\}, G_e); n \mapsto en. \tag{3.3.1}$$

If (X, G) is k-transitive, for $k > 1$, then G_e is a subgroup of the automorphism group of N that is $(k-1)$-transitive on $N - \{1\}$ (Exercise 3K). This forces N to be a p-group for some prime p:

Proposition 3.3.4. *Let A be a subgroup of the automorphism group of a non-trivial finite group N.*

(a) *If $(N - \{1\}, A)$ is 1-transitive, then N is a direct power of $(\mathbb{Z}_p, +)$ for some prime p.*
(b) *If $(N - \{1\}, A)$ is 2-transitive, then $p = 2$ or N is isomorphic to $(\mathbb{Z}_3, +)$.*
(c) *If $(N - \{1\}, A)$ is 3-transitive, then N is isomorphic to $(\mathbb{Z}_3, +)$ or the Vierergruppe V.*
(d) *It is impossible for $(N - \{1\}, A)$ to be 4-transitive.*

Proof. (a) Since N is non-trivial, some prime p divides its order. By Cauchy's Theorem (Corollary 3.3.2(a)), N has an element of order p. Since orders are preserved by automorphisms, each non-identity element of N has order p. By Corollary 3.3.2(b), the order of N is some power p^r of p. By Corollary 2.5.5, the center $Z(N)$ of N is non-trivial. Since $Z(N) - \{1\}$ is invariant under the transitive group A of automorphisms (Exercise 2.5G), it follows that $N = Z(N)$ is abelian. Let H be a subgroup of N with $x \in N - H$. Then H and $\langle x \rangle$ are normal subgroups of $\langle H, x \rangle = H\langle x \rangle$ with $H \cap \langle x \rangle = \{1\}$. By Section O 4.4, the permutable (Proposition 2.2.3) congruences on $\langle H, x \rangle$ determined by the normal subgroups H and $\langle x \rangle$ decompose $\langle H, x \rangle \cong H \times \langle x \rangle$. Since $\langle x \rangle \cong (\mathbb{Z}_p, +)$, it follows by induction that $N \cong (\mathbb{Z}_p, +)^r$.

(b) Since N is abelian, the inversion map J of (1.3.2) is an automorphism of N. Moreover, the subsets $\langle J \rangle$ and A of Aut N are permutable (cf. Exercise 2.1G). Consider the $\langle J \rangle$-set $(N - \{1\}, \langle J \rangle)$. The kernel of the projection $(N - \{1\}) \to (N - \{1\})/\langle J \rangle$ is an A-subset of $(N - \{1\})^2$. Since it is reflexive, it is either the diagonal, in which case the $\langle J \rangle$-orbits are trivial and $p = 2$, or else it is $(N - \{1\})^2$, in which case there is only one $\langle J \rangle$-orbit on $N - \{1\}$, necessarily of order 2, and N is isomorphic to $(\mathbb{Z}_3, +)$.

(c) If $p = 2$, then N contains an (abstract) Vierergruppe $V = \{1, x, y, xy\}$ as a subgroup. The stabilizer A_x acts 2-transitively on $N - \langle x \rangle$ (cf. Exercise 3K). Let α be the kernel of the projection $N \to \langle x \rangle \backslash N$. Then $\alpha \cap (N - \langle x \rangle)^2$ is a reflexive A_x-subset of $(N - \langle x \rangle)^2$ containing (y, xy). It thus coincides with $(N - \langle x \rangle)^2$, in which case $N - \langle x \rangle = \{y, xy\}$ and $N = V$.

(d) Neither $(\mathbb{Z}_3, \mathrm{Aut}(\mathbb{Z}_3, +))$ nor $(V, \mathrm{Aut} V)$ is 4-transitive. □

Corollary 3.3.5. *Let N be a regular normal subgroup of a permutation group G of a set X of order n.*

(a) *If (X, G) is 2-transitive, then n is a prime power p^r and $N \cong (\mathbb{Z}_p, +)^r$.*
(b) *If (X, G) is 3-transitive, then $n = 2^r$ or $n = 3$.*
(c) *If (X, G) is 4-transitive, then $n = 4$ or $n = 3$.*
(d) *It is impossible for (X, G) to be 5-transitive.*

Proof. If e is an element of the k-transitive G-set X, then $(X - \{e\}, G_e)$ is $(k - 1)$-transitive (cf. Exercise 3K). The isomorphism (3.3.1) then yields a $(k - 1)$-transitive action of G_e on $N - \{1\}$. The disjoint union with the singleton G_e-set $\{1\}$ then yields G_e as a subgroup of the automorphism group of N. □

Theorem 3.3.6. *For $n > 4$, the group A_n is simple.*

Proof. Let N be a non-trivial normal subgroup of the $(n - 2)$-transitive permutation group A_n on $\mathbb{Z}_n$. Now A_n acts on $\mathbb{Z}_n/N$ by $g : xN \mapsto xNg = xgg^{-1}Ng = xgN$. Thus the projection $\mathbb{Z}_n \to \mathbb{Z}_n/N$ is an A_n-homomorphism whose kernel is a non-trivial A_n-congruence. By Exercises 3I and 3J, the action A_n is simple. Thus the kernel of the projection is all of $\mathbb{Z}_n^2$, whence $(\mathbb{Z}_n, N)$ is transitive and n divides $|N|$.

If $n = 5$, then $|A_n| = 5!/2 = 2^2 \cdot 3 \cdot 5$. By Proposition 3.2.2(b), there are $5!/5 = 24$ permutations of type $1^0 2^0 3^0 4^0 5^1$ in S_5, and hence in A_5. Each Sylov 5-subgroup of A_5 contains 4 such permutations (along with the identity), so there are 6 Sylov 5-subgroups of A_5. By Sylov's Theorem 3.3.3(b), they are all conjugate. Now N has order a multiple of 5, so by Cauchy's Theorem it contains at least one of these Sylov subgroups. But since N is normal, it contains all 6, i.e. $|\mathrm{Syl}_5(N)| = 6$. By Sylov's Theorem 3.3.3(b), it follows that 6 also divides $|N|$, whence $|N|$ is $2 \cdot 3 \cdot 5$ or $2^2 \cdot 3 \cdot 5$. By Cauchy's Theorem again, N contains at least one Sylov 3-subgroup of A_5. There are $5!/3 \cdot 2 = 20$ permutations of type $1^0 2^0 3^1 4^0 5^0$ in S_5 and A_5, pairing off with the identity to make 10 Sylow 3-subgroups. All are contained in the normal subgroup N, so N contains 20 cycles of length 3 and 24 of length 5. In particular, $30 < |N|$, so $|N| = 60$ and N is improper. In other words, A_5 is simple.

Now for $n > 5$, assume by induction that A_{n-1} is simple. Let N be minimal amongst the non-trivial normal subgroups of A_n. Recall that A_{n-1} is the stabilizer of n in A_n. If $N \cap A_{n-1}$ were trivial, the stabilizer N_n of n in N would be trivial. By Proposition 2.4.5, all the stabilizers N_i in the transitive action $(\mathbb{Z}_n, N)$ would then be trivial, whence N would become a regular normal subgroup of the $(n - 2)$-transitive permutation group A_n. By Corollary 3.3.5(d), this would force $n = 6$. But since 6 is not a prime power, the demand of Corollary 3.3.5(a) would be violated. Thus $N \cap A_{n-1}$ is a non-trivial normal subgroup of A_{n-1}. By the induction assumption, $N \cap A_{n-1} = A_{n-1}$, so that $N_n = A_{n-1}$. Since $(\mathbb{Z}_n, N)$ is transitive, Proposition 1.3.2 yields $|N| = |\mathbb{Z}_n| \cdot |N_n| = n \cdot (n - 1)! = n! = |A_n|$, so that $N = A_n$ and A_n is simple. □

EXERCISES

3.3A. Let N be a normal subgroup of a group G. For a prime number p, show that G is a p-group if and only if both N and G/N are p-groups.

3.3B. Show that the product $\mathbb{Z} \to \prod_{r=1}^{\infty} \mathbb{Z}^{\langle 2^r \rangle}$ of the maps $n \mapsto n^{\langle 2^r \rangle}$ injects.

3.3C. By considering the factor 6 of $|A_4|$, show that Cauchy's Theorem does not extend to composite divisors of the order of a finite group.

3.3D. By considering (2.2), show that Proposition 3.3.1 does not extend to quasigroups.

3.3E. Determine the number of Sylov 2-subgroups of S_4. Show that one of them is equal to D_4.

3.3F. Show that the action of a Sylov p-subgroup of S_{p^2}, for prime p, is similar to the wreath product action $C_p \wr C_p$.

3.3G. Let the order of a group G be the product pq of primes p and q, with $p > q$. Show that G has a normal subgroup of order p.

3.3H. For each odd prime p, determine the Sylov p-subgroups of the dihedral group D_n.

3.3I. Show that there is no simple group of order 56.

3.3J. (a) Determine the orders of the (group) conjugacy classes of the group A_5.

(b) Show that no proper non-singleton Inn A_5-subset of A_5 containing 1 has order dividing 60.

(c) Deduce that A_5 is simple (independently of the proof of Theorem 3.3.6).

4. LOOPS, NETS, AND ISOTOPY

In Example 2.4, the quasigroup Q with multiplication table (2.2) was obtained from the Latin square in (2.1) by an arbitrary assignment of the elements of Q as labels for the rows and columns of the table. Rearranging the row and column labels would yield the multiplication table of a different quasigroup, not necessarily isomorphic to the original. For example, the quasigroup Q with table (2.2) has three idempotents, namely A, B, and C. If the column headings are changed by application of the cycle $(ABCV)$, then the new quasigroup Q' [as in (4.2) below] only has a single idempotent, namely V. Isomorphism of quasigroups is too strong a relationship to describe the intrinsic structure of a Latin square. The appropriate relationship is known as "isotopy."

Recall that a quasigroup homomorphism $f:(Q,\cdot,/,\backslash) \to (P,\cdot,/,\backslash)$ is a set map $f:Q \to P$ with $xf \cdot yf = (xy)f$ for all x, y in Q (Exercise 2.2E). A *quasigroup homotopy* (f_1, f_2, f_3): $(Q,\cdot,/,\backslash) \to (P,\cdot,/,\backslash)$ is a triple of set maps $f_i : Q \to P$ with

$$xf_1 \cdot yf_2 = (xy)f_3 \tag{4.1}$$

for all x, y in Q. Thus a homomorphism is a homotopy in which the three components agree. A *quasigroup isotopy* is a homotopy in which each component bijects. Two quasigroups Q_1, Q_2 are said to be *isotopic*, written $Q_1 \sim Q_2$, if there is an isotopy $Q_1 \to Q_2$. One also says that Q_2 is an *isotope* of Q_1. Considering the quasigroups Q and Q' on the set $\{A, B, C, V\}$ of Example 2.4, with multiplication tables

(4.2)

Q	A	B	C	V
A	A	V	B	C
B	C	B	A	V
C	V	A	C	B
V	B	C	V	A

,

Q′	A	B	C	V
A	C	A	V	B
B	V	C	B	A
C	B	V	A	C
V	A	B	C	V

,

one sees that Q and Q' are isotopic via the isotopy $(1, (ABCV), 1)\colon\ Q \to Q'$. The second component of the isotopy records the permutation of the column labels in the original table of Q. Similarly, the first component of an isotopy records symmetries of the row labels, while the third component records symmetries of the set of table entries. The arithmetic mean quasigroup $(\mathbb{R}, \circ)$ of Example 2.3 is isotopic to the abelian group $(\mathbb{R}, +)$ via the isotopy $(1, 1, 2)\colon (\mathbb{R}, \circ) \to (\mathbb{R}, +)$ whose third component doubles its argument.

An isotopy with equal domain and codomain is said to be *principal* if its third component is the identity map of its domain. Note that any isotopy (f_1, f_2, f_3) factorizes as a product $(f_1 f_3^{-1}, f_2 f_3^{-1}, 1)(f_3, f_3, f_3)$ of a principal isotopy with an isomorphism, or vice versa as a product $(f_3, f_3, f_3)(f_3^{-1} f_1, f_3^{-1} f_2, 1)$ of an isomorphism with a principal isotopy. Now a *loop* $(Q, \cdot, /, \backslash, 1)$ is a quasigroup $(Q, \cdot, /, \backslash)$ with an *identity* element 1 of Q satisfying

$$1 \cdot x = x = x \cdot 1 \tag{4.3}$$

for all x in Q. Let $(Q, \cdot, /, \backslash)$ be a quasigroup with element e. Define a new quasigroup structure $(Q, *, //, \backslash\backslash)$ on Q by

$$\begin{cases} x * y = (x/(e \backslash e)) \cdot (e \backslash y), \\ x//y = (x/(e \backslash y)) \cdot (e \backslash e), \\ x \backslash\backslash y = (e/e) \cdot ((x/e) \backslash y). \end{cases} \tag{4.4}$$

Then $(Q, *, //, \backslash\backslash, e)$ is a loop. Moreover, there is a principal isotopy

$$(R(e \backslash e), L(e), 1) : (Q, \cdot, /, \backslash) \to (Q, *, //, \backslash\backslash).$$

Thus:

Proposition 4.1. *Let Q be a quasigroup with element e. Then Q is principally isotopic to a loop with identity element e.* □

One often wishes to study properties of a quasigroup that are shared by its isotopes. To this end, a relational structure is associated with a quasigroup in such a way that isotopic quasigroups correspond to isomorphic structures.

Definition 4.2. Let k be an integer bigger than 2. Then a *k-net* is a relational structure $(N, \langle \alpha_i | 1 \leq i \leq k \rangle)$ consisting of a k-element multiset of equivalence relations such that $\forall 1 \leq i \neq j \leq k$, $\alpha_i \cap \alpha_j = \hat{N}$ and $\alpha_i \circ \alpha_j = N^2$.

Proposition 4.3. *Let* $(N, \langle \alpha_k | 1 \leq i \leq k \rangle)$ *be a k-net.*

(a) *The set* $\{\alpha_i | 1 \leq i \leq k\}$ *is permutable.*
(b) *For* $1 \leq i \neq j \leq k$, *the map* $N \to N^{\alpha_i} \times N^{\alpha_j}$; $n \mapsto (n^{\alpha_i}, n^{\alpha_j})$ *bijects.*
(c) *The sets* N^{α_i} *are all isomorphic for* $1 \leq i \leq k$.
(d) *If* $|N| > 1$, *then* $i \neq j \Rightarrow \alpha_i \neq \alpha_j$.
(e) *If* $1 < |N| < \infty$, *then* $(k-1)^2 \leq |N|$.

Proof. (a) By the definition, $\alpha_i \circ \alpha_j = N^2 = \alpha_j \circ \alpha_i$.

(b) Cf. Section O 4.4.

(c) If N is empty, so is each N^{α_i}. Otherwise, fix e in N. Consider $|\{h, i, j\}| = 3$ with $1 \leq h, i, j \leq k$. Then there is a bijection

$$T(e, h, i, j) : N^{\alpha_i} \to N^{\alpha_j} \tag{4.5}$$

given by $n^{\alpha_i} \mapsto \bigcup \{t^{\alpha_j} | \{t\} = n^{\alpha_i} \cap e^{\alpha_h}\}$.

(d) Suppose $i \neq j$ and $\alpha_i = \alpha_j$. Then $\hat{N} = \alpha_i \cap \alpha_j = \alpha_i \circ \alpha_j = N^2$, whence $|N| \leq 1$.

(e) Firstly, note $|N^{\alpha_i}| > 1$ for each $1 \leq i \leq k$: otherwise, (c) and (b) would imply $|N| \leq 1$. Pick e and f in N with $e^{\alpha_1} \neq f^{\alpha_1}$. For $1 < i \leq k$, pick t_i in $e^{\alpha_i} \cap f^{\alpha_1}$. Then $t_2, \ldots, t_k$ are distinct, for $t_i = t_j$ would imply $(e, t_i) = (e, t_j) \in \alpha_i \cap \alpha_j = \hat{N}$, whence $f^{\alpha_1} = t_i^{\alpha_1} = e^{\alpha_1}$. Now $t_2^{\alpha_2}, \ldots, t_k^{\alpha_2}$ are distinct, for $t_i^{\alpha_2} = t_j^{\alpha_2}$ would imply $(t_i, t_j) \in \alpha_1 \cap \alpha_2 = \hat{N}$. Thus $|N^{\alpha_2}| \geq k - 1$ and $|N| = |N^{\alpha_1}| \cdot |N^{\alpha_2}| = |N^{\alpha_2}|^2 \geq (k-1)^2$, as required. □

In view of Proposition 4.3(b) and Section O 4.4, one may loosely describe a k-net as a set that is the product of any pair of a k-element multiset of quotients.

A quasigroup Q determines a 3-net $Q^2 = \text{Net}(Q)$. The underlying set of the net is the Cartesian square $N = Q^2$, equipped with the projections $\pi_i : Q^2 \to Q$; $(x_1, x_2) \mapsto x_i$. Then $\alpha_i = \ker \pi_i$ for $i = 1, 2$, while $\alpha_3 = \ker (Q^2 \to Q; (x_1, x_2) \mapsto x_1 x_2)$. Note, for example, that $(x_1, x_2)\, \alpha_1\, (x_1, x_1 \backslash y_1 y_2)\, \alpha_3\, (y_1, y_2)$, using the notational convention introduced following Proposition 2.2.2, so that $\alpha_1 \circ \alpha_3 = N^2$. Also $(x_1, x_2)\, \alpha_1 \cap \alpha_3\, (x_1, y_2) \Rightarrow x_2 = x_1 \backslash x_1 x_2 = x_1 \backslash x_1 y_2 = y_2$, so that $\alpha_1 \cap \alpha_3 = \hat{N}$. Similarly $\alpha_2 \cap \alpha_3 = \hat{N}$ and $\alpha_2 \circ \alpha_3 = N^2$. Especially for a finite quasigroup Q, the 3-net $\text{Net}(Q)$

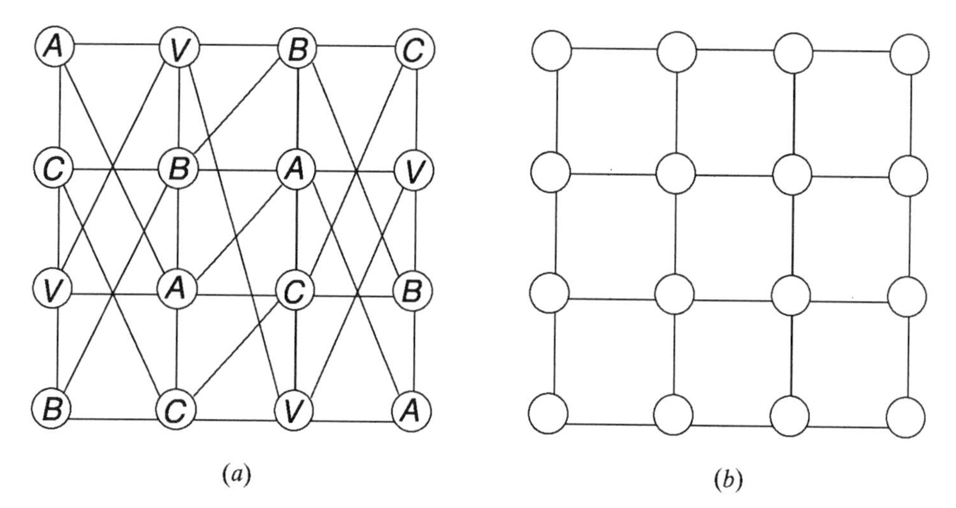

Figure 1.2. (*a*) The Latin square (1.2.2). (*b*) The corresponding net.

may be viewed as an unlabeled version of the Latin square obtained from the multiplication table by removing the borders. The α_1-classes correspond to the rows, so the elements of an α_1-class may be joined by chains of horizontal lines. Similarly, the elements of an α_2-class, representing a column in the table, may be joined by chains of vertical lines. Finally, the elements of an α_3-class, representing table locations with the same entry, may be joined by chains of diagonal lines. Figure 1.2(a) shows the Latin square determined by the multiplication table of the quasigroup Q of (2.2) or (4.2), while Figure 1.2(b) displays the corresponding 3-net Net (Q) using the conventions described. Because of these conventions, the α_i-classes of a 3-net are often called the *horizontal*, *vertical*, and *diagonal* lines of the net, or sometimes just the *i-lines*, for $i = 1, 2, 3$, respectively. Two 3-nets $(N, \langle \alpha_1, \alpha_2, \alpha_3 \rangle)$ and $(N', \langle \alpha_1', \alpha_2', \alpha_3' \rangle)$ are *isomorphic* if there is a bijection $f : N \to N'$ such that $(n_1, n_2) \in \alpha_i \Rightarrow (n_1 f, n_2 f) \in \alpha_i'$ for each i. Thus an isomorphism of 3-nets carries horizontals to horizontals, verticals to verticals, and diagonals to diagonals.

Now given a 3-net $(N, \langle \alpha_1, \alpha_2, \alpha_3 \rangle)$, one may construct a quasigroup on a set $\sqrt{N}$ that is in bijection with each of the isomorphic sets $N^{\alpha_i} (1 \leq i \leq 3)$ of Proposition 4.3(c). For $1 \leq i \leq 3$, define $p_i : N \to \sqrt{N}$ with $\ker p_i = \alpha_i$. Then a quasigroup structure $N(p_1, p_2, p_3) = (\sqrt{N}, \cdot)$ is defined by

$$x_1 \cdot x_2 = x_3 \Leftrightarrow \varnothing \neq \bigcap_{1 \leq i \leq 3} p_i^{-1}\{x_i\}. \tag{4.6}$$

The following result shows how isotopic quasigroups correspond to isomorphic 3-nets.

Theorem 4.5.

(a) *Let* $(Q, \cdot)$ *and* $(P, \circ)$ *be isotopic quasigroups. Then the 3-nets* $\mathrm{Net}(Q, \cdot)$ *and* $\mathrm{Net}(P, \circ)$ *are isomorphic.*

(b) *Let* $p_i : N \to \sqrt{N}$ *and* $p'_i : N \to \sqrt{N}$ *(for* $1 \leq i \leq 3$*) be two triples of maps from a 3-net N with* $\alpha_i = \ker p_i = \ker p'_i$*. Then the quasigroups* $N(p_1, p_2, p_3) = (\sqrt{N}, \cdot)$ *and* $N(p'_1, p'_2, p'_3) = (\sqrt{N}, \circ)$ *are isotopic.*

Proof. (a) Suppose there is a isotopy $(f_1, f_2, f_3) : (Q, \cdot) \to (P, \circ)$, so that $x_1 f_1 \circ x_2 f_2 = (x_1 x_2) f_3$ for x_1, x_2 in Q. Then $f : Q^2 \to P^2$; $(x_1, x_2) \mapsto (x_1 f_1, x_2 f_2)$ bijects. It will be shown that f is an isomorphism from $\mathrm{Net}(Q, \cdot)$ to $\mathrm{Net}(P, \circ)$. By construction, f carries horizontals to horizontals and verticals to verticals. Suppose $(x_1, x_2)\, \alpha_3\, (y_1, y_2)$ in Q^2. Then $x_1 f_1 \circ x_2 f_2 = (x_1 x_2) f_3 = (y_1 y_2) f_3 = y_1 f_1 \circ y_2 f_2$, as required.

(b) For $1 \leq i \leq 3$, the First Isomorphism Theorem O 3.3.1 shows that $p_i : N \to \sqrt{N}$ factorizes through an isomorphism $\overline{p_i} : N^{\alpha_i} \to \sqrt{N}$ and that p'_i factorizes through an isomorphism $\overline{p'_i} : N^{\alpha_i} \to \sqrt{N}$. Define bijections $f_i = \overline{p_i}^{-1} \overline{p'_i} : \sqrt{N} \to \sqrt{N}$. Then for x_1, x_2, x_3 in $\sqrt{N}$, one has

$$
\begin{aligned}
& x_1 \cdot x_2 = x_3 \\
\Leftrightarrow\ & \exists n \in N.\ \forall 1 \leq i \leq 3,\ n(\mathrm{nat}\ \alpha_i)\overline{p_i} = n p_i = x_i \\
\Leftrightarrow\ & \exists n \in N.\ \forall 1 \leq i \leq 3,\ n p'_i = n(\mathrm{nat}\ \alpha_i)\overline{p'_i} = x_i f_i \\
\Leftrightarrow\ & x_1 f_1 \circ x_2 f_2 = x_3 f_3 .
\end{aligned}
$$

Thus $x_1 f_1 \circ x_2 f_2 = (x_1 x_2) f_3$, yielding the required isotopy. □

EXERCISES

4A. If $(f_1, f_2, f_3) : Q \to Q'$ and $(g_1, g_2, g_3) : Q' \to Q''$ are quasigroup homotopies, prove that their composite $(f_1 g_1, f_2 g_2, f_3 g_3) : Q \to Q''$ is one too.

4B. Prove that a quasigroup homotopy $(f_1, f_2, f_3) : Q \to P$ is an isotopy if and only if f_i bijects for one i in $\{1, 2, 3\}$.

4C. (a) Prove that isotopy is an equivalence relation on any set of quasigroups.

(b) Prove that principal isotopy is an equivalence relation on the set of quasigroup structures on a given set.

4D. Prove that isotopic groups are isomorphic.

4E. Prove that a non-empty quasigroup Q is a loop if and only if $x/x = y \backslash y$ for all x, y in Q.

4F. Prove that the multiplication group of the loop (4.4) is a subgroup of the multiplication group of the quasigroup $(Q, \cdot, /, \backslash)$. Give an example of a quasigroup $(Q, \cdot, /, \backslash)$ for which this subgroup is proper.

4G. Determine a principal isotope of the quasigroup of (2.4) having 1 as its identity element. Display the multiplication table of the principal isotope.

4H. Determine the right and left divisions in the quasigroup $(\sqrt{N}, \cdot)$ of (4.6).

4I. In the notation of Exercise O 4.4C, consider the 3-net $N = (\mathbb{R}^2, \langle \alpha_i | \ 1 \leq i \leq 3 \rangle)$.

(a) Find maps $p_i : \mathbb{R}^2 \to \mathbb{R}$ with $\ker p_i = \alpha_i$, for $1 \leq i \leq 3$.

(b) Identify the quasigroup $N(p_1, p_2, p_3)$.

4J. Let $(f_1, f_2, f_3) : (Q, \cdot, /, \backslash) \to (P, \cdot, /, \backslash)$ be a quasigroup homotopy. For each element g of S_3, determine quasigroup structures $Q(g)$ on Q and $P(g)$ on P such that $(f_{1g}, f_{2g}, f_{3g}) : Q(g) \to P(g)$ is a homotopy.

4.1. Inverse Properties and Moufang Loops

A quasigroup Q is said to have the *left inverse property* if there is a (necessarily bijective) map $l : Q \to Q$ such that $L(x)^{-1} = L(x^l)$ for each x in Q. Note $L(x^{l^2}) = L(x^l)^{-1} = L(x)$, so $l^2 = 1$. Similarly, the *right inverse property* is the existence of $r : Q \to Q$ with $R(x)^{-1} = R(x^r)$ for each x in Q. Again in this case, one has $r^2 = 1$. If Q has both right and left inverse properties, then it is said to have the *inverse property*. For example, the quasigroup of (2.4) has the inverse property, with $r = l = 1$. Any group has the inverse property, with $r = l = J$ as in (1.3.2). Suppose that Q has the inverse property. Then $x = xy \cdot y^r \Rightarrow (xy)^l \ x = y^r \Rightarrow (xy)^l = y^r x^r$. Similarly, $(xy)^r = y^l x^l$. If Q is a loop with identity element 1, then $xx^l = x^{l^2} \cdot x^l 1 = 1 = 1x \cdot x^r = xx^r$, so that $x^l = x^r$. Thus in an inverse property loop Q, there is a map $J : Q \to Q;\ x \mapsto x^J = x^r = x^l = x^{-1}$ with

$$(xy)^J = y^J x^J \tag{4.1.1}$$

for x, y in Q. In particular, $R(x)^J = L(x^J) = L(x)^{-1}$ and $L(x)^J = R(x^J) = R(x)^{-1}$ for x in Q. Moreover, $x/y = xy^{-1}$ and $x \backslash y = x^{-1}y$.

An *autotopy* of a quasigroup $(Q, \cdot, /, \backslash)$ is an isotopy (f_1, f_2, f_3): $(Q, \cdot /, \backslash) \to (Q, \cdot, /, \backslash)$ from Q to itself. The set Atp Q of autotopies of Q forms a group under composition (cf. Exercise 4A), containing the automorphism group Aut Q as a subgroup. If Q is an inverse property loop, then there is an action of S_3 on Atp Q.

Proposition 4.1.1. *Let Q be an inverse property loop. Then* Atp Q *is an S_3-set under the actions* $(12) : (f_1, f_2, f_3) \mapsto (f_2^J, f_1^J, f_3^J)$ *and* $(23) : (f_1, f_2, f_3) \mapsto (f_1^J, f_3, f_2)$. *Moreover, a bijection appearing in one component of an autotopy appears in the other components in certain other autotopies.*

Proof. For (f_1, f_2, f_3) in Atp Q and x,y in Q, (4.1.1) yields $x^{Jf_2J} \cdot y^{Jf_1J} = (y^{Jf_1} \cdot x^{Jf_2})^J = (y^J x^J)^{f_3J} = (xy)^{Jf_3J}$, so that (f_2^J, f_1^J, f_3^J) lies in Atp Q. Moreover, $(f_1, f_2, f_3)(12)^2 = (f_2^J, f_1^J, f_3^J)(12) = (f_1, f_2, f_3)$. Now $y^{f_3} = (x^J \cdot xy)^{f_3} = x^{Jf_1}(xy)^{f_2}$, whence $x^{Jf_1J} \cdot y^{f_3} = (xy)^{f_2}$ and $(f_1^J, f_3, f_2) \in$ Atp Q. Moreover,

$(f_1, f_2, f_3)(23)^2 = (f_1^J, f_3, f_2)(23) = (f_1, f_2, f_3)$. Finally, $(f_1, f_2, f_3)((12)(23))^3 = (f_2, f_3^J, f_1^J)((12)(23))^2 = (f_3^J, f_1, f_2^J)((12)(23)) = (f_1, f_2, f_3)$. Thus an S_3-action is obtained (cf. Exercise 3.2I). Moreover, if the bijection f_1 appears in any one component of a certain autotopy α, then it appears in each of the other two components of other autotopies, namely the images of α under various elements of S_3. □

In view of Proposition 4.1.1, one may define a *topomorphism* of an inverse property loop Q to be a bijection of Q appearing as a component of an autotopy of Q. The set Tpm Q of topomorphisms of Q forms a subgroup of $Q!$. An element of Q is a *Moufang element* if and only if it lies in the orbit of the identity element of Q under the action of Tpm Q. Moufang elements are characterized by the following:

Proposition 4.1.2. *Let Q be an inverse property loop. The set $M =$ Mfg Q of Moufang elements of Q forms a subloop of Q. For a Moufang element m, $L(m)R(m) = R(m)L(m)$. Moreover, there is a map*

$$(4.1.2) \qquad \mathrm{Mfg}Q \to \mathrm{Atp}Q\,;\, m \mapsto (L(m), R(m), L(m)R(m)),$$

so that Mlt M *is a subgroup of* Tpm Q.

Proof. Let m be a Moufang element of Q, say $m = 1f_1$ with $(f_1, f_2, f_3) \in$ Atp Q. Now for x in Q, one has $xf_2L(m) = 1^{f_1}x^{f_2} = xf_3$, so that $L(m) = f_2^{-1}f_3$. Thus $(f_1^{-1}f_1^J, L(m), L(m)^{-1}) = (f_1^{-1}, f_2^{-1}, f_3^{-1})(f_1^J, f_3, f_2) = (f_1, f_2, f_3)^{-1}(f_1, f_2, f_3)^{(23)} \in$ Atp Q, whence $(L(m)^{-1}, L(m)^J, f_1^{-1}f_1^J) = (f_1^{-1}f_1^J, L(m), L(m)^{-1})^{(13)} \in$ Atp Q and also $(L(m), R(m), (f_1^{-1})^Jf_1) = (L(m)^{-1}, L(m)^J, f_1^{-1}f_1^J)^{-1} \in$ Atp Q. Then for x in Q, one has $xR(m)L(m) = m \cdot xm = (1 \cdot x)(f_1^{-1})^Jf_1 = x(f_1^{-1})^Jf_1 = (x \cdot 1)(f_1^{-1})^Jf_1 = mx \cdot m = xL(m)R(m)$.Finally, if m and n are Moufang elements, then so are $mn = 1L(n)L(m)$ and $m^{-1} = 1L(m)^{-1}$, so that $M =$ Mfg Q is a subloop of Q □

An inverse property loop is said to be a *Moufang loop* if every element is a Moufang element. By Proposition 4.1.2, Moufang loops satisfy the *flexible identity*

$$(4.1.3) \qquad xy \cdot x = x \cdot yx$$

and the *Third Moufang Identity*

$$(4.1.4) \qquad zx \cdot yz = (z \cdot xy)z.$$

Any group is a Moufang loop. Here are two examples of Moufang loops that are not associative: The first is commutative, so that (4.1.4) may be rewritten in the more readily verifiable form $xz \cdot yz = (xy \cdot z)z$.

Example 4.1.3 (Zassenhaus's Commutative Moufang Loop). Let Q be the set $\mathbb{Z}_3^4$. Define a new multiplication $\circ$ on $\mathbb{Z}_3^4$, for $\underline{x} = (x_1, x_2, x_3, x_4)$ and $\underline{y} = (y_1, y_2, y_3, y_4)$, by

$$\underline{x} \circ \underline{y} = \underline{x} + \underline{y} + (0, 0, 0, (x_3 - y_3)(x_1 y_2 - x_2 y_1)). \tag{4.1.5}$$

Then $(Q, \circ)$ is a commutative Moufang loop that is not associative. □

Example 4.1.4 (Zorn's Vector-Matrix Algebra). Let M be the set of matrices $\begin{bmatrix} \alpha & \mathbf{a} \\ \mathbf{b} & \beta \end{bmatrix}$ with real scalars α, β and real 3-vectors $\mathbf{a}, \mathbf{b}$. Define the product of two elements of M by

$$\begin{bmatrix} \alpha & \mathbf{a} \\ \mathbf{b} & \beta \end{bmatrix} \cdot \begin{bmatrix} \gamma & \mathbf{c} \\ \mathbf{d} & \delta \end{bmatrix} = \begin{bmatrix} \alpha\gamma + \mathbf{a} \cdot \mathbf{d} & \alpha\mathbf{c} + \delta\mathbf{a} - \mathbf{b} \times \mathbf{d} \\ \gamma\mathbf{b} + \beta\mathbf{d} + \mathbf{a} \times \mathbf{c} & \mathbf{b} \cdot \mathbf{c} + \beta\delta \end{bmatrix} \tag{4.1.6}$$

using the scalar product $\mathbf{x} \cdot \mathbf{y}$ and cross product $\mathbf{x} \times \mathbf{y}$ of 3-vectors. Define the "determinant" of a matrix by

$$\left\| \begin{bmatrix} \alpha & \mathbf{a} \\ \mathbf{b} & \beta \end{bmatrix} \right\| = \alpha\beta - \mathbf{a} \cdot \mathbf{b}. \tag{4.1.7}$$

Let Q be the subset of M consisting of matrices whose determinant is 1. Set

$$I = \begin{bmatrix} 1 & \mathbf{0} \\ \mathbf{0} & 1 \end{bmatrix}$$

Then $(Q, \cdot, I)$ is a Moufang loop that is neither commutative nor associative. □

There are other ways to characterize Moufang loops. Define the *First* or *Left Moufang Identity*

$$(zy \cdot z)x = z(y \cdot zx) \tag{4.1.8}$$

and the *Second* or *Right Moufang Identity*

$$x(z \cdot yz) = (xz \cdot y)z. \tag{4.1.9}$$

Note that each implies the flexible identity on setting $x = 1$. The first may be written in the form $L(z)L(y)L(z) = L(zy \cdot z)$, involving *left* multiplications, while the second may be written as $R(z)R(y)R(z) = R(z \cdot yz)$, involving *right* multiplications.

Proposition 4.1.5. *In a loop $(Q, \cdot, 1)$, the three Moufang identities are equivalent. Each implies the inverse property, and thus that Q is a Moufang loop.*

Proof. To begin, suppose that the First Moufang Identity holds. Setting $y = z \backslash 1$ yields $zx = z((z \backslash 1)(zx))$, whence $x = (z \backslash 1) \cdot zx$. Thus the left inverse property holds, with $z^l = z \backslash 1$. Using the First Moufang Identity again, $(uz)z^l = (z(z^l u) \cdot z)z^l = z(z^l u \cdot zz^l) = z(z^l u) = u$, so that the right inverse property also holds. By (4.1.1), the map $J = l: Q \to Q$ gives $x(z \cdot yz) = ((z^J y^J \cdot z^J)x^J)^J = (z^J(y^J \cdot z^J x^J))^J = (xz \cdot y)z$, so that the Second Moufang Identity holds. Moreover, $(zy \cdot z)(z^J x) = z(y \cdot z(z^J x)) = z(yx)$, so that each $L(z)$ is a topomorphism and each element $z = 1L(z)$ is a Moufang element. In particular, the Third Moufang Identity holds. A similar argument shows that the Second Moufang Identity implies the inverse property and the other Moufang identities.

Conversely, suppose that the Third Moufang Identity (4.1.4) holds. Setting $z = 1/x$ yields $y(1/x) = (1/x)x \cdot y(1/x) = ((1/x) \cdot xy)(1/x)$, whence $y = (1/x) \cdot xy$. Thus the left inverse property holds, with $x^l = 1/x$. Setting $y = 1$ in (4.1.4) yields the flexible identity (4.1.3). Then $zx = zx.z^l z = (z.xz^l)z = z(xz^l.z)$, whence $x = xz^l.z$ and the right inverse property holds. Thus Q becomes a Moufang loop: (4.1.4) shows that each $L(z)$ is a topomorphism and each element $z = 1L(z)$ is a Moufang element. Finally, since $(L(z)^J, L(z)R(z), R(z)) = (L(z), R(z), L(z)R(z))^{(23)}$ is an autotopy by Proposition 4.1.1, one has $x(z \cdot yz) = (xz \cdot z^J)(zy \cdot z) = (z \cdot (xz)^J)^J(zy \cdot z) = (xz \cdot y)z$, so the Second Moufang Identity also holds. □

EXERCISES

4.1A. Give an example of an inverse property quasigroup with $r \neq l$.

4.1B. Give an example of a quasigroup having the right inverse property, but not the left.

4.1C. For a quasigroup Q, prove Aut $Q \triangleleft$ Atp Q.

4.1D. If Q is an inverse property loop, under what conditions is Aut Q an S_3-subset of Atp Q?

4.1E. Let (f_1, f_2, f_3) be an autotopy of an inverse property loop Q. List all 6 elements of the orbit $(f_1, f_2, f_3)\ S_3$ in the S_3-set Atp Q.

4.1F. Show that a Moufang loop satisfies the *left* and *right alternative laws* $xx \cdot y = x \cdot xy$ and $y \cdot xx = yx \cdot x$.

4.1G. Verify the claims of Example 4.1.3.

4.1H. Verify the claims of Example 4.1.4. Hints: Show that the "determinant" of a product of vector-matrices is the product of their "determinants". Also show that

$$\begin{bmatrix} \alpha & \mathbf{a} \\ \mathbf{b} & \beta \end{bmatrix} \begin{bmatrix} \beta & -\mathbf{a} \\ -\mathbf{b} & \alpha \end{bmatrix} = \left\| \begin{bmatrix} \alpha & \mathbf{a} \\ \mathbf{b} & \beta \end{bmatrix} \right\| \cdot I.$$

4.1I. Show that the following conditions on a loop $(Q, \cdot, /, \backslash, 1)$ are equivalent:

(a) Q has the right inverse property;

(b) Q satisfies $yx \backslash y = zx \backslash z$;

(c) Q satisfies $yx \backslash y = x \backslash 1$.

4.1J. Consider a vector-matrix algebra as in Example 4.1.4, but with complex scalars and vectors. Set $\mathbf{u}_1 = (1,0,0)$, $\mathbf{u}_2 = (0,1,0)$ and $\mathbf{u}_3 = (0,0,1)$. Define

$$e_0 = \begin{bmatrix} 1 & \mathbf{0} \\ \mathbf{0} & 1 \end{bmatrix}, \quad e_j = \begin{bmatrix} 0 & -\mathbf{u}_j \\ \mathbf{u}_j & 0 \end{bmatrix} \quad \text{for } 1 \le j \le 3,$$

$$e_4 = \begin{bmatrix} i & \mathbf{0} \\ \mathbf{0} & -i \end{bmatrix},$$

and

$$e_{4+j} = \begin{bmatrix} 0 & i\mathbf{u}_j \\ i\mathbf{u}_j & 0 \end{bmatrix} \quad \text{for } 1 \le j \le 3.$$

(a) Show that $Q = \{\pm e_j | 0 \le j \le 7\}$ forms a 16-element Moufang loop under the vector-matrix product.

(b) For $1 \le j, k \le 7$, show that $e_j e_k = \pm e_{j \cdot k}$, where $j \cdot k$ is calculated in the quasigroup (2.4).

4.1K. (a) **(Minkowski spacetime in the vector-matrix algebra.)** Show that a 4-vector $(t, \mathbf{r})$ in Minkowski spacetime may be identified with the element $\begin{bmatrix} t & -\mathbf{r} \\ \mathbf{r} & -t \end{bmatrix}$ of the real vector-matrix algebra, so that the "determinant" becomes the Lorentz metric. (Hint: Pick units so that the speed of light $c = 1$.)

(b) **(Maxwell's equations in the vector-matrix algebra.)** Pick units so that the dielectric constant and the permeability take the value 1. Defining the differential operator matrix

$$D = \begin{bmatrix} -\dfrac{\partial}{\partial t} & \nabla \\ \nabla & -\dfrac{\partial}{\partial t} \end{bmatrix},$$

the field matrix

$$F = \begin{bmatrix} 0 & -\mathbf{E} + \mathbf{B} \\ \mathbf{E} + \mathbf{B} & 0 \end{bmatrix},$$

and the 4-current

$$J = \begin{bmatrix} \rho & -\mathbf{j} \\ \mathbf{j} & -\rho \end{bmatrix},$$

show that Maxwell's equations may be written as the single equation $DF = J$ in the vector-matrix algebra.

4.1L. Let $(Q, \cdot, 1)$ be a group in which each non-identity element has order 3. Define a new multiplication on Q by $x \circ y = y^2 x y^2$. Show that $(Q, \circ, 1)$ is a commutative Moufang loop.

4.1M. In the multiplication group of a Moufang loop, set $P(x) = L(x)^{-1}R(x)^{-1}$. Derive the *triality relations*

$$P(xyx) = P(x)P(y)P(x), \quad R(xyx) = R(x)R(y)R(x),$$
$$L(xyx) = L(x)L(y)L(x)$$

and

$$P(x)^{R(y)} = P(xy)P(y^{-1}), \quad P(x)^{L(y)} = P(yx)P(y^{-1}),$$
$$R(x)^{L(y)} = R(xy)R(y^{-1}), \quad R(x)^{P(y)} = R(yx)R(y^{-1}),$$
$$L(x)^{P(y)} = L(xy)L(y^{-1}), \quad L(x)^{R(y)} = L(yx)L(y^{-1}).$$

4.2. Loop Isotopes and Bol Loops

For a loop $(Q, \cdot, 1)$, the right inverse property may be displayed graphically within the 3-net $Q^2 =$ Net$(Q, \cdot, 1)$ determined by Q. Consider Figure 1.3, in which part of the net is drawn as a sort of optical bench. The vertical line forming the column labeled 1 represents a lens refracting horizontals to diagonals and vice versa. The vertical line forming the column labeled x represents an object mounted on the bench. The point (y, x) in the object has the point $(yx, yx \backslash y)$ in the vertical labeled $yx \backslash y$ as its image. Similarly, the point (z, x) in the object has $(zx, zx \backslash z)$ as its image. In general, the image points of different points of the vertical object line x lie in different vertical lines: The image is unfocused [Fig. 1.3(a)]. However, the right inverse property in the loop Q means precisely (Exercise 4.1I) that $yx \backslash y = zx \backslash z = x \backslash 1 = x^r$. Graphically, this means that the image points of the vertical object line x all lie on a vertical image line x^r: The image is focused (Fig. 1.3(b)).

This interpretation of the right inverse property in a loop Q as a property of the corresponding net enables one to specify the stronger property that the loop Q has to possess in order for every loop isotopic to Q also to have the right inverse property. Each vertical z in the net may end up being labeled

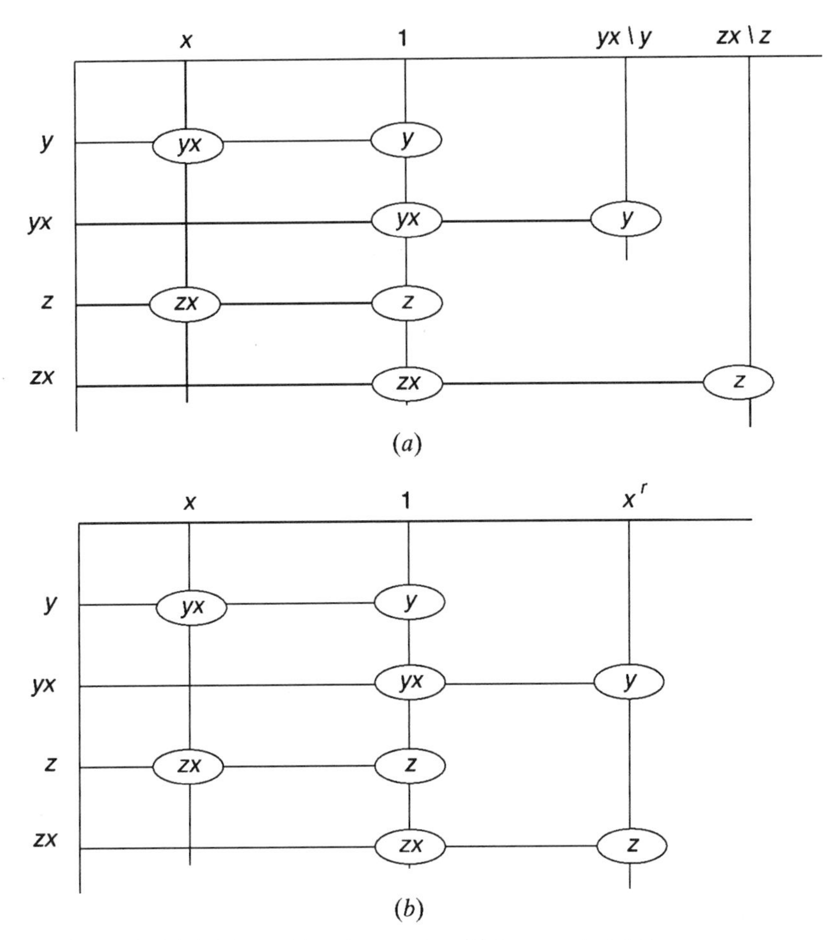

Figure 1.3. (*a*) "Unfocused": $yx \backslash y \neq zx \backslash z$. (*b*) "Focused": right inverse property.

by the identity element of some loop isotope Q' of Q, and the right inverse property in Q' then means that the given vertical z, behaving as a lens, has to focus other object verticals sharply onto an image vertical. As shown in Figure 1.4, this condition reduces to the so-called *Second* or *Right Bol Identity*

$$x(zy \cdot z) = (xz \cdot y)z. \tag{4.2.1}$$

A loop Q satisfying (4.2.1) is called a (*right*) *Bol loop*. Note the subtle distinction from the Second or Right Moufang Identity (4.1.9), namely in the association within the parentheses on the left hand side of the identity. Setting $x = 1$ in (4.1.9) yields the flexible law, while setting $x = 1$ in (4.2.1) yields nothing. Thus the Right Moufang Identity implies the Right Bol Identity. On the other hand, the Right Bol Identity is strictly weaker than the Right Moufang Identity:

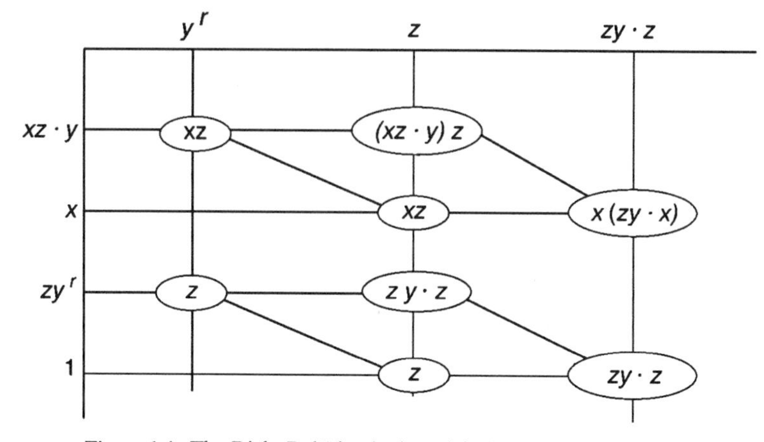

Figure 1.4. The Right Bol Identity in a right inverse property loop.

Example 4.2.3. Let Q be the set $\mathbb{Z}_2^3$. Define a new multiplication $\circ$ on $\mathbb{Z}_2^3$, for $\mathbf{x} = (x_1, x_2, x_3)$ and $\mathbf{y} = (y_1, y_2, y_3)$, by

$$\mathbf{x} \circ \mathbf{y} = \mathbf{x} + \mathbf{y} + (0, 0, x_2 y_1 y_2). \tag{4.2.2}$$

Then $(Q, \circ)$ is a right Bol loop that is not Moufang. □

In the positive direction, the Right Bol Identity implies the right alternative law (cf. Exercise 4.1F) on setting $y = 1$. Moreover, one has the following characterization.

Theorem 4.2.4. *A loop is a right Bol loop if and only if each of its loop isotopes has the right inverse property.*

Proof. If each loop isotope of a loop Q has the right inverse property, then Figure 1.4 shows that Q is a right Bol loop. Conversely, let $(Q, \cdot, 1)$ be a right Bol loop, so that (4.2.1) holds. Then $xz = x(z(z \backslash 1) \cdot z) = (xz)(z \backslash 1) \cdot z$, whence $x = (xz)(z \backslash 1)$. Thus $(Q, \cdot, 1)$ is a right inverse property loop, with $z^r = z \backslash 1$, and each vertical in $\mathrm{Net}(Q, \cdot, 1)$ has the focusing property of Figure 1.4. Let $(P, \circ, e)$ be a loop isotopic to $(Q, \cdot, 1)$. By Theorem 4.5(a), Net $(P, \circ, e)$ is isomorphic to $\mathrm{Net}(Q, \cdot, 1)$. Thus $\mathrm{Net}(P, \circ, e)$ also has the focussing property, showing that $(P, \circ, e)$ has the right inverse property. □

Corollary 4.2.5. *Each loop isotope of a right Bol loop is itself a right Bol loop.* □

Interchanging right and left, i.e. interchanging loops with their opposites, one obtains the *First* or *Left Bol Identity*

$$(z \cdot yz)x = z(y \cdot zx). \tag{4.2.3}$$

A loop satisfying (4.2.3) is a *(left) Bol loop*. Thus each loop isotope of a left Bol loop has the left inverse property. Again, the Left Bol Identity is weaker than the Left Moufang Identity. However:

Proposition 4.2.6. *Let* $(Q, \cdot, 1)$ *be a left Bol loop. Then the following conditions are equivalent*:

(a) *Q is a Moufang loop*;

(b) *Q has the right inverse property*;

(c) *Q satisfies the right alternative law*;

(d) *Q is a right Bol loop.*

Proof. If Q is a Moufang loop, then it satisfies the flexible law, and thus is a right Bol loop. In particular, it then has the right inverse property and satisfies the right alternative law. Conversely, it will be shown that each of (b) and (c) implies (a). Since (d) suffices for (b) and (c), the proof will then be complete.

(b) $\Rightarrow$ (a): As a left Bol loop, Q already has the left inverse property. Thus (b) means that Q is an inverse property loop. Setting $x = wL(z)^{-1}$ in (4.2.3) yields that $(R(z)L(z), L(z)^{-1}, L(z))$ is an autotopy, so that $L(z)$ is a topomorphism. Thus each element of Q is a Moufang element.

(c) $\Rightarrow$ (a): By the right alternative law, (4.2.3) yields $(z \cdot yz)z = z(y \cdot zz) = z(yz \cdot z)$. Substituting $yz = x$, one obtains the flexible law $zx \cdot z = z \cdot xz$ that suffices to transform the left Bol identity into the left Moufang identity. □

Corollary 4.2.7. *Each loop isotope of a Moufang loop is itself a Moufang loop.*

Proof. A loop is Moufang if and only if it is both a left and a right Bol loop. By Corollary 4.2.5, loop isotopes of left and right Bol loops are themselves left and right Bol loops. □

The right Bol identity is strong enough to guarantee the good behavior of exponentiation in a loop. Let $(Q, \cdot, 1)$ be a right Bol loop, so that in particular Q has the right inverse property $xy \cdot y^r = x$. Define the integral powers y^n of an element y of Q by

$$y^n = 1R(y)^n. \tag{4.2.4}$$

Theorem 4.2.8. *In a right Bol loop* $(Q, \cdot, 1)$,

$$R(y^n) = R(y)^n \tag{4.2.5}$$

holds.

Proof. The equation (4.2.5) will first be proved by induction for non-negative n. It is true for $n = 0$, $n = 1$. Then for $n > 1$, one has

$xR(y^n) = x(yR(y)^{n-2} \cdot y) = x(yR(y^{n-2}) \cdot y) = x(yy^{n-2} \cdot y) = (xy \cdot y^{n-2})y = (xy)R(1R(y)^{n-2})R(y) = xR(y)^n$. Thus (4.2.5) holds for all non-negative n. For negative n, one then has $R(1R(y)^n) = R(1R(y^r)^{|n|}) = R(y^r)^{|n|} = R(y)^n$, so that (4.2.5) holds for all integers n. □

EXERCISES

4.2A. Verify the assertions of Example 4.2.3.

4.2B. Prove Corollary 4.2.5 equationally, without recourse to nets.

4.2C. Draw and label the analogue of Figure 1.4 corresponding to the left Bol identity in a left inverse property loop.

4.2D. Show that a right Bol loop satisfying $(xy)^r = y^r x^r$ (for $x^r = 1/x$) is a Moufang loop.

4.2E. Show that a commutative Bol loop is Moufang.

4.2F. Give an example of an element y of a loop Q such that $R(yy) \neq R(y)^2$.

4.2G. Let $(Q, \cdot, 1)$ be a group in which each non-identity element has order 5. Define a new multiplication on Q by $x \circ y = y^3 x y^3$. Show that $(Q, \circ, 1)$ is a right Bol loop. (Cf. Exercise 4.1L.)

4.2H. (a) **(The Laplacian in the vector-matrix algebra.)** For D as in Exercise 4.1K (b) and for

$$\overline{D} = \begin{bmatrix} -\dfrac{\partial}{\partial t} & -\nabla \\ -\nabla & -\dfrac{\partial}{\partial t} \end{bmatrix},$$

show that $D\overline{D} = \Delta \cdot I$ with the Laplacian $\Delta = \partial^2/\partial t^2 - \nabla^2$.

(b) **(Electromagnetic waves in the vector-matrix algebra.)** Defining the electromagnetic potential matrix

$$P = \begin{bmatrix} \varphi & -\mathbf{A} \\ \mathbf{A} & -\varphi \end{bmatrix},$$

with $F = \overline{D}P$, obtain the wave equation $\Delta P = J$. (Hint: Consider the left inverse property, or the left-handed version of Theorem 4.2.8, in the Moufang loop of invertible elements of the vector-matrix algebra.)

4.3. Right Loops and Loop Transversals

In Section 2.2, a quasigroup Q was specified as a set-with-structure $(Q, \cdot, /, \backslash)$ satisfying (2.2.3). This specification breaks up naturally into left- and right-handed parts that share the multiplication. Taking the right-hand part alone, a *right quasigroup* $(Q, \cdot, /)$ is a set equipped with a binary *multiplication* $\cdot$ and *right division* $/$ such that (2.2.3)IR and (2.2.3)SR are satisfied. Recall that (2.2.3)IR yields the injectivity of each right multiplication, while (2.2.3)SR yields its surjectivity. The right division x/y is then expressed in the form $xR(y)^{-1}$ of (2.2.1).

Example 4.3.1. Let Q be a set. Define $x.y = x = x/y$ for x, y in Q. Then $(Q, \cdot, /)$ is a right quasigroup, with $R(y) = 1$ for all y in Q. □

The extreme Example 4.3.1 shows that the right multiplication map $R: Q \to Q!$ of a right quasigroup Q need not inject. Nevertheless, it is still convenient to study the *(combinatorial) right multiplication group* G, or RMlt $(Q, \cdot, /)$, of a right quasigroup $(Q, \cdot, /)$, namely the subgroup of $Q!$ generated by the image $R(Q)$ of the right multiplication map. The right quasigroups of Example 4.3.1 have trivial right multiplication groups.

Restriction from right quasigroups to right loops eliminates trivialities such as those inherent in Example 4.3.1. A *right loop* $(Q, \cdot, /, 1)$ is a right quasigroup $(Q, \cdot, /)$ with an *identity* element 1 satisfying (4.3). For a right loop Q, the right multiplication map $R: Q \to Q!$ injects, since $R(x) = R(y)$ implies $x = 1R(x) = 1R(y) = y$. A *right quasigroup homomorphism* is a set map between right quasigroups that preserves the multiplication and right division. A *right loop homomorphism* is a right quasigroup homomorphism between right loops mapping the identity of the domain to the identity of the codomain. A *right subquasigroup* of a right quasigroup $(Q, \cdot, /)$ is a subset S of Q that is closed under multiplication and right division; a *right subloop* of a right loop is a right subquasigroup that contains the identity element. A *congruence (relation)* on a right quasigroup or right loop $(Q, \cdot, /)$ is an equivalence relation α on Q that is a right subquasigroup of the direct square right quasigroup $(Q^2, \cdot, /)$ with componentwise operations. For a right loop $(Q, \cdot, /, 1)$, define

$$(x, y, z)P = (x/y) \cdot z. \tag{4.3.1}$$

Then $(y, y, z)P = (y/y)z = 1 \cdot z = z$ and $(z, y, y)P = (z/y) \cdot y = z$ for all y, z in Q. In other words, the operation P of (4.3.1) in a right loop Q has the properties of the operation P of (2.2.5) in a quasigroup Q. Consequently, one obtains the following result.

Proposition 4.3.2. *Let $(Q, \cdot, /, 1)$ be a right loop.*

(a) *The congruence relations on Q are permutable.*
(b) *A right subloop α of Q^2 is a congruence if and only if it contains the diagonal right subloop $\hat{Q}$.*

Proof. (a) As for Proposition 2.2.3.
(b) As for Proposition 2.2.4. □

Of course, one may also study oppositely-handed versions of the above, namely left quasigroups, left loops, etc.

The primary sources of right loops are right transversals to subgroups of groups. Let H be a subgroup of a group $(G, \cdot, /, \backslash, 1)$, and let T be a right transversal to H in G such that 1 represents H. (Transversals having the identity as the representative for the subgroup are often described as *normalized.*) Thus $G = \dot{\bigcup}_{t \in T} Ht$. Define a map $\varepsilon : G \to T;\ g \mapsto g^{\varepsilon}$ by

$$g \in Hg^{\varepsilon}, \tag{4.3.2}$$

so that g^{ε} or $g\varepsilon$ is the unique representative in T for the right coset of H that contains g. It is also convenient to define a map $\delta : G \to H;\ g \mapsto g^{\delta}$ by

$$g = g^{\delta} g^{\varepsilon}. \tag{4.3.3}$$

Note that $1^{\delta} = 1^{\varepsilon} = 1$. Moreover $h^{\delta} = h$ and $h^{\varepsilon} = 1$ for h in H, while $t^{\delta} = 1$ and $t^{\varepsilon} = t$ for t in T. Now define a binary multiplication $*$ and a binary right division $\|$ on T by

$$t * u = (tu)\varepsilon, t\|u = (t/u)\varepsilon \tag{4.3.4}$$

for t, u in T, i.e. by $tu \in H(t * u)$ and $t/u = tu^{-1} \in H(t\|u)$. Since $H(t\|u)u \ni (t/u) \cdot u = t \in Ht$ and $H(t * u)/u \ni (tu)/u = t \in Ht$, one has $(t\|u) * u = t$ and $(t * u)\|u = t$ for all t, u in T. Moreover $1 * t = (1t)\varepsilon = t\varepsilon = t = t\varepsilon = (t1)\varepsilon = t * 1$. Summarizing:

Proposition 4.3.3. *Let T be a normalized right transversal from a group G to a subgroup H. Then $(T, *, \|, 1)$ is a right loop.* □

To within right loop isomorphism, every right loop may be obtained by the construction of Proposition 4.3.3 (cf. Exercise 4.3I). Note also that the set bijection $T \to H \backslash G;\ t \mapsto Ht$ may be used to transfer the right loop structure from the normalized right transversal T to the homogeneous space $H \backslash G$.

In certain circumstances, the right loop of Proposition 4.3.3 will become a loop. If this happens, the normalized right transversal T is called a *loop transversal.* Thus T is a loop transversal if and only if, for each ordered pair (t, u) of elements of T, the equation

$$t * x = u \tag{4.3.5}$$

has a unique solution. The solution x is the result of t dividing u from the left in the loop.

Example 4.3.4. Let N be a normal subgroup of the group G. Then a normalized right transversal T from G to N is a loop transversal. Indeed, the set bijection $T \to N \backslash G;\ t \mapsto Nt$ becomes a right loop isomorphism

$(T, *, \|, 1) \to (N \setminus G, \cdot, /, N) = (G/N, \cdot, /, N)$, since $N(t * u) = Ntu = NNtu = Nt \cdot t^{-1}Ntu = Nt \cdot Nu$ by Corollary 2.4.7. □

Example 4.3.5. Let Q be a quasigroup with element e and multiplication group G. Then the normalized right transversal $T = \{\rho(e, y) | y \in Q\}$ from G to the stabilizer G_e [cf. (2.2.4)] is a loop transversal. Indeed, one may identify T with Q via the mutually inverse maps $T \to Q$; $t \mapsto et$ and $Q \to T$; $y \mapsto \rho(e, y)$. Given $\varepsilon : G \to T$; $g \mapsto \rho(e, eg)$, this yields the right loop $(Q, *, e)$ with $x * y = e(\rho(e, x) * \rho(e, y)) = e\rho(e, x\rho(e, y)) = x\rho(e, y) = xR(e \setminus e)^{-1}R(e \setminus y)$. This is just the loop $(Q, *, e)$ of (4.4). □

Proposition 4.3.6. *Let T be a normalized right transversal from a group G to a subgroup H. Then T is a loop transversal if and only if it is a right transversal to each conjugate H^g of H in G.*

Proof. Suppose first that T is a loop transversal. Note that $H^g = g^{-1}Hg = (g^\delta g^\varepsilon)^{-1}Hg^\delta g^\varepsilon = H^{g^\varepsilon}$. Then for x in T and a in G, $a \in H^g x \Leftrightarrow a \in H^{g^\varepsilon}x \Leftrightarrow g^\varepsilon \cdot a \in Hg^\varepsilon \cdot x \Leftrightarrow (g^\varepsilon \cdot a)\varepsilon = (g^\varepsilon \cdot x)\varepsilon \Leftrightarrow g^\varepsilon * x = (g^\varepsilon \cdot a)\varepsilon$. Since $(T, *)$ is a loop, there is a unique solution x to the latter equation. Thus T is a right transversal to H^g in G.

Conversely, suppose that T is a right transversal to each conjugate of H in G. It must be shown that (4.3.5) has a unique solution x. But $t * x = u \Leftrightarrow (tx)\varepsilon = u \Leftrightarrow Hu = H(tx)^\varepsilon = Htx \Leftrightarrow u \in Htx \Leftrightarrow t^{-1}u \in H^t x$. Since T is a right transversal to H^t, there is a unique x in T for which $t^{-1}u \in H^t x$, and thus for which $t * x = u$. □

EXERCISES

4.3A. Consider the set $(P, \cdot)$ of real polynomials with multiplication $p(x) \cdot q(x) = p(x) + q'(x)$ as in Example 2.2.1.
  (a) Define a right division $/$ on P so that $(P, \cdot, /)$ becomes a right quasigroup.
  (b) Show that there is no element e of P for which $(P, \cdot, /, e)$ would be a right loop.

4.3B. (a) Let $(Q, \cdot, /)$ be a right quasigroup. Show that $(Q, / \cdot)$ is a right quasigroup, with $\text{RMlt}(Q, /, \cdot) = \text{RMlt}(Q, \cdot, /)$.
  (b) Give an example of a right quasigroup $(Q, \cdot, /)$ for which $(Q, \cdot, /)$ is different from $(Q, /, \cdot)$.

4.3C. Let $T = \{(1), (123), (132)\}$ and $H = \{(1), (23)\}$ in the group S_3. Determine the right loop $(T, *, \|, 1)$ given by Proposition 4.3.3 in this case. Is it a loop?

4.3D. Let $(Q, \cdot, /, \setminus)$ be a quasigroup. Let $//$ denote the opposite of right division, so that $x//y = y/x$ for x, y in Q. Prove that $(Q, \setminus, //)$ is a right quasigroup.

4.3E. Formulate and prove First Isomorphism Theorems for right quasigroups and right loops.

4.3F. Show that congruence relations on a right quasigroup need not be permutable.

4.3G. Give an example of a right subquasigroup α of the square Q^2 of a right quasigroup $(Q, \cdot, /)$ such that α contains the diagonal $\hat{Q}$, but nevertheless fails to be a congruence relation on Q.

4.3H. Suppose that a binary multiplication $\cdot$ is defined on a set Q. Show that there is a right division $/$ yielding a right quasigroup $(Q, \cdot, /)$ if and only if, for all a, b in Q, there is a unique solution x to the equation $x \cdot a = b$. For finite Q, show that this happens if and only if the multiplication is *left cancellative*: $c \cdot a = d \cdot a \Rightarrow c = d$ for all a, c, d in Q.

4.3I. Let $(Q, \cdot, /, 1)$ be a right loop, with right multiplication group G. Let H be the stabilizer of 1 in G.

(a) Show that $R(Q)$ is a normalized right transversal to H in G.

(b) Let $(R(Q), *, \|, 1)$ be the right loop on $R(Q)$ given by Proposition 4.3.3. Show that $R: (Q, \cdot, /, 1) \to (R(Q), *, \|, 1)$ is a right-loop isomorphism.

4.3J. Let T be a normalized right transversal to a subgroup H of a group G. If T is an Inn G-subset of $(G, \text{Inn } G)$, show that $(T, *, 1)$ is a loop.

4.4. Loop Transversal Codes

The concept of a loop transversal offers a quick and elementary introduction to the subject known as algebraic coding theory. Algebraic coding theory addresses certain aspects of the problem of transmitting information through channels that are subject to interference. The effect of the interference is to corrupt the signals being transmitted. Nevertheless, algebraic coding theory offers methods of encoding the original information into a signal for transmission, in such a way that the original information may be recovered from a corrupt received signal, or at least so that a signal may be recognized as being corrupt. The information transmission may be taking place through space, sending a message from one physical location to another. On the other hand, it may also be taking place through time, recording a message in a memory and then reading it back later.

The usual scheme of algebraic coding theory may be summarized as follows. A finite alphabet A is given, and the set of words to be encoded is the uniform code A^k for some k (cf. Exercise O 4.1B). The information channel carries words from the uniform code A^n for some $n \geq k$. The integer n is known as the *length* of the channel. A subset C of A^n is chosen. This subset C is known as the *code* (or a *block code*, to avoid confusion with the concept of Definition O 4.1.1, which in turn may be described as a *source code*). The *encoding* is an embedding $A^k \to A^n$ with image C, restricting to a bijection $\eta : A^k \to C$. Thus $|C| = |A|^k$. The integer k is known as the *dimension* of the code. If a word c from the code C is transmitted through the channel without corruption, then it is received as the same word c. The original encoded word from A^k may then be recovered as $c\eta^{-1}$. However, the emitted codeword c may have been subject to interference in the channel, being received as a corrupted word x in A^n. A *decoding* map

$$\delta : A^n \to C \tag{4.4.1}$$

assigns a codeword x^δ to the received word x. Provided that the received word x was not corrupted excessively from the emitted codeword c, one should expect that $x^\delta = c$. In particular, one should have $c^\delta = c$ for c in C.

Example 4.4.1 (Repetition Codes). Let $A = \{0,1\}$ and $k = 1$. Consider a channel length of 3. Define $0\eta = 000$ and $1\eta = 111$. Thus $C = \{000, 111\}$. Define the decoding (4.4.1) by $\delta^{-1}\{000\} = \{000, 001, 010, 100\}$ and $\delta^{-1}\{111\} = \{111, 110, 101, 011\}$ ("majority vote"). Provided that at most one letter of the emitted codeword gets corrupted in the channel, the decoder is able to recover the codeword. One may extend this scheme to channels of greater odd length. □

For further analysis, it is convenient to put an abelian group structure $(A, +, 0)$ on the alphabet A. Usually, for $|A| = l$, one takes A to be the cyclic group C_l. The channel A^n is the n-th direct power of A, the i-th projection $A^n \to A$ simply selecting the i-th letter of a word (cf. Exercise O 4.1J). Thus the channel A^n becomes the abelian group $(A^n, +, 0)$, or more pedantically $(A^n, +, 00\ldots0)$. This abelian group structure may be used to describe the interference taking place in the channel. If an emitted codeword c is received as the corrupted word x, one says that the *error* $x - c$ was added to c during passage through the channel. The decoder $\delta : x \mapsto c$ is then said to *correct* the error $x - c$. To measure the seriousness of the error, one may define the *Hamming weight* $|x|$ of a channel word x in A^n to be the number of non-zero letters in x. The *Hamming distance* between two words x, y is then $|x - y|$. Note that the triangle inequality

$$|x + y| \leq |x| + |y| \tag{4.4.2}$$

is satisfied. Indeed, $|x + y| > |x| + |y|$ is impossible, since $x + y$ can only have a non-zero letter in a certain slot if at least one of x and y has a non-zero letter in that slot. Moreover, $|x| = 0 \Leftrightarrow x = 0$.

The decoding may be analyzed using the abelian group structure. An *error map*

$$\varepsilon : A^n \to A^n \tag{4.4.3}$$

determines that a received word x was the result of an error x^{ε}. Thus

$$x = x^{\delta} + x^{\varepsilon} \tag{4.4.4}$$

for each x in A^n. The key idea behind loop transversal codes is the observation that (4.4.4) may just be an instance of (4.3.3). Thus the code C is defined to be *linear* if it is a subgroup of the channel A^n. Since A^n is abelian, such a subgroup C is normal. As in Example 4.3.4, any normalized right transversal T to C in A^n is then a loop transversal. Taking the error map ε as in (4.3.2), one obtains the loop transversal T as the set of errors corrected by the code. Note that the loop $(T, *, 0)$ defined by (4.3.4) is an abelian group, since the map $T \to A^n/C;\ t \mapsto C + t$ of Example 4.3.4 is a right loop isomorphism of T with the abelian group A^n/C. Nevertheless, it is often convenient to continue to refer to the operation $*$ as a loop multiplication, in order to distinguish it from the abelian group operation $+$ on A^n.

Example 4.4.2. Consider the length 3 binary repetition code C of Example 4.4.1. Interpret A as $\mathbb{Z}_2$. Then C becomes linear, and the normalized right transversal $T = \{000, 001, 010, 100\}$ is the set of errors corrected by C. The abelian group multiplication $*$ on T given by (4.3.4) has the table

*	000	001	010	100
000	000	001	010	100
001	001	000	100	010
010	010	100	000	001
100	100	010	001	000

Note that the table may be summarized by the specification that the map

$$s : (T, *) \to (A^2, +);\ 001 \mapsto 01, 010 \mapsto 10, 100 \mapsto 11$$

is an abelian group homomorphism. □

If one knows a linear code C in a channel A^n, one may determine a loop transversal T to C by selecting representatives of the various cosets of C. Typically, one picks *coset leaders*—representatives having minimal Hamming weight within their cosets. On the other hand, one of the major problems of algebraic coding theory is to determine a suitable code C to begin with, for a given channel A^n. If the loop $(T, *, 0)$ is known, then the code C may be obtained from T by the so-called *Principle of Local Duality*. To formulate this principle, it is convenient to establish some notation. For elements $t_1, t_2, \ldots$ of T, define $\Sigma_{i=1}^{m} t_i$ inductively by

$$\sum_{i=1}^{0} t_i = 0 \quad \text{and} \quad \sum_{i=1}^{m} t_i = t_m + \sum_{i=1}^{m-1} t_i.$$

Define $\prod_{i=1}^{m} t_i$ inductively by

$$\prod_{i=1}^{0} t_i = 0 \quad \text{and} \quad \prod_{i=1}^{m} t_i = t_m * \prod_{i=1}^{m-1} t_i.$$

In compound expressions involving loop operations $*$, $\|$ and abelian group operations $+$, $-$, the loop operations will bind more strongly than the group operations. For example, $t + u - t * u = t + u - (t * u)$.

Proposition 4.4.3 (Principle of Local Duality). Let T be a loop transversal to a linear code C in a channel A^n over a finite abelian group alphabet A. Suppose that T is a set of generators for A^n. Then

$$C = \left\{ \sum_{i=1}^{m} t_i - \prod_{i=1}^{m} t_i \,\middle|\, \langle t_1, \ldots, t_m \rangle \in T^{*\kappa} \right\}.$$

Proof. Recall that $t^{\varepsilon} = t$ for t in T. Induction on m using (4.3.4) then shows that $(\sum_{i=1}^{m} t_i)\varepsilon = \prod_{i=1}^{m} t_i$ for $t_1, \ldots, t_m$ in T. Since T generates A^n and A is finite, each channel word x may be written in the form $x = \sum_{i=1}^{m} t_i$ for some multisubset $\langle t_1, \ldots, t_m \rangle$ of T. Then

$$C = \{x^{\delta} | x \in A^n\} = \{x - x^{\varepsilon} | x \in A^n\} = \left\{ \sum_{i=1}^{m} t_i - \prod_{i=1}^{m} t_i \,\middle|\, \langle t_1, \ldots, t_m \rangle \in T^{*\kappa} \right\}.$$

□

The full force of the Principle of Local Duality comes into play when it is not even known in advance that there is some code C to which a loop $(T, *, 0)$ in A^n is transversal. For simplicity, the case $A = \mathbb{Z}_2 = \{0, 1\}$ will be discussed here. (For a more general version, see Exercise II 2.4T.) Given a channel A^n, one normally has a list of the errors one would like to correct (e.g. the commonest errors), and this list usually includes the n-element set B of errors of Hamming weight 1. Let T be a 2^{n-k}-element set of errors to be corrected, with $T \supseteq \{0\} \cup B$. Suppose that T carries a loop structure $(T, *, 0)$ given by an isomorphism

$$s : (T, *, 0) \to (A^{n-k}, +, 0) \tag{4.4.5}$$

(e.g. as in Example 4.4.2). Let $t_1, \ldots, t_m$ be elements of T. By the closure of $(T, *)$, the loop product $\prod_{i=1}^{m} t_i$ always lies in T. On the other hand, the sum $\sum_{i=1}^{m} t_i$ may only lie in T for certain choices of $t_1, \ldots, t_m$. The isomorphism (4.4.5) is said to be a *partial homomorphism* $s : (T, +) \to (A^{n-k}, +)$ if $(\sum_{i=1}^{m} t_i)s = \sum_{i=1}^{m} t_i^s$ whenever $\sum_{i=1}^{m} t_i \in T$. Of course, this means that $\sum_{i=1}^{m} t_i = \prod_{i=1}^{m} t_i$ in such cases, since the two sides of the equation have the same image under the isomorphism (4.4.5).

Theorem 4.4.4. *Let T be a 2^{n-k}-element subset of the length n binary channel A^n, such that T contains 0 and the n-element set B of errors of Hamming weight 1. Suppose that T carries a loop structure $(T, *, 0)$ given by an isomorphism (4.4.5) such that $s:(T, +) \to (A^{n-k}, +)$ is a partial homomorphism. Then there is a linear code C of dimension k in A^n to which $(T, *, 0)$ is a loop transversal. Moreover, T is precisely the set of errors corrected by C.*

Proof. Note that each element x of A^n has a unique expression $x = \Sigma\{b_i | i \in X\}$ for a subset X of B. Define the *syndrome*

$$s : A^n \to A^{n-k}; \sum_{i \in X} b_i \mapsto \sum_{i \in X} b_i^s. \tag{4.4.6}$$

Since (4.4.5) is a partial homomorphism, it is the restriction of the syndrome to T. Now for x, y in A^n, with $x = \Sigma_{i \in X} b_i$ and $y = \Sigma_{i \in Y} b_i$, one has

$$x^s + y^s = \sum_{i \in X} b_i^s + \sum_{i \in Y} b_i^s = \sum \{b_i^s | i \in (X \cup Y) - (X \cap Y)\} = (x + y)s.$$

Thus the syndrome is an abelian group homomorphism. Let $C = \text{Ker } s$ be its group kernel $s^{-1}\{0\}$. Note that $|C| = |A|^k$. For $x = \Sigma_{i \in X} b_i$ in A, define $x^\delta = \Sigma_{i \in X} b_i - \Pi_{i \in X} b_i$ and $x^\varepsilon = \Pi_{i \in X} b_i$. Then $x^\delta \in C$ and $x^\varepsilon \in T$, with $x = x^\delta + x^\varepsilon$. Thus $A^n = C + T$. But $|A^n| = |C| \cdot |T|$, so T is a loop transversal to C in A^n. Moreover, $\delta : A^n \to C;\ x \mapsto x^\delta$ and $\varepsilon : A^n \to T;\ x \mapsto x^\varepsilon$ surject, indeed $\varepsilon|_T = 1_T$, so T is precisely the set of errors corrected by C. □

Example 4.4.5. For the alphabet $\mathbb{Z}_2 = \{0, 1\}$, consider the set $\mathbb{Z}_2^4 = \{0 = 0000, 1 = 0001, 2 = 0010, \ldots, 9 = 1001, A = 1010, B = 1011, \ldots, F = 1111\}$ of hexadecimal digits. This set is to be encoded for transmission through a binary channel in such a way that errors of single Hamming weight may be corrected. Let b_i, for $1 \leq i \leq 7$, denote the binary word of length 7 and Hamming weight 1 with its unique non-zero letter in the i-th slot. Thus $b_1 = 1000000, \ldots, b_3 = 0010000$, etc. Set $B = \{b_i | 1 \leq i \leq 7\}$ and $T = \{0000000\} \cup B$. Define $s : T \to \mathbb{Z}_2^3 = \mathbb{Z}_2^{7-4}$ as a partial homomorphism by sending b_i to the binary representation of i, e.g. $b_3^s = 011$. This sets up an isomorphism (4.4.5), e.g. $b_1 * b_3 = (b_1^s + b_3^s)s^{-1} = (001 + 011)s^{-1} = 010s^{-1} = b_2$. By Theorem 4.4.4, the loop transversal $(T, *, 1)$ then determines a code C of dimension 4. The 2^4 hexadecimal digits may be encoded by bijection with C. The elements of C may be determined by the Principle of Local Duality, e.g. $b_1 + b_3 - b_1 * b_3 = 1000000 + 0010000 - 0100000 = 1110000 \in C$. □

EXERCISES

4.4A. Describe the repetition code C of length 5 for the alphabet $A = \{0, 1\}$, by analogy with Example 4.4.1. Decode the received words 10110 and 01010.

4.4B. Taking A as $\mathbb{Z}_2$ in Exercise 4.4A, determine a loop transversal T from A^5 to C comprising the errors corrected by C. Exhibit an abelian group homomorphism $s:(T, *) \to (A^4, +)$.

4.4C. Use Exercise 4.3J (instead of Example 4.3.4) to show that a normalized right transversal to a linear code is a loop transversal.

4.4D. Write out the cosets of the linear code C of Example 4.4.2, and determine a coset leader for each coset. Are these leaders unique?

4.4E. Write out the cosets of the linear code C of Exercise 4.4B, and determine a coset leader for each coset. Are these leaders unique?

4.4F. For $A = \mathbb{Z}_3 = \{0, 1, 2\}$, consider the linear repetition code $C = \{000, 111, 222\}$ of length 3. Write out the cosets of the linear code C, and determine a coset leader for each coset. Hence determine a loop transversal T to C. Describe the abelian group $(T, *)$, either by a table or by an isomorphism with A^2. Are the coset leaders unique? If not, repeat the exercise with a different loop transversal U.

4.4G. Verify $\varepsilon|_T = 1_T$ in the proof of Theorem 4.4.4.

4.4H. (a) Determine all 16 elements of C in Example 4.4.5.

(b) Set up an encoding bijection $\eta : \mathbb{Z}_2^4 \to C$.

(c) If the hexadecimal digit D is encoded, and subjected to the error b_4 during passage through the channel $\mathbb{Z}_2^7$, show that it may be recovered.

(d) If the hexadecimal digit D is encoded, and subjected to the error $b_2 + b_4$, to which hexadecimal digit is the received word decoded?

4.4I. Consider the binary channel $\mathbb{Z}_2^{15}$ of length 15. Let B be the set of channel words of Hamming weight 1, and let T be the subset $\{0\} \cup B$ of $\mathbb{Z}_2^{15}$. By analogy with Example 4.4.5, determine a map $s:T \to \mathbb{Z}_2^4 = \mathbb{Z}_2^{15-11}$ yielding a code C in $\mathbb{Z}_2^{15}$ of dimension 11.

4.4J. Consider the binary channel $\mathbb{Z}_2^7$ of length 7, as in Example 4.4.5. One wishes to correct errors of Hamming weight 1, and also the so-called *burst* errors of Hamming weight 2, namely those channel words of weight 2 in which the two non-zero entries are adjacent, e.g. 0011000. Starting with the assignments $b_1 \mapsto 0001, b_2 \mapsto 0010, b_3 \mapsto 0100, b_4 \mapsto 1000, b_5 \mapsto 1111, b_6 \mapsto 0101$ and, $b_7 \mapsto 1011$, construct a map $s:T \to \mathbb{Z}_2^4$ with domain containing the burst and single-weight errors, by means of which Theorem 4.4.4 builds a code C of dimension 3 correcting the full set T of errors.

II

LINEAR ALGEBRA

1. GENERAL ALGEBRA AND LINEAR ALGEBRA

General algebra is the study of sets with structure. Thus its basic objects are sets, and the maps between them are just set functions. Against this background, linear algebra (in its widest sense) may be described as the study of abelian groups with structure. Thus its basic objects are abelian groups, and the maps between them are group homomorphisms. This analogy between general algebra and linear algebra is only one aspect of a very complex relationship. A second aspect is the obvious fact that linear algebra is just a specialized part of general algebra. For instance, linear algebra provides a copious collection of examples of many general algebraic phenomena. A third aspect of the relationship is that linear algebra represents a kind of ideal mathematical world, where calculations are easy and everything is well-behaved. From this point of view, general algebra is often concerned with measuring how far non-linear systems have diverged from the ideal. A major goal of general algebra is then to develop concepts that help to bring non-linear systems back closer to the ideal.

In linear algebra, the underlying abelian groups are usually "written additively". In the language of right loops (Section I 4.3), this means that the "multiplication" is addition $(+)$, the "right division" is subtraction $(-)$, and the "identity element" is the zero element (0). Ironically, the non-associative operation of subtraction is the most fundamental. Indeed, in an abelian group A, the zero element is $a - a$ for any a in A, while the sum $a + b$ of two elements a,b is $a - ((a - a) - b)$. Thus a non-empty subset S of an abelian group A is a subgroup iff it is closed under subtraction (cf. Exercise I 2F).

The major difference between a set function $f: X \to Y$ and an abelian group homomorphism $h: A \to B$ is that a set function may map any element of X to any element of Y, while an abelian group homomorphism is constrained to map the zero element of A to the zero element of B. The special role of zero elements accounts for much of the distinction between general algebra and linear algebra. For example, recall that the kernel $\ker f$

of the set function $f : X \to Y$ is the equivalence relation $\{(x_1, x_2) | x_1 f = x_2 f\}$ on X. The First Isomorphism Theorem O 3.3.1 factorizes $f : X \to Y$ as the composite of the natural projection nat $\ker f : X \to X^{\ker f}$; $x \mapsto x^{\ker f}$, the isomorphism $X^{\ker f} \to Xf$; $x^{\ker f} \mapsto xf$, and the injection $Xf \hookrightarrow Y$. Forgetting the abelian group structure, this First Isomorphism Theorem for sets applies to the homomorphism $h : A \to B$. The kernel $\ker h$ of h is a binary relation on A. However, it may be described entirely by the group kernel $\operatorname{Ker} h = \{a \in A | ah = 0\}$, a normal subgroup of A. The congruence class $x^{\ker h}$ of an element x of A is then just the coset $x + \operatorname{Ker} h = \{x + a | ah = 0\}$. Thus $A^{\ker h} = A/\operatorname{Ker} h$. The construction of the binary relation $\ker f$ of an arbitrary set function f is a typical general algebraic contrivance that attempts to approximate a simpler construct of linear algebra, the unary group kernel relation $\operatorname{Ker} h$.

For abelian groups, the First Isomorphism Theorem may readily be extended. It is customary to write $\operatorname{Im} h$ for the image Ah of an abelian group homomorphism $h : A \to B$. The *cokernel* $\operatorname{Coker} h$ of h is then defined to be the quotient group $B/\operatorname{Im} h$, usually taken with the natural projection $B \to B/\operatorname{Im} h$; $b \mapsto b + \operatorname{Im} h$. A sequence

$$(1.1) \quad A_0 \longrightarrow \cdots \xrightarrow{f_{n-2}} A_{n-1} \xrightarrow{f_{n-1}} A_n \xrightarrow{f_n} A_{n+1} \xrightarrow{f_{n+1}} \cdots \longrightarrow A_N$$

of abelian groups and homomorphisms is said to be *exact at* A_n if $\operatorname{Im} f_{n-1} = \operatorname{Ker} f_n$. It is said to be *exact* if it is exact at each intermediate group $A_1, A_2, \dots, A_{N-1}$. The singleton abelian group $\{0\}$ is often denoted just as 0. The extended First Isomorphism Theorem for Abelian Groups may then be formulated as follows.

Proposition 1.1. *Let $h : A \to B$ be a homomorphism of abelian groups. Then there is an exact sequence*

$$(1.2) \qquad 0 \longrightarrow \operatorname{Ker} h \longrightarrow A \xrightarrow{h} B \longrightarrow \operatorname{Coker} h \longrightarrow 0. \quad \square$$

Note that the homomorphism $0 \to \operatorname{Ker} h$ is not labeled, since the only possibility sends the unique element 0 of the singleton $\{0\}$ to the zero element 0 of $\operatorname{Ker} h$. The exactness at $\operatorname{Ker} h$ says that the group kernel of $\operatorname{Ker} h \to A$ is $\{0\}$, or in other words that $\operatorname{Ker} h$ embeds into A. At the other end, the homomorphism $\operatorname{Coker} h \to 0$ is not labeled, since the only possibility sends each element of $\operatorname{Coker} h$ to 0 in $\{0\}$. The exactness at $\operatorname{Coker} h$ says that the image of $B \to \operatorname{Coker} h$ is the group kernel of $\operatorname{Coker} h \to 0$, namely all of $\operatorname{Coker} h$. In other words, $B \to \operatorname{Coker} h$ surjects. The exactness at A corresponds to the definition of $\operatorname{Ker} h$, while the exactness at B corresponds to the definition of $\operatorname{Coker} h$.

If X and Y are sets, then the collection of all set functions $f : X \to Y$ forms a new set Y^X. There is a parallel construction in linear algebra. Given two abelian groups A and B, let $\operatorname{Hom}(A, B)$ denote the set of all abelian group homomorphisms $h : A \to B$. This set is nonempty, since it contains the *zero homomorphism* $0 : A \to B;\ a \mapsto 0$, i.e. the composite $A \to 0 \to B$ of the unique homomorphisms $A \to 0$ and $0 \to B$. The abelian group structure on B then defines an abelian group structure on $\mathrm{Hom}(A, B)$ via the subtraction

$$(1.3) \qquad - : \mathrm{Hom}(A,B)^2 \longrightarrow \mathrm{Hom}(A,B);\ (f,g) \mapsto (f-g : a \mapsto af - ag).$$

For a single abelian group A, the abelian group $\mathrm{Hom}(A, A)$ is often denoted as $\mathrm{End}(A,+)$ or just $\mathrm{End}\,A$. Its elements are called *endomorphisms* of A. Thus $\mathrm{End}\,A$ is the linear algebra counterpart of the set X^X of maps from a set X into itself. Since composites of homomorphisms are homomorphisms, $\mathrm{End}\,A$ forms a submonoid of the monoid $(A^A, \cdot, 1)$. Moreover, there is an abelian group homomorphism $j : \mathbb{Z} \to \mathrm{End}\,A;\ 1 \mapsto 1$. The image of an integer n under this homomorphism is also denoted n, thus

$$(1.4) \qquad n : A \to A;\ a \mapsto an.$$

Often one writes na instead of an, and $-a$ instead of $(-1)a$.

EXERCISES

1A. Describe the inversion map (I 1.3.2) of an abelian group A purely in terms of the binary subtraction operation.

1B. (a) Prove that an abelian group homomorphism $h : A \to B$ is injective iff $\mathrm{Ker}\,h = 0$.

(b) Prove that an abelian group homomorphism $h : A \to B$ is surjective iff $\mathrm{Coker}\,h = 0$.

1C. Prove that an abelian group homomorphism $h : A \to B$ is an isomorphism iff the sequence $0 \to A \xrightarrow{h} B \to 0$ is exact.

1D. Given an abelian group homomorphism $h : A \to B$, establish the exact sequence $0 \to \mathrm{Ker}\,h \to A \to A/\mathrm{Ker}\,h \to 0$.

1E. Let $0 \to A \to B \to C \to 0$ be an exact sequence of finite abelian groups. Prove that $\log|A| - \log|B| + \log|C| = 0$.

1F. Let $0 \to A_0 \to A_1 \to \cdots \to A_n \to 0$ be an exact sequence of finite abelian groups. Prove that $\Sigma_{i=0}^{n}(-1)^i \log|A_i| = 0$.

1G. For f, g in $\mathrm{Hom}(A, B)$, verify that $f - g$ as defined by (1.3) really is a homomorphism.

1H. Give an example of two group homomorphisms $f : G \to H$ and $g : G \to H$ such that $f/g : G \to H;\ x \mapsto (xf) \cdot (xg)^{-1}$ is not a group homomorphism.

1I. Let P and Q be quasigroups, with Q entropic. Given quasigroup homomorphisms $f : P \to Q$ and $g : P \to Q$, define $f \cdot g : P \to Q;\ p \mapsto (pf) \cdot (pg)$ and $f/g : P \to Q;\ p \mapsto (pf)/(pg)$. Prove that $f \cdot g$ and f/g

are quasigroup homomorphisms. Show that the set of all quasigroup homomorphisms $f : P \to Q$ is a quasigroup with $\cdot$ as multiplication.

1J. For an abelian group A, does Aut A form a subgroup of End A?

1K. Determine the abelian groups End$(\mathbb{Z}_n, +)$ for each positive integer n.

1L. Let T be a loop transversal to a linear code C in an abelian group channel A^n (cf. Section I 4.4). Exhibit an exact sequence $0 \to C \to A^n \xrightarrow{\varepsilon} T \to 0$.

1M. Let X, Y, Z be subgroups of an abelian group A or $(A, +)$. Write $X - Y = \{x - y | x \in X, y \in Y\}$. Verify the commutative law

$$X - Y = Y - X,$$

the associative law

$$(X - Y) - Z = X - (Y - Z),$$

and the idempotent law

$$X - X = X$$

for subtraction. Conclude that, under subtraction, the set Sb$(A, +)$ of subgroups of the abelian group $(A, +)$ forms a semilattice. What order relation $\leq$ on the set Sb$(A, +)$ is determined via O 4.3.3 by the semilattice (Sb$(A, +)$, $-$)?

1.1. Products and Coproducts of Abelian Groups

Given two abelian groups A, B, their (*direct*) *product* is the Cartesian product (O 3.2.1), equipped with the componentwise structure $(a_1, b_1) - (a_2, b_2) = (a_1 - a_2, \ b_1 - b_2)$. This structure may also be defined by the middle arrow in the diagram

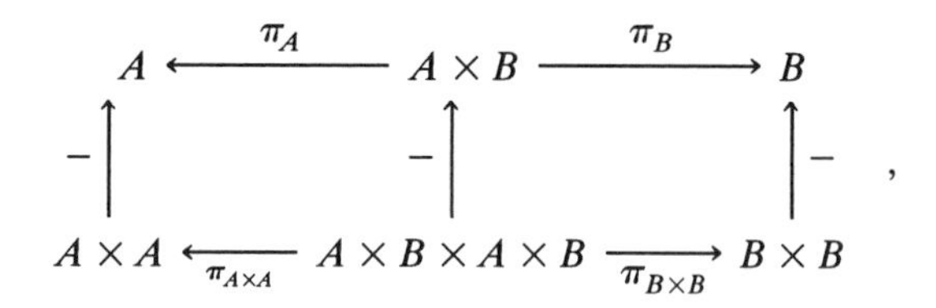

an instance of (O 3.2.4). Here the bottom arrows are given by $\pi_{A \times A} : (a_1, b_1, a_2, b_2) \mapsto (a_1, a_2)$, and similarly for $\pi_{B \times B}$. The relation (O 3.2.4) for products of sets then specializes to an isomorphism

$$\text{(1.1.1)} \qquad \text{Hom}\,(C, A \times B) \cong \text{Hom}\,(C, A) \times \text{Hom}\,(C, B)$$

of abelian groups. More generally, given a whole multiset $\langle A_i | i \in I \rangle$ of abelian groups, the *product* $\prod_{i \in I} A_i$ is the set product with subtraction (and hence abelian group structure) given by the left-hand side of

$$\begin{array}{ccc} \prod A_i & \xrightarrow{\pi_i} & A_i \\ \uparrow - & & \uparrow - \\ \prod A_i \times \prod A_i & \xrightarrow[(\pi_i,\, \pi_i)]{} & A_i \times A_i \end{array},$$

in which the bottom arrow projects from the first product to the first factor, and from the second product to the second factor. One then obtains an abelian group isomorphism

$$\operatorname{Hom}\Big(B, \prod_{i \in I} A_i\Big) \cong \prod_{i \in I} \operatorname{Hom}(B, A_i). \tag{1.1.2}$$

For each i in I, the projection $\operatorname{Hom}(B, \prod_{i \in I} A_i) \to \operatorname{Hom}(B, A_i)$; $f \mapsto f\pi_i$ may be used to verify that $\operatorname{Hom}(B, \prod_{i \in I} A_i)$ satisfies the requirements for $\prod_{i \in I} \operatorname{Hom}(B, A_i)$.

The group kernel of the projection $\pi_B : A \times B \to B$; $(a, b) \mapsto b$ is the subgroup $A \times 0 = \{(a, 0) \in A \times B | a \in A\}$. The map $\iota_A : A \to A \times B$; $a \mapsto (a, 0)$ then yields an exact sequence $0 \to A \xrightarrow{\iota_A} A \times B \xrightarrow{\pi_B} B \to 0$. Similarly, there is an exact sequence $0 \to B \xrightarrow{\iota_B} A \times B \xrightarrow{\pi_A} A \to 0$. Consider a third abelian group C, the common codomain of group homomorphisms $f : A \to C$ and $g : B \to C$. Define $f \oplus g : A \times B \to C$; $(a, b) \mapsto af + bg$. Then $f \oplus g$ is again a group homomorphism, uniquely specified by the equations $\iota_A(f \oplus g) = f$ and $\iota_B(f \oplus g) = g$. Considering the diagram

$$\begin{array}{ccccc} A & \xrightarrow{\iota_A} & A \times B & \xleftarrow{\iota_B} & B \\ {\scriptstyle f}\downarrow & & \downarrow{\scriptstyle f \oplus g} & & \downarrow{\scriptstyle g} \\ C & = & C & = & C \end{array},$$

it becomes apparent that the abelian group $A \times B$ with the *insertions* ι_A and ι_B plays a similar role to the disjoint union of sets, or of M-sets for a monoid M. Taking account of this special role, the product $A \times B$ of abelian groups is often written as $A \oplus B$, and called the (*external*) *direct sum* or *coproduct* of the abelian groups A and B. Since it serves simultaneously as both a product and a coproduct, it may also be described as the *biproduct* of A and B. (The biproduct is a by-product of the product.) Finally, note that the whole group $A \oplus B$ is generated by the union $A\iota_A \cup B\iota_B$ of the images of A and B under the insertions ι_A and ι_B. The external direct sum $A \oplus B$ of A and B is often described as the *internal direct sum* of the images $A\iota_A$ and $B\iota_B$. However, it is also customary to identify the direct summands A and B with their images in $A \oplus B$, thus suppressing the distinction between internal and external direct sums.

The biproduct construction extends to an arbitrary finite number of summands. Let $A_1, \ldots, A_n$ be abelian groups. For $1 \le j \le n$, define

$$\pi_{\hat{j}} : \prod_{i=1}^{n} A_i \longrightarrow \prod_{i=1}^{j-1} A_i \times \prod_{i=j+1}^{n} A_i ; (a_1, \ldots, a_n) \mapsto (a_1, \ldots, a_{j-1}, a_{j+1}, \ldots, a_n)$$

and

$$\iota_j : A_j \longrightarrow \prod_{i=1}^{n} A_i ; a_j \mapsto (0, \ldots, 0, a_j, 0, \ldots, 0),$$

yielding an exact sequence

(1.1.3)
$$0 \longrightarrow A_j \xrightarrow{\iota_j} \prod_{i=1}^{n} A_i \xrightarrow{\pi_{\hat{j}}} \prod_{i \ne j} A_i \to 0.$$

Now consider an abelian group C, the common codomain of homomorphisms $f_i : A_i \to C$ for each i. Define $f_1 \oplus \cdots \oplus f_n : \prod_{i=1}^n A_i \to C$; $(a_1, \ldots, a_n) \mapsto \sum_{i=1}^n a_i f_i$. Then $f_1 \oplus \cdots \oplus f_n$ is again a group homomorphism, uniquely specified by the equations $\iota_j (f_1 \oplus \cdots \oplus f_n) = f_j$ for $j = 1, \ldots, n$. Thus the product $\prod_{i=1}^n A_i$, equipped with the insertions $\iota_j : A_j \to \prod_{i=1}^n A_i$ for $j = 1, \ldots, n$, becomes the *coproduct* $\coprod_{i=1}^n A_i$ of $A_1, \ldots, A_n$, the analogue of the disjoint union of sets or M-sets. The analogy is expressed by the fact that there is an isomorphism

(1.1.4)
$$\mathrm{Hom}\left(\coprod_{i=1}^{n} A_i , C \right) \cong \prod_{i=1}^{n} \mathrm{Hom}(A_i, C)$$

of abelian groups. Since the same construction works for both product and coproduct, one often describes the group $\prod_{i=1}^n A_i = \coprod_{i=1}^n A_i$ as the *biproduct* $\bigoplus_{i=1}^n A_i$ of the *direct summands* $A_1, \ldots, A_n$. The biproduct is also described as the *external direct sum* of the A_j, or the *internal direct sum* of the $A_j \iota_j$. It is generated by $\bigcup \{ A_j \iota_j | 1 \le j \le n \}$, i.e. by $\sum_{i=1}^n A_j \iota_j$.

If a given multiset $\langle A_i | i \in I \rangle$ of abelian groups is infinite, then the biproduct construction no longer works: the product and coproduct separate. For $j \in I$, define

$$\pi_{\hat{j}} : \prod_{i \in I} A_i \to \prod_{j \ne i \in I} A_i \text{ by}$$

$$\pi_{\hat{j}} \left(\pi_i : \prod_{j \ne i \in I} A_i \to A_i \right) = \left(\pi_i : \prod_{i \in I} A_i \to A_i \right) \quad \text{for } j \ne i \in I.$$

Define $\iota_j : A_j \to \prod_{i \in I} A_i$ to yield an exact sequence

(1.1.5)
$$0 \to A_j \xrightarrow{\iota_j} \prod_{i \in I} A_i \xrightarrow{\pi_{\hat{j}}} \prod_{j \ne i \in I} A_i \to 0,$$

the analogue of (1.1.3). Define the *coproduct* $\coprod_{i\in I} A_i$ to be the subgroup of $\prod_{i\in I} A_i$ generated by $\{A_j\iota_j | j \in I\}$. If there is an infinite number of non-zero summands A_i, then the coproduct is a proper subgroup of the product $\prod_{i\in I} A_i$. The coproduct $\coprod_{i\in I} A_i$, equipped with the insertions $\iota_j : A_j \to \coprod_{i\in I} A_i$ for each j in I, possesses the analogue of the defining property for a disjoint union of sets or M-sets, namely the existence of an isomorphism

$$\text{(1.1.6)} \qquad \mathrm{Hom}\Big(\coprod_{i\in I} A_i, C\Big) \cong \prod_{i\in I} \mathrm{Hom}(A_i, C).$$

Given homomorphisms $f_i : A_i \to C$ for each i in I, a homomorphism $\Sigma_{i\in I} f_i : \coprod_{i\in I} A_i \to C;\ a_{i_1}\iota_{i_1} + \cdots + a_{i_n}\iota_{i_n} \mapsto a_{i_1} f_{i_1} + \cdots + a_{i_n} f_{i_n}$ is specified uniquely by the equations $\iota_j(\Sigma_{i\in I} f_i) = f_j$ for j in I. The coproduct $\coprod_{i\in I} A_i$ is also known as the (*external*) *direct sum* of the A_j, and as the *internal direct sum* of the $A_j\iota_j$.

EXERCISES

1.1A. Verify that (1.1.1) and (1.1.2) really do give isomorphisms of abelian groups.

1.1B. Let A be an abelian group.

(a) Verify that subtraction gives a homomorphism $A \times A \xrightarrow{-} A$.

(b) With the diagonal $\Delta : A \to A \times A;\ a \mapsto (a, a)$, show that there is an exact sequence $0 \to A \xrightarrow{\Delta} A \times A \xrightarrow{-} A \to 0$.

1.1C. Prove that an abelian group C is the internal direct sum of subgroups A and B if and only if $C = A + B$ and $A \cap B = \{0\}$.

1.1D. Let $0 \to A \xrightarrow{j} E \xrightarrow{p} Q \to 0$ be an exact sequence of abelian groups. (Such an exact sequence, with three inside groups sandwiched between zero groups, is called a *short exact sequence*. Then E is called an *extension* of A by Q.) Prove that the following three conditions are equivalent:

(a) The surjection $p : E \to Q$ has a section $s : Q \to E$ that is a homomorphism;

(b) The injection $j : A \to E$ has a retraction $r : E \to A$ that is a homomorphism;

(c) E is the internal direct sum of $\mathrm{Im}\, j$ and $\mathrm{Coker}\, j$.

(If these conditions obtain, then the extension and the exact sequence are described as *split*.)

1.1E. Prove $\mathrm{Hom}(A \oplus B, C) \cong \mathrm{Hom}(A, C) \times \mathrm{Hom}(B, C)$.

1.1F. (a) Let $A_1, \dots, A_n$ be finite sets. Prove that $\left| \bigcup_{i=1}^n A_i \right| = \Sigma_{i=1}^n |A_i|$.

(b) Let $A_1, \dots, A_n$ be finite abelian groups. Prove that $\log\left| \bigoplus_{i=1}^n A_i \right| = \Sigma_{i=1}^n \log|A_i|$.

1.1G. Verify that (1.1.4) and (1.1.6) really do give isomorphisms of abelian groups.

1.1H. For abelian groups A and B, can you exhibit the disjoint union $A \uplus B$ as a subset of the underlying set of $A \oplus B$?

1.1I. Prove that an abelian group A is the internal direct sum of a finite set $\{A_1, \ldots, A_n\}$ of subgroups iff the following two conditions are satisfied:

(a) $A = \Sigma_{i=1}^{n} A_i$;

(b) $\forall 1 \leq j \leq n,\ A_j \cap \Sigma_{i \neq j} A_i = \{0\}$.

1.1J. Let A be a finite abelian group. Prove that A is the internal direct sum of its Sylov subgroups.

1.1K. Let $A, A_1, \ldots, A_n$ be abelian groups. Prove that A is the biproduct of $A_1, \ldots, A_n$ if and only if there are homomorphisms $p_i : A \to A_i$ and $j_i : A_i \to A$ for each i such that:

(a) $\forall 1 \leq h, i \leq n,\ j_h p_i = \delta_{hi} 1_{A_i}$;

(b) $1_A = \Sigma_{i=1}^{n} p_i j_i$ in $\operatorname{Hom}(A, A)$.

1.1L. Let X be a set, and let A be an abelian group. Define an abelian group structure on A^X by $x(f - g) = xf - xg$. For x in X, define the *evaluation map* $\pi_x : A^X \to A; f \mapsto xf$. Prove that A^X, together with the evaluation maps, becomes the direct product of $|X|$ copies of the group A.

1.1M. An element f of A^X (as in Exercise 1.1L) is *almost zero* (or "has finite support") if $\{x \in X | xf \neq 0\}$ is finite. For an element x of X, define the *insertion* $\iota_x : A \to A^X;\ a \mapsto (y \mapsto$ **if** $y = x$ **then** a **else** $0)$. Prove that the set A_0^X of almost zero functions, together with the insertions, becomes the coproduct of $|X|$ copies of the group A.

1.1N. Continuing the notation of Exercises 1.1L, and 1.1M, let XK be the submonoid of $(\mathbb{Z}_0^X, +, 0)$ generated by $\{\delta_x | x \in X\}$, where δ_x is the *delta function* $1\iota_x$. Prove that XK, together with the injection

$$\iota : X \to XK;\ x \mapsto \delta_x$$

[cf. (I 1.5.1)], becomes the free commutative monoid on X. For $m \in XK$ and $x \in X$, prove that $m\pi_x$ is the multiplicity of occurrence of x in the multiset m.

1.2. Matrices

A key feature of linear algebra is the use of matrices to describe homomorphisms between biproducts of abelian groups. Given abelian groups $A_1, \ldots, A_l, B_1, \ldots, B_m$, the isomorphisms (1.1.4) and (1.1.2) yield a composite isomorphism

$$\mathrm{Hom}\left(\bigoplus_{i=1}^{l} A_i, \bigoplus_{j=1}^{m} B_j\right) \cong \bigoplus_{i=1}^{l} \mathrm{Hom}\left(A_i, \bigoplus_{j=1}^{m} B_j\right)$$
$$\cong \bigoplus_{i=1}^{l} \bigoplus_{j=1}^{m} \mathrm{Hom}(A_i, B_j); f \mapsto \sum_{i=1}^{l} \sum_{j=1}^{m} f_{ij}$$

with

$$f_{ij} = \iota_i f \pi_j \in \mathrm{Hom}(A_i, B_j). \tag{1.2.1}$$

It is convenient to display the various f_{ij} of (1.2.1) in an array:

$$\begin{bmatrix} f_{11} & \cdots & f_{1j} & \cdots & f_{1m} \\ \vdots & & \vdots & & \vdots \\ f_{i1} & \cdots & f_{ij} & \cdots & f_{im} \\ \vdots & & \vdots & & \vdots \\ f_{l1} & \cdots & f_{lj} & \cdots & f_{lm} \end{bmatrix} = [f_{ij}]_{l \times m} = [f_{ij}]. \tag{1.2.2}$$

This array is the *matrix* of f. The map f_{ij} is the (i, j)*-component* or (i, j)*-entry* of the matrix. The element $[f_{i1} \dots f_{ij} \dots f_{im}]$ or $(f_{i1}, \dots, f_{ij}, \dots, f_{im})$ of $\prod_{j=1}^{m} \mathrm{Hom}(A_i, B_j)$ is the i-th *row* of the matrix. The element

$$\begin{bmatrix} f_{1j} \\ \vdots \\ f_{ij} \\ \vdots \\ f_{lj} \end{bmatrix}$$

or $(f_{1j}, \dots, f_{ij}, \dots, f_{lj})$ of $\prod_{i=1}^{l} \mathrm{Hom}(A_i, B_j)$ is the j-th *column* of the matrix. A matrix such as (1.2.2), with l rows and m columns, is described as an $l \times m$-matrix. If $l = m$ and $A_i = B_i$ for $1 \le i \le l$, the matrix is *square*.

Example 1.2.1. The $l \times m$ matrix of $\pi_i \iota_j \in \mathrm{Hom}(\bigoplus_{i=1}^{l} A, \bigoplus_{j=1}^{m} A)$ has the identity map $1 \in \mathrm{Hom}(A, A)$ as its (i, j)-entry, and the zero map $0 \in \mathrm{Hom}(A, A)$ as its (i', j')-entry for $i' \neq i$ or $j' \neq j$. This matrix is known as the (i, j)-th *elementary matrix* E_{lm}^{ij} or E^{ij}. □

Example 1.2.2. The matrix of the identity map $1 \in \mathrm{Hom}(\bigoplus_{i=1}^{l} A_i, \bigoplus_{j=1}^{l} A_j)$ has the identity map $1 \in \mathrm{Hom}(A_i, A_i)$ as its (i, i)-entry for $1 \le i \le l$, and the zero map $0 \in \mathrm{Hom}(A_i, A_j)$ as its (i, j)-entry for $i \neq j$. This square matrix is known as the $l \times l$ *identity matrix* I_l or I. □

Example 1.2.3. The matrix of the zero map $0 \in \mathrm{Hom}(\bigoplus_{i=1}^{l} A_i, \bigoplus_{j=1}^{m} B_j)$ has the zero map $0 \in \mathrm{Hom}(A_i, B_j)$ as its (i, j)-entry for $1 \leq i \leq l$ and $1 \leq j \leq m$. This matrix is known as the $l \times m$ *zero matrix* $0_{l \times m}$ or 0. □

Example 1.2.4. The group $(\mathbb{Z}_{12}, +)$ is the internal direct sum $\mathbb{Z}_4 \oplus \mathbb{Z}_3$ of its non-trivial Sylov subgroups (cf. Exercise 1.1J), via insertions $\iota_1 : \mathbb{Z}_4 \to \mathbb{Z}_{12}; x^{\langle 4 \rangle} \mapsto 9x^{\langle 12 \rangle}$, $\iota_2 : \mathbb{Z}_3 \to \mathbb{Z}_{12}; y^{\langle 3 \rangle} \mapsto 4y^{\langle 12 \rangle}$ and projections $\pi_1 : \mathbb{Z}_{12} \to \mathbb{Z}_4; z^{\langle 12 \rangle} \mapsto z^{\langle 4 \rangle}$, $\pi_2 : \mathbb{Z}_{12} \to \mathbb{Z}_3; z^{\langle 12 \rangle} \mapsto z^{\langle 3 \rangle}$. For integers r and n with $n > 1$, consider the abelian group endomorphism $r : (\mathbb{Z}_n, +) \to (\mathbb{Z}_n, +); x^{\langle n \rangle} \mapsto (rx)^{\langle n \rangle}$ [cf. (1.4)]. What is the matrix of $5 : \mathbb{Z}_{12} \to \mathbb{Z}_{12}$? For $x^{\langle 4 \rangle}$ in $\mathbb{Z}_4$, equation (1.2.1) yields $x^{\langle 4 \rangle} 5_{11} = x^{\langle 4 \rangle} \iota_1 5 \pi_1 = (45x)^{\langle 4 \rangle} = x^{\langle 4 \rangle}$ and $x^{\langle 4 \rangle} 5_{12} = x^{\langle 4 \rangle} \iota_1 5 \pi_2 = (45x)^{\langle 3 \rangle} = 0$. Similarly, for $y^{\langle 3 \rangle}$ in $\mathbb{Z}_3$, one obtains $y^{\langle 3 \rangle} 5_{21} = y^{\langle 3 \rangle} \iota_2 5 \pi_1 = (20y)^{\langle 4 \rangle} = 0$ and $y^{\langle 3 \rangle} 5_{22} = y^{\langle 3 \rangle} \iota_2 5 \pi_2 = (20y)^{\langle 3 \rangle} = 2y^{\langle 3 \rangle}$. Thus the matrix of 5 is $\begin{bmatrix} 1 & 0 \\ 0 & 2 \end{bmatrix}$. □

Example 1.2.5. For $A = (\mathbb{Z}_2, +)$, let C be a linear code of dimension k in the binary channel $A^n = \bigoplus_{i=1}^{n} A$ of length n, given by the construction of Theorem I 4.4.4. As noted in the proof of Theorem I 4.4.4, the syndrome $s : \bigoplus_{i=1}^{n} A \to \bigoplus_{i=1}^{n-k} A$ of (4.4.6) is an abelian group homomorphism. Its $n \times (n-k)$-matrix is known as the *parity check matrix* of the code C. For $1 \leq i \leq n$, the insertion $\iota_i : A \to \bigoplus_{i=1}^{n} A$ maps 1 to the binary word b_i of length n and Hamming weight 1 with its unique non-zero letter in the i-th slot. Thus the i-th row of the parity check matrix is the binary word b_i^s of length $n-k$. For example, the parity check matrix of the code of Example 4.4.5 is

$$\begin{bmatrix} 0 & 0 & 1 \\ 0 & 1 & 0 \\ 0 & 1 & 1 \\ 1 & 0 & 0 \\ 1 & 0 & 1 \\ 1 & 1 & 0 \\ 1 & 1 & 1 \end{bmatrix}. \quad \square$$

Now consider further abelian groups $C_1, \ldots, C_n$ and the isomorphism

$$\mathrm{Hom}\left(\bigoplus_{j=1}^{m} B_j, \bigoplus_{k=1}^{n} C_k \right) \cong \bigoplus_{j=1}^{m} \bigoplus_{k=1}^{n} \mathrm{Hom}(B_j, C_k);\ g \mapsto \sum_{j=1}^{m} \sum_{k=1}^{n} g_{jk}.$$

For $1 \leq i \leq l$ and $1 \leq k \leq n$, the (i, k)-component of the matrix of

$fg \in \mathrm{Hom}(\bigoplus_{i=1}^{l} A_i, \bigoplus_{k=1}^{n} C_k)$ is given via (1.2.1) as

$$(fg)_{ik} = \iota_i fg\pi_k = \iota_i f 1 g\pi_k = \iota_i f\left(\sum_{j=1}^{m} \pi_j\iota_j\right) g\pi_k = \sum_{j=1}^{m} \iota_i f\pi_j\iota_j g\pi_k = \sum_{j=1}^{m} f_{ij}g_{jk}$$

[for the third equality, cf. Exercise 1.1K(b)]. Summarizing,

$$(1.2.3) \qquad (fg)_{ik} = \sum_{j=1}^{m} f_{ij}g_{jk}.$$

This is the equation by means of which the $l \times m$-matrix of f is multiplied by the $m \times n$-matrix of g to yield the $l \times n$-matrix of fg.

EXERCISES

1.2A. Determine the (square) matrix of $\pi_k\iota_l \in \mathrm{Hom}(\bigoplus_{i=1}^{n} A, \bigoplus_{i=1}^{n} A)$.

1.2B. Give an example of a 2×2-matrix that is not square.

1.2C. Writing $(\mathbb{Z}_{45}, +)$ as the internal direct sum $\mathbb{Z}_9 \oplus \mathbb{Z}_5$ of its non-trivial Sylov subgroups (cf. Exercise 1.1J and Example 1.2.4), determine the 2×2 matrix of the homomorphism $13 : (\mathbb{Z}_{45}, +) \to (\mathbb{Z}_{45}, +);\ z^{\langle 45 \rangle} \mapsto (13z)^{\langle 45 \rangle}$.

1.2D. Determine the parity check matrix of the code of Exercise I 4.4I.

1.2E. Derive a formula for the product $E^{ij}E^{i'j'}$ of two elementary square $l \times l$-matrices $E^{ij}, E^{i'j'}$ as in Example 1.2.1.

1.2F. Consider the elements of Hom $(\mathbb{Z}_2 \oplus \mathbb{Z}_2, \mathbb{Z}_2 \oplus \mathbb{Z}_2)$ represented by the matrices $I_2, \begin{bmatrix} 0 & 1 \\ 1 & 0 \end{bmatrix}, \begin{bmatrix} 0 & 1 \\ 1 & 1 \end{bmatrix}, \begin{bmatrix} 1 & 0 \\ 1 & 1 \end{bmatrix}, \begin{bmatrix} 1 & 1 \\ 0 & 1 \end{bmatrix}, \begin{bmatrix} 1 & 1 \\ 1 & 0 \end{bmatrix}$. Prove that they form a group under multiplication that is isomorphic with S_3. (Cf. Section O 2.)

1.2G. A square matrix (1.2.2) is *symmetric* if $f_{ij} = f_{ji}$ for all i, j. Is the product of symmetric matrices symmetric?

1.2H. A square matrix (1.2.2) is *diagonal* if $f_{ij} = 0$ for $i \neq j$. Which endomorphisms of $(\mathbb{Z}_{12}, +)$, considered as endomorphisms of $\mathbb{Z}_4 \oplus Z_3$, have diagonal matrices?

1.2I. Let A be a non-trivial abelian group. For $T \in S_n$, define

$$\rho_T = \sum_{i=1}^{n} \iota_i\pi_{iT} \in \mathrm{End} \bigoplus_{i=1}^{n} A.$$

(a) Prove that $\rho : S_n \to \mathrm{End} \bigoplus_{i=1}^{n} A;\ T \mapsto \rho_T$ is a monoid homomorphism.

(b) The matrices of the ρ_T are called *permutation matrices*. Prove that an $n \times n$ square matrix is a permutation matrix if and only if it has 1_A as the unique non-zero entry in each row and each column.

1.2J. Let $(\mathbb{Z}_n, \cdot, /, \backslash)$ be a quasigroup on the set $\mathbb{Z}_n$. With notation as in Exercise 1.2I, and $A = \mathbb{Z}_n$, prove that the matrix of $\sum_{i=0}^{n-1} i \rho_{R(i)}$ is a Latin square that yields the left division table when bordered by $0, 1, \ldots, n-1$ in order.

1.2K. For $\theta \in \mathbb{R}$, define

$$\exp \theta := \begin{bmatrix} \cos\theta & \sin\theta \\ -\sin\theta & \cos\theta \end{bmatrix} \in \mathbb{R}_2^2.$$

Show that the equation $\exp\theta \exp\varphi = \exp(\theta + \varphi)$ yields the usual addition formulae for the sine and cosine functions.

1.3. Unital and Non-Unital Rings

Let A be an abelian group. The set $\operatorname{End} A$ or $\operatorname{Hom}(A, A)$ of endomorphisms of A carries two algebraic structures: the abelian group structure given by (1.3), and the monoid structure it inherits as a submonoid of A^A under composition. These two structures are connected by the *right distributive law*

$$(x - y)z = xz - yz, \tag{1.3.1}$$

a consequence of the definition of $-$ and the fact that z is an endomorphism. On the other side, they are connected by the *left distributive law*

$$x(y - z) = xy - xz, \tag{1.3.2}$$

a consequence of the definition of subtraction. In compound expressions such as the right hand sides of (1.3.1) and (1.3.2), the multiplication binds more strongly than addition and subtraction. Abstracting the properties of the set of endomorphisms of an abelian group, one obtains the following definitions.

Definition 1.3.1. A *ring*, or *unital ring*, or *ring-with-a-one*, or *$\mathbb{Z}$-algebra*, is a set R equipped with an abelian group structure and a monoid structure, connected by the right and left distributive laws. □

Definition 1.3.2. A *ring*, or *non-unital ring*, or *ring-without-one*, is a set R equipped with an abelian group structure and a semigroup structure, connected by the right and left distributive laws. □

The ambiguity implicit in these definitions—does a ring have a multiplicative identity or not?—is a perennial source of confusion. As a very vague and unreliable rule of thumb, it is more natural to consider rings with one when focussing on individual rings and ways of constructing them from other individual rings, while rings without one tend to arise naturally when studying whole classes of rings together. In practice, one needs to determine from the context whether the word "ring" is referring to a unital or non-unital ring.

Example 1.3.3. Let A be an abelian group. Then $(\mathrm{End}A, +, \cdot, 1_A)$ is a unital ring. Defining $x * y = 0$ for x, y in A, one obtains a non-unital ring $(A, +, *)$. Rings of such form $(A, + *)$ for an abelian group $(A, +)$ are known as *zero rings*. □

Example 1.3.4. Let B be a set. Define the operation **xor** of *symmetric difference* or *exclusive or* on the power set $\mathscr{P}(B)$ by X **xor** $Y = (X - Y) \cup (Y - X)$. Then $(\mathscr{P}(B), \mathbf{xor}, \cap, B)$ is a unital ring. Let $\mathscr{P}_{<\omega}(B)$ be the set of finite subsets of B. Then$(\mathscr{P}_{<\omega}(B), \mathbf{xor}, \cap)$ is a non-unital ring. □

A ring R is *commutative* if the semigroup R is commutative. The rings of Example 1.3.4 are commutative. On the other hand, $\mathrm{End}(\mathbb{Z} \oplus \mathbb{Z})$ is not commutative, since $(\pi_1 \iota_2)(\pi_2 \iota_1) = \pi_1 \iota_1$, while $(\pi_2 \iota_1)(\pi_1 \iota_2) = \pi_2 \iota_2$.

Example 1.3.5. The set $\mathbb{Z}$ of integers forms a unital ring under addition and multiplication. The set $2\mathbb{Z}$ of even integers forms a non-unital ring under addition and multiplication. □

A set function $f: R \to S$ between unital rings is a (*unital*) *ring homomorphism* if it is a homomorphism both of abelian groups and of monoids, for the respective parts of the ring structure of R and S. Similarly, a set function between non-unital rings is a (*non-unital*) *ring homomorphism* if it is both an abelian group homomorphism and a semigroup homomorphism. A ring *isomorphism* is a bijective ring homomorphism. Rings R and S are *isomorphic*, written $R \cong S$, if there is a ring isomorphism between them. A subset R of a unital ring S is a (*unital*) *subring* if it is both an abelian subgroup and a submonoid. A subset R of a non-unital ring S is a (*non-unital*) *subring* if it is both a subgroup and a subsemigroup. Thus $2\mathbb{Z}$ is a non-unital subring of $\mathbb{Z}$, but not a unital subring. As a trickier example, $(\mathscr{P}(\{0,1\}), \mathbf{xor}, \cap, \{0,1\})$ is a unital ring that is a non-unital subring of the unital ring $(\mathscr{P}(\mathbb{N}), \mathbf{xor}, \cap, \mathbb{N})$, but not a unital subring.

Now consider a ring $(S, + \cdot)$. For each element s of S, there are various elements of S^S determined by s:

$$(1.3.3) \qquad \begin{cases} R_+(s) & : S \to S;\ x \mapsto x + s; \\ L(s) & : S \to S;\ x \mapsto sx; \\ R(s) & : S \to S;\ x \mapsto xs. \end{cases}$$

Thus $R_+(s)$ is the right (or left) "multiplication" by s in the quasigroup $(S, +)$. In the present context, it is called *translation* by s. The map $L(s)$ is called *left multiplication* by s. The left distributive law says that it is an endomorphism of $(S,+)$. Similarly, the right distributive law says that the *right multiplication* $R(s)$ is an endomorphism of $(S, +)$. The right multiplications feature in the linear algebraic version of the general algebraic Theorem O 4.3.

Theorem 1.3.6. *Let* $(S, +, \cdot, 1)$ *be a unital ring. Then* $R: S \to End(S, +)$; $s \mapsto R(s)$ *gives a ring isomorphism* $S \to SR$ *of* $(S, + \cdot)$ *with a ring of endomorphisms of the abelian group* $(S, +)$.

Proof. Theorem O 4.3 shows that R gives a monoid isomorphism. The left distributive law $xR(s - t) = x[R(s) - R(t)]$ shows that R also gives an abelian group isomorphism, and hence a ring isomorphism. □

The map R of Theorem 1.3.6 is called the *right regular representation* of the ring S (cf. Section I 1).

Given a family $\langle S_i | i \in I \rangle$ of rings, the Cartesian product $\prod_{i \in I} S_i$, along with the projections $\pi_j : \prod_{i \in I} S_i \to S_j$, yields a *product of rings*, since it is simultaneously the product of abelian groups and monoids or semigroups. Specifically, the π_j are ring homomorphisms, and given ring homomorphisms $f_i : T \to S_i$ with common domain for each i, there is a unique ring homomorphism $f : T \to \prod_{i \in I} S_i$ with $f\pi_j = f_j$ for each j. This works equally well for unital as for non-unital rings and homomorphisms.

Theorem 1.3.6 and the product construction for rings suggest consideration of matrices with ring elements as entries. Let $(S, +, \cdot)$ be a (non-unital) ring. For positive integers l,m, let S_l^m denote the set of $l \times m$ arrays or matrices (1.2.2) with entries from S. For $1 \leq i \leq l$ and $1 \leq j \leq m$, the map $\pi_{ij} : S_l^m \to S$; $f \mapsto f_{ij}$ sending a matrix to its (i, j)-entry is a projection making S_l^m the product of lm copies of the ring S. Thus S_l^m has componentwise ring structure. The componentwise product of two matrices is called their *Hadamard product*. If S is a unital ring, then S_l^m becomes a unital ring with the "all-ones" matrix J or

$$J_l^m = \sum_{i=1}^{l} \sum_{j=1}^{m} E_{lm}^{ij}$$

as identity. Now for an additional positive integer n, define a *matrix product*

$$S_l^m \times S_m^n \to S_l^n; (f, g) \mapsto fg = \left[\sum_{j=1}^{m} f_{ij} g_{jk} \right] \tag{1.3.4}$$

as in (1.2.3). In particular, the abelian group S_l^l of $l \times l$ square matrices becomes a ring under the matrix product. If S is unital, then S_l^l is unital, with I_l as identity element. In this case Theorem 1.3.6 extends to show that

$$S_l^m \to \mathrm{Hom}\left(\bigoplus_{i=1}^{l} S, \bigoplus_{j=1}^{m} S \right); [f_{ij}] \mapsto \sum_{i=1}^{l} \sum_{j=1}^{m} R(f_{ij}) \tag{1.3.5}$$

is an abelian group embedding, and a unital ring embedding if $l = m$. For an element f of a non-unital ring S, define the $l \times l$ *scalar matrix* $f = [f_{ij}]_{l \times l}$ by $f_{ij} =$ **if** $i = j$ **then** f **else** 0. Then there is a ring embedding

$$S \to S_l^l;\ f \mapsto f. \tag{1.3.6}$$

For scalar matrices, the matrix product and Hadamard product coincide.

EXERCISES

1.3A. Let $(S, +, \cdot, 1)$ be a unital ring.

(a) Under the sum $x \oplus y = x + y - 1$ and the product $x \circ y = x + y - xy$, verify that $(S, \oplus, \circ, 0)$ is a unital ring. What is its zero element?

(b) Writing $' : S \to S;\ x \mapsto 1 - x$, verify the *De Morgan Law* $(xy)' = x' \circ y'$.

(c) Interpret $'$ in the unital ring of Example 1.3.4.

1.3B. Let $(S, +, \cdot)$ be a non-unital ring. Define a product on the abelian group $\mathbb{Z} \oplus S$ by $(m + s)(n + t) = mn + sn + mt + st$ [cf. (1.4)]. Prove that $(\mathbb{Z} \oplus S, +, \cdot, 1)$ is a unital ring, and that the abelian group insertion $\iota_S : S \to \mathbb{Z} \oplus S$ is a non-unital ring homomorphism.

1.3C. Prove that every non-unital ring is isomorphic with a ring of endomorphisms of an abelian group. Contrast with Exercise O 4I.

1.3D. Let $\langle S_i | i \in I \rangle$ be a multiset of non-unital rings. Prove that the group coproduct $\coprod_{i \in I} S_i$ forms a subring of the product ring $\prod_{i \in I} S_i$. Does this result extend to unital rings?

1.3E. Let S be a commutative ring. Define the *trace* $\mathrm{tr} : S_l^l \to S;\ [f_{ij}] \mapsto \sum_{i=1}^{l} f_{ii}$. Prove that the trace is an abelian group homomorphism, and that $\mathrm{tr}(fg) = \mathrm{tr}(gf)$.

1.3F. Let S be a unital ring. Show that there is a unique unital ring homomorphism $\mathbb{Z} \to S$.

1.3G. (a) Prove that (1.3.4) defines an action of the monoid S_m^m on the set S_1^m.

(b) If S is unital, show that the action is faithful.

(c) Give an example of a ring S and a positive integer m for which the action is not faithful.

(d) Show that the $m \times m$ permutation matrices [cf. Exercise 1.2I (b)] form a subgroup S_m of the monoid S_m^m, for S unital.

(e) Let $X = \{E_{1m}^{1j} | 1 \le j \le m\}$ be the set of elementary $1 \times m$-matrices. Show that X is an S_m-subset of the S_m-set S_1^m.

(f) Prove that the trace of a permutation matrix is the number of fixed points it has in X.

1.3H. Let S be a ring and l a positive integer. An $l \times l$-matrix $[f_{ij}]$ in S_l^l is *diagonal* if $f_{ij} = 0$ for $i \neq j$ (cf. Exercise 1.2H). Prove that the set of $l \times l$ diagonal matrices with matrix product forms a subring of the ring S_l^l, under matrix product, and a subring of the ring S_l^l under Hadamard product.

1.3I. Let S be a unital ring. Prove that an $l \times l$ diagonal matrix $[f_{ij}]$ is invertible in the monoid $(S_l^l, \cdot, I_l)$—cf. Exercise I 1.3M—iff each of its diagonal entries f_{ii} (for $1 \leq i \leq l$) is invertible in the monoid $(S, \cdot, 1)$.

1.3J. Let S be a non-trivial unital ring. Set $X = \{1, 2, \ldots, n\}$. For a relation ρ on X, the *incidence matrix* A_ρ of ρ is the $n \times n$ matrix $\Sigma_{(i,j) \in \rho} E_{nn}^{ij}$. Verify the following:

(a) $A_{X \times X} = J$;

(b) $A_{\hat{X}} = I$;

(c) $E^{1i} A_\rho A_\sigma E^{j1} \neq 0 \Rightarrow (i, j) \in \rho \circ \sigma$.

Describe $A_{\rho \cap \sigma}$ in terms of the Hadamard product on S_n^n.

1.4. Ideals, Fields, and Domains

A subset K of a semigroup S is a *sink* if $\forall k \in K, \forall s \in S, ks \in K$ and $sk \in K$. If $f: S \to T$ is a non-unital ring homomorphism, then the group kernel $\operatorname{Ker} f$ is a non-unital subring of the ring S and a sink in the semigroup S. This follows since $\{0\}$ is a sink in the semigroup T. In general, a non-unital subring K of a ring S is said to be an *ideal* of S, written $K \lhd S$, if K is a sink in the semigroup S. This means that the quotient group S/K becomes a ring under the well-defined semigroup product

$$(x + K)(y + K) = xy + K. \tag{1.4.1}$$

If S is unital, then S/K is unital with identity element $1 + K$. In any case, K is the kernel of the natural projection $R_+(K): S \to S/K;\ x \mapsto x + K$, and the exact sequence $0 \to K \to S \to S/K \to 0$ of groups becomes an exact sequence of non-unital rings. Each ring S possesses the *trivial* ideal $\{0\}$ and the *improper* ideal S. Note that no proper ideal of a unital ring S can contain the identity element. A ring S is *simple* if it has no proper, non-trivial ideals.

Let $\operatorname{Id} S$ denote the set of ideals of a ring S. The set is partially ordered by containment. If $\{K_i | i \in I\}$ is a set of ideals of S, then the intersection $\bigcap_{i \in I} K_i$ is also an ideal. For a subset X of S, the *ideal* $\langle X \rangle$ or $\langle X \rangle_S$ *generated* by X is $\bigcap\{K \lhd S | X \subseteq K\}$. For a finite subset $X = \{x_1, \ldots, x_r\}$, write $\langle X \rangle = \langle x_1, \ldots, x_r \rangle$. If S is commutative and unital, then $\langle x \rangle = xS$ for x in S. In any ring S, $\langle x \rangle$ is called the *principal ideal* generated by the element x of S.

Let S be a unital ring. Denote the group of units or invertible elements of the monoid S by S^*, i.e. $x \in S^* \Leftrightarrow \exists x^{-1} \in S.\ xx^{-1} = 1 = x^{-1}x$. If $S^* = S - \{0\}$, then the ring S is a *skewfield* or *division ring*, and a *field* if it is commutative.

Proposition 1.4.1. *Let S be a commutative, unital ring. Then S is a field if and only if it is simple and non-trivial.*

Proof. Let x be a non-zero element of simple, non-trivial S. Then xS, as a non-trivial ideal of S, contains the identity element, i.e. $\exists x^{-1} \in S.\ xx^{-1} = 1 = x^{-1}x$. Conversely, suppose each non-zero element is invertible. Then $\{0\} < K \lhd S$, say $0 \neq x \in K$, implies $1 = xx^{-1} \in xS \subseteq K$, whence $K = S$ and S is simple. □

Corollary 1.4.2. *A non-zero ring homomorphism, whose domain is a field, injects.*

Proof. For S a field and $0 \neq f: S \to T$ a ring homomorphism, the group kernel of f is a proper ideal of S. Thus $\operatorname{Ker} f = \{0\}$, i.e. f injects. □

Let Prop S denote the set of proper ideals of a non-trivial ring S. Thus (Prop S, $\subseteq$) is a non-empty poset. Since the union $\bigcup_{i \in I} K_i$ of a chain $\{K_i | i \in I\}$ of ideals is an ideal, the poset is inductive. Then by Zorn's Lemma, the ring S has maximal (proper) ideals. Indeed, a similar argument shows that any given proper ideal K of S is contained in a maximal ideal of S. Write Max S for the set of maximal ideals of S.

Proposition 1.4.3. *Let S be a non-trivial, commutative unital ring. For $K \lhd S$,*

$$K \in \operatorname{Max} S \Leftrightarrow S/K \text{ is a field.}$$

Proof. K is maximal $\Leftrightarrow S/K$ is non-trivial and simple (cf. Exercise 1.4J) $\Leftrightarrow S/K$ is a field. □

A commutative, unital ring S is a field iff $S - \{0\}$ is a subgroup of the monoid S. Analogously, one has the following:

Definition 1.4.4. Let S be a commutative, unital ring. Then S is a *domain* or *integral domain* iff $S - \{0\}$ is a submonoid of the monoid S. □

Note that a field is certainly a domain. On the other hand, the ring $\mathbb{Z}$ of integers is an integral domain (whence the name) which is not a field. By Proposition 1.4.3, an ideal M in a commutative, unital ring S is maximal iff S/M is a field. One thus defines an ideal P in a commutative, unital ring S to be *prime* iff S/P is a domain. The *spectrum* of S is the set Spec S of prime ideals of S. [The rationale for this terminology becomes apparent later

(Exercise 3.4L).] Since fields are domains, maximal ideals are prime. On the other hand, $\{0\}$ is a prime ideal of $\mathbb{Z}$ which is not maximal (being contained in the ideal $2\mathbb{Z}$, for example). For a positive integer n, the ideal $n\mathbb{Z}$ is prime $\Leftrightarrow \mathbb{Z}/n\mathbb{Z} = \mathbb{Z}_n$ is an integral domain $\Leftrightarrow n$ is prime. In general, prime ideals P in a commutative, unital ring S are characterized by each of the following equivalent conditions:

$$(1.4.2) \qquad \begin{cases} P \text{ prime} \\ \Leftrightarrow S/P \quad \text{is a domain} \\ \Leftrightarrow (\forall x, y \in S, x \notin P \text{ and } y \notin P \Rightarrow xy \notin P) \\ \Leftrightarrow (\forall x, y \in S, xy \in P \Rightarrow x \in P \text{ or } y \in P). \end{cases}$$

EXERCISES

1.4A. Show that the product (1.4.1) is well-defined.

1.4B. Verify the claims of the second paragraph.

1.4C. Determine the group of units of the ring $\mathbb{Z}_{24}$.

1.4D. Determine the group of units of the ring $(\mathbb{Z}_2)_2^2$ of 2×2 matrices over the ring $\mathbb{Z}_2$ (cf. Exercise 1.2F).

1.4E. Formulate the First Isomorphism Theorem for rings.

1.4F. If S is a commutative, non-unital ring, show that the principal ideal generated by an element x is $x\mathbb{Z} + xS$.

1.4G. If S is a unital, non-commutative ring, show that the principal ideal generated by an element x is $SxS := \{\Sigma_{i=1}^r s_i x t_i | r \in \mathbb{N};\ s_i,\ t_i \in S\}$.

1.4H. If S is a non-unital, non-commutative ring, show that the principal ideal generated by an element x is $x\mathbb{Z} + xS + Sx + SxS$, the latter summand as in Exercise 1.4G.

1.4I. Prove $\{K_i | i \in I\} \subseteq \mathrm{Id}\, S \Rightarrow \Sigma_{i \in I} K_i \lhd S$.

1.4J. For $K \lhd S$, prove that there is a bijection $\{J \lhd S | K \subseteq J\} \to \mathrm{Id}(S/K)$; $J \mapsto J/K$.

1.4K. (a) For ideals K, L of a commutative ring S, prove

$$KL := \left\{ \sum_{i=1}^{r} k_i l_i | r \in \mathbb{N};\ k_i \in K;\ l_i \in l \right\} \lhd S.$$

(b) Prove $KL \subseteq K \cap L$.

(c) For $K, L \in \mathrm{Spec}(\mathbb{Z})$, prove $KL = K \cap L$.

(d) Give an example of ideals K, L of a commutative ring S with $KL \subset K \cap L$.

1.4L. For positive integers m,n, prove m divides n iff $n\mathbb{Z} \subseteq m\mathbb{Z}$.

1.4M. A non-zero element x of a ring S is a *zero divisor* if there is a non-zero element y of S with $yx = 0$ or $xy = 0$. If S is unital and commutative, prove that S is an integral domain iff it has no zero divisors.

1.4N. Let F be a field. For $n > 1$, show that the matrix ring F_n^n is a simple ring with zero divisors. (Hint: show that a non-zero ideal contains an elementary matrix. Show that no proper ideal can contain an elementary matrix.)

1.4O. For a prime number p, show that the zero ring having $\mathbb{Z}_p$ as the underlying abelian group is simple.

1.4P. Let

$$S = \left\{ \begin{bmatrix} a & -b \\ b & a \end{bmatrix} \middle| a, b \in \mathbb{Z}_3 \right\}.$$

(a) Prove that S is a subring of the ring $(\mathbb{Z}_3)_2^2$ of 2×2 matrices over the ring $\mathbb{Z}_3$.

(b) Prove that S is a field.

1.4Q. Let I be an interval in $\mathbb{R}$. Let $C(I)$ be the set of continuous functions $f : I \to \mathbb{R}$, considered as a ring under pointwise addition and multiplication. For x in I, show that $M_x = \{f \in C(I) | xf = 0\}$ is a maximal ideal of $C(I)$.

1.4R. Let $\langle S_i | i \in I \rangle$ be a family of non-unital rings. Prove that the group coproduct $\coprod_{i \in I} S_i$ is an ideal in the product ring $\prod_{i \in I} S_i$.

1.4S. Let $f : R \to S$ be a commutative, unital ring homomorphism. Show that there is a well-defined map f^{-1} : Spec $S \to$ Spec R; $P \mapsto f^{-1}(P)$.

2. VECTOR SPACES AND MODULES

The general-algebraic concept of a right S-set X for a monoid S was defined in (I 1.1) via the representation $R : S \to X^X$, a monoid homomorphism. Given a ring S, the linear-algebraic concept of a *right S-module* X or X_S or $(X, +, S)$ is defined analogously: it is an abelian group X, together with a ring homomorphism (*representation*)

$$R : S \to \text{End } X;\ s \mapsto (x \mapsto xs). \tag{2.1}$$

If S and R are non-unital, the module X is *non-unital*. If S and R are unital, the module X is *unital*. Note that a unital S-module X is an S-set for the monoid S. A right S-module X is *faithful* if the representation (2.1) injects. For a faithful module, one often identifies elements of the ring S with their images under the representation R. If S is a field, Corollary 1.4.2 shows that a non-trivial unital S-module X is faithful. Unital modules over a field S are known as *vector spaces* over the field. The elements of S are called *scalars* in this context. The image s or $R(s): X \to X$ of a scalar s under the representation is called a *homothety* or *scalar multiplication* by the scalar s. This terminology is occasionally extended to modules over more general rings S.

Example 2.1. For a unital ring S, the right regular representation of Theorem 1.3.6 is a representation in the sense of (2.1), making S a unital right S-module. For reasons that become clear later, this module is described as the *free monogenic (right) S-module.* □

Example 2.2. For a ring S and a positive integer m, the action (1.3.4) of the monoid S_m^m on the set S_1^m of *row matrices* [cf. Exercise 1.3G(a)] makes the abelian group S_1^m a right S_m^m-module, faithful if S is unital [cf. Exercise 1.3G(b)]. □

Example 2.3. Let B be a subset of a set A. Then the semigroup action of the semilattice $(\mathscr{P}(B), \cap)$ on the set $\mathscr{P}(A)$ given in Exercise I 1A makes the abelian group $(\mathscr{P}(A), \mathbf{xor})$ a right module over the ring $(\mathscr{P}(B), \mathbf{xor}, \cap)$ of Example 1.3.4. □

Most of the general algebraic constructions for monoid actions studied in Section I 1 have their counterparts in linear algebra. For instance, given a ring homomorphism $f: S \to T$ and a right T-module X with representation $R: T \to \text{End } X$, the composite $fR: S \to \text{End } X$ is a representation making X a right S-module.

Example 2.4. For a ring S and a positive integer m, the ring homomorphism (1.3.6) mapping S to the scalar matrices makes a right S-module S_1^m out of the right S_m^m-module of Example 2.2. The scalar multiplications in the right S-module S_1^m are "entrywise": $(s_1, \ldots, s_m)s = (s_1 s, \ldots, s_m s)$. In analogy with the terminology of Example 2.1, the right S-module S_1^m is described as the *free (right) S-module on m generators.* □

Given two right S-modules $(X, +, S)$ and $(Y, +, S)$, a *right S-module homomorphism* $f:(X, +, S) \to (Y, +, S)$ is an abelian group homomorphism $f:(X, +) \to (Y, +)$ and an S-homomorphism $f:(X, S) \to (Y, S)$. If the context is clearly linear-algebraic—S a ring and X,Y modules—an S-module homomorphism may just be described as an S-homomorphism. Denote the set of S-homomorphisms from X to Y by $S(X,Y)$. Now $S(X,Y)$ is a subgroup of $\text{Hom}(X,Y)$. If S is commutative, then $S(X,Y)$ is an S-subset of the S-set Y^X with $s: Y^X \to Y^X; f \mapsto fs$ for $s \in S$. Thus $S(X,Y)$ is itself an S-module in this case.

In an S-module $(X,+,S)$, an S-invariant subgroup K of X is a *submodule* of $(X, +, S)$, a relationship denoted by $K \leq X$ (and $K < X$ for proper containment). Corestriction of the representation $R: S \to \text{End } X$ to a representation $S \to \operatorname{End} K$ makes K an S-module and the embedding $K \hookrightarrow X$ an S-homomorphism. Given a submodule K of X, the group quotient X/K is an S-module via $(x + K)s = xs + K$ for $x \in X, s \in S$. Then $(X/K, S)$ is known as the *quotient module*. Given an S-homomorphism $f:(X, +, S) \to (Y, +, S)$, the group kernel $\operatorname{Ker} f$ is a submodule of X and the image $Xf = \operatorname{Im} f$ is a submodule of Y. It is customary to carry over the exact sequence notation and terminology from groups to modules. Thus an S-homomorphism $f: X \to Y$ yields an exact sequence

$$(2.2) \qquad 0 \longrightarrow \operatorname{Ker} f \longrightarrow X \xrightarrow{f} Y \longrightarrow \operatorname{Coker} f \longrightarrow 0,$$

the analogue of (1.2). In particular, (2.2) defines the *cokernel* $\operatorname{Coker} f$ as the quotient $Y/\operatorname{Im} f$.

Given a family $\langle X_i | i \in I \rangle$ of S-modules, the Cartesian product $\prod_{i \in I} X_i$ is the product both of the groups X_i and the S-sets X_i. In other words, given S-homomorphisms $f_i : Y \to X_i$ with common domain Y, there is a unique S-homomorphism $f: Y \to \prod_{i \in I} X_i$ with $f\pi_i = f_i$ for each i. Now the abelian group coproduct $\coprod_{i \in I} X_i$ is an S-subset of the S-set $\prod_{i \in I} X_i$, and thus an S-module. It is the *coproduct* of the S-modules X_i: Given S-homomorphisms $f_i : X_i \to Y$ with common codomain Y, there is a unique S-homomorphism $f: \coprod_{i \in I} X_i \to Y$ with $\iota_i f = f_i$ for each i. If there are only finitely many non-trivial modules X_i, then the product and coproduct coincide to give the *biproduct* $\bigoplus_{i \in I} X_i$. For a finite biproduct $\bigoplus_{i=1}^n X_i$ of S-modules, and an element s of S, the matrix of the scalar multiplication $\left(s: \bigoplus_{i=1}^n X_i \to \bigoplus_{i=1}^n X_i\right)$ in $\operatorname{Hom}\left(\bigoplus_{i=1}^n X_i, \bigoplus_{i=1}^n X_i\right)$ is the scalar matrix s of (1.3.6).

EXERCISES

2A. (a) Let $(X, +, S)$ be a unital S-module. For each element x of X, show that there is a unique S-homomorphism $L(x):(S, +, S) \to (X, +, S)$ to X from the free monogenic S-module with $1L(x) = x$.

(b) Show that $x \mapsto L(x)$ and $p \mapsto 1p$ are mutually inverse maps yielding an abelian group isomorphism $X \cong S(S, X)$. Under what conditions does this become a module isomorphism?

2B. Verify that the action $(x + K)s = xs + K$ (for $x \in X, s \in S$) on the quotient X/K of an S-module X by a submodule K is well-defined.

2C. Formulate and prove a First Isomorphism Theorem for S-Modules (over a given ring S).

2D. Let $(X, +, S)$ be a unital S-module (over a unital ring S). Let $(Y, +, S)$ be a non-unital S-module, and let $f:(X, +, S) \to (Y, +, S)$ be an S-module homomorphism. Show that $\operatorname{Im} f$ is a unital S-module inside the non-unital S-module Y.

2E. Let A and B be abelian groups.

(a) Show that A is a $\mathbb{Z}$-module under the representation $j:\mathbb{Z} \to \operatorname{End} A$ of (1.4).

(b) Show that an abelian group homomorphism $f: A \to B$ is a $\mathbb{Z}$-homomorphism.

2F. How to make a module unital. Let X be a non-unital module over a non-unital ring S, with representation $R: S \to \operatorname{End} X$. Let $U:\mathbb{Z} \to \operatorname{End} X$ be the unique unital ring homomorphism of Exercise 1.3F. Let $U + R:\mathbb{Z} \oplus S \to \operatorname{End} X$ be the abelian group coproduct of the maps U and R. Show that, if $\mathbb{Z} \oplus S$ is the ring of Exercise 1.3B, then $U + R$ is a representation making X a unital module over the unital ring $\mathbb{Z} \oplus S$.

2G. How to make a module faithful. Let X be an S-module that is not necessarily faithful, with representation $R: S \to \operatorname{End} X$. Define the *annihilator* $\operatorname{An} X$ or $\operatorname{An}_S X$ to be the group kernel $\operatorname{Ker} R$ in S. Factorize $R: S \to \operatorname{End} X$ through $\overline{R}: S/\operatorname{An} X \to \operatorname{End} X$ using the First Isomorphism Theorem for Rings. Show that X is a faithful $S/\operatorname{An} X$-module.

2H. For a family $\langle X_i | i \in I\rangle$ of S-modules, exhibit abelian group isomorphisms

$$S\Big(Y, \prod_{i\in I} X_i\Big) \cong \prod_{i\in I} S(Y, X_i)$$

and

$$S\Big(\coprod_{i\in I} X_i, Y\Big) \cong \prod_{i\in I} S(X_i, Y)$$

for a given S-module Y.

2I. Let U be a subring of a ring S. Show that the concatenation of the embedding $U \hookrightarrow S$ with the right regular representation $S \to \operatorname{End} S$ of S makes S a right U-module.

2J. Let X be an S-module. Let $\operatorname{Sb}(X, +, S)$ or $\operatorname{Sb}X_S$ denote the set of S-submodules of X. Show that $(\operatorname{Sb}X_S, +)$, with $K + L = \{k + l | k \in K, l \in L\}$, is a semilattice, and a monoid with identity element $\{0\}$.

2K. (a) Let $f: Y \to Z$ be an S-module homomorphism. For an S-module X, show that there is an abelian group homomorphism $R(f) = S(X, f): S(X, Y) \to S(X, Z);\ \theta \mapsto \theta f$.

(b) Given a second S-module homomorphism $g: Z \to W$, show that $S(X, f)S(X, g) = S(X, fg)$.

(c) If Y and Z are unital, show that there is a commutative diagram

$$\begin{array}{ccc} Y & \cong & S(S,Y) \\ {\scriptstyle f}\downarrow & & \downarrow{\scriptstyle S(S,f)} \\ Z & \cong & S(S,Z) \end{array}$$

whose rows are the isomorphisms of Exercise 2A(b).

2.1. Duality and Transposed Matrices

Let S be a non-commutative ring, with abelian group $(S, +)$ and semigroup $(S, \cdot)$ connected by the right and left distributive laws (1.3.1) and (1.3.2). Then the *opposite* ring S^{op} is the ring with abelian group $(S, +)$ and semigroup $(S, \check{\circ})$ opposite to $(S, \cdot)$, as in (I 1.4). The right distributive law for S yields the left distributive law for S^{op}. A *left S-module* ${}_SX$ or $(X, S, +)$ is a right S^{op}-module $X_{S^{\mathrm{op}}}$ or $(X, +, S^{\mathrm{op}})$. The scalar multiplications in a left S-module are usually written on the left of their arguments, following (I 1.5) and (I 1.6).

Example 2.1.1. Let S be a ring. Then S is a right S-module under the *right regular representation* $R: S \to \mathrm{End}(S, +)$; $s \mapsto (x \mapsto xR(s) = xs)$ and a left S-module under the *left regular representation* $L: S^{\mathrm{op}} \to \mathrm{End}(S, +)$; $s \mapsto (x \mapsto xL(s) = sx)$. □

Example 2.1.2. For a ring S and a positive integer m, the action (1.3.4) of the monoid S_m^m on the set S_m^1 of *column matrices* (cf. Example 2.2) makes S_m^1 a left S_m^m-module. □

Given left S-modules $(X, S, +)$ and $(Y, S, +)$, a *left S-module homomorphism* $f: (X, S, +) \to (Y, S, +)$ is a right S^{op}-module homomorphism $f: (X, +, S^{\mathrm{op}}) \to (Y, +, S^{\mathrm{op}})$. Thus f is an abelian group homomorphism, and $s(xf) = (sx)f$ holds for s in S and x in X. If the ring S is commutative, then there is no distinction between right and left S-modules. In this case the scalar multiplications may be written on either side of their arguments.

For an abelian group $(X, +)$ and right S-module $(Y, +, S)$ the abelian group $\mathrm{Hom}(X, Y)$ of abelian group homomorphisms $f: X \to Y$ has the structure of a right S-module, with scalar multiplications defined by

$$x(ft) = (xf)t \tag{2.1.1}$$

for $x \in X$, $f \in \mathrm{Hom}(X, Y_S)_S$, $t \in S$, via the right S-module structure of Y. One sometimes refers to the definition (2.1.1) as a *mixed associative law*: an associative law because of its form, but mixed because its arguments are from the various sets X, $\mathrm{Hom}(X, Y)$, and S in order. Now suppose that $(X, +, S)$ is a right S-module, while $(Y, +)$ is just an abelian group. Then $\mathrm{Hom}(X, Y)$ has the structure of a left S-module, with scalar multiplications defined by the mixed associative law

$$x(sf) = (xs)f \tag{2.1.2}$$

for $x \in X, s \in S, f \in {}_S\mathrm{Hom}(X_S, Y)$ via the right S-module structure of X. Note the verification that $\mathrm{Hom}(X, Y)$ is a left S-set: $x((s_1s_2)f) = (x(s_1s_2))f = ((xs_1)s_2)f = (xs_1)(s_2f) = x((s_1(s_2f))$ for $x \in X, s_i \in S, f \in \mathrm{Hom}(X, Y)$. Given a right S-module homomorphism $f: U \to V$, a left S-module homomorphism $L(f) = \mathrm{Hom}(f, Y): \mathrm{Hom}(V, Y) \to \mathrm{Hom}(U, Y)$; $\theta \mapsto f\theta$ is defined.

Indeed, $u(s(f\theta)) = (us)f\theta = ((uf)s)\theta = (uf)(s\theta) = u(f(s\theta))$ for $u \in U$, $s \in S, \theta \in$ Hom (V, Y). The definition may be displayed by the following commutative diagram:

(2.1.3)

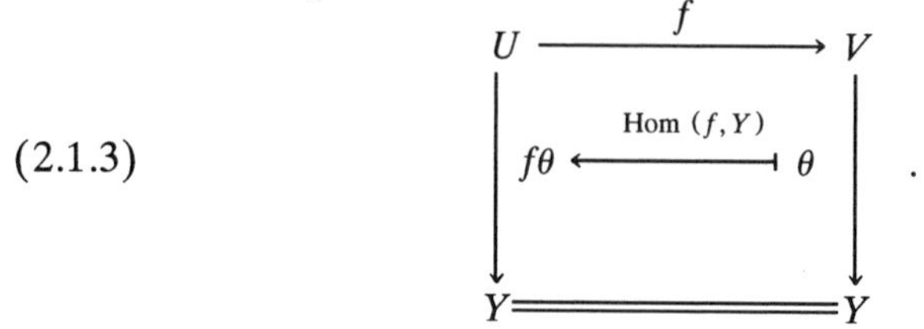

Given a second right S-module homomorphism $g : V \to W$, one obtains

$$\text{Hom}(g, Y)\text{Hom}(f, Y) = \text{Hom}(fg, Y). \tag{2.1.4}$$

Indeed, for φ in Hom(W, Y), the definition (2.1.3) yields φ Hom(g, Y) Hom$(f, Y) = (g\varphi)$ Hom$(f, Y) = f(g\varphi) = (fg)\varphi = \varphi$ Hom(fg, Y). Compare this with Exercise 2K(b).

Now fix a non-unital ring S and an abelian group G. For a right S-module X, set X^T to be the left S-module Hom(X, G). For an S-homomorphism f, set $f^T =$ Hom(f, G). Then (2.1.3) becomes $\theta f^T = f\theta$, while (2.1.4) becomes $g^T f^T = (fg)^T$. The left S-module ${}_S X^T$ is known as the *dual* of the right S-module X_S, and the left S-module homomorphism f^T is known as the *dual* of the right S-module homomorphism f. In general, *duality* in linear algebra means applying the construction Hom(__, G)–viz. "fill in the blank"—to modules and homomorphisms between them. (Cf. Section O 3.1.)

Example 2.1.3. Let S be a unital ring, and let G be the abelian group $(S, +)$. Consider the right S-module S_S given by the right regular representation of S. Define $R : S \to S_S^T; t \mapsto (R(t) : S \to S; x \mapsto xt)$. For x and s in S, one has $x(sR(t)) = (xs)R(t) = (xs)t = x(st) = xR(st)$. Note also $t = 1R(t)$. Thus R is a left S-module homomorphism embedding the left S-module ${}_S S$ given by the left regular representation of S (cf. Example 2.1.1) as a submodule of the dual S_S^T. □

Example 2.1.4. Let $\langle A_i | i \in I \rangle$ be a family of right S-modules. By (1.1.6), there is an abelian group isomorphism $\theta : \text{Hom}(\coprod_{i \in I} A_i, G) \to \prod_{i \in I} \text{Hom}(A_i, G)$, under which the image $(\Sigma_{i \in I} f_i)\theta$ of an element $\Sigma_{i \in I} f_i : \coprod_{i \in I} A_i \to G$ is determined by its projections $(\Sigma_{i \in I} f_i)\theta\pi_j = f_j : A_j \to G$ for $j \in I$. Now for s in S, one has $(s\Sigma_{i \in I} f_i)\theta\pi_j = (\Sigma_{i \in i} sf_i)\theta\pi_j = sf_j$, whence $(s\Sigma_{i \in I} f_i)\theta = s((\Sigma_{i \in I} f_i)\theta)$, so that θ becomes a left S-module isomorphism. In other words, the dual $(\coprod_{i \in I} A_i)^T$ of the coproduct is (isomorphic to) the product $\prod_{i \in I} A_i^T$ of the duals. □

Example 2.1.5. Let I be a finite interval in $\mathbb{R}$, and let $C(I)$ be the set of continuous functions $f : I \to \mathbb{R}$. Then $C(I)$ is an $\mathbb{R}$-module under $(f - g)(x) = f(x) - g(x)$ and $(rf)(x) = rf(x)$ for $x \in I, r \in \mathbb{R}$, and $f, g \in C(I)$. (Cf.

Exercise 1.4Q.) Let G be the abelian group $\mathbb{R}$. Then for $g \in C(I)$, the Riemann integral $J_g : C(I) \to \mathbb{R}; f \mapsto \int_I f(x)g(x)\,dx$ is an element of the dual $C(I)^T$. Moreover, the map $C(I) \to C(I)^T;\ g \mapsto J_g$ injects (Exercise 2.1I). For a point y in I, the Dirac δ-function δ_y is not a function $I \to \mathbb{R}$, let alone a continuous function, but it determines an element $J_y : C(I) \to \mathbb{R}; f \mapsto f(y)$ of the dual $C(I)^T$. It is thus convenient to define $\int_I f(x)\delta_y(x)\,dx = f(y)$, and to think of δ_y as a "generalized function" or "distribution" $\delta_y : I \to \mathbb{R}$. □

For right S-modules $A_1, \ldots, A_l, B_1, \ldots, B_m$, consider an abelian group homomorphism $f \in \mathrm{Hom}(\bigoplus_{i=1}^l A_i, \bigoplus_{j=1}^m B_j)$. Recall that f has a matrix $[\iota_i f \pi_j]_{l \times m}$ as in (1.2.2). By Example 2.1.4, one obtains $(\bigoplus_{i=1}^l A_i)^T = \mathrm{Hom}(\bigoplus_{i=1}^l A_i, G) = \mathrm{Hom}(\coprod_{i=1}^l A_i, G) \cong \prod_{i=1}^l \mathrm{Hom}(A_i, G) = \bigoplus_{i=1}^l \mathrm{Hom}(A_i, G) = \bigoplus_{i=1}^l A_i^T$ and $(\bigoplus_{j=1}^m B_j)^T \cong \bigoplus_{j=1}^m B_j^T$, where the isomorphisms are left S-module isomorphisms. Thus f^T, a priori an element of $\mathrm{Hom}((\bigoplus_{j=1}^m B_j)^T, (\bigoplus_{i=1}^l A_i)^T)$, becomes an element of $\mathrm{Hom}(\bigoplus_{j=1}^m B_j^T, \bigoplus_{i=1}^l A_i^T)$. Two questions then arise: what is the $m \times l$ matrix of f^T, and what relationship does it bear to the $l \times m$ matrix of f? To answer these questions, consider $\beta_j \in B_j^T$ for $1 \le j \le m$ in the following commutative diagram:

$$\begin{array}{ccccccc}
A_i & \xrightarrow{\iota_i} & \bigoplus_{h=1}^{l} A_h & \xrightarrow{f} & \bigoplus_{k=1}^{m} B_k & \xrightarrow{\pi_j} & B_j \\
\downarrow{\scriptstyle \iota_i f \pi_j \beta_j} & & \downarrow{\scriptstyle f \pi_j \beta_j} & & \downarrow{\scriptstyle \pi_j \beta_j} & & \downarrow{\scriptstyle \beta_j} \\
G & = & G & = & G & = & G
\end{array} .$$

Then under the composite

$$B_j^T \xrightarrow{\iota_j} \bigoplus_{k=1}^{m} B_k^T \xrightarrow{f^T} \bigoplus_{h=1}^{l} A_h^T \xrightarrow{\pi_i} A_i^T,$$

the element β_j maps to $\iota_i\, f \pi_j\, \beta_j = f_{ij}\, \beta_j = \beta_j f_{ij}^T$. Thus:

Theorem 2.1.6. *If $f \in \mathrm{Hom}\left(\bigoplus_{i=1}^l A_i, \bigoplus_{j=1}^m B_j\right)$ has $l \times m$-matrix $[f_{ij}]_{l \times m}$, then the dual homomorphism $f^T \in \mathrm{Hom}\left(\bigoplus_{j=1}^m B_j^T, \bigoplus_{i=1}^l A_i^T\right)$ has an $m \times l$ matrix whose (j, i)-entry is the dual f_{ij}^T of the (i, j)-entry of the matrix of f.* □

The $m \times l$ matrix

$$\begin{bmatrix}
f_{11}^T & \cdots & f_{21}^T & \cdots & f_{l1}^T \\
\vdots & & \vdots & & \vdots \\
f_{1j}^T & \cdots & f_{ij}^T & \cdots & f_{lj}^T \\
\vdots & & \vdots & & \vdots \\
f_{1m}^T & \cdots & f_{im}^T & \cdots & f_{lm}^T
\end{bmatrix}$$

is called the *transpose*

$$\begin{bmatrix} f_{11} & \cdots & f_{1j} & \cdots & f_{1m} \\ \vdots & & \vdots & & \vdots \\ f_{i1} & \cdots & f_{ij} & \cdots & f_{im} \\ \vdots & & \vdots & & \vdots \\ f_{l1} & \cdots & f_{lj} & \cdots & f_{lm} \end{bmatrix}^T$$

of the matrix (1.2.2).

EXERCISES

2.1A. Let S be a unital ring. Show that the left regular representation $L: S^{\text{op}} \to \text{End}(S, +)$ of Example 2.1.1 makes S a faithful unital left S-module, i.e. a faithful unital right S^{op}-module.

2.1B. Let $0 \longrightarrow A \xrightarrow{f} B \xrightarrow{g} C \longrightarrow 0$ be an exact sequence of right S-modules over a ring S.

(a) Show that, for an abelian group G, the sequence

$$\text{Hom}(A,G) \xleftarrow{\text{Hom}(f,G)} \text{Hom}(B,G) \xleftarrow{\text{Hom}(g,G)} \text{Hom}(C,G) \longleftarrow 0$$

is exact.

(b) Give an example of an exact sequence $0 \longrightarrow A \xrightarrow{f} B \xrightarrow{g} C \longrightarrow 0$ and a ring S for which

$$0 \longleftarrow \text{Hom}(A,G) \xleftarrow{\text{Hom}(f,G)} \text{Hom}(B,G) \xleftarrow{\text{Hom}(g,G)} \text{Hom}(C,G) \longleftarrow 0$$

is not exact.

(c) Show that, if $0 \longrightarrow A \xrightarrow{f} B \xrightarrow{g} C \longrightarrow 0$ is a split exact sequence, then so is

$$0 \longleftarrow \text{Hom}(A,G) \xleftarrow{\text{Hom}(f,G)} \text{Hom}(B,G) \xleftarrow{\text{Hom}(g,G)} \text{Hom}(C,G) \longleftarrow 0.$$

2.1C. Let G be the abelian group $\mathbb{Z}$. Let m be an integer. Determine the dual m^T of the $\mathbb{Z}$-homomorphism $m: \mathbb{Z} \to \mathbb{Z}; x \mapsto xm$.

2.1D. Let X and Y be (right) modules over a commutative ring S. Show that $S(X,Y)$ is a submodule of the (right) S-module $\text{Hom}(X, Y_S)$ given by (2.1.1) and of the (left) S-module Hom (X_S, Y) given by (2.1.2).

2.1E. (a) Let ${}_S X$ be a left S-module over a ring S. For an abelian group G, write elements f of $\mathrm{Hom}(X, G)$ on the left of their arguments, i.e. $f : X \to G;\ x \mapsto f(x)$. Show that $\mathrm{Hom}(X, G)$ is a right S-module ${}_S X^T$ via the action specified by the mixed associative law $(fs)(x) = f(sx)$ with $f \in X^T, s \in S, x \in X$.

(b) For a right S-module X, show that *evaluation* gives a map $X \to X^{TT};\ x \mapsto (f \mapsto xf)$. For what structures on X and X^{TT} is this map a homomorphism?

2.1F. For a ring S, let G be the abelian group $(S, +)$. For a positive integer m, show that matrix multiplication $S_1^m \times S_m^1 \to S_1^1 = S$ as in (1.3.4) yields a left S_m^m-module homomorphism from the left S_m^m-module S_m^1 of Example 2.1.2 to the dual $(S_1^m)^T$ of the right S_m^m-module S_1^m of Example 2.2.

2.1G. Consider the abelian group $\mathbb{Z}_n$ as a $\mathbb{Z}$-module. Let G be the *circle group*, the group $\{z \in \mathbb{C} \mid z\bar{z} = 1\}$ of complex numbers of unit modulus, under multiplication. Show that $\mathbb{Z}_n$ is isomorphic with its dual $\mathbb{Z}_n^T$.

2.1H. Let y be a point of the interval I of Example 2.1.5. Define the *Heaviside function* $H_y : I \to \mathbb{R}$ by $H_y(x) = \textbf{if } x < y \textbf{ then } 0 \textbf{ else } 1$.

(a) Show that H_y, while not itself an element of $C(I)$, determines an element $f \mapsto \int_I f(x) H_y(x)\, dx = \int_{I \cap [y, \infty)} f(x)\, dx$ of $C(I)^T$.

(b) If formal "integration by parts" is to hold, how should the "derivative" H'_y be defined?

2.1I. Prove that the function $C(I) \to C(I)^T;\ g \mapsto J_g$ of Example 2.1.5 injects.

2.2. Solving Linear Equations

Let X and Y be abelian groups, and let $f \in \mathrm{Hom}(X, Y)$. For a fixed element y of Y, an equation of the form

$$xf = y \tag{2.2.1}$$

with $x \in X$ is called an *affine equation* or an (*inhomogeneous*) *linear equation*. The equation

$$xf = 0 \tag{2.2.2}$$

is then called the *corresponding homogeneous equation*. In general, an equation of the form (2.2.2) is called a *linear equation* or a *homogeneous linear equation*. Of the alternative terminologies, the former is more correct algebraically, while the latter is more widely used. The term y of (2.2.1), especially if it is non-zero, is called the *inhomogeneous term* of (2.2.1).

Classically, given f and y in (2.2.1), one is interested in determining the *solution set* $f^{-1}\{y\}$. The endeavor is known as *solving* (2.2.1). The key issues are the dual notions of *existence* and *uniqueness* of solutions, i.e. the respective truth values $[y \in \operatorname{Im} f]$ and $[\operatorname{Ker} f = \{0\}]$. If solutions exist, i.e. $y \in \operatorname{Im} f$, then any element x of $f^{-1}\{y\}$ is called a *particular solution* of (2.2.1). The coset $x + \operatorname{Ker} f$ is called the set of *general solutions* of (2.2.1). Thus a general solution of the inhomogeneous linear equation (2.2.1) is the sum of a particular solution and a solution to the corresponding homogeneous equation (Exercise 2.2B). Note that the homogeneous equation (2.2.2) always has at least one solution, viz. 0. There are three possibilities for (2.2.1):

$$(2.2.3) \qquad \begin{cases} (0) & \text{no solutions} - y \notin \operatorname{Im} f; \\ (1) & \text{a unique solution} - y \in \operatorname{Im} f \text{ and } \operatorname{Ker} f = \{0\}; \\ (\infty) & \text{many solutions} - y \in \operatorname{Im} f \text{ and } \operatorname{Ker} f > \{0\}. \end{cases}$$

The equation (2.2.1) is described as *inconsistent* or *overdetermined* in case (0), *well-posed* in case (1), and *underdetermined* in case (∞).

Example 2.2.1. Consider the syndrome homomorphism $s : (A^n, +) \to (A^{n-k}, +)$ of the proof of Theorem I 4.4.4. Suppose that a word w has been received. The decoding problem may be construed as an example of solving the inhomogeneous equation $xs = w^s$ for x. Since (I 4.4.5) surjects, the equation has a particular solution $w^\varepsilon \in T$. The codewords, as elements of $\operatorname{Ker} s$, are solutions to the corresponding homogeneous equation $xs = 0$. The solution x of the inhomogeneous equation $xs = w^s$ chosen by the decoder is then the sum $w = w^\varepsilon + w^\delta$ of the particular solution w^ε and the solution w^δ of the corresponding homogeneous equation. □

Example 2.2.2. A vehicle suspension system maintains a variable vertical displacement s between an axle and a mounting point on the body of the vehicle. The suspension system includes two key components: a spring and a damper. The spring exerts a net force ks proportional to the displacement s, while the damper exerts a force ls' proportional to the vertical velocity s' (the time derivative of s) of the body with respect to the axle. The suspension system has to control an inertial force due to vertical movement of the body that is given as ms'' by Newton's Second Law, where the constant m is proportional to the mass of the body. If the axle is subjected to a net vertical force of F (the effect of the weight of the vehicle having been subtracted both from this term and from the spring force ks), then the behavior of the suspension system is described by the differential equation

$$(2.2.4) \qquad ks + ls' + ms'' = F.$$

Let X be the abelian group $C^2(0,1)$ of real-valued twice-differentiable functions $s : (0,1) \to \mathbb{R}$ from the time interval $(0,1)$, with pointwise addition.

Let Y be the abelian group $\mathbb{R}^{(0,1)}$ of real-valued functions on $(0, 1)$, again with pointwise addition. Then the function $f: X \to Y;\ s \mapsto ks + ls' + ms''$ is an abelian group homomorphism, and the differential equation (2.2.4) becomes an instance of (2.2.1). It is customary to check the performance of a suspension system (e.g. to see whether the damper constant l is still good) on a stationary vehicle by displacing the body, and then releasing it at time 0. Over the time interval $(0, 1)$, one is then observing a solution to the homogeneous equation corresponding to (2.2.4). The validity of this procedure as an indication of the dynamic behavior of the system, described by a solution to the inhomogeneous equation (2.2.4), rests on the fact that this solution is the sum of a particular solution to (2.2.4)—dependent on F, and thus on the irregularities of the surface over which the vehicle is traveling—and of a solution to the homogeneous equation that is likely to be similar to the solution observed during the stationary test. □

Now let S be a unital ring. By virtue of Theorem 1.3.6, S is identified with its image under the right regular representation. Given positive integers l and m, a *system of m linear equations in l unknowns* is an equation of the form

$$xA = b \tag{2.2.5}$$

with $x = (x_1, \dots, x_l) \in S_1^l$, $A = [a_{ij}] \in S_l^m$, and $b = (b_1, \dots, b_m) \in S_1^m$. The components of x are the *unknowns*, and A is called the *coefficient matrix* of the system. By (1.3.5), the coefficient matrix A in S_l^m determines an element of $\mathrm{Hom}(S_1^l, S_1^m)$, while b determines an element of $\mathrm{Hom}\,(S_1^1, S_1^m)$. One identifies matrices with their images under (1.3.5). Thus (2.2.5) becomes an instance of (2.2.1). The $(l+1) \times m$-matrix of $A \oplus b \in \mathrm{Hom}(S_1^l \oplus S_1^1, S_1^m) = \mathrm{Hom}(S_1^{l+1}, S_1^m)$, i.e. the matrix A with b added as an extra bottom row, is called the *augmented matrix* of the system (2.2.5). Note that the augmented matrix describes the system (2.2.5) completely.

Two systems $xA = b$, $xA_0 = b_0$ of m linear equations in l unknowns are *equivalent* if their solution sets agree, i.e. $A^{-1}(b) = A_0^{-1}(b_0)$ as subsets of S_1^l. The general technique for solving (2.2.5) involves passage from the original system, through a series of intermediate equivalent systems, to a final equivalent system whose solution set is immediately apparent. Such a final system is described as *reduced*, while the passage to a reduced system is described as *reduction*. During the reduction, it suffices to record the augmented matrices of the intermediate systems, since a system $xA = b$ is determined completely by its augmented matrix $A \oplus b$. Reduction is achieved by post-multiplying $(l + 1) \times m$ augmented matrices by invertible $m \times m$ square matrices.

Proposition 2.2.3. *If there is an invertible matrix E in S_m^m with $(A \oplus b)E = A_0 \oplus b_0$, then $xA = b$ is equivalent to $xA_0 = b_0$.*

Proof. Firstly, note $(A \oplus b)E = A_0 \oplus b_0$ implies $(A_0 + b_0)E^{-1} = A \oplus b$, so the relationship between the two systems is symmetrical. Then $x \in A^{-1}(b) \subseteq S_1^l \Rightarrow xA = b \Rightarrow xA_0 = xAE = xbE = xb_0 \Rightarrow x \in A_0^{-1}(b_0) \subseteq S_1^l$, i.e. $A^{-1}(b) \subseteq A_0^{-1}(b_0)$. Similarly, $A_0^{-1}(b_0) \subseteq A^{-1}(b)$, whence equality of the solution sets. □

To perform a reduction in easy steps, one selects particular kinds of invertible matrices: permutation matrices (cf. Exercise 1.2I), diagonal matrices with invertible entries (cf. Exercises 1.2H, 1.3H), and matrices of the form $I + sE^{ij}$ for $i \neq j$ and $s \in S$, where E^{ij} is the (i, j)-th elementary matrix (cf. Example 1.2.1). Consider a system (2.2.5) of 3 equations in 2 unknowns, i.e.

$$\begin{cases} x_1 a_{11} + x_2 a_{21} = b_1 \\ x_1 a_{12} + x_2 a_{22} = b_2 \\ x_1 a_{13} + x_2 a_{23} = b_3 \end{cases} \tag{2.2.6}$$

Permutation. Post-multiplying the augmented matrix of the system by the permutation matrix $\rho_{(123)}$ yields

$$\begin{bmatrix} a_{11} & a_{12} & a_{13} \\ a_{21} & a_{22} & a_{23} \\ b_1 & b_2 & b_3 \end{bmatrix} \begin{bmatrix} 0 & 1 & 0 \\ 0 & 0 & 1 \\ 1 & 0 & 0 \end{bmatrix} = \begin{bmatrix} a_{13} & a_{11} & a_{12} \\ a_{23} & a_{21} & a_{22} \\ b_3 & b_1 & b_2 \end{bmatrix}.$$

In general, post-multiplication of an augmented matrix by a permutation matrix ρ_T yields the augmented matrix of a new system such that the i-th equation of the old system becomes the iT-th equation of the new system. The i-th column of the old matrix becomes the iT-th column of the new matrix. The matrix is said to have undergone the *elementary column operation of permutation.*

Rescaling. Post-multiplying the augmented matrix of (2.2.6) by the invertible diagonal matrix

$$\begin{bmatrix} c_1 & 0 & 0 \\ 0 & c_2 & 0 \\ 0 & 0 & c_3 \end{bmatrix}$$

yields

$$\begin{bmatrix} a_{11} & a_{12} & a_{13} \\ a_{21} & a_{22} & a_{23} \\ b_1 & b_2 & b_3 \end{bmatrix} \begin{bmatrix} c_1 & 0 & 0 \\ 0 & c_2 & 0 \\ 0 & 0 & c_3 \end{bmatrix} = \begin{bmatrix} a_{11}c_1 & a_{12}c_2 & a_{13}c_3 \\ a_{21}c_1 & a_{22}c_2 & a_{23}c_3 \\ b_1c_1 & b_2c_2 & b_3c_3 \end{bmatrix}.$$

In general, post-multiplication of a matrix by a diagonal matrix with unit i-th diagonal entry c_i yields a new matrix whose i-th column is the i-th column of the old matrix postmultiplied by the scalar c_i. The matrix is said to have undergone the *elementary column operation of rescaling*.

Shear. Post-multiplying the augmented matrix of (2.2.6) by the invertible matrix $I + sE^{12}$ yields

$$\begin{bmatrix} a_{11} & a_{12} & a_{13} \\ a_{21} & a_{22} & a_{23} \\ b_1 & b_2 & b_3 \end{bmatrix} \begin{bmatrix} 1 & s & 0 \\ 0 & 1 & 0 \\ 0 & 0 & 1 \end{bmatrix} = \begin{bmatrix} a_{11} & a_{11}s + a_{12} & a_{13} \\ a_{21} & a_{21}s + a_{22} & a_{23} \\ b_1 & b_1 s + b_2 & b_3 \end{bmatrix}.$$

In general, post-multiplication of an augmented matrix by a matrix $I + sE^{ij}$ with $i \neq j$ postmultiplies the i-th equation of the system by s and adds the result to the j-th equation. The matrix itself, having had its i-th column post-multiplied by s and added to its j-th column, is said to have undergone the *elementary column operation of shear*. The name comes from the geometric action $\square \mapsto$ ▱ of $I + sE^{21}$ on the Cartesian plane $\mathbb{R}^2_1$.

An *elementary column operation* is of one of these three kinds: permutation, rescaling, or shear. The submonoid of S^m_m generated by the elementary column operations is called the monoid of *column operations*.

If the unital ring S is a field, then the reduction and ultimate solution of a system (2.2.5) may be achieved by the method known in the West as *Gaussian elimination* (but already known to the Chinese of the Former Han dynasty two millenia before Gauss' time). Consider a system over the field $\mathbb{R}$ of real numbers with augmented matrix

$$\begin{bmatrix} 1 & 2 & 1 \\ -1 & 0 & 1 \\ 1 & 2 & 1 \\ -1 & 0 & 1 \\ 1 & 2 & 1 \end{bmatrix}.$$

Initially, the entries of the coefficient matrix are described as *unused*, and are not underlined. An invertible entry of the coefficient matrix is then selected as a so-called *pivot*, and underlined:

$$\begin{bmatrix} 1 & \underline{2} & 1 \\ -1 & 0 & 1 \\ 1 & 2 & 1 \\ -1 & 0 & 1 \\ 1 & 2 & 1 \end{bmatrix}.$$

By rescaling the column of the pivot, the pivot element becomes 1:

$$\begin{bmatrix} 1 & \underline{2} & 1 \\ -1 & 0 & 1 \\ 1 & 2 & 1 \\ -1 & 0 & 1 \\ 1 & 2 & 1 \end{bmatrix} \begin{bmatrix} 1 & 0 & 0 \\ 0 & \frac{1}{2} & 0 \\ 0 & 0 & 1 \end{bmatrix} = \begin{bmatrix} 1 & \underline{1} & 1 \\ -1 & 0 & 1 \\ 1 & 1 & 1 \\ -1 & 0 & 1 \\ 1 & 1 & 1 \end{bmatrix}.$$

Using shears, the other elements of the row of the pivot are cut down to zero, and all elements of the row and column of the pivot element in the coefficient matrix are underlined and considered as *used*:

$$\begin{bmatrix} 1 & \underline{1} & 1 \\ -1 & 0 & 1 \\ 1 & 1 & 1 \\ -1 & 0 & 1 \\ 1 & 1 & 1 \end{bmatrix} \begin{bmatrix} 1 & 0 & 0 \\ -1 & 1 & -1 \\ 0 & 0 & 1 \end{bmatrix}$$

$$= \begin{bmatrix} 1 & \underline{1} & 1 \\ -1 & 0 & 1 \\ 1 & 1 & 1 \\ -1 & 0 & 1 \\ 1 & 1 & 1 \end{bmatrix} \begin{bmatrix} 1 & 0 & 0 \\ -1 & 1 & 0 \\ 0 & 0 & 1 \end{bmatrix} \begin{bmatrix} 1 & 0 & 0 \\ 0 & 1 & -1 \\ 0 & 0 & 1 \end{bmatrix}$$

$$= \begin{bmatrix} \underline{0} & \underline{1} & \underline{0} \\ -1 & \underline{0} & 1 \\ 0 & \underline{1} & 0 \\ -1 & \underline{0} & 1 \\ 0 & \underline{1} & 0 \end{bmatrix}.$$

At the next stage, an unused invertible entry of the coefficient matrix is chosen as pivot and rescaled to 1. Again, the other elements of the row of this new pivot are cut down to zero with shears (during which process no zero elements of rows of earlier pivots ever become non-zero), and at the end the elements of the row and column of the current pivot in the coefficient matrix are underlined and considered as used:

$$\begin{bmatrix} \underline{0} & \underline{1} & \underline{0} \\ -1 & \underline{0} & \underline{1} \\ 0 & \underline{1} & 0 \\ -1 & \underline{0} & \underline{1} \\ 0 & \underline{1} & 0 \end{bmatrix} \begin{bmatrix} 1 & 0 & 0 \\ 0 & 1 & 0 \\ 1 & 0 & 1 \end{bmatrix} = \begin{bmatrix} \underline{0} & \underline{1} & \underline{0} \\ \underline{0} & \underline{0} & \underline{1} \\ \underline{0} & \underline{1} & \underline{0} \\ \underline{0} & \underline{0} & \underline{1} \\ \underline{0} & \underline{1} & \underline{0} \end{bmatrix}.$$

This step is repeated until, as now in the example, no unused non-zero elements remain in the coefficient matrix. An $l \times m$-matrix is said to be *column-reduced* of *column rank r* if it has the form

$$\begin{bmatrix} I_r & 0_{r\times(m-r)} \\ B_{(l-r)\times r} & 0_{(l-r)\times(m-r)} \end{bmatrix}$$

for some $0 \leq r \leq \min\{n, m\}$ and an arbitrary $(l-r) \times r$-matrix $B_{(l-r)\times r}$. The final step in Gaussian elimination is a permutation to bring the coefficient matrix into column-reduced form:

$$\begin{bmatrix} 0 & 1 & 0 \\ 0 & 0 & 1 \\ 0 & 1 & 0 \\ 0 & 0 & 1 \\ 0 & 1 & 0 \end{bmatrix} \begin{bmatrix} 0 & 0 & 1 \\ 1 & 0 & 0 \\ 0 & 1 & 0 \end{bmatrix} = \begin{bmatrix} 1 & 0 & 0 \\ 0 & 1 & 0 \\ 1 & 0 & 0 \\ 0 & 1 & 0 \\ 1 & 0 & 0 \end{bmatrix}.$$

If the augmented matrix is not column-reduced, then the system (2.2.5) is inconsistent, since one of the $(r+1)$-st, $\dots, m$-th columns of the final augmented matrix $A' \oplus b'$ corresponds to an equation $x_1 0 + \cdots + x_l 0 = b'_j$ with non-zero b'_j. Otherwise, one may pick $x_{r+1}, \dots, x_m$ to be arbitrary members of the field S (so the system is under-determined if $r < m$), and then use the first r equations of the reduced system to obtain $x_1, \dots, x_r$ in terms of $b'_1, \dots, b'_r$ and possibly $x_{r+1}, \dots, x_m$. In the example, one obtains $x_1 = 1 - x_3$ and $x_2 = -x_4$. Thus the solution set is $(1, 0, 0, 0) + (-1, 0, 1, 0)\mathbb{R} + (0, -1, 0, 1)\mathbb{R}$.

EXERCISES

2.2A. (a) Show that the solution set of (2.2.1) is always closed under the operation P of (I 2.2.5). [Hint: For $f \in \operatorname{Hom}(X, Y)$ and $a, b, c \in X$, one has $(a, b, c)Pf = (af, bf, cf)P$.]

(b) Show that the solution set of (2.2.1) is a subgroup of X iff the equation is homogeneous.

2.2B. If $pf = y$ in (2.2.1), prove $p + \operatorname{Ker} f = \{x \in X | xf = y\}$.

2.2C. Give an example of abelian groups X, Y, an element $y \in Y$, and $f \in \operatorname{Hom}(X, Y)$ such that (2.2.1) has exactly two solutions.

2.2D. Let $C^2[0, 2\pi]$ be the ring of twice-differentiable functions $\alpha : [0, 2\pi] \to \mathbb{R}$, with pointwise operations.

(a) Show that $C^2[0, 2\pi] \to \mathbb{R}^{[0, 2\pi]}$; $\alpha \mapsto \alpha'' + \alpha$ is an abelian group homomorphism.

(b) Show that $\alpha = x^3 - 6x$ is a particular solution of $\alpha'' + \alpha = x^3$.

(c) Find the general solution of $\alpha'' + \alpha = x^3$.

2.2E. For elements f and y of a field, consider the equation $xf = y$. Show that the truth values of uniqueness and existence of solutions are given respectively by $[\text{Ker } f = \{0\}] = [f \neq 0]$ and $[y \in \text{Im } f] = [f = 0][y = 0] + [f \neq 0]$.

2.2F. Solve the systems of linear equations over $\mathbb{R}$ with the following augmented matrices:

$$\begin{bmatrix} 1 & 2 & 3 & 2 \\ 1 & 3 & 9 & 1 \\ 5 & -3 & 3 & 1 \\ -7 & 14 & 15 & 2 \end{bmatrix}; \begin{bmatrix} 1 & 2 & 1 & -1 \\ 1 & 1 & 2 & -1 \\ 1 & -2 & -3 & 3 \\ 3 & 1 & 1 & 1 \end{bmatrix};$$

$$\begin{bmatrix} 1 & -1 & 0 & 0 \\ 3 & 2 & -1 & -7 \\ 0 & -3 & 1 & 3 \\ -3 & 4 & -1 & 1 \\ 1 & -4 & 3 & -3 \end{bmatrix}.$$

2.2G. Solve the system of linear equations over $\mathbb{Z}$ with augmented matrix

$$\begin{bmatrix} 5 & 21 & 7 \\ 2 & 9 & 3 \\ 0 & 5 & 2 \\ 2 & 0 & 1 \end{bmatrix}.$$

2.2H. A system (2.2.5) of linear equations over a field S has 5 solutions. Show that S is isomorphic to $\mathbb{Z}_5$.

2.3 Bases and Free Modules

Let S be a unital ring, and let V or V_S be a unital right S – module. Consider the left S-module S or ${}_SS$. As in Exercise 2.1E, the group $\text{Hom}(S, V)$ of abelian group homomorphisms $\theta\colon\ S \to V;\ r \mapsto \theta(r)$ becomes a right S-module $\text{Hom}(S, V)_S$ with actions specified by the mixed associative law $(\theta s)(r) = \theta(sr)$ for $\theta \in \text{Hom}(S, V)$ and $r, s \in S$. The group $S(S, V)$ or $S(S_S, V_S)$ of right S-module homomorphisms $p : S \to V$[cf. Exercise 2A(a)] is a subgroup of Hom (S, V). Moreover, it is right S-submodule $S(S_S, V_S)_S$, since for $p \in S(S, V)$ and $v, s, t \in S$ one has $(ps)(rt) = p(srt) = p(sr)t = (ps)(r)t$. The middle equality holds since p is an S-homomorphism, while the other two are instances of the mixed associative law. As in Exercise 2A(b), there are mutually inverse abelian group homomorphisms $L(1) : S(S, V) \to V$; $p \mapsto p(1)$ and $L : V \to S(S, V)$; $v \mapsto (L(v)\colon\ r \mapsto vr)$. Now for $p \in S(S, V)$ and $s \in S$, one has $L(1) : ps \mapsto ps(1) = p(s1) = p(1s) = p(1)s$, so that $L(1)$ —and hence also its inverse L—are S-homomorphisms. Thus for any unital right S-module V, the modules V and $S(S, V)$ are isomorphic. It is often convenient to identify them.

For an index set I, an *I-linear combination* is an element of $\coprod_{i \in I} S \cong S(S, \coprod_{i \in I} S)$. One usually realizes the coproduct $\coprod_{i \in I} S$ as the module S_0^I of almost-zero functions $I \to S$ (cf. Exercise 1.1M). In particular, if $I = \{1, \ldots, m\}$, then $\coprod_{i \in I} S$ is the module S_1^m of $1 \times m$ matrices, as in Example 2.4. For a right S-module V, an *I-tuple of elements of V* is an element of the product (cf. Exercise 1.1L) $V^I = \prod_{i \in I} V \cong \prod_{i \in I} S(S, V) \cong S(\coprod_{i \in I} S, V)$, the latter isomorphism as in Exercise 2H. Composition of S-homomorphisms yields a map

$$(2.3.1) \quad S\Big(S, \coprod_{i \in I} S\Big) \times S\Big(\coprod_{i \in I} S, V\Big) \to S(S, V) \cong V; (\xi, f) \mapsto \xi f.$$

Example 2.3.1. For a finite index set $I = \{1, \ldots, m\}$, an I-linear combination is called an *m-linear combination.* In particular, the element $(1, 1, \ldots, 1) \in S^m = \coprod_{i=1}^m S$ is usually written as Σ or $\Sigma_{i=1}^m$. Then for $v = (v_1, \ldots, v_m) \in V^m = V^I \cong S(\coprod_{i=1}^m S, V)$, (2.3.1) specializes to $(\Sigma, v) \mapsto \Sigma v = \Sigma_{i=1}^m v_i$. □

In general, given $\xi \in S_0^I$ and $f: I \to V$, one has $\xi f = \Sigma_{i\xi \neq 0}\, if \cdot i\xi$. The sum is finite since ξ is almost zero. Note also $(\varnothing \to S)(\varnothing \to V) = 0$.

Now consider a fixed I-tuple $f: I \to V$. For such f, (2.3.1) yields an S-homomorphism

$$(2.3.2) \qquad R(f): S\Big(S, \coprod_{i \in I} S\Big) \to V; \xi \mapsto \xi f.$$

Properties of the I-tuple $f: I \to V$, or of the multiset $\langle if | i \in I \rangle$ (cf. Exercise 2.3A), are defined by properties of the map (2.3.2).

Definition 2.3.2. The I-tuple $f: I \to V$ or multiset $\langle if | i \in I \rangle$:

(S) *spans* (V) if $R(f)$ surjects;
(I) is *independent* if $R(f)$ injects;
(B) is a *basis* if $R(f)$ bijects. □

(The alliteration is a useful mnemonic.) Note that $f: I \to V$ is independent, i.e. $R(f)$ injects, iff $\operatorname{Ker} R(f) = 0$ [cf. Exercise 1B(a)], i.e. iff the homogeneous linear equation

$$(2.3.3) \qquad \xi R(f) = 0$$

has a unique solution ξ, the zero linear combination $0 \in S(S, \coprod_{i \in I} S)$. In particular, the $\varnothing$-tuple $\varnothing \to V$ is independent. If (2.3.3) is underdetermined, i.e. f is not independent, then f is said to be *dependent.* If the unital ring S is non-zero, then $if = jf$ for $i \neq j \in I$ implies that (2.3.3) has the non-zero solution $\xi = (1 \mapsto 1\iota_i) - (1 \mapsto 1\iota_j)$. Thus a multiset $\langle if | i \in I\rangle$ with any element of multiplicity more than 1 is necessarily dependent in this case. For arbitrary $f: I \to V$, the image Im $R(f)$ of (2.3.2) is a submodule of V. This submodule is called the *span* of f or $\langle if | i \in I\rangle$. Thus f spans in the sense of Definition 2.3.2(S) iff Im $R(f) = V$.

Example 2.3.3. Let S be a unital commutative ring, considered as a right S-module S_S.

(a) The singleton $\{s\} \subseteq S$ is dependent iff s is zero or a zero divisor (cf. Exercise 1.4M).
(b) The singleton $\{s\}$ spans iff s is invertible.
(c) The singleton $\{1\}$ is a basis.
(d) In $\mathbb{Z}$, the subset (2-element multiset) $\{2, 3\}$ is dependent, and spans. The subsets $\{2\}$ and $\{3\}$ are each independent, while neither spans. □

Example 2.3.4. The right $\mathbb{Z}$-module $\mathbb{Z}_2$ has no basis. In particular, the singleton $\{1\}$ spans, but is dependent. □

The existence of the isomorphism $L: V \to S(S, V)$; $v \mapsto (L(v): s \mapsto vs)$ may be reformulated as a freeness property, analogous to (O 4.1.2), Exercise I 1U, or Section I 1.4. Given a set map $\theta: \{1\} \to V$; $1 \mapsto v$, there is a unique S-homomorphism $\bar{\theta} = L(v)$: $S \to V$ with $1\bar{\theta} = 1\theta$. This property is expressed by saying that the right S-module S_S is the *free S-module on the generating set* $\{1\}$ (cf. Example 2.1). Now consider the index set I as the disjoint union $\Sigma_{j \in I}\{j\}$ of its singleton subsets $\{j\}$. The disjoint union of the composites $\{j\} \to \{1\} \hookrightarrow S \xrightarrow{\iota_j} \coprod_{i \in I} S$, with the insertions ι_j as in (1.1.5), defines an insertion $\iota: I \to \coprod_{i \in I} S$. Suppose a set map $\theta: I \to W$ to an S-module W is given. For j in I, define $\theta_j: \{1\} \to W$; $1 \mapsto j\theta$. The freeness of S on $\{1\}$ yields a unique S-homomorphism $\bar{\theta}_j: S \to W$ with $1\bar{\theta}_j = 1\theta_j$, i.e. with $1\bar{\theta}_j = j\theta$. There is then a unique S-homomorphism $\bar{\theta}: \coprod_{i \in I} S \to W$ with $j\iota\bar{\theta} = j\theta$ for each j in I, namely $\bar{\theta} = \Sigma_{i \in I}\bar{\theta}_i$. Indeed $j\iota\bar{\theta} = 1\iota_j\bar{\theta} = 1\bar{\theta}_j = j\theta$. Summarizing: there is an insertion $\iota: I \to \coprod_{i \in I} S$ such that, given a set map θ: $I \to W$, there is a unique S-homomorphism $\bar{\theta}: \coprod_{i \in I} S \to W$ with $\iota\bar{\theta} = \theta$. One says that $\coprod_{i \in I} S$ is a *free S-module on the generating set I*. (Cf. Example 2.4 for the case of finite I.) The property may be expressed diagrammatically as:

(2.3.4)

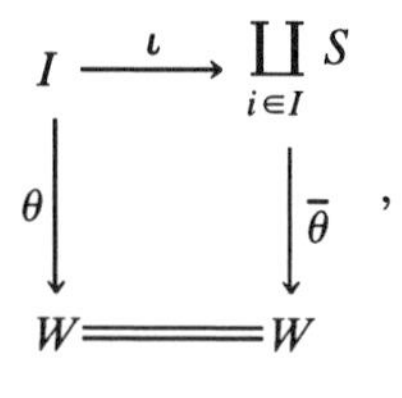

an analogue of (O 4.1.3).

Now suppose that an S-module V has a basis $f : I \to V$, so that (2.3.2) yields an isomorphism

$$\coprod_{i\in I} S \xrightarrow{L} S\Big(S, \coprod_{i\in I} S\Big) \xrightarrow{R(f)} V \tag{2.3.5}$$

of S-modules. Note that $\iota LR(f) = f : I \to V$. Given a set map $\theta : I \to W$ from I to an S-module W, there is a unique S-homomorphism $\overline{\theta} : \coprod_{i\in I} S \to W$ with $\iota\overline{\theta} = \theta$. Thus there is a unique S-homomorphism $\tilde{\theta} = (LR(f))^{-1}\overline{\theta} : V \to W$ with $f\tilde{\theta} = \iota LR(f)\tilde{\theta} = \iota\overline{\theta} = \theta$. Diagrammatically, one has the analogue

$$\begin{array}{ccc} I & \xrightarrow{f} & V \\ {\scriptstyle\theta}\downarrow & & \downarrow{\scriptstyle\tilde{\theta}} \\ W & = & W \end{array} \tag{2.3.6}$$

of (2.3.4) and (O 4.1.3), so one is also justified in calling V a *free S-module*, with basis $f : I \to V$. In this sense, $\coprod_{i\in I} S$ is a free S-module with basis $\iota : I \to \coprod_{i\in I} S$ (cf. Exercise 2.3G).

If V is a free S-module with basis $f : I \to V$, then the inverse $R(f)^{-1}$ of the isomorphism $R(f) : S_0^I \to V$ of (2.3.5) yields an isomorphism

$$R(f)^{-1} : V \to S_0^I;\ v \mapsto \big(i \mapsto i\big(vR(f)^{-1}\big)\big). \tag{2.3.7}$$

For i in I, the scalar $i(vR(f)^{-1}) \in S$ is called the i-th *coefficient of v with respect to the basis $f : I \to V$.*

Example 2.3.5. For an I-linear combination ξ in the free module S_0^I with basis $\iota : I \to S_0^I;\ i \mapsto \iota_i$, the i-th coefficient of ξ is the element $i\xi$ of S. □

Example 2.3.4 shows that, for an arbitrary ring S, an S-module V need not possess a basis. In other words : not all modules are free. For a vector space V over a field K, however, the following theorem shows that V does have a basis. Thus all vector spaces are free.

Theorem 2.3.6. *Let V be a vector space over a field K. Then V has a basis.*

Proof. Let $(A, \subseteq)$ be the poset of independent subsets of V, with order induced from the power set $(\mathscr{P}(V), \subseteq)$. The poset $(A, \subseteq)$ is inductive, since the union $U = \bigcup C$ of a chain $(C, \subseteq)$ of independent sets is independent. Indeed, if $\xi R(1_U) = 0$ with $\xi = (\eta : X \to K^*) \cup ((U - X) \to \{0\})$ and finite, non-empty X, then X would be a subset of an element W of C. But then $((\eta : X \to K^*) \cup ((W - X) \to \{0\}))R(1_W) = 0$ would contradict the independence of W. Now by Zorn's Lemma O 3.5.3, $(A, \subseteq)$ has a maximal element B. As an element of A, the (multi-)set B is independent, i.e. $R(1_B)$ injects. It remains to be shown that $R(1_B)$ surjects. For b in B, one has $((\{b\} \to \{1\}) \dot\cup ((B - \{b\}) \to \{0\}))R(1_B) = b$. For v in $V - B$, one has $B \subset B \cup \{v\} = B \dot\cup \{v\}$. Since B is maximal in A, the set $B \cup \{v\}$ cannot lie in A, and must thus be dependent. Consider a non-zero solution $\xi = (J \to K^*) \dot\cup ((B - J) \to \{0\}) \dot\cup (\{v\} \to \{k\})$ to $\xi R(1_{B \cup \{v\}}) = 0$, with J a finite subset of B. Since B is independent, $k \neq 0$. Then $v = \sum_{j \in J} j \cdot (-j\xi \cdot k^{-1}) \in \operatorname{Im} R(1_J) \subseteq \operatorname{Im} R(1_B)$. Thus $R(1_B)$ surjects. □

Corollary 2.3.7. *A linearly independent subset I of a vector space V is a subset of a basis B of V.*

Proof. The element I of A is contained in a maximal element B of A, since Zorn's Lemma applies to the inductive subposet of A whose elements contain I. As in the proof of Theorem 2.3.6, the (multi-)set B is then a basis for V. □

Determination of a basis B containing a given independent set I as a subset is known as *extending* the independent set I to the basis B. The process is used, for example, in proving the final theorem of the next section.

EXERCISES

2.3A. Let $b : I \to I$ be a bijection of an index set I. Show that an I-tuple $f : I \to V$ has each of the properties of Definition 2.3.2 iff the I-tuple $bf : I \to V$ has the corresponding property.

2.3B. Verify the statements of Example 2.3.3.

2.3C. Verify the statement of Example 2.3.4.

2.3D. For each element α of an index set A, let $f_\alpha : I_\alpha \to V_\alpha$ be a basis for a right S-module V_α. Show that the disjoint union

$$\dot{\bigcup_{\alpha \in A}} \left(I_\alpha \xrightarrow{f_\alpha} V_\alpha \xrightarrow{\iota_\alpha} \coprod_{\beta \in A} V_\beta \right),$$

with ι_α as in (1.1.5), is a basis for $\coprod_{\alpha \in A} V_\alpha$.

2.3E. Let V be the real vector space $\mathbb{R}^3 = \mathbb{R}^3_1$. An l-tuple f of elements of V is an element $[f_{ij}]$ of $V^l \cong \mathbb{R}(\mathbb{R}, \mathbb{R}^3)^l \cong \mathbb{R}(\bigoplus_{i=1}^l \mathbb{R}, \bigoplus_{i=1}^3 \mathbb{R}) \cong \mathbb{R}^3_l$. An l-linear combination ξ is an element $[\xi_1, \ldots, \xi_l]$ of $\mathbb{R}^l_1 \cong \mathbb{R}(\mathbb{R}, \bigoplus_{i=1}^l \mathbb{R})$. Using e.g. Gaussian elimination to solve the homogeneous system $[\xi_1, \ldots, \xi_l][f_{ij}] = [0, 0, 0]$ of 3 equations in l unknowns [viz. (2.3.3)], determine for each of the following $l \times 3$-matrices $[f_{ij}]_{l \times 3}$

whether $f:\{1,\ldots,l\} \to \mathbb{R}_1^3$ is linearly independent:

$$(a)\begin{bmatrix}1 & 2 & 0\\ 2 & 2 & -2\\ 0 & 1 & 1\end{bmatrix};\quad (b)\begin{bmatrix}1 & 2 & 0\\ 2 & 2 & 1\\ 0 & 1 & 1\end{bmatrix};\quad (c)\begin{bmatrix}1 & 2 & 0\\ 2 & 2 & 1\\ 0 & 1 & 1\\ 2 & 0 & 1\end{bmatrix}.$$

2.3F. For $J \subseteq I$, let $(g:J \to V) = (f:I \to V)|_J$.
(a) If g spans, prove that f does also.
(b) If f is independent, prove that g is also. (Hint: Cf. Exercise O 3.1F.)

2.3G. For the I-tuple $\iota: I \to \coprod_{i\in I} S$, verify that $R(\iota): S(S, \coprod_{i\in I} S) \to \coprod_{i\in I} S$ [an instance of (2.3.2)] bijects. Hence verify that $\iota: I \to \coprod_{i\in I} S$ is a basis of $\coprod_{i\in I} S$.

2.3H. Let V be a vector space over a field K. Show that an I-tuple $f:I \to V$ or multisubset $\langle if \mid i \in I\rangle$ of V is dependent iff $\exists i \in I.\ if \in \operatorname{Im} R(f|_{I-\{i\}})$, i.e. iff one element of the multiset lies in the span of its complement.

2.3I. Let S be a unital ring. Show that for each (unital) S-module V, there is an exact sequence

$$0 \to R \to F \to V \to 0$$

of S-modules with F free.

2.3J. Let S be a unital ring. A (unital) right S-module V is *projective* if, for each surjective S-homomorphism $g: W \to X$ and $h \in S(V, X)$, there is an S-homomorphism $\bar{h}: V \to W$ with $\bar{h}g = h$. [The homomorphism $h \in S(V, X)$ is said to have been *lifted* to the homomorphism $\bar{h} \in S(V, W)$.] Show that free S-modules are projective.

2.3K. Show that an S-module V is projective iff it is a direct summand of a free module.

2.3L. For a field K, consider the ring $S = K^2$. Show that the S-submodule $K = K \times \{0\}$ of the S-module S_S is projective, but not free.

2.3M. Prove that an S-module Q is projective iff each short exact sequence

$$0 \to A \to E \to Q \to 0$$

of S-modules splits (cf. Exercise 1.1D).

2.3N. (a) The concept of an *injective* right S-module V is dual to the concept of a projective right S-module. Use this information to formulate an exact definition of what it means for an S-module V to be injective.
(b) Verify the validity of the definition given in answer to (a) by proving that an S-module A is injective iff each short exact sequence $0 \to A \to E \to Q \to 0$ of S-modules splits.

2.3O. Show that the abelian group $(\mathbb{Q}, +, 0)$ of rational numbers is an injective $\mathbb{Z}$-module.

2.4 Determinants and Dimension

Throughout this section, let K be a commutative unital ring. If V is a K-module, then the *conjugate* (or "dual") $V' := K(V, K)$ or $K(V; K)$ or $K^1(V; K)$ is again a K-module (cf. Exercise 2.1D). Elements of V' are known as *functionals* on V. (For example, Exercise 2.1H exhibits the Heaviside function as a functional on the real vector space $C(I)$ of continuous functions on the interval I.) Now let U and W be further (unital) K-modules. Recall the Currying bijection

$$(2.4.1)\quad (U^W)^V \to U^{V \times W}; (g : v \mapsto (f_v : w \mapsto u)) \mapsto (h : (v, w) \mapsto u)$$

of (O 3.4.1). The image of the subset $K(V, K(W, U))$ under this bijection is called the set $K(V, W; U)$ of *bilinear functions* from $V \times W$ to U. Elements of $K(V, W; K)$ are called *bilinear pairings* of V with W. Linearity of f_v in (2.4.1) corresponds to linearity of h in its second slot. Linearity of g in (2.4.1) corresponds to linearity of h in its first slot. The restriction of the bijection (2.4.1) to the subset $K(V, K(W, U))$ gives a bijection from $K(V, K(W, U))$ to $K(V, W; U)$. This bijection transfers the K-module structure on $K(V, K(W, U))$ (cf. Exercise 2.1D) to $K(V, W; U)$. In the case $V = W$ and $U = K_K$, the module $K(V, V; K)$ is called the module $K^2(V; K)$ of *bilinear functionals on* V. For example, the usual Euclidean inner product $(x, y) \mapsto xy^T$ on $\mathbb{R}_1^3$ is a bilinear functional on $\mathbb{R}^3$. The signed area $x_{11}x_{22} - x_{12}x_{21}$ of the parallelogram $(0, 0), (x_{11}, x_{12}), (x_{21}, x_{22}), (x_{11} + x_{21}, x_{12} + x_{22})$ spanned by an ordered pair $((x_{11}, x_{12}), (x_{21}, x_{22}))$ of vectors in $\mathbb{R}^2$ yields a bilinear functional $((x_{11}, x_{12}), (x_{21}, x_{22})) \mapsto x_{11}x_{22} - x_{12}x_{21}$ on $\mathbb{R}^2$. The signed area vanishes when the two vectors are equal. In general, a bilinear functional on a K-module V is said to *alternate* if its restriction to the diagonal $\hat{V}$ in V^2 is zero, i.e. if it vanishes on the complement of the binary diversity relation on V. The set $\tilde{K}^2(V; K)$ of *2-forms* or alternating bilinear functionals on V is a submodule of $K^2(V; K)$.

For a natural number n, consider the bijection $((\ldots(K^V)\ldots)^V)^V \to K^{V \times V \times \cdots \times V}$, with n copies of V, an instance of the extension of (O 3.4.1) or (O 3.4.3) as in Exercise O 3.4D. The image of $K(V, K(V, \ldots, K(V, K)\ldots))$ under the bijection is the module $K^n(V; K)$ of *n-linear functionals* on V. Elements of the submodule $\tilde{K}^n(V; K)$ of n-linear functionals vanishing on the complement of the n-ary diversity relation on V, i.e. vanishing whenever at least two of their arguments agree, are known as *n-forms* or *alternating* n-linear functionals. Note that the semigroup $\underline{\underline{\mathrm{Set}}}(\mathbb{Z}_n, \mathbb{Z}_n)$ and symmetric group S_n have left actions on $K^n(V; K)$ given by

$$(2.4.2)\qquad T : h \mapsto (Th : (V_1, \ldots, V_n) \mapsto h(V_{1T}, \ldots, V_{nT}))$$

for $T : \mathbb{Z}_n \to \mathbb{Z}_n$, with $\tilde{K}^n(V; K)$ as an invariant subset. If h is an n-form and T is the transposition (ij), then $0 = h(v_1, \ldots, v_i + v_j, \ldots, v_i + v_j, \ldots, v_n) = h(v_1, \ldots, v_i, \ldots, v_j, \ldots, v_n) + h(v_1, \ldots, v_j, \ldots, v_i, \ldots, v_n) = h(v_1, \ldots, v_i, \ldots, v_j, \ldots, v_n) + Th\,(v_1, \ldots, v_i, \ldots, v_j, \ldots, v_n)$ implies $Th = (-1)h$. Since $\tilde{K}^n(V; K)$ is an S_n-set, it follows more generally that $Th = h$ for even T and $Th = -h$ for odd T in S_n. Thus for h in $\tilde{K}^n(V; K)$ and $T \in \underline{\underline{\mathrm{Set}}}(\mathbb{Z}_n, \mathbb{Z}_n)$, one has

$$(2.4.3) \qquad Th = \textbf{if } T \in S_n \textbf{ then } (\operatorname{sgn} T)h \textbf{ else } 0.$$

In a different direction, elements of $\mathrm{Fix}(K^n(V; K), S_n)$ are described as *symmetric* n-linear functionals.

A K-homomorphism $\theta : V \to W$ induces a K-homomorphism $\sum_{i=1}^n \theta : V^n \to W^n$. One then obtains a K-homomorphism

$$(2.4.4) \qquad K^n(\theta; K) : K^n(W; K) \to K^n(V; K);\ h \mapsto \left(\sum_{i=1}^{s} \theta \right) h.$$

In other words, $K^n(\theta; K)h(v_1, \ldots, v_n) = h(v_1\theta, \ldots, v_n\theta)$ for $(v_1, \ldots, v_n) \in V^n$. If h is an n-form on W, and $v_i = v_j$, then $K^n(\theta; K)h(\ldots, v_i, \ldots, v_j, \ldots) = h(\ldots, v_i\theta, \ldots, v_j\theta, \ldots) = 0$. Thus $K^h(\theta; K)$ restricts to a K-homomorphism $\tilde{K}^n(\theta; K) = \tilde{K}^n(W; K) \to \tilde{K}^n(V; K)$. If $\varphi : U \to V$ is also a K-homomorphism, then $\tilde{K}^n(\varphi\theta; K) = \tilde{K}^n(\theta; K)\tilde{K}^n(\varphi; K)$: $\tilde{K}^n(W; K) \to \tilde{K}^n(U; K)$. Also note that $\tilde{K}^n(1_V; K)$ is the identity map on $\tilde{K}^n(V; K)$.

An r-linear functional h on the free K-module $K^n = K_1^n$ may be construed as a function $h : K_r^n \to K;\ f \mapsto fh = h([f_{11}, \ldots, f_{1n}], \ldots, [f_{r1}, \ldots, f_{rn}])$.

Proposition 2.4.1. *For each element k of K, there is a unique element k* det *of $\tilde{K}^n(K^n, K)$ assigning the value k to the identity matrix I.*

Proof. If such a function k det exists, it satisfies

$$\begin{aligned}
k \det(f) &= k \det\left(\sum_{j_1=1}^{n} f_{1j_n} E_{1n}^{1j_1}, \ldots, \sum_{j_n=1}^{n} f_{nj_n} E_{1n}^{1j_n} \right) \\
&= \sum_{j_1=1}^{n} \cdots \sum_{j_n=1}^{n} f_{1j_1} \cdots f_{nj_n} k \det\left(E_{1n}^{1j_1}, \ldots, E_{1n}^{1j_n} \right) \\
&= \sum_{T \in \underline{\underline{\mathrm{Set}}}(\mathbb{Z}_n, \mathbb{Z}_n)} f_{1,1T} \cdots f_{n,nT} Tk \det\left(E_{1n}^{11}, \ldots, E_{1n}^{1n} \right) \\
&= \sum_{T \in S_n} f_{1,1T} \cdots f_{n,nT} (\operatorname{sgn} T) k \det(I) \\
&= k \sum_{T \in S_n} (\operatorname{sgn} T) f_{1,1T} \cdots f_{n,nT},
\end{aligned}$$

and is thus specified uniquely. On the other hand, defining

$$|f| = \sum_{U \in S_n} (\operatorname{sgn} U) \prod_{i \in \mathbb{Z}_n} f_{i, iU} \tag{2.4.5}$$

does give an n-linear functional on K_1^n, taking the value 1 on I. Now suppose that the i-th and j-th rows of f agree, for $i \neq j$. Then

$$\begin{aligned}
|f| &= \sum_{U \in S_n} (\operatorname{sgn} U) \prod_{i \in \mathbb{Z}_n} f_{i, iU} \\
&= \sum_{U \in A_n} (\operatorname{sgn} U) \prod_{k \in \mathbb{Z}_n} f_{k, kU} + \sum_{U \in (ij)A_n} (\operatorname{sgn} U) \prod_{k \in \mathbb{Z}_n} f_{k, kU} \\
&= \sum_{T \in A_n} (\operatorname{sgn} T) f_{1,1T} \cdots f_{i,iT} \cdots f_{j,jT} \cdots f_{n,nT} \\
&\quad - \sum_{T \in A_n} (\operatorname{sgn} T) f_{1,1T} \cdots f_{i,jT} \cdots f_{j,iT} \cdots f_{n,nT} = 0
\end{aligned}$$

since $f_{i,k} = f_{j,k}$ for $1 \leq k \leq n$. Thus defining $k \det : K_n^n \to K; f \mapsto k|f|$, one obtains the unique element of $\tilde{K}^n(K^n; K)$ with $k \det I = k$, as required. □

Definition 2.4.2. For an $n \times n$-matrix $f \in K_n^n$, the scalar $\det f$ or $|f|$ given by (2.4.5) is called the *determinant* of f. □

Theorem 2.4.3 (Properties of the Determinant). *For $f, g \in K_n^n$:*

(a) $|fg| = |f| \cdot |g|$;
(b) $|f^T| = |f|$.

Proof. (a) For $R(g) : K_1^n \to K_1^n; v \mapsto vg$, one has $\tilde{K}^n(R(g); K)\det I = \det IR(g) = \det g$. Thus $\tilde{K}^n(R(g);K) \det = \det g \cdot \det$ by Proposition 2.4.1, whence $\det fg = \tilde{K}^n(R(g); K)\det f = \det g \cdot \det f$.

(b) By (2.4.5), one has

$$\begin{aligned}
|f^T| &= \sum_{U \in S_n} (\operatorname{sgn} U) \prod_{i \in \mathbb{Z}_n} f^T_{i, iU} \\
&= \sum_{U \in S_n} (\operatorname{sgn} U) \prod_{i \in \mathbb{Z}_n} f_{iU, i} \\
&= \sum_{U^{-1} \in S_n} (\operatorname{sgn} U^{-1}) \prod_{j \in \mathbb{Z}_n} f_{j, jU^{-1}} = |f|. \quad \square
\end{aligned}$$

Theorem 2.4.4. *For each natural number n, the module $\tilde{K}^n(K^n; K)$ of n-forms on the free module K^n is the free module with singleton basis* {det}.

Proof. For $h \in \tilde{K}^n(K^n; K)$, one has $h = h(I)\det$ by Proposition 2.4.1. □

Proposition 2.4.5. *If $r > n$, then $\tilde{K}^r(K^n; K) = 0$.*

Proof. For $h \in \tilde{H}^r(K^n; K)$ and $f \in K_r^n$, one has

$$fh = h\left(\sum_{j_1=1}^{n} f_{1j_1} E_{1n}^{1j_1}, \ldots, \sum_{j_r=1}^{n} f_{rj_r} E_{1n}^{1j_r}\right)$$

$$= \sum_{j_1=1}^{n} \cdots \sum_{j_r=1}^{n} f_{1j_1} \cdots f_{rj_r} h\left(E_{1n}^{1j_1}, \ldots, E_{1n}^{1j_r}\right)$$

$$= \sum_{T \in \underline{\underline{\mathrm{Set}}}(\mathbb{Z}_r, \mathbb{Z}_n)} f_{1,1T} \cdots f_{r,rT} h\left(E_{1n}^{1,1T}, \ldots, E_{1n}^{1,rT}\right) = 0,$$

since $r > n$ implies $\forall T \in \underline{\underline{\mathrm{Set}}}(\mathbb{Z}_r, \mathbb{Z}_n)$, $\exists 1 \leq i < j \leq r.\ \ iT = jT$. □

Theorem 2.4.6. *Let K be a non-zero ring. Let V be a free K-module with a finite basis $\beta : \mathbb{Z}_n \to V$. Then the cardinality n of the basis is uniquely determined by V.*

Proof. By (2.3.2), there is a K-isomorphism $R(\beta) : K^n \to V$. Suppose that V also had a basis $\gamma : \mathbb{Z}_r \to V$ with $r > n$, yielding a K-isomorphism $R(\gamma) : K^r \to V$. Then $R(\gamma)R(\beta)^{-1}R(\beta)R(\gamma)^{-1} = 1_{K^r}$, whence the identity map $\tilde{K}^r(1_{K^r}; K)$ on $\tilde{K}^r(K^r; K)$ would factor as $(\tilde{K}^r(R(\beta)R(\gamma)^{-1}; K) : \tilde{K}^r(K^r; K) \to \tilde{K}^r(K^n; K))(\tilde{K}^r(R(\gamma)R(\beta)^{-1}; K) : \tilde{K}^r(K^n; K) \to \tilde{K}^r(K^r; K))$. But this is impossible, since $\tilde{K}^r(K^n; K) = 0$ by Proposition 2.4.5, while $\tilde{K}^r(K^r; K) \cong K_K \neq 0$ by Theorem 2.4.4 and the hypothesis. □

Definition 2.4.7. If a free K-module V has a basis $\beta : \mathbb{Z}_n \to V$, then V is said to have *rank* or *dimension* $\dim V = n$. □

Consider an endomorphism $\theta : V \to V$ of a free module with bases $\beta : \mathbb{Z}_n \to V$ and $\gamma : \mathbb{Z}_n \to V$. Note that $R(\gamma)R(\beta)^{-1}.\ R(\beta)R(\gamma)^{-1} = R(\gamma)1_V R(\gamma)^{-1} = 1_{K^n}$. Identifying $K(K^n, K^n) \cong K_n^n$, Theorem 2.4.3(a) yields $|R(\gamma)\theta R(\gamma)^{-1}| = |R(\gamma)R(\beta)^{-1} \cdot R(\beta)\theta R(\beta)^{-1} \cdot R(\beta)R(\gamma)^{-1}| = |R(\gamma)R(\beta)^{-1}| \cdot |R(\beta)\theta R(\beta)^{-1}| \cdot |R(\beta)R(\gamma)^{-1}| = |R(\gamma)R(\beta)^{-1}| \cdot |R(\beta)R(\gamma)^{-1}| \cdot |R(\beta)\theta R(\beta)^{-1}| = |R(\beta)\theta R(\beta)^{-1}|$. One may thus define the *determinant of* the *endomorphism* θ unambiguously by $\det \theta = |\theta| := |R(\beta)\theta R(\beta)^{-1}|$, independently of the choice of basis $\beta : I \to V$. In particular, $\det 1_V = 1$. Moreover, for a second endomorphism φ, one has $|\theta\varphi| = |R(\beta)\theta\varphi R(\beta)^{-1}| = |R(\beta)\theta R(\beta)^{-1}R(\beta)\varphi R(\beta)^{-1}| = |R(\beta)\theta R(\beta)^{-1}| \cdot |R(\beta)\varphi R(\beta)^{-1}| = |\theta| \cdot |\varphi|$. Thus invertible endomorphisms have invertible determinants. Conversely, it will be shown that matrices or endomorphisms with invertible determinants are invertible. Given a matrix $f = [f_{ij}]_{n \times n}$ in K_n^n, define a new matrix $\hat{f} = [\hat{f}_{ji}]_{n \times n}$ by

$$\hat{f}_{ji} = \sum_{U \in (in)S_{n-1}(nj)} \mathrm{sgn}\, U \prod_{i \neq h \in \mathbb{Z}_n} f_{h,hU}. \tag{2.4.6}$$

Now the (i, i)-entry of the product matrix $f\hat{f}$ is

$$\sum_{j=1}^{n} f_{ij}\hat{f}_{ji} = \sum_{j=1}^{n} f_{ij} \sum_{U \in (in)S_{n=1}(nj)} \operatorname{sgn} U \prod_{i \neq h \in \mathbb{Z}_n} f_{h,hU}$$
$$= \sum_{U \in S_n} \operatorname{sgn} U \prod_{i \in \mathbb{Z}_n} f_{i,iU} = |f|,$$

since $S_n = \bigcup_{j \in \mathbb{Z}_n} (in)S_{n-1}(nj)$. For $i \neq k$, let g_k be the matrix obtained from f by replacing f_{ij} with f_{kj}. Then, as above, one has $\Sigma_{j=1}^{n} f_{kj}\hat{f}_{ji} = |g_k| = |g_k^T| = 0$. In other words, $\Sigma_{j=1}^{n} f_{kj}\hat{f}_{ji} = \delta_{ik}. |f|$, or

$$f\hat{f} = |f| \cdot I_n \tag{2.4.7}$$

in K_n^n.

Proposition 2.4.8. *For f in K_n^n, the following conditions are equivalent:*

(a) $\exists g.\ fg = 1$;
(b) $\exists g.\ gf = 1$;
(c) *f is invertible;*
(d) *$|f|$ is invertible.*

The conditions are also equivalent when applied to endomorphisms of a free K-module of finite dimension.

Proof. If (a) or (b) hold, or (c) holds such that f has inverse g, then $|fg| = |f|.|g| = |gf| = 1$, so (d) holds. Conversely, if (d) holds for a matrix f, then (2.4.7) shows that $f(|f|^{-1}\hat{f}) = f\hat{f} \cdot |f|^{-1} = I$ and $(|f|^{-1}\widehat{f^T}^T)f = |f|^{-1}\widehat{f^T}^T \cdot f^{TT} = |f|^{-1}(f^T \cdot \widehat{f^T}) = |f|^{-1} \cdot |f^T| \cdot I = I$, so that f is invertible. Finally, if $\beta : \mathbb{Z}_n \to V$ is a basis for a K-module V, then the various conditions on endomorphisms θ translate to the corresponding conditions on the matrices $R(\beta)\theta R(\beta)^{-1}$. □

If K is a field, then part (c) of Theorem 2.4.9 below gives an alternative approach to the equivalence of (a) and (b) in Proposition 2.4.8 above.

Theorem 2.4.9. *Let K be a field, and let $\theta : V \to W$ be a K-homomorphism.*

(a) *If V has finite dimension, then* $\dim V = \dim \operatorname{Ker} \theta + \dim \operatorname{Im} \theta$.
(b) *If W has finite dimension, then* $\dim W = \dim \operatorname{Coker} \theta + \dim \operatorname{Im} \theta$.
(c) *If* $\dim V = \dim W$ *is finite, then θ injects iff it surjects.*

Proof. (a) Pick a basis $\kappa : J \to \operatorname{Ker} \theta$ for $\operatorname{Ker} \theta$, and extend to a basis $(\kappa \cup \lambda) : J \cup L \to V$ of V. Then $\lambda\theta : L \to V\theta$ is a basis for $\operatorname{Im} \theta$. Indeed, for $v \in V$, one has $v\theta = (vR(\kappa \uplus \lambda)^{-1})((0: \coprod_J S \to \coprod_J S) + (1: \coprod_L S \to \coprod_L S))R(\lambda\theta)$, so $\lambda\theta$ spans Im θ. On the other hand, for $\eta \in S(S, \coprod_L S) \leq S(S, \coprod_{J \uplus L} S)$, the equation $\eta R(\lambda\theta) = 0$ implies $\eta R(\lambda) \in \operatorname{Ker}\theta$, i.e. $\exists \xi \in S(S, \coprod_J S) \leq S(S, \coprod_{J \uplus L} S)$. $\xi R(\kappa) = \eta R(\lambda)$, whence $(\xi - \eta)R(\kappa \uplus \lambda) = 0$. Since $\kappa \uplus \lambda$ is a basis for V, it follows that $\xi - \eta = 0$ or $\xi = \eta \in (\coprod_J S) \cap (\coprod_L S) = \{0\}$, i.e. $\eta = 0$. Thus $\lambda\theta$ is independent, and so a basis for Im θ. Then $\dim V = |J \uplus L| = |J| + |L| = \dim \operatorname{Ker}\theta + \dim \operatorname{Im}\theta$.

(b) The proof of (b) is dual to (a). Pick a basis $\kappa : J \to \operatorname{Im}\theta$ of Im θ, and extend to a basis $(\kappa \uplus \lambda): J \uplus L \to W$ of W. Then the composite of λ: $L \to W$ with the natural projection $W \to \operatorname{Coker}\theta$; $w \mapsto w + \operatorname{Im}\theta$ is a basis for Coker θ (Exercise 2.4O).

(c) If θ injects, then $\dim \operatorname{Ker}\theta = 0$. Thus $\dim \operatorname{Im}\theta = \dim V = \dim W$. It follows that Im $\theta = W$, for otherwise a basis for Im θ would have a proper extension to a basis for W, violating the equality of the dimensions. Thus θ surjects. In the other direction, suppose that θ surjects. Then $\dim \operatorname{Ker}\theta = \dim V - \dim \operatorname{Im}\theta = 0$, so that θ injects. □

EXERCISES

2.4A. For a K-module X, show that the evaluation map of Exercise 2.1E, with $G = K$, furnishes a bilinear pairing of X with its conjugate $X' = K(X, K)$.

2.4B. (a) Let f be a fixed element of K_m^n. Show that matrix multiplication gives a bilinear pairing $(\xi, \eta) \mapsto \xi f \eta$ of K_1^m with K_n^1.

(b) Show that $f \mapsto ((\xi, \eta) \mapsto \xi f \eta)$ gives a K-isomorphism between K_m^n and $K(K_1^m, K_n^1; K)$.

2.4C. Show that the Riemann integral $C(I) \times C(I) \to \mathbb{R}$; $(f, g) \mapsto \int_I f(x)g(x)\,dx$ gives a bilinear functional on the $\mathbb{R}$-module $C(I)$ of Example 2.1.5.

2.4D. For a fixed 2-linear combination $\xi \in K^2$, show that the function $V^2 \to V$; $f \mapsto \xi f$ [as in (2.3.1)] is bilinear.

2.4E. Verify that (2.4.2) makes $K^n(V; K)$ a left $\underline{\underline{\mathrm{Set}}}(\mathbb{Z}_n, \mathbb{Z}_n)$-set and a left S_n-set. Verify that $\tilde{K}^n(V;K)$ is an invariant subset of each.

2.4F. Express each bilinear functional on a real vector space as the sum of a symmetric bilinear functional and an alternating bilinear functional.

2.4G. (a) Show that the determinant of a diagonal matrix is the product of its diagonal entries.

(b) Show that the determinant of a shear matrix $I_n + kE_{nn}^{ij}$ ($i \neq j$, and $k \in K$) is 1.

2.4H. For $T \in S_n$, let ρ_T be the corresponding permutation matrix as in Exercise 1.2I, with $A = K$. Prove that $\det \rho_T = \operatorname{sgn} T$.

2.4I. For what values of n is the determinant of an $n \times n$-matrix $f \in K_n^n$ given by *Sarrus' Rule* $\det f = \Sigma_{T \in D_n} (\operatorname{sgn} T) \prod_{i \in \mathbb{Z}_n} f_{i, iT}$?

2.4J. Give an example of a unital, non-commutative ring S and distinct natural numbers m, n such that the S-modules S^m and S^n are isomorphic.

2.4K. Let $\mathrm{GL}_n(K)$, the *general linear group* of *dimension n* over K, be the group of invertible elements of the monoid $(K_n^n, \cdot, I_n)$. Show that matrices in the same (group) conjugacy class have the same determinant.

2.4L. Let $\mathrm{SL}_n(K)$, the *special linear group* of *dimension n* over K, be the set of elements of $\mathrm{GL}_n(K)$ whose determinant is 1_K. Show that $\mathrm{SL}_n(K)$ is a normal subgroup of $\mathrm{GL}_n(K)$. What is the quotient group $\mathrm{GL}_n(K)/\mathrm{SL}_n(K)$?

2.4M. Exhibit a group isomorphism $\mathrm{SL}_2(\mathbb{Z}_2) \cong S_3$.

2.4N. Let $0 \to A_0 \to A_1 \to \cdots \to A_n \to 0$ be an exact sequence of finite dimensional vector spaces. Prove that $\Sigma_{i=0}^{n}(-1)^i \dim A_i = 0$. (Cf. Exercise 1F.)

2.4O. Complete the proof of Theorem 2.4.9(b).

2.4P. For vector spaces U and V over a field K, prove that $U \leq V$ and $\dim U = \dim V < \infty$ implies $U = V$.

2.4Q. Let $\mathbb{C}$ denote the subset

$$\left\{ \begin{bmatrix} x & y \\ -y & x \end{bmatrix} \middle| x, y \in \mathbb{R} \right\}$$

of $\mathbb{R}_2^2$. Elements of $\mathbb{C}$ are known as *complex numbers*.

(a) Show that $\mathbb{C}$ is a commutative subring of $\mathbb{R}_2^2$, containing the ring of scalar matrices as a subring isomorphic to, and identified with, the ring of real numbers $\mathbb{R}$.

(b) Set $i = \begin{bmatrix} 0 & 1 \\ -1 & 0 \end{bmatrix}$. Show that $i^2 = -1$.

(c) Show that each complex number z can be expressed uniquely in the form $z = x + iy$, with *real part* $x = \frac{1}{2}\,\mathrm{tr}\, z$ (cf. Exercise 1.3E) and *imaginary part* $y = -i(z - \frac{1}{2}\mathrm{tr}\, z)$.

(d) The transpose z^T of a complex number is called its *complex conjugate* $\bar{z}$. Show that transposition or *conjugation* yields an automorphism of the ring $\mathbb{C}$.

(e) Prove $\det z = z\bar{z}$ and $\mathrm{tr}\, z = z + \bar{z}$.

(f) Prove that the determinant of a complex number z is non-negative. The *absolute value* or *norm* or *modulus* of z is $\sqrt{\det z}$. It is denoted by $|z|$ (not to be confused with the notation $|z|$ for $\det z$). Prove $|zw| = |z| \cdot |w|$.

(g) Show that $\mathbb{C}$ is a field.

2.4R. Let m and n be positive integers that can be expressed as the sum of two squares of positive integers, e.g. $10 = 3^2 + 1^2$ and $20 = 4^2 + 2^2$. Prove that the product mn can be expressed similarly. [Hint: Use Exercise 2.4Q(f).]

2.4S. Let $\mathbb{H}$ denote the subset

$$\left\{ \begin{bmatrix} t & x & y & z \\ -x & t & -z & y \\ -y & z & t & -x \\ -z & -y & x & t \end{bmatrix} \middle| \, t, x, y, z \in \mathbb{R} \right\}$$

of $\mathbb{R}^4_4$. Elements of $\mathbb{H}$ are known as *quaternions*.

(a) Show that $\mathbb{H}$ is a subring of $\mathbb{R}^4_4$, containing the ring of scalar matrices as a subring isomorphic to, and identified with, the ring of real numbers $\mathbb{R}$.

(b) Set

$$i = \begin{bmatrix} 0 & 1 & 0 & 0 \\ -1 & 0 & 0 & 0 \\ 0 & 0 & 0 & -1 \\ 0 & 0 & 1 & 0 \end{bmatrix}, \quad j = \begin{bmatrix} 0 & 0 & 1 & 0 \\ 0 & 0 & 0 & 1 \\ -1 & 0 & 0 & 0 \\ 0 & -1 & 0 & 0 \end{bmatrix},$$

$$k = \begin{bmatrix} 0 & 0 & 0 & 1 \\ 0 & 0 & -1 & 0 \\ 0 & 1 & 0 & 0 \\ -1 & 0 & 0 & 0 \end{bmatrix}.$$

Show that $i^2 = j^2 = k^2 = -1$, $ij = k$, $jk = i$, $ki = j$, $ji = -k$, $kj = -i$, and $ik = -j$. (Compare Exercise I 2.5E.)

(c) The transpose q^T of a quaternion q is called its *conjugate* $\bar{q}$. Show that transposition yields an isomorphism of $\mathbb{H}$ with $\mathbb{H}^{\mathrm{op}}$.

(d) Prove that $\mathbb{H}$ is a skewfield.

2.4T. **(Loop Transversal Codes over Finite Ring Alphabets.)** Let A be a commutative unital finite ring, regarded as an alphabet. Let T be a $|A|^{n-k}$-element subset of the free A-module A^n such that T contains 0 and a basis B for A^n. Suppose that T carries a unital A-module structure $(T, *, A)$ given by an A-isomorphism $s:(T, *, A) \to (A^n, +, A)$ such that $s:(T, +, A) \to (A^n, +, A)$ is a *partial A-homomorphism*. In other words, for $t, u \in T$ and $a \in A$, one has $t + u \in T \Rightarrow (t + u)^s = t^s + u^s$ and $ta \in T \Rightarrow (ta)^s = t^s a$. Show that there is a linear code C, a free submodule of dimension k in A^n, to which $(T, *, 0)$ is a loop transversal. Moreover, show that T is precisely the set of errors corrected by C. (Hint: Generalize Theorem I 4.4.4.)

2.4U. Express $\hat{f}_{ji}$ from (2.4.6) in terms of the determinant of the $(n-1) \times (n-1)$-matrix obtained from f by removing its i-th row and j-th column.

3. COMMUTATIVE ALGEBRA

A module over a ring is the linear-algebraic analogue of a set with a monoid action. Linear algebraic analogues of monoids are known as "algebras." (The rather grandiose name probably arose because early twentieth-century algebra was almost exclusively devoted to the study of "algebras" in this sense.) If K is a unital commutative ring, then a *K-algebra* $(A_K, \cdot, 1)$ is a unital K-module A_K and a monoid $(A, \cdot, 1)$ such that the monoid multiplication $A \times A \to A$; $(a, b) \mapsto ab$ is bilinear. In other words, the right regular representation $R: A \to K(A, A)$; $b \mapsto (x \mapsto xb)$ and the map $L : A \to K(A, A)$; $a \mapsto (x \mapsto ax)$ are K-homomorphisms. Note that a $\mathbb{Z}$-algebra is just a unital ring. Note that K itself is a K-algebra $(K_K, \cdot, 1)$. The following proposition describes a copious source of K-algebras.

Proposition 3.1. *Let $f: K \to S$ be a unital ring homomorphism from K to a unital ring $(S, \cdot, 1)$ such that*

$$\forall k \in K, \forall s \in S, \ k^f s = sk^f. \tag{3.1}$$

Then the composite of $f: K \to S$ with the right regular representation $R: S \to$ End S of S makes S a right K-module S_K and a K-algebra $(S_K, \cdot, 1)$.

Proof. For s in S, one has $R(s) \in K(S_K, S_K)$, since $\forall x \in S$, $\forall k \in K$, $(xk)R(s) = xk^f \cdot s = xs \cdot k^f = (xR(s))k$ by (3.1). Then $R: S \to K(S_K, S_K)$ is a K-homomorphism, since $\forall x \in S$, $\forall k \in K$, $xR(sk) = x \cdot sk^f = xs \cdot k^f = (xR(s))k = x(R(s)k)$. Similarly $L(s) \in K(S_K, S_K)$, since $\forall x \in S$, $\forall k \in K$, $(xk)L(s) = s \cdot xk^f = sx \cdot k^f = (xL(s))k$. Finally, $L: S \to K(S_K, S_K)$ is a K-homomorphism, since $\forall x \in S$, $\forall k \in K$, $xL(sk) = sk^f \cdot x = sx \cdot k^f = (xL(s))k = x(L(s)k)$ by (3.1). □

Example 3.2. For a commutative unital ring K and a positive integer n, the unital ring homomorphism $f: K \to K_n^n$ embedding K as the ring of scalar matrices makes K_n^n into a K-algebra according to Proposition 3.1. In this context, one speaks of the *matrix algebra* K_n^n. □

Example 3.3. For an interval I in the real line $\mathbb{R}$, let $C(I)$ be the set of continuous functions $I \to \mathbb{R}$. Note that $C(I)$ is a commutative unital ring under pointwise operations (cf. Exercise 1.4Q). Let $f: \mathbb{R} \to C(I)$; $r \mapsto (x \mapsto r)$ embed $\mathbb{R}$ as the ring of constant functions. Then Proposition 3.1 makes $C(I)$ into an $\mathbb{R}$-algebra. □

Example 3.4. The complex numbers $\mathbb{C}$ (Exercise 2.4Q) and the quaternions $\mathbb{H}$ (Exercise 2.4S) form $\mathbb{R}$-algebras according to Proposition 3.1, taking the homomorphism f in each case that embeds $\mathbb{R}$ as the ring of scalar matrices. $\square$

Given two K-algebras $(A_K, \cdot, 1)$ and $(B_K, \cdot, 1)$, a *K-algebra homomorphism* $h:(A_K, \cdot, 1) \to (B_K, \cdot, 1)$ is a function $h: A \to B$ that is a K-homomorphism $h: A_K \to B_K$ and a monoid homomorphism $h:(A, \cdot, 1) \to (B, \cdot, 1)$.

For a given unital commutative ring K, the most fundamental K-algebra is the *polynomial ring* $K[T]$ *in the indeterminate* T. Let $T^{\mathbb{N}}$ be the free monoid on the singleton T (cf. Exercise O 4.2C). Then the K-module $K[T]_K$ is the coproduct $\coprod_{T^{\mathbb{N}}} K$, with insertion $\iota_{T^n}: K \to K[T]$; $k \mapsto kT^n$. In particular, ι_1 embeds K in $K[T]$. Since $T^{\mathbb{N}}$ is a basis for $K[T]_K$, each *polynomial* $p(T)$, i.e. element of $K[T]$, has a unique expression $p(T) = p_n T^n + \cdots + p_1 T + p_0$, with $p_n \neq 0$ unless $p(T) = 0$. For non-zero $p(T)$, the natural number n is called the *degree* $\deg p(T)$ of $p(T)$, and p_n is called the *leading coefficient* of $p(T)$. If $n > 0$ and $p_n = 1$, then $p(T)$ is said to be *monic*. Further, $\deg 0 := -\infty$. Note $\deg(p(T) + q(T)) \leq \max\{\deg p(T), \deg q(T)\}$. The element $1\iota_{T^n} = T^n$ of $K\iota_{T^n}$ is called the *monomial* of degree n. Elements of $K = K\iota_1$ are called *constants*. Note that the degree of a monomial T^n is just its length as an element T^n of the free monoid $T^{\mathbb{N}}$. Now the right regular representation $R(T^n): T^{\mathbb{N}} \to T^{\mathbb{N}}$ of an element T^n of $T^{\mathbb{N}}$ concatenates with the embedding $T^{\mathbb{N}} \hookrightarrow K[T]$ of monic monomials to yield a set mapping $R(T^n): T^{\mathbb{N}} \to K[T]$. Since $K[T]$ is the free K-module on $T^{\mathbb{N}}$, there is then a unique K-homomorphism $R(T^n)$: $K[T] \to K[T]$ restricting to $R(T^n): T^{\mathbb{N}} \to K[T]$ on $T^{\mathbb{N}}$. Finally, the set map $R: T^{\mathbb{N}} \to K(K[T]_K, K[T]_K)$; $T^n \mapsto R(T^n)$ extends to a unique K-homomorphism

$$R: K[T] \to K(K[T]_K, K[T]_K), \tag{3.2}$$

the *right regular representation* of $K[T]$.

Proposition 3.5. *The multiplication* $p(T) \cdot q(T) := p(T)R(q(T))$ *makes the polynomial ring* $K[T]$ *into a commutative K-algebra* $(K[T]_K, \cdot, 1)$.

Proof. For natural numbers m and n, both $R(T^{m+n})$ and $R(T^m)R(T^n)$ in $K(K[T], K[T])$ restrict to the same map $T^{\mathbb{N}} \to K[T]$. Thus $R: T^{\mathbb{N}} \to K(K[T]_K, K[T]_K)$ is a monoid homomorphism, with commutative image. Now for $p(T)$ and $q(T)$ in $K[T]$, one has

$$p(T)q(T) = p(T)R(q(T)) = \left(\sum_{i=0}^{\deg p} p_i T^i\right) R\left(\sum_{j=0}^{\deg q} q_j T^j\right) = \sum_{i=0}^{\deg p} \sum_{j=0}^{\deg q} p_i q_j T^{i+j}$$

$$= q(T)p(T),$$

so that $K[T]$ has a commutative, bilinear multiplication. Also

$$\begin{aligned} R(p(T)q(T)) &= R\left(\sum_{i=0}^{\deg p}\sum_{j=0}^{\deg q} p_i q_j T^{i+j}\right) \\ &= \sum_{i=0}^{\deg p}\sum_{j=0}^{\deg q} p_i q_j R(T^{i+j}) \\ &= \sum_{i=0}^{\deg p}\sum_{j=0}^{\deg q} p_i q_j R(T^i)R(T^j) \\ &= \left[\sum_{i=0}^{\deg p} p_i R(T^i)\right]\left[\sum_{j=0}^{\deg q} q_j\, R(T^j)\right] \\ &= R(p(T))R(q(T)), \end{aligned}$$

so the multiplication in $K[T]$ is associative. □

Make $\{-\infty\} \cup \mathbb{N}$ a semigroup under $+$ with $\{-\infty\}$ as a sink. Then $\deg(p(T)q(T)) \leq \deg p(T) + \deg q(T)$, with equality guaranteed if either of $p(T), q(T)$ is zero, or otherwise if neither of the leading coefficients of $p(T)$ or $q(T)$ is a zero divisor. In particular, equality always holds if K is an integral domain, so that $K[T]$ is then itself an integral domain. Furthermore,

(3.3) $$K[T]^* = K^*$$

in this case, since $p(T)q(T) = 1 \Rightarrow 0 \leq \deg p(T) + \deg q(T) = \deg p(T)q(T) = \deg 1 = 0$, whence $\deg p(T) = \deg q(T) = 0$ and $p_0 q_0 = 1$.

The importance of the polynomial algebra $K[T]$ rests on the following *universality property*: Given an element t of a K-algebra S, there is a unique K-algebra homomorphism

(3.4) $$K[T] \to S;\ p(T) \mapsto p(t)$$

mapping T to t. Indeed, for $p(T) = \sum_{i=0}^{\deg p} p_i T^i$, one has $p(t) = \sum_{i=0}^{\deg p} p_i\ t^i$. A first application of the universality property takes the K-algebra S to be the set $\underline{\underline{\mathrm{Set}}}(K, K)$ of set functions from K to K, with the pointwise unital ring operations of unit $1 : x \mapsto 1$, subtraction $g - h : x \mapsto x^g - x^h$, and product $g \cdot h : x \mapsto x^g x^h$. Let $f : K \to \underline{\underline{\mathrm{Set}}}(K, K);\ k \mapsto (x \mapsto k)$ embed K as the subring of constant functions. By Proposition 3.1, $\underline{\underline{\mathrm{Set}}}(K, K)$ becomes a K-algebra. For the element $1_K : x \mapsto x$ of $\underline{\underline{\mathrm{Set}}}(K, K)$, the universality property (3.4) gives a unique K-algebra homomorphism

(3.5) $$K[T] \to \underline{\underline{\mathrm{Set}}}(K, K);\ p(T) \mapsto (k \mapsto p(k)).$$

The image of a polynomial $p(T)$ under (3.5) is described as the *polynomial function* $p(x) : K \to K;\ x \mapsto p(x)$. Elements of the preimage $\{x \in K \mid p(x) = 0\}$

of 0 under $p(x)$ are called *roots* of the polynomial $p(T)$ in $K[T]$. Note that distinct polynomials in $K[T]$ may yield the same polynomial function. For example, $T^3 + 2T$ is a non-zero polynomial in $\mathbb{Z}_3[T]$ yielding the zero polynomial function $\mathbb{Z}_3 \to \mathbb{Z}_3$. A useful tool for designing polynomial functions over fields is provided by *Lagrange interpolation.*

Proposition 3.6. *Let K be a field.*

(a) **(Lagrange Interpolation.)** *Let $X = \{x_1, \ldots, x_n\}$ be a finite, non-empty subset of K. For $1 \le i \le n$, define the delta polynomial*

$$\delta_{x_i}(T) = \prod_{j \ne i} \frac{T - x_j}{x_i - x_j}. \tag{3.6}$$

Let $f: X \to K$ be a function. Then the Lagrange interpolant polynomial $\Sigma_{i=1}^{n} f(x_i)\delta_{x_i}(T)$ yields a polynomial function extending f.

(b) *If K is finite, then each element of $\underline{\underline{\text{Set}}}(K, K)$ may be realized as a polynomial function.*

Proof. (a) The delta polynomial (3.6) yields a polynomial function $K \to K$ restricting to the delta function $X \to K$; $x \mapsto$ **if** $x = x_i$ **then** 1 **else** 0 on X. The value of the Lagrange interpolant polynomial function at x_i in X is then $\Sigma_{j=1}^{n} f(x_j)\delta_{x_j}(x_i) = f(x_i)$, agreeing with the value of f there.

(b) Take X in (a) to be K. □

Since the polynomial ring construction applies to arbitrary commutative unital rings, it applies to polynomial rings themselves. In particular, given a commutative unital ring K and indeterminates T, U, one may form the polynomial $K[T]$-algebra $K[T][U]$. Consider a commutative K-algebra S and an element (t, u) of S^2. By (3.4) for $K[T]$, there is a unique K-algebra homomorphism $f: K[T] \to S$; $p(T) \to p(t)$ mapping T to t. By Proposition 3.1, S is a $K[T]$-algebra. By (3.4) for $K[T][U]$, there is then a unique $K[T]$-algebra homomorphism $K[T][U] \to S$; $\Sigma_j p_j(T)U^j \mapsto \Sigma_j p_j(T)^f u^j = \Sigma_j(\Sigma_{i=1}^{\deg p_j} p_{ji}t^i)u^j$ mapping U to u and restricting to f on $K[T]$, i.e. mapping (T, U) to (t, u). Continuing by induction, one obtains the following:

Theorem 3.7. *Let K be a commutative, unital ring and let n be a natural number. Then there is a commutative K-algebra $K[T_1, \ldots, T_n]$ containing a subset $\{T_1, \ldots, T_n\}$ such that, given a commutative K-algebra S and an element $(t_1, \ldots, t_n)$ of S^n, there is a unique K-algebra homomorphism*

$$K[T_1, \ldots, T_n] \to S; \; p(T_1, \ldots, T_n) \mapsto p(t_1, \ldots, t_n) \tag{3.7}$$

mapping each T_i to t_i. □

Consider $\underline{\underline{\text{Set}}}(K^n, K)$ as a unital commutative ring with pointwise operations. Let $f: K \to \underline{\underline{\text{Set}}}(K^n, K)$ embed K as the subring $\{K^n \to K; (x_1, \ldots, x_n) \mapsto k | k \in K\}$ of constant functions. Proposition 3.1 makes

$\underline{\underline{\text{Set}}}(K^n, K)$ into a commutative K-algebra. For the element $((x_1, \ldots, x_n) \mapsto x_1, \ldots, (x_1, \ldots, x_n) \mapsto x_n)$ of $\underline{\underline{\text{Set}}}(K^n, K)^n$, the universality property of Theorem 3.7 gives a unique K-algebra homomorphism

$$K[T_1, \ldots, T_n] \to \underline{\underline{\text{Set}}}(K^n, K);$$
$$p(T_1, \ldots, T_n) \mapsto ((k_1, \ldots, k_n) \mapsto p(k_1, \ldots, k_n)), \tag{3.8}$$

the multivariate analogue of (3.5).

EXERCISES

3A. Let A be a K-algebra.

(a) Show that $f : K \to A;\ k \mapsto 1k$ is a ring homomorphism.

(b) Show that

$$\forall k \in K, \forall a \in A,\ k^f a = ak^f$$

[cf. (3.1)].

(c) If K is a field and $A \neq \{0\}$, show that f injects.

(d) If $K = \mathbb{Z}$ and A is a unital ring, show that there is a non-negative integer n, the *characteristic* of A, such that $\operatorname{Ker} f = n\mathbb{Z}$.

(e) Show that the characteristic of a field is zero or prime.

3B. (a) Let A be a K-module. Let M be the subset of $K(A, A; A)$ consisting of those bilinear maps $m : A \times A \to A$ for which $(A_K, m, 1)$ is a K-algebra. Show that M is a submodule of $K(A, A; A)$.

(b) If A is a vector space of dimension n over a field K, determine $\dim K(A, A; A)$ and $\dim M$.

3C. (a) Show that

$$f : \mathbb{C} \to \mathbb{H};\ \begin{bmatrix} t & x \\ -x & t \end{bmatrix} \mapsto \begin{bmatrix} t & x & 0 & 0 \\ -x & t & 0 & 0 \\ 0 & 0 & t & -x \\ 0 & 0 & x & t \end{bmatrix}$$

is a unital ring homomorphism.

(b) Verify the need for hypothesis (3.1) in Proposition 3.1 by showing that the method of Proposition 3.1 applied to the ring homomorphism $f : \mathbb{C} \to \mathbb{H}$ does not make $\mathbb{H}$ a $\mathbb{C}$-algebra.

3D. Polly and Queeny are waitresses. In a typical week, Polly takes r cents in tips with probability p_r, while Queeny takes s cents with probability q_s. Consider the polynomials $p(T) = \Sigma p_i\, T^i$ and $q(T) = \Sigma q_j T^j$ in $\mathbb{R}[T]$. Show that the probability that the two waitresses' total combined tips in the week are t cents is given by the coefficient of T^t in the product polynomial $p(T)q(T)$.

3E. Give a direct verification of the associative law $p(T)q(T) \cdot r(T) = p(T) \cdot q(T)r(T)$ in the polynomial ring $K[T]$.

3F. Verify the universality property (3.4).

3G. Show that the universality property (3.4) of $K[T]$ is equivalent to the following freeness property: Given a set map $\theta : \{T\} \to A$ to a K-algebra A, there is a unique K-algebra homomorphism $\bar{\theta}: K[T] \to A$ with $T\bar{\theta} = t\theta$. (Remark: This interpretation shows that $K[T]$ is the linear-algebraic analogue of the free monoid $T^{\mathbb{N}}$ on the singleton $\{T\}$, as in Exercise O 4.2C.)

3H. Show that $K[T]$ has the following universal property: Given a unital ring homomorphism $f: K \to S$, and an element t of the (not necessarily commutative) ring S such that $k^f \cdot t = t \cdot k^f$ for all k in K, there is a unique ring homomorphism $K[T] \to S$; $p(T) \mapsto p^f(t)$ mapping T to t and restricting to f on the subring K of $K[T]$.

3I. Give an example of a cubic (i.e. degree 3) polynomial in $\mathbb{Z}_3[T]$ yielding a non-zero homomorphism of the abelian group $(\mathbb{Z}_3, -, 0)$.

3J. **(Simpson's Rule.)** Let $f:[-1,1] \to \mathbb{R}$ be a continuous, real-valued function on the interval $[-1,1]$. Simpson's Rule approximates the Riemann integral $\int_{[-1,1]} f(x)\,dx$ as the integral of the Lagrange interpolant polynomial function determined by the function $\{-1,0,1\} \to \mathbb{R}; -1 \mapsto f(-1), 0 \mapsto f(0), 1 \mapsto f(1)$. Find the quadratic interpolant, and evaluate its Riemann integral over $[-1,1]$.

3K. **(Permutation polynomials.)** Let K be a finite field. A polynomial in $K[T]$ is a *permutation polynomial* if its polynomial function is a permutation of K.

(a) Show that a polynomial of degree 1 is a permutation polynomial.

(b) Show that the symmetric group S_3 is the set of degree 1 polynomial functions over the field $\mathbb{Z}_3$.

(c) Show that the degree 1 polynomial functions over the field $\mathbb{Z}_5$ generate a proper subgroup of the symmetric group S_5.

(d) Show that the degree 1 polynomial functions together with the degree 3 permutation polynomial functions generate the symmetric group S_5.

(e) Give an example of a cubic polynomial over the field $\mathbb{Z}_5$ which is not a permutation polynomial.

3L. Write out a careful proof of Theorem 3.7.

3M. Theorem 3.7 builds $K[T_1, \ldots, T_n]$ inductively as $K[T_1][T_2]\ldots[T_n]$. Give an alternative construction of the K-algebra $K[T_1, \ldots, T_n]$ using the coproduct $\coprod_{X^{**}} K$ of K over the free commutative monoid on the set $X = \{T_1, \ldots, T_n\}$ of indeterminates. (Hint: Generalize Proposition 3.5.)

3N. Let U be a subset of $\mathbb{R}^n$. Prove that $\{p(T_1, \ldots, T_n) \mid \forall (u_1, \ldots, u_n) \in U^n, p(u_1, \ldots, u_n) \neq 0\}$ is a submonoid of the monoid $(K[T_1, \ldots, T_n], \cdot, 1)$.

3O. If K is an integral domain, prove that $K[T_1, \ldots, T_n]$ is an integral domain.

3.1. Fractions

The field $\mathbb{Q}$ of fractions is probably the most sophisticated algebraic construct common to the experience of every educated person. Identifying 51/17 with 531/177 is no mean feat, and many students succumb to the temptation of linearity by believing that the sum of $\frac{3}{5}$ and $\frac{4}{7}$ should be $(3 + 4)/(5 + 7)$. Some perverse consolation may be derived from the realization that one is dealing with the universal $\mathbb{Z}$-algebra $\mathbb{Q}$ in which the elements of the submonoid $\mathbb{Z} - \{0\}$ of $\mathbb{Z}$ have been made invertible. This section studies a general version of the construction of fractions. Throughout, K is a commutative unital ring.

Let M be a submonoid of a commutative K-algebra A. Consider the direct product monoid $M \times A$. Define a relation α on $M \times A$ by

$$(m_1, a_1)\alpha(m_2, a_2) \Leftrightarrow \exists m \in M.\ m_1 m a_2 = a_1 m m_2. \tag{3.1.1}$$

Then α is a monoid congruence on $M \times A$. It is reflexive since M is non-empty, and symmetric since A is commutative. Suppose $(m_1, a_1)\alpha(m_2, a_2)\alpha(m_3, a_3)$, say $m_1 m a_2 = a_1 m m_2$ and $m_2 n a_3 = a_2 n m_3$ with m, n in M. Then $m_1(m m_2 n)a_3 = m_1 m(m_2 n a_3) = m_1 m(a_2 n m_3) = (m_1 m a_2)n m_3 = (a_1 m m_2)n m_3 = a_1(m m_2 n)m_3$, so that $(m_1, a_1)\alpha(m_3, a_3)$ and α is transitive. Finally, the commutativity of A readily yields that α is a submonoid of $(M \times A)^2$. The quotient monoid $(M \times A)^\alpha$ is denoted by $M^{-1}A$, and an equivalence class $(m, a)^\alpha$ is denoted by $m^{-1}a$ or a/m or $\frac{a}{m}$. Note the "cancellation" $an/mn = a/m$ for n in M. Now $M \times A$ is a K-set under the action $M \times A \times K \to M \times K$; $(m, a, k) \mapsto (m, ak)$, and α is a K-congruence, so $M^{-1}A$ becomes a K-set with action $M^{-1}A \times K \to M^{-1}A$; $(a/m, k) \mapsto ak/m$. The well-defined subtraction

$$\frac{a_1}{m_1} - \frac{a_2}{m_2} = \frac{a_1 m_2 - m_1 a_2}{m_1 m_2} \tag{3.1.2}$$

then makes $M^{-1} A$ into a K-algebra (cf. Exercise 3.1A). The K-algebra $M^{-1}A$ has the universality property given by the following:

Proposition 3.1.1. *There is a K-algebra homomorphism*

$$\eta : A \to M^{-1}A;\ a \mapsto a/1 \tag{3.1.3}$$

such that, given a K-algebra homomorphism $f : A \to B$ with $Mf \leq B^*$, there is a unique K-algebra homomorphism

$$\bar{f} : M^{-1}A \to B;\ m^{-1}a \mapsto (mf)^{-1} \cdot af \tag{3.1.4}$$

with $\eta\bar{f} = f$. Furthermore, $\operatorname{Ker} \eta = \{a \in A | \exists m \in M.\ ma = 0\}$ *and* $M\eta \leq (M^{-1}A)^*$.

Proof. Verification that $\bar{f}$ is well-defined and has the claimed properties is left as Exercise 3.1B. Now $\operatorname{Ker}\eta = \eta^{-1}\{0\} = \{a \in A \mid 0/1 = a/1\} = \{a \in A \mid (1,0)\alpha(1,a)\} = \{a \in A \mid \exists m \in M.\ 1ma = 0m1 = 0\}$. Finally, for m in M, one has $(m/1)(1/m) = 1/1 = 1$. □

Example 3.1.2 (The Rationals). If M is the submonoid $\mathbb{Z} - \{0\}$ of the $\mathbb{Z}$-algebra $\mathbb{Z}$, then $(\mathbb{Z} - \{0\})^{-1}\mathbb{Z}$ is the field $\mathbb{Q} = \{m^{-1}n | n \in \mathbb{Z},\ m \in \mathbb{Z} - \{0\}\}$. □

Example 3.1.3 (The Field of Fractions of an Integral Domain). Generalizing Example 3.1.2, let S be an integral domain (and in particular a commutative $\mathbb{Z}$-algebra). Then $(S - \{0\})^{-1}S$ is a field, since each non-zero element a/b has an inverse b/a. By Proposition 3.1.1, $\operatorname{Ker}(\eta : S \to (S - \{0\})^{-1}S) = \{0\}$, so η embeds S in $(S - \{0\})^{-1}S$. The field $(S - \{0\})^{-1}S$ extending S is called the *field of fractions* of the integral domain S. □

Example 3.1.4 (Rationals over an Integral Domain). For an integral domain K, the polynomial algebra $K[T]$ is an integral domain. The field of fractions of $K[T]$ has the form

$$K(T) = \{p(T)/q(T) | q(T) \neq 0\}. \tag{3.1.5}$$

Its elements are known as *rationals*, or more colloquially as "rational functions" (cf. the usage of Example 3.1.6 and Exercises 3.1F and 3.1K). □

Example 3.1.5 (Division by Zero). If M contains 0, then $\operatorname{Ker}\eta = A$ by Proposition 3.1.1. □

Example 3.1.6 (Localization). Let P be a prime ideal of a commutative unital ring A. By (1.4.2), the complement $A - P$ of P is a submonoid of A. The ring $(A - P)^{-1}A$ is called the *localization* of A at P. For instance, consider the polynomial ring $\mathbb{R}[T]$ identified via (3.5) with a subring of the ring $C(\mathbb{R})$ of Example 3.3. For a real number x, the set $P_x = \{p(T) | p(x) = 0\}$ is a prime (indeed maximal) ideal of $\mathbb{R}[T]$, namely the kernel of the ring homomorphism $\mathbb{R}[T] \to \mathbb{R};\ p(T) \to p(x)$ of (3.4) onto the field $\mathbb{R}$. Then the localization $(\mathbb{R}[T] - P_x)^{-1}\mathbb{R}[T]$ comprises those rational functions that are well-defined in a neighborhood of the point x, i.e. "locally" around x (cf. Exercise 3.1F). □

It is instructive to study the invertible elements of the result $(A - P)^{-1}A$ of the localization process of Example 3.1.6 applied to a prime ideal P of the commutative, unital ring A. For $m^{-1}u \in (A - P)^{-1}A$ with $u \notin P$, one has $(m^{-1}u)(u^{-1}m) = 1$, so $m^{-1}u \in (A - P)^{-1}(A - P) \Rightarrow m^{-1}u \in ((A - P)^{-1}A)^*$. Conversely, suppose $m^{-1}u \in ((A - P)^{-1}A)^*$, say $(u/m) \cdot (v/n) = (1/1)$ or $\exists l \in A - P.\ uvl = mnl \in A - P$. Now $u \in P$ would imply $uvl \in P$, a contradiction. Thus $u \notin P$, and $((A - P)^{-1}A)^* = (A - P)^{-1}(A - P)$. In other words, the non-invertible elements of the localization $(A - P)^{-1}A$ form the ideal $(A - P)^{-1}P$ of $(A - P)^{-1}A$. In general, a commutative, unital ring is said to be *local* if its non-invertible elements form an ideal. Since no multiple of a non-invertible element can be invertible, this is equivalent to saying that the non-invertible elements form a non-empty semigroup under addition. An alternative characterization is given by:

Proposition 3.1.7. *Let A be a commutative, unital ring. Then A is local iff* $|\text{Max } A| = 1$.

Proof. Let x be an element of a proper ideal I of a local ring A. Then x is an element of the ideal $A - A^*$, for otherwise $1 \in xA \subseteq I$ and $I = A$. In particular, the unique maximal ideal of the local ring A is its ideal $A - A^*$ of non-invertible elements. Conversely, suppose Max $A = \{I\}$, and $x, y \in A - A^*$. Then $xA \subseteq I$ and $yA \subseteq I$, since the proper principal ideals xA and yA are contained in some maximal ideal. Thus $(x + y)A \subseteq xA + yA \subseteq I \subset A$, so that $x + y \in A - A^*$. Finally, note $\varnothing \subset I \subseteq A - A^*$. Thus the non-invertible elements form a non-empty subsemigroup of $(A, +)$, i.e. A is local. □

If A is a local ring, then the field $A/(A - A^*)$ is called the *residue field* of A. For example, the residue field of a localization $(A - P)^{-1}A$ of a ring A at a prime ideal P is $((A - P)^{-1}A)/((A - P)^{-1}P)$. Now

$$\eta : A/P \to \big((A - P)^{-1}A\big)/\big((A - P)^{-1}P\big); \qquad (3.1.6)$$
$$a + P \mapsto (a/1) + (A - P)^{-1}P$$

satisfies the universality property of Proposition 3.1.1 for the field of fractions of the integral domain A/P, so the residue field $((A - P)^{-1}A)/((A - P)^{-1}P)$ is this field of fractions. As a trivial example, note that each field is a local ring, and thus is its own residue field.

EXERCISES

3.1A. Verify that the subtraction (3.1.2) is well-defined, and yields a K-module structure $M^{-1}A_K$ on $M^{-1}A$ so that $(M^{-1}A_K, \cdot, 1)$ is a K-algebra. [Hint: For a non-computational verification of the associative law for addition, use the injection $\exp : M \times A \to A_2^2; (m, a) \mapsto \begin{bmatrix} m & a \\ 0 & m \end{bmatrix}$ to give a monoid homomorphism $(M \times A, +, 0) \to (A_2^2, \cdot, 1)$.]

3.1B. Verify that $\bar{f}$ in (3.1.4) is a well-defined K-algebra homomorphism with $\eta\bar{f} = f$.

3.1C. Let Z be the set consisting of zero and the zero divisors of a commutative, unital ring A (cf. Exercise 1.4M).

(a) Show that $A - Z$ is a submonoid of A.

(b) Show that the ring homomorphism $\eta : A \to (A - Z)^{-1}A$ of (3.1.3) injects. [The extension $(A - Z)^{-1}A$ of A is called the *total ring of fractions* of the ring A.]

(c) If $A - Z$ is a proper submonoid of a submonoid M of A, show that the corresponding ring homomorphism $\eta : A \to M^{-1}A$ of (3.1.3) does not inject.

3.1D. Let

$$A = \left\{ \begin{bmatrix} m & n \\ -n & m \end{bmatrix} \middle| m, n \in \mathbb{Z} \right\}.$$

(a) Show that A is a subring of the ring $\mathbb{Z}_2^2$.

(b) Show that A is an integral domain.

(c) Identify the field of fractions of A.

3.1E. For a prime number p, show that the localization $(\mathbb{Z} - p\mathbb{Z})^{-1}\mathbb{Z}$ is the subring $\{m^{-1}n | m \notin p\mathbb{Z}\}$ of $\mathbb{Q}$.

3.1F. Let x be a real number, and let $f(T)/g(T)$ be a rational function (quotient of polynomials) with $g(x) \neq 0$. Give a detailed proof that there is an open interval I containing x such that $I \to \mathbb{R}; t \mapsto f(t)/g(t)$ is a continuous function.

3.1G. Let M be the submonoid $\{(1,1),(0,1)\}$ of $\mathbb{R}^2$. Show that $\eta : \mathbb{R}^2 \to M^{-1}\mathbb{R}^2$ is the projection $\mathbb{R}^2 \to \mathbb{R}; (x, y) \mapsto y$.

3.1H. For $I \lhd A$ and $(M, \cdot, 1) \leq (A, \cdot, 1)$, prove $M^{-1}I \lhd M^{-1}A$.

3.1I. Verify that (3.1.6) satisfies the universality property for the field of fractions of A/P.

3.1J. (a) For $\eta : A \to M^{-1}A$ as in (3.1.3), consider the map η^{-1}: Spec $(M^{-1}A) \to$ Spec A of Exercise 1.4S. Show that η^{-1} is an injection, with image $\{Q \in \text{Spec } A | Q \cap M = \varnothing\}$.

(b) For $P \in$ Spec A, consider the map $\eta_P : A \to (A - P)^{-1}A$ of (3.1.3). Show that η_P^{-1} : Spec $((A - P)^{-1}A) \to$ Spec A is an injection, with image $\{Q \in \text{Spec} A | Q \leq P\}$.

3.1K. Let x be a real number.

(a) Show that the ring $R_x = \{f(T)/g(T) | f(T), g(T) \in \mathbb{R}[T], g(x) \neq 0\}$ of rational functions defined at x is a local ring.

(b) Identify the unique maximal ideal and the residue field of the ring R_x.

3.1L. Let A be the ring $C(\mathbb{R})$ as in Example 3.3. Let x be a real number. Define a relation α on $S_x = C(\mathbb{R}) \times [-\infty, x) \times (x, \infty]$ by $(f, a, b)\alpha(g, c, d) \Leftrightarrow \exists l, m \in \mathbb{R} \cup \{-\infty, \infty\}.\ \max\{a, c\} \leq l < x < m \leq \min\{b, d\}$ and $f|_{(l,m)} = g|_{(l,m)}$.

(a) Show that α is an equivalence relation on S_x.

(b) Defining $\mathbb{R} \to S_x^\alpha;\ r \mapsto ((t \mapsto r), -\infty, \infty)^\alpha$, show that S_x^α is an $\mathbb{R}$-algebra with $(f, a, b)^\alpha + (g, c, d)^\alpha = (f + g, \max\{a,c\}, \min\{b,d\})^\alpha$.

(c) Show that S_x^α is a local ring, and identify its unique maximal ideal. (S_x^α is known as the "algebra of germs of continuous real-valued functions at x.")

3.2 Factors

Let A be a commutative, unital ring. The group A^* of invertible elements or units of A acts on A by (right) multiplication. Elements common to an A^*-orbit are described as *associates*. For example, $\mathbb{Z}^* = \{1, -1\} \cong C_2$, and each integer n is associated with its negative. Non-negative integers are usually chosen as representatives for the $\mathbb{Z}^*$-orbits. The association relation in the ring A is a congruence on the monoid $(A, \cdot, 1)$, since A is commutative. Thus the set A/A^* of orbits carries the quotient monoid structure, with A^* as identity. Now suppose that A is an integral domain, so that $A - \{0\}$ is a submonoid of $(A, \cdot, 1)$. Note that $\{0\}$ is a singleton A^*-orbit. Then $(A - \{0\})/A^*$ is a submonoid of A/A^*. A non-zero, non-unit element a of A is said to be *irreducible* if $aA^* = bA^* \cdot cA^* \Rightarrow bA^* = A^*$ or $cA^* = A^*$. The set of irreducible elements of A is denoted as Irr A. In particular, if $(A - \{0\})/A^*$ is a free commutative monoid, then it is free on the set $(\mathrm{Irr}\ A)/A^*$ of orbits of irreducible elements. [Interpreting $(A - \{0\})/A^*$ as a set of finite multisets, the elements of $(\mathrm{Irr}\ A)/A^*$ correspond to the one-element multisets.]

As motivation for the concepts of this section, consider the prototypical integral domain—the ring $\mathbb{Z}$ of integers. The Fundamental Theorem of Arithmetic says that $(\mathbb{Z} - \{0\})/\mathbb{Z}^*$ is the free commutative monoid on the set of orbits of prime numbers (cf. Exercise I 1.5F). The proof of the Fundamental Theorem of Arithmetic depends on the Division Algorithm. The Division Algorithm shows that, given a fixed non-zero integer d known as the *divisor*, the set $\mathbb{Z}_d = \{0, 1, \ldots, d - 1\}$ of non-negative integers less than d is a loop transversal to the subgroup $d\mathbb{Z}$ of $(\mathbb{Z}, +, 0)$. Thus a given integer n (known in this context as the *dividend*) has a decomposition $n = n\delta_d + n\varepsilon_d$ as in (I 4.3.3), with the element $n\delta_d = dq \in d\mathbb{Z}$ and $n\varepsilon_d \in \mathbb{Z}_d$. The element $n\varepsilon_d$ of $\mathbb{Z}_d$ is traditionally called the *remainder* and the integer q is called the *quotient*.

Definition 3.2.1. Let A be an integral domain.

(a) A is said to be a *Euclidean domain*, or ED, if there is a *length function* $A \to \mathbb{N};\ a \mapsto |a|$ such that:

  (i) $|a| = 0 \Leftrightarrow a = 0$;

  (ii) for each non-zero element d of A, there is a subset A_d of the set $\{a \in A \mid |a| < |d|\}$ of elements shorter than d such that A_d is a loop transversal in $(A, +, 0)$ to the principal ideal dA of A.

(b) A is said to be a *principal ideal domain*, or PID, if each element of Id A is principal.
(c) A is said to be a *unique factorization domain*, or UFD, if $(A - \{0\})/A^*$ is a free commutative monoid. □

It is sometimes helpful to use the language of loop transversal codes in dealing with Euclidean domains. Consider a *divisor*, a non-zero element d of a Euclidean domain A. Imagine that A represents a noisy channel through which one wishes to transmit elements of the alphabet A. The elements are first encoded by the embedding $L(d)\colon A \to A;\ a \mapsto da$, with image the code dA. Note that $L(d)$ injects since A is an integral domain: $da_1 = da_2 \Rightarrow d(a_1 - a_2) = 0 \Rightarrow a_1 - a_2 = 0$. Now suppose that a *dividend*, an element n of A, is received as output from the channel. The errors to be corrected form the loop transversal A_d. The received word n is determined to have been subject to the *error* or *remainder* $n\varepsilon_d$ with $n \in dA + n\varepsilon_d$, so one has the *error map* $\varepsilon_d : A \to A_d$. Then n is decoded to $n\delta_d = n - n\varepsilon_d$. One obtains the *decoding* map $\delta_d : A \to dA$. Finally, the code element $n\delta_d = dq$ determines the *quotient* $q = n\delta_d L(d)^{-1}$ as the original letter from A that had been encoded as dq.

In any integral domain A, a non-zero element p is said to be *prime* iff pA is prime. This generalizes the familiar terminology from the ring $\mathbb{Z}$, since a positive integer p is prime iff $p\mathbb{Z} \in \operatorname{Spec}\mathbb{Z}$. Note that, for non-zero elements a, a' of an integral domain A,

$$aA = a'A \Leftrightarrow aA^* = a'A^*. \tag{3.2.1}$$

Certainly $aA^* = a'A^* \Rightarrow aA = aA^*A = a'A^*A = a'A$. Conversely, suppose $aA = a'A$, say $a = a'u$ and $a' = av$. Then $a = avu$ or $a(1 - vu) = 0$. Since a is a non-zero and A is an integral domain, it follows that $1 - vu = 0$, i.e. $vu = 1$. Thus u and v are invertible, so $a \in a'A^*$ and $a' \in aA^*$.

Proposition 3.2.2. (a) *In any integral domain A, prime elements are irreducible.*
(b) *In a unique factorization domain A, irreducible elements are prime.*

Proof. (a) Suppose $pA \in \operatorname{Spec} A - \{\{0\}\}$ and $pA^* = bA^* \cdot cA^*$. Certainly $pA \subseteq bA$ and $pA \subseteq cA$. Also $bc \in pA \Rightarrow b \in pA$ or $c \in pA$, i.e. $bA \subseteq pA$ or $cA \subseteq pA$. Thus $bA = pA$ or $cA = pA$. If $bA = pA$, then $pA^* = bA^* \cdot cA^* \Rightarrow pA = bA \cdot cA = pA \cdot cA$, so $p = pca$ for some a in A. Then $p(1 - ca) = 0 \Rightarrow ca = 1$, so $cA^* = A^*$. If $cA = pA$, then similarly $bA^* = A^*$.

(b) For $p \in \operatorname{Irr} A$ and $bc \in pA$, say $bc = pa$ or $bA^* \cdot cA^* = pA^* \cdot aA^*$, one has $bA^* \subseteq pA^* \cdot A$ or $cA^* \subseteq pA^* \cdot A$, i.e. $b \in pA$ or $c \in pA$. □

Let $K[T]$ be a polynomial algebra. Define $K[T] \to \mathbb{N};\ p(T) \mapsto |p(T)| =$ **if** $p(T) = 0$ **then** 0 **else** $2^{\deg p(T)}$. The following result shows that, for a field K, the polynomial ring $K[T]$ is a Euclidean domain with $p(T) \mapsto |p(T)|$ as length function. Recall that $K[T]$ is already an integral domain if its coefficient ring K is.

Proposition 3.2.3. *Let K be an integral domain. Let $d(T)$ in $K[T]$ have invertible leading coefficient d_m. Then $V = \{r(T) | \deg r(T) < m\}$ is a loop transversal to $d(T) \cdot K[T]$.*

Proof. By induction on the length $|f(T)|$ of the dividend $f(T)$, it will be shown that V contains a representative for the coset $f(T) + d(T) \cdot k[T]$. If $|f(T)| < 2^m$, then $f(T) = d(T) \cdot 0 + f(T) \in d(T)K[T] + V$. Suppose V already represents cosets of polynomials of length less than the positive integer $|f(T)| = 2^n \geq 2^m$. Then $f(T) - f_n d_m^{-1} \ T^{n-m} d(T) \in d(T)K[T] + V \Rightarrow f(T) \in d(T)K[T] + V$. Finally, to show that V contains a unique representative for $f(T) + d(T) \cdot K[T]$, suppose $f(T) = d(T)q_1(T) + r_1(T) = d(T)q_2(T) + r_2(T)$. Then $m > \deg[r_1(T) - r_2(T)] = \deg\{d(T)[q_2(T) - q_1(T)]\} = \deg d(T) + \deg[q_2(T) - q_1(T)] = m + \deg[q_2(T) - q_1(T)] \Rightarrow \deg[r_1(T) - r_2(T)] = \deg[q_2(T) - q_1(T)] = -\infty \Rightarrow q_1(T) = q_2(T)$ and $r_2(T) = r_1(T)$. □

For an element k of an integral domain K, consider the monic polynomial $(T - k)$. Proposition 3.2.3 shows that K is a loop transversal to $(T - k)K[T]$. Indeed, for any polynomial $f(T)$ in $K[T]$, one has $f(T) = (T - k)q(T) + f(k)$. Thus the principal ideal $(T - k)K[T]$ is the kernel of the ring homomorphism $k[T] \to K$; $f(T) \mapsto f(k)$ onto the integral domain K. This shows that $T - k$ is a prime element of $K[T]$.

The relationship between the various domain properties of Definition 3.2.1 is given by the following:

Theorem 3.2.4. (a) *Every Euclidean domain is a principal ideal domain.*

(b) *Every principal ideal domain is a unique factorization domain.*

Proof. (a) Let I be a non-zero ideal of a Euclidean domain A. The set $|I - \{0\}|$ of lengths of non-zero elements of I, as a subset of the positive integers, contains a least element $|d|$ for some fixed (but not necessarily unique) d in $I - \{0\}$. Certainly $dA \subseteq I$. Conversely, for n in I, one has a quotient $q \in A$ and a remainder $r \in A_d$ such that $n = dq + r$. But $r = n - dq \in I$, so $|r| < |d|$ implies $r = 0$. Thus $n = dq \in dA$, and $I \subseteq dA$.

(b) Let A be a principal ideal domain. Each maximal ideal of A is of the form pA for a prime element p of A. A mutually inverse pair μ, φ of monoid homomorphisms

$$(A - \{0\})/A^* \underset{\mu}{\overset{\varphi}{\rightleftarrows}} (\max A)^{*\kappa} \tag{3.2.2}$$

between $(A - \{0\})/A^*$ and the free commutative monoid $(\max A)^{*\kappa}$ of multisubsets of Max A will be exhibited: φ is called *factorization* and μ is called *multiplying out*. For each one-element multisubset $\langle pA \rangle$ of Max A, define $\langle pA \rangle \mu = pA^*$. Note that μ is well-defined by (3.2.1). Since (max $A)^{*\kappa}$ is the free commutative monoid on these one-element multisubsets, μ extends to a unique monoid homomorphism (also called μ) from (Max $A)^{*\kappa}$ to the commutative monoid $(A - \{0\})/A^*$. Injectivity of μ, namely the implication $(p_1 A \cdot p_2 A \ldots p_m A)^\kappa \mu = (q_1 A \cdot q_2 A \ldots q_n A)^\kappa \mu \Rightarrow (p_1 A \cdot p_2 A \ldots p_m A)^\kappa = (q_1 A \cdot q_2 A \ldots q_n A)^\kappa$, is proved by induction on $m + n$. The premise of the implication yields $q_1 q_2 \ldots q_n + p_1 A = p_1 A$ in the field $A/p_1 A$, whence $\exists 1 \leq i \leq n.\ q_i A = p_1 A$. Then $(p_2 A \ldots p_m A)^\kappa \mu = (\ldots q_{i-1} A \cdot q_{i+1} A \ldots)^\kappa \mu$ implies $(p_2 A \ldots p_m A)^\kappa = (\ldots q_{i-1} A \cdot q_{i+1} A \ldots)^\kappa$ by the induction hypothesis, whence $(p_1 A \cdot p_2 A \ldots p_m A)^\kappa = (q_1 A \ldots q_i A \ldots q_n A)^\kappa$, as required.

Now consider a non-identity element aA^* of $(A - \{0\})/A^*$. Since the ideal aA is proper, it is contained in a maximal ideal $p_1 A$, say $a = p_1 a_1$. If $a_1 A$ is not proper, one has $a_1 = p_2 a_2$ with $p_2 A \in \text{Max } A$ and $aA \subset a_1 A \subset a_2 A$. Continue thus. The process stops, since $aA \subset a_1 A \subset a_2 A \subset \cdots \subset a_n A \subset \cdots$ would imply $\bigcup_{i=1}^{\infty} a_i A = bA$ for an element b of an ideal $a_n A$, whence $a_n A \subseteq bA \subseteq a_n A$, a contradiction. Thus $a = p_1 p_2 \ldots p_m u$ for prime p_i and invertible u. Define $aA^* \varphi = (p_1 A \cdot p_2 A \ldots p_m A)^\kappa$. Then $aA^* \varphi \mu = aA^*$, so that μ surjects. Thus φ is the inverse of the monoid isomorphism μ. □

Example 3.2.5. The subring $\left\{ \left[\begin{smallmatrix} a & 23b \\ b & a \end{smallmatrix}\right] \middle| a, b \in \mathbb{Z} \right\}$ of $\mathbb{Z}_2^2$ is a principal ideal domain which is not Euclidean. [See W. Narkiewicz, *Elementary and Analytic Theory of Algebraic Numbers*, PWN, Warsaw, 1974 (1st ed.), 1990 (2nd ed.), Chapter 3, Section 4.] □

Example 3.2.6. If K is a field, then the polynomial ring $K[T]$ is a Euclidean domain, by Proposition 3.2.3. Theorem 3.2.4(a) then shows that $K[T]$ is a principal ideal domain. However, $K[T, U]$ is not a principal ideal domain. Consider the ideal $\langle T, U \rangle_{K[T,U]}$ of $K[T, U]$, the ring kernel of the homomorphism $K[T, U] \to K;\ p(T, U) \mapsto p(0, 0)$ of (3.7). If $\langle T, U \rangle_{K[T,U]} = f(T, U)K[T, U]$, then $T = f(T, U)g(T, U)$ and $U = f(T, U)h(T, U)$. It follows that $f(T, U)$ would have degree 1 both as an element of $(K[U])[T]$ and as an element of $(K[T])[U]$, so $f(T, U) = kT + lU$ with coefficients $k, l \in K$. But $T = (kT + lU)g(T, U)$ implies $l = 0$, and similarly $k = 0$, whence the contradiction $f(T, U) = 0$. Thus polynomial rings over principal ideal domains are not necessarily principal ideal domains. However, Theorem 3.2.8 below shows that polynomial rings over unique factorization domains are themselves unique factorization domains. In particular, $K[T, U]$ is a unique factorization domain which is not a principal ideal domain. □

A polynomial $p[T] = \Sigma_{i=0}^{\deg p} p_i T^i$ in the polynomial ring $K[T]$ over an integral domain K is said to be *primitive* if its set $\{p_i \mid 0 \leq i \leq \deg p\}$ of coefficients is not contained in any maximal ideal of K.

Proposition 3.2.7 (Gauss' Lemma). *The set X of primitive polynomials forms a submonoid of* $(K[T], \cdot, 1)$.

Proof. Suppose that the set of coefficients of the product of primitive polynomials $p(T)$ and $g(T)$ lies in a maximal ideal M of K. Consider the K-algebra $(K/M)[T]$. By (3.4), there is a unique K-algebra homomorphism $\pi: K[T] \to (K/M)[T]; T \mapsto T$. Then $(p(T)q(T))^\pi = 0$, but $p(T)^\pi \neq 0$ and $q(T)^\pi \neq 0$. This contradicts the fact that the polynomial ring $(K/M)[T]$ over the field K/M is an integral domain. □

For the rest of this section, let K be a unique factorization domain, i.e. $(K - \{0\})/K^* = (\text{Irr}\, K/K^*)^{*\kappa}$. Let X be the monoid of primitive polynomials of $K[T]$, and let $L = (K - \{0\})^{-1}K$ be the field of fractions of K. The polynomial ring $L[T]$ is a Euclidean domain, by Proposition 3.2.3, and thus a UFD, by Theorem 3.2.4. Recall (3.3): $K[T]^* = K^*$ and $L[T]^* = L^*$.

Theorem 3.2.8. *The polynomial ring $K[T]$ of a UFD K is itself a UFD. Moreover,*

$$\text{Irr}\, K[T] = \text{Irr}\, K \cup (X \cap \text{Irr}\, \text{L[T]}) \tag{3.2.3}$$

for $L = (K - \{0\})^{-1}K$ and X the set of primitive polynomials of $K[T]$.

Proof. Consider $K \subset K[T] \subseteq L[T]$. The elements of $\text{Irr}\, K$ have degree 0 in $K[T]$; since they have no non-trivial factorization in K, they can have none in $K[T]$. Suppose that, for an element $f(T)$ of $X \cap \text{Irr}\, L[T]$, one has $f(T)K^* = g(T)K^* \cdot h(T)K^*$ in $K[T]$. Then $f(T)L^* = g(T)L^* \cdot h(T)L^*$ implies (without loss of generality) $g(T)L^* = L^*$, by the irreducibility of $f(T)$ in $L[T]$, so $g(T) \in K - \{0\}$. Since $f(T)$ is primitive, one then has $g(T)K^* = K^*$, as required. Thus $\text{Irr}\, K \cup (X \cap \text{Irr}\, L[T]) \subseteq K[T]$.

The set X is a K^*-set (X, K^*). Consider the construction of the free L^*-set $(X, K^*)\uparrow_{K^*}^{L^*} = ((X \times L^*)/K^*, L^*)$ over (X, K^*), as in Section I 2.3. Elements $(f(T), k_1/k_2)K^*$ of $(X \times L^*)/K^*$ may be identified with elements $(k_1/k_2)f(T)$ of $L[T]$, and each non-zero element of $L[T]$ appears in this form. The bijection $(X, K^*)/K^* \to ((X \times L^*)/K^*)/L^* = (L[T] - \{0\})/L^* = (L[T] - \{0\})/(L[T])^*$ of Exercise I 2.3F becomes a monoid isomorphism. Since $L[T]$ is a UFD, it follows that X/K^* is a free commutative monoid and $(X - K^*)/K^*$ is a free commutative semigroup on the set $(X \cap \text{Irr}\, L[T])/K^*$. Also $(K - K^*)/K^*$ is a free commutative semigroup on the set $(\text{Irr}\, K)/K^*$, since K is a UFD. Thus $[\text{Irr}\, K \cup (X \cap \text{Irr}\, L[T])]/K^*$ generates a free commutative subsemigroup of $(K[T]/K^*, \cdot)$—note that products of elements of $\text{Irr}\, K$ all have degree 0, while products of elements of $X \cap \text{Irr}\, L[T]$ all have bigger degree. Finally, the subsemigroup is in fact all of $K[T] - K^*$ [and in particular, equality holds in (3.2.3)]. Indeed, consider an element $f(T) = k_1 \cdot f^1(T)$ of $K[T]$. If the coefficients

$\{f_0^1, \ldots, f_{\deg f^1}^1\}$ have a common prime factor p_2, set $k_2 = k_1 p_2$ and $f(T) = k_2 f^2(T)$. Continuing thus, one eventually obtains $f(T)$ in the form $f(T) = k_r f^r(T)$ with $k_r \in K$ and $f^r(T) \in X$. □

Corollary 3.2.9. *For a UFD K, each multivariate polynomial ring $K[T_1, \ldots, T_n]$ is a UFD.*

Proof. By induction on n, using Theorem 3.2.8. □

EXERCISES

3.2A. Prove that association is a congruence relation on the monoid $(A, \cdot, 1)$, for a commutative unital ring A.

3.2B. Determine the loop structure (I 4.3.4) given on the loop transversal $\mathbb{Z}_d$ to $d\mathbb{Z}$ in $(\mathbb{Z}, +, 0)$.

3.2C. For an odd integer d, show that $\{r \in \mathbb{Z} \| r| < |d/2|\}$ is a loop transversal to $d\mathbb{Z}$ in $(\mathbb{Z}, +, 0)$.

3.2D. Determine the quotient and remainder when the dividend $2T^5 - 9T^3 + 9T - 2$ is divided in $\mathbb{Z}[T]$ by the divisor $T^3 - 2T + 1$.

3.2E. Let K be an integral domain.

(a) Prove, by induction on the degree of a non-zero polynomial $f(T)$ in $K[T]$, that $f(T)$ has at most $\deg f(T)$ roots in K.

(b) If K is infinite, prove that the ring homomorphisms (3.5) and (3.8) are injective.

(c) Let K be a field. Can you prove the following:
  (i) K infinite $\Leftrightarrow$ (3.5) injects;
  (ii) K finite $\Leftrightarrow$ (3.5) surjects?

3.2F. Show that the ring A of Exercise 3.1D is a Euclidean domain, with the determinant as length function.

3.2G. Let a be a non-zero element of a principal ideal domain A. For each maximal ideal P of A, let e_P be the multiplicity of P in the factorization $aA^*\varphi$ of a. Prove $|\mathrm{Id}\,(A/aA)| = \prod_{P \in \mathrm{Max}\, A}(1 + e_P)$.

3.2H. Write a short essay discussing the use of Zorn's Lemma in the proof of Theorem 3.2.4(b). (Hints: Can you prove the result without Zorn's Lemma? Can you factorize any 1000-digit integer?)

3.2I. In a principal ideal domain A, show that each prime ideal is maximal.

3.2J. In the unique factorization domain $\mathbb{Z}$, show that 12 has 12 factorizations as a product $p_1 p_2 p_3$ of prime elements.

3.2K. Let m be a positive integer. Set

$$\mathbb{Z}[\sqrt{-m}] = \left\{ \begin{bmatrix} a & -mb \\ b & a \end{bmatrix} \middle| a, b \in \mathbb{Z} \right\}.$$

(a) Show that $\mathbb{Z}[\sqrt{-m}]$ is a subring of $\mathbb{Z}_2^2$.

(b) Show that

$$J : \mathbb{Z}[\sqrt{-m}] \to \mathbb{C}_2^2; \begin{bmatrix} a & -mb \\ b & a \end{bmatrix} \mapsto \begin{bmatrix} a & mib \\ ib & a \end{bmatrix}$$

is a unital ring homomorphism.

(c) Show that $N : (\mathbb{Z}[\sqrt{-m}], \cdot, I) \to (\mathbb{N}, \cdot, 1); f \mapsto \det f^J$ is a monoid homomorphism.

(d) Show that $\mathbb{Z}[\sqrt{-m}]$ is an integral domain.

(e) For $m = 2$, show that $\mathbb{Z}[\sqrt{-m}$ is a Euclidean domain.

(f) For $m = 5$, show that $\mathbb{Z}[\sqrt{-m}]$ is not a unique factorization domain.

3.2L. Show that the polynomial ring $\mathbb{Z}[T]$ is a UFD, but not a PID.

3.2M. Express the polynomial $12T_1^2 + 6T_1T_2 - 36T_2^2$ as a product of primes in $\mathbb{Z}[T_1, T_2]$.

3.3. Modules over Principal Ideal Domains

A module V over a ring S is said to be *finitely generated* if it is spanned by a finite multisubset. The main result of this section, Theorem 3.3.4, identifies the isomorphism classes of finitely generated modules over a PID A. Specifically, for a positive integer m, an $(m-1)$*-simplex* in a poset $(X, \leq)$ is an m-element multisubset $\langle x_1, \ldots, x_m \rangle$ of X with $x_1 \leq x_2 \leq \ldots \leq x_m$. The set of $(m-1)$-simplices in $(X, \leq)$ is denoted by $(X, \leq)_{m-1}$. Then the isomorphism classes of A-modules spanned by m-element multisubsets correspond to $(m-1)$-simplices in $(\text{Id } A, \supseteq)$. For Euclidean domains A, an algorithm for determining the simplex of a module is given: The algorithm essentially consists of Gaussian elimination. Taking A as the ring $\mathbb{Z}$ of integers, the algorithm classifies finitely generated abelian groups. In the following section, the algorithm classifies finite-dimensional linear dynamical systems by taking A as the polynomial ring $K[T]$ over a field K.

The special property of principal ideal domains that facilitates the classification of finitely generated modules is given by the following characterization.

Theorem 3.3.1. *A commutative, unital ring S is a principal ideal domain iff each submodule of a free S-module of finite rank is a free module of equal or lesser rank.*

Proof. First assume that submodules of finitely generated free modules are free modules of equal or lesser rank. Since $\{0\} = \{0\}^{274}$ is a free $\{0\}$-submodule of the free $\{0\}$-module $\{0\}^{97}$, the ring S is not the zero ring $\{0\}$. Moreover, a zero divisor s of S would yield an S-submodule sS of S_S that was not free [cf. Example 2.3.3 (a)], so S is an integral domain. Finally, a non-zero ideal I of S, as a submodule I_S of the free S-module S_S of rank 1, is a free module of rank 1, and thus a principal ideal.

Conversely, suppose that S is a principal ideal domain. Then submodules of free modules of rank 1 are free modules of rank 1 or 0. Fix a positive integer n. Suppose, by induction, that submodules of free modules of rank $r < n$ are free of rank at most r. Let U be a submodule of S^n. Consider the projection $p : S^n \to S;\ (s_1, \dots, s_n) \mapsto s_n$, with kernel $\operatorname{Ker} p = S^{n-1}$. If $U \leq \operatorname{Ker} p$, then U is free of rank at most $n - 1$, by the induction hypothesis. Otherwise, Up is a rank 1 free submodule of S, again by the induction hypothesis. Also $\operatorname{Ker}(p|_U)$, as a submodule of the free rank $(n - 1)$ module $\operatorname{Ker} p$, is free of rank at most $n - 1$. Since Up is free, the exact sequence $0 \to \operatorname{Ker}(p|_U) \longrightarrow U \xrightarrow{p} Up \longrightarrow 0$ splits, and then U, being isomorphic to the direct sum $\operatorname{Ker}(p|_U) \oplus Up$, is free of rank at most $(n - 1) + 1 = n$. □

Let U_S and V_S be right modules over a ring S, with corresponding groups $\operatorname{Aut} U_S$ and $\operatorname{Aut} V_S$ of automorphisms (invertible module endomorphisms) under composition. Thus U is a right $(\operatorname{Aut} U)$-set and V is a right $(\operatorname{Aut} V)$-set. Now consider the set $S(U, V)$ of S-homomorphisms from U to V. This set is a right $(\operatorname{Aut} U)$-set under the action $\theta : f \mapsto \theta^{-1} f$ and a right $(\operatorname{Aut} V)$-set under the action $\varphi : f \mapsto f\varphi$. Thus it is a right $(\operatorname{Aut} U) \times (\operatorname{Aut} V)$-set under the action $(\theta, \varphi) : f \mapsto \theta^{-1} f \varphi$. The $(\operatorname{Aut} U) \times (\operatorname{Aut} V)$-orbits on $S(U, V)$ are known as *equivalence* classes: Two S-homomorphisms f_1, f_2 are *equivalent* if $\exists \theta \in \operatorname{Aut} U.\ \exists \varphi \in \operatorname{Aut} V.\ f_1 = \theta^{-1} f_2 \varphi$. From now on, fix a principal ideal domain A. The first goal is to show that $(\operatorname{Id} A, \supseteq)_{n-1}$, for $n = \min\{l, m\}$, classifies the equivalence classes of A-homomorphisms from a free A-module U of rank l to a free A-module V of rank m, for positive integers l and m. This will be done by induction on l and m. The base case $l = m = 1$ is immediate from (3.2.1) and the observation $\operatorname{Aut} A_A = \{A \to A;\ 1 \mapsto u | u \in A^*\}$: The equivalence class of the typical A-homomorphism $L(d) : A \to A;\ 1 \mapsto d$ then corresponds to the 0-simplex $\langle dA \rangle$ in $(\operatorname{Id}\ A, \supseteq)$. Note that an $(n - 1)$-simplex $d_1 A \supseteq d_2 A \supseteq \cdots \supseteq d_n A$ determines the (well-defined) equivalence class of the A-homomorphism

$$\left(\sum_{i=1}^{n} L(d_i)\right) \oplus 0 : A^l = \left(\bigoplus_{i=1}^{n} A\right) \oplus \left(\bigoplus_{j=1}^{l-n} A\right) \to \left(\bigoplus_{i=1}^{n} A\right)$$

$$= \left(\bigoplus_{i=1}^{n} A\right) \oplus 0 \leq \left(\bigoplus_{i=1}^{n} A\right) \oplus \left(\bigoplus_{j=1}^{m-n} A\right) = A^m.$$

Proposition 3.3.2. *Let U be a module and let V be a free module over the principal ideal domain A. Then*:

$$\forall 0 \neq f \in A(U, V),\ \exists d_1 \in A.\ \exists f_1 \in A(U, V).\ \exists \theta_1 \in V'.$$

$$f = L(d_1) f_1 \text{ and } \operatorname{Im}(f_1 \theta_1) = A.$$

Proof. Recall the factorization φ of (3.2.2). Consider the union $\bigcup_{\theta \in V'} \mathrm{Im}(f\theta)$ of the images of composites of f with functionals θ on V. This union is a subset of A. Pick an element d_1 of $[\bigcup_{\theta \in V'} \mathrm{Im}(f\theta)] - \{0\}$ such that $d_1 A^* \varphi$ in $(\mathrm{Max}\, A)^{*\kappa}$ has minimal length as a (commutative) word over the alphabet Max A. Suppose $d_1 = u_1 f \theta_1$ for $u_1 \in U$ and $\theta_1 \in V'$. Then

$$(3.3.1) \qquad \forall \theta_2 \in V',\ u_1 f \theta_2 A \leq d_1 A.$$

Indeed, suppose $dA = u_1 f\theta_1 A + u_1 f \theta_2 A$, say $d = u_1 f \theta_1 a_1 + u_1 f \theta_2 a_2 = u_1 f(\theta_1 a_1 + \theta_2 a_2) \in \bigcup_{\theta \in V'} \mathrm{Im}(f\theta)$. Then $d_1 \in dA$ implies that $dA^*\varphi$ is a multisubset of $d_1 A^* \varphi$, whence $dA^*\varphi = d_1 A^* \varphi$ by the minimality of $d_1 A^* \varphi$. Thus $dA^* = dA^* \varphi\mu = d_1 A^* \varphi \mu = d_1 A^*$, and then $dA = d_1 A$ by (3.2.1). The containment (3.3.1) follows. Similarly,

$$(3.3.2) \qquad \forall u_2 \in U,\ u_2 f \theta_1 A \leq d_1 A.$$

Indeed, suppose $cA = u_1 f \theta_1 A + u_2 f \theta_1 A$, say $c = u_1 f \theta_1 a_1 + u_2 f \theta_1 a_2 = (u_1 a_1 + u_2 a_2) f \theta_1 \in \bigcup_{\theta \in V'} \mathrm{Im}(f\theta)$. Then $cA = d_1 A$ by the argument yielding (3.3.1). Finally,

$$(3.3.3) \qquad \forall u_2 \in U, \forall \theta_2 \in V',\ u_2 f \theta_2 A \leq d_1 A.$$

Certainly $\exists b \in A.\ u_1 f \theta_2 = u_1 f \theta_1 b$, by (3.3.1). Consider $\theta_1' = \theta_2 - \theta_1 b + \theta_1$. Then $d_1 = u_1 f \theta_1 = u_1 f \theta_1'$, and $u_2 f \theta_2 = u_2(b-1) f \theta_1 + u_2 f \theta_1'$. By (3.3.2), one has $u_2(b-1) f\theta_1 A \leq d_1 A$. By (3.3.2) with θ_1' replacing θ_1, one has $u_2 f \theta_1' A \leq d_1 A$. The containment (3.3.3) follows. Let $\beta : I \to V$ be a basis for V. For each element i of I, the isomorphism (2.3.7) yields a functional

$$\beta_i : V \to A;\ v \mapsto i\left(vR(\beta)^{-1}\right)$$

selecting the i-th coefficient of a module element v with respect to the basis β. By (3.3.3), one may then define $f_1 \in A(U, V)$ such that, for $u \in U$, one has $uf\beta_i = d_1(uf_1 \beta_i)$ for $i \in I$. Thus $f = L(d_1) f_1$. Also $d_1 = u_1 f \theta_1 = u_1 d_1 f_1 \theta_1 = d_1(u_1 f_1 \theta_1)$ implies $u_1 f_1 \theta_1 \in A^*$, so $\mathrm{Im}(f_1 \theta_1) = A$. □

Proposition 3.3.2 is used to complete the inductive proof by showing that each A-homomorphism $f : U \to V$ from the free module U of rank l to the free module V of rank m is equivalent to $\Sigma_{i=1}^{n} L(d_i)$ for some $(n-1)$-simplex $d_1 A \supseteq d_2 A \supseteq \cdots \supseteq d_n A$. If $f = 0$, then $d_1 A = \cdots = d_n A = \{0\}$. Otherwise, Proposition 3.3.2 yields $u_1 \in U, f_1 \in A(U, V)$ and $\theta_1 \in V'$ such that $L(u_1) f_1 \theta_1 = 1_A$:

$$(3.3.4) \qquad \begin{array}{ccc} U & \xrightarrow{f_1} & V \\ {\scriptstyle L(u_1)}\uparrow & & \downarrow{\scriptstyle \theta_1} \\ A & = & A \end{array} .$$

Note that the exact sequences $0 \to \operatorname{Ker} f_1\theta_1 \longrightarrow U \xrightarrow{f_1\theta_1} A \longrightarrow 0$ and $0 \longrightarrow \operatorname{Ker}\theta_1 \longrightarrow V \xrightarrow{\theta_1} A \longrightarrow 0$ split, namely $U = AL(u_1) \oplus \operatorname{Ker} f_1\theta_1$ and $V = AL(u_1)f_1 \oplus \operatorname{Ker}\theta_1$. Moreover, f_1 restricts to an A-homomorphism $f_1' : \operatorname{Ker} f_1\theta_1 \to \operatorname{Ker}\theta_1$. By induction, there is an $(n-2)$-simplex $e_2A \supseteq e_3A \supseteq \cdots \supseteq e_nA$ such that f_1' is equivalent to $\Sigma_{i=2}^n L(e_i)$. Also, f_1 restricts to an isomorphism $f_1'' : AL(u_1) \to AL(u_1)f_1$. Thus f_1 is equivalent to $L(1) \oplus \Sigma_{i=2}^n L(e_i): U \to V$, whence $f = L(d_1)f_1$ is equivalent to $\Sigma_{i=1}^n L(d_i)$ with $d_i = d_1e_i$ for $2 \le i \le n$ and with $d_1A \supseteq d_2A \supseteq \cdots \supseteq d_nA$. Summarizing:

Theorem 3.3.3. *Let A be a principal ideal domain. Let U and V be finitely generated non-zero free modules over A, of respective ranks l and m. Then for $n = \min\{l, m\}$, the map*

$$(\operatorname{Id} A, \supseteq)_{n-1} \to A(U,V)/(\operatorname{Aut} U \times \operatorname{Aut} V);$$

$$(d_1A \supseteq \cdots \supseteq d_nA) \mapsto \left(\sum_{i=1}^{n} L(d_i)\right)(\operatorname{Aut} U \times \operatorname{Aut} V)$$

bijects. □

Now let W be a non-zero finitely generated module over the principal ideal domain A. Suppose that W is spanned by $\beta : \mathbb{Z}_m \to W$, so the A-homomorphism $R(\beta): A^m \to W$ [an instance of (2.3.2)] surjects. By Theorem 3.3.1, $\operatorname{Ker} R(\beta)$ is a free A-module of rank l, say, with $0 \le l \le m$. If $l = 0$, then β is a basis for W, which becomes free of rank m. In this case $W \cong A^m \cong \oplus_{i=1}^n (A/d_iA)$ with $d_1A = d_2A = \cdots = d_mA = \{0\}$. Otherwise, $\operatorname{Ker} R(\beta)$ is a non-zero free submodule U of the free module $V = A^m$, and (2.3.2) yields the exact sequence

$$0 \longrightarrow U \xrightarrow{f} V \xrightarrow{R(\beta)} W \longrightarrow 0. \tag{3.3.5}$$

In general, a *free resolution* of a module W is an exact sequence $\cdots \to A_n \to \cdots \to A_2 \to A_1 \to W \to 0$ of modules with $A_1, A_2, \ldots, A_n, \cdots$ all free. By the First Isomorphism Theorem applied to $R(\beta)$ in (3.3.5), one has $W \cong V/\operatorname{Ker} R(\beta) \cong V/Uf$, so the isomorphism class of W only depends on the equivalence class of f in (3.3.5). By Theorem 3.3.3, f is equivalent to $\Sigma_{i=1}^l L(d_i)$ for an $(l-1)$-simplex $d_1A \supseteq \cdots \supseteq d_lA$ in $(\operatorname{Id} A, \supseteq)$. Thus $W \cong A^m/(d_1A \oplus \cdots \oplus d_lA) \cong \left(\oplus_{i=1}^l (A/d_iA)\right) \oplus A^{m-l}$. Suppose $d_{k+1}A = d_{k+2}A = \cdots = d_lA = \{0\}$ for some k with $1 \le k \le l$ (by convention $k = l$ means $d_lA \supset \{0\}$). Then $W \cong \oplus_{i=1}^k (A/d_iA) \oplus A^{m-k}$. If $d_1A = A$, then $W \cong \oplus_{i=2}^k (A/d_iA) \oplus A^{m-k}$ could have been spanned by some $\beta' = \mathbb{Z}_{m-1} \to W$. One thus obtains the following classification of isomorphism classes of finitely generated A-modules.

Theorem 3.3.4. *Let A be a principal ideal domain. Consider the poset $(\operatorname{Prop} A, \supseteq)$ of proper ideals of A. Then finitely generated non-zero A-modules*

W are classified up to isomorphism by the set $\bigcup_{n\in\mathbb{N}}$ (Prop A, $\supseteq$)$_n$ *of simplices in* (Prop A, $\supseteq$), *namely*

$$W \cong \bigoplus_{i=1}^{m} (A/d_iA) \tag{3.3.6}$$

for $A \supset d_1A \supseteq d_2A \supseteq \cdots \supseteq d_mA$. $\square$

In the context of (3.3.6), suppose $d_kA \supset d_{k+1}A = 0$ for some $k \in \{0, 1, \ldots, m\}$. Then the natural number $m - k$ is called the *rank* of W. The ideals $d_1A, \ldots, d_mA$ are called the *invariant factors* of W. A module of rank 0, and the zero module itself, are called *torsion modules*.

Suppose that d is a non-zero, non-unit element of A, such that dA^* has the factorization $dA^*\varphi = ((p_1A)^{e_1} \ldots (p_rA)^{e_r})^\kappa \in (\text{Max } A)^{*\kappa}$ with $|\{p_1A, \ldots, p_rA\}| = r$. Then the A-homomorphism $A/dA \to \bigoplus_{i=1}^r (A/p_i^{e_i}A)$; $a + dA \mapsto (a + p_1^{e_1}A, \ldots, a + p_r^{e_r}A)$ bijects. For a non-zero torsion A-module W with $(m-1)$-simplex $d_1A \supseteq d_2A \supseteq \cdots \supseteq d_mA$, suppose $d_mA^*\varphi = ((p_1A)^{e_{1m}} \ldots (p_rA)^{e_{rm}})^\kappa$ with $|\{p_1A, \ldots, p_rA\}| = r$. Then for $1 \le j \le m$, one has $d_jA^*\varphi = ((p_1A)^{e_{1j}} \ldots (p_rA)^{e_{rj}})^\kappa$ with $0 \le e_{i1} \le \cdots \le e_{ij} \le \cdots \le e_{im}$ for $1 \le i \le r$, and

$$W \cong \bigoplus_{i=1}^{r} \bigoplus_{j=1}^{m} (A/p_i^{e_{ij}}A). \tag{3.3.7}$$

The decomposition (3.3.7) is called the *primary decomposition* of the torsion module W. The proper ideals $p_i^{e_{ij}}A$ in (3.3.7), i.e. with $e_{ij} > 0$, are called the *elementary divisors* of W. If $r = 1$, then W is said to be a *primary module*, or more specifically a p_1A-primary module. If $A = \mathbb{Z}$, so that W is a finite abelian group, then the p_iA-primary modules $\bigoplus_{j=1}^m (A/p_i^{e_{ij}}A)$ in (3.3.7) are the Sylov subgroups of W. More generally, abelian groups of prime-power order are primary $\mathbb{Z}$-modules.

For the rest of the section, assume that A is a Euclidean domain. Given an element $[a\ b]$ of $A \times (A - \{0\})$, a generator d of the principal ideal $dA = aA + bA$ may be found by the *Euclidean Algorithm*

begin $[n_0\ d_0] := [a\, b]$, $i := 0$

  while $n_i \varepsilon_{d_i} \neq 0$ **do**

  $$[n_{i+1}\ d_{i+1}] := [n_i\ d_i] \begin{bmatrix} 1 & 0 \\ -n_i\delta_{d_i}L(d_i)^{-1} & 1 \end{bmatrix} \begin{bmatrix} 0 & 1 \\ 1 & 0 \end{bmatrix}$$

  $i := i + 1$

  end while

output $d := d_i$

described using the loop transversal code terminology introduced after Definition 3.2.1. Note that $n_{i+1} = d_i$ and $d_{i+1} = n_i - d_i(n_i\delta_{d_i}L(d_i)^{-1})$, so $n_{i+1}A + d_{i+1}A \leq n_iA + d_iA$. Conversely, $n_i = n_{i+1}(n_i\delta_{d_i}L(d_i)^{-1}) + d_{i+1}$ and $d_i = n_{i+1}$, so $n_iA + d_iA \leq n_{i+1}A + d_{i+1}A$. If $n_i\varepsilon_{d_i} = 0$, then $n_iA + d_iA = d_iA$. Also $aA + bA = n_0A + d_0A$. Thus $aA + bA = dA$. The algorithm terminates in at most $|b|$ steps, since $d_{i+1} \in A_{d_i}$ implies $0 < |d| < \cdots < |d_{i+1}| < |d_i| < \cdots < |d_0| = |b|$.

Given an element f of A_l^m or $A(A^l, A^m)$ with $n = \min\{l, m\}$, the Euclidean Algorithm combines with Gaussian elimination to determine the $(n-1)$-simplex $d_1A \supseteq \cdots \supseteq d_nA$ such that f is equivalent to $\Sigma_{i=1}^n L(d_i)$. Note that column operations are part of the action of Aut A^m on $A(A^l, A^m)$. Similarly, one obtains *row operations* as part of the action of Aut A^l on $A(A^l, A^m)$ by premultiplying by permutation, rescaling, and shear matrices. Now the 2×2 matrices in the **while** loop of the Euclidean Algorithm are a shear matrix and a permutation matrix respectively. Using the Euclidean Algorithm, one may apply column operations to reduce f to an equivalent matrix f' with zero off-diagonal elements in the first row, such that $f'_{11}A = f_{11}A + \cdots + f_{1m}A$. Similarly, one may then apply row operations to reduce f' to an equivalent matrix f'' with zero off-diagonal elements in the first row and first column, such that $f''_{11}A = f_{11}A + \cdots + f_{1m}A + f_{21}A + \cdots + f_{l1}A$. If $\forall 1 \leq i \leq l, \forall 1 \leq j \leq m, f''_{ij} \in f''_{11}A$, one may then take $d_1 = f''_{11}$. Otherwise, if $f''_{ab} \notin f''_{11}A$, one may use a shear to add the b-th row of f'' to the first row, and then reiterate the use of the Euclidean Algorithm. Eventually f is reduced to an equivalent matrix $R(d_1) + g$ with g in $A(A^{l-1}, A^{m-1})$, and such that $\forall 1 \leq i \leq l-1, \forall 1 \leq j \leq m-1, g_{ij} \in d_1A$. By induction, g is equivalent to $\Sigma_{i=2}^n L(e_i)$ for an $(n-1)$-simplex $e_2A \supseteq \cdots \supseteq e_nA$. Setting $d_i = d_1e_i$ for $2 \leq i \leq n$, one thus obtains f as equivalent to $\Sigma_{i=1}^n L(d_i)$ for the $(n-1)$-simplex $d_1A \supseteq d_2A \supseteq \cdots \supseteq d_nA$. This procedure determines the simplex of a non-zero, finitely generated module W by operating on the homomorphism f in a corresponding free resolution (3.3.5).

Example 3.3.5. Consider

$$\begin{bmatrix} 2 & 8 & -4 & 4 \\ -2 & -2 & 4 & -4 \\ 6 & 24 & -12 & 12 \end{bmatrix} \in \mathbb{Z}_3^4.$$

It is equivalent to

$$\begin{bmatrix} 2 & 8 & -4 & 4 \\ -2 & -2 & 4 & -4 \\ 6 & 24 & -12 & 12 \end{bmatrix} \begin{bmatrix} 1 & -4 & 2 & -2 \\ 0 & 1 & 0 & 0 \\ 0 & 0 & 1 & 0 \\ 0 & 0 & 0 & 1 \end{bmatrix}$$

$$= \begin{bmatrix} 2 & 0 & 0 & 0 \\ -2 & 6 & 0 & 0 \\ 6 & 0 & 0 & 0 \end{bmatrix},$$

and then to

$$\begin{bmatrix} 1 & 0 & 0 \\ 1 & 1 & 0 \\ -3 & 0 & 1 \end{bmatrix} \begin{bmatrix} 2 & 0 & 0 & 0 \\ -2 & 6 & 0 & 0 \\ 6 & 0 & 0 & 0 \end{bmatrix} = \begin{bmatrix} 2 & 0 & 0 & 0 \\ 0 & 6 & 0 & 0 \\ 0 & 0 & 0 & 0 \end{bmatrix}. \quad \square$$

EXERCISES

3.3A. Show that the submodule $2\mathbb{Z}[T] + T\mathbb{Z}[T]$ of the free $\mathbb{Z}[T]$-module $\mathbb{Z}[T]_{\mathbb{Z}[T]}$ is not free.

3.3B. In Proposition 3.3.2, suppose that $\beta : \mathbb{Z}_l \to U$ and $\gamma : \mathbb{Z}_m \to V$ are bases, so that $R(\beta)fR(\gamma)^{-1} \in A(A^l, A^m) = A_l^m$. Show that each entry of the matrix $R(\beta)fR(\gamma)^{-1}$ lies in the ideal d_1A of A.

3.3C. Let U be a submodule of a (not necessarily finitely generated) free module V over the principal ideal domain A. Let $\beta : I \to V$ be a basis for V. Let $X = \{(J, S) | J \supseteq I, S \subseteq U, \ S$ a basis for $U \cap \operatorname{Im} R(\beta|_J)\}$. Define a partial order $\leq$ on X by $(J_1, S_1) \leq (J_2, S_2) \Leftrightarrow J_1 \supseteq J_2$ and $S_1 \subseteq S_2$.

(a) Show that $(X, \leq)$ is an inductive poset.

(b) Use Zorn's Lemma to obtain a maximal element (J, S) of $(X, \leq)$. If $\exists i \in I - J$, use the embedding $(U \cap \operatorname{Im} R(\beta|_{J \cup \{i\}}))/(U \cap \operatorname{Im} R(\beta|_J)) \to \operatorname{Im} R(\beta|_{J \cup \{i\}})/\operatorname{Im} R(\beta|_J) \cong A$ to obtain a contradiction to the maximality of (J, S).

(c) Deduce that every submodule of a free A-module is free.

3.3D. (a) Let W be a module over the principal ideal domain A. An element x of W is a *torsion element* if the annihilator $\operatorname{An}_A xA$ is non-zero. Prove that the set of torsion elements of W is a submodule $\operatorname{Tor} W$ of W.

(b) Let W be a non-zero, non-free, finitely generated A-module, with simplex $d_1A \supseteq \cdots \supseteq d_mA$, where $d_kA \supset d_{k+1}A = 0$ for some $k \in \{1, \ldots, m\}$. Prove that $\operatorname{Tor} W$ is a non-zero, non-free, finitely generated A-module, with simplex $d_1A \supseteq \cdots \supseteq d_kA$.

3.3E. Determine the invariant factors of the finite abelian group with multiset $\langle 4\mathbb{Z}, 4\mathbb{Z}, 25\mathbb{Z}, 27\mathbb{Z}, 27\mathbb{Z}, 29\mathbb{Z} \rangle$ of elementary divisors.

3.3F. (a) Show that each finite abelian group is the direct product of its Sylov subgroups.

(b) Show that

$$H = \left\{ \begin{bmatrix} 1 & a & b \\ 0 & 1 & c \\ 0 & 0 & 1 \end{bmatrix} \middle| \, a, b, c \in \mathbb{Z}_6 \right\}$$

is a non-abelian subgroup of $\mathrm{SL}_3(\mathbb{Z}_6)$ (cf. Exercise 2.4L). Show that H is the direct product of its Sylov subgroups.

(c) Exhibit a finite group that is not the direct product of its Sylov subgroups.

3.3G. Prove the **Chinese Remainder Theorem**: Let A be a principal ideal domain. Let $p_1, \ldots, p_r$ be distinct prime elements of A. Then for natural numbers $e_1, \ldots, e_r$ and for $d = p_1^{e_1} \ldots p_r^{e_r}$, there is an A-algebra isomorphism

$$A/dA \cong \prod_{i=1}^{r} A/p_i^{e_i}A$$

(cf. Exercise O 4.4J).

3.3H. Determine the primary decompositions of the groups of units of the rings $\mathbb{Z}_{21}$ and $\mathbb{Z}_{105}$.

3.3I. For the abelian group $\mathbb{Z}^4/([1\ -1\ 1\ -1]\mathbb{Z} + [0\ 0\ 2\ -2]\mathbb{Z})$, determine the invariant factors.

3.3J. Show that there are 7 isomorphism classes of abelian groups of order 32. (Hint: Cf. Exercise I 1.5G.)

3.3K. Exhibit a free subgroup of $\prod_{n \in \mathbb{N}}(\mathbb{Z}/2^n\mathbb{Z})$.

3.3L. Are the groups $\mathbb{Z}_{50} \oplus \mathbb{Z}_{80}$ and $\mathbb{Z}_{40} \oplus \mathbb{Z}_{100}$ isomorphic?

3.3M. (a) Set $B_0 = \prod_{n \in \mathbb{N}}(\mathbb{Z}/4\mathbb{Z})$, $B_1 = \prod_{n \in \mathbb{N}}((\mathbb{Z}/4\mathbb{Z}) \oplus (\mathbb{Z}/2\mathbb{Z}))$, and $B_2 = \prod_{n \in \mathbb{N}}(\mathbb{Z}/2\mathbb{Z})$. Show that $B_0 \times B_1 \cong B_0 \times B_2$, but $B_1 \not\cong B_2$.

(b) If B_0, B_1, B_2 are finitely generated abelian groups, prove $B_0 \times B_1 \cong B_0 \times B_2 \Rightarrow B_1 \cong B_2$.

3.4. Linear Dynamical Systems

According to Section O 4.2, a dynamical system is a set X with a function $T: X \to X$, the evolution operator. In other words, X is a $\langle T \rangle$-set for the free monoid $\langle T \rangle$ on the single generator T. Analogously, a linear dynamical system (U, t) is an S-module U together with an S-homomorphism $t: U_S \to U_S$, for a ring S. The evolution operator t is an element of the ring $\mathrm{End}_S\, U$ of S-endomorphisms of the S-module U. Two such endomorphism t_1, t_2 are said to be *similar* if there is an action morphism $(\theta, t_1 \mapsto t_2)$ from the $\langle t_1 \rangle$-set $(U, \langle t_1 \rangle)$ to the $\langle t_2 \rangle$-set $(U, \langle t_2 \rangle)$, with the special property that the isomorphism $\theta: U \to U$ is an S-homomorphism. Thus

$$\begin{array}{ccc} U & \xrightarrow{\theta} & U \\ {\scriptstyle t_1}\downarrow & & \downarrow{\scriptstyle t_2} \\ U & \xrightarrow[\theta]{} & U \end{array} :$$

$\theta t_2 = t_1\theta$ in $\mathrm{End}_S U$. Alternately, one may consider the action $\theta : t \mapsto \theta^{-1}t\theta$ of $\mathrm{Aut}\, U_S$ on $\mathrm{End}\, U_S$. Then t_1 and t_2 are similar iff they lie in the same (Aut U)-orbit. The fundamental problem in the study of linear dynamical systems is to classify their similarity classes. The ideal is to have an algorithm deciding whether or not two given linear dynamical systems are similar.

For a commutative unital ring K, consider a linear dynamical system (V, t) in which V is a unital K-module via the representation $K \to \mathrm{End}\, V_{\mathbb{Z}}$. This representation makes $\mathrm{End}\, V_K$ a K-algebra. The universality property (3.4) for the polynomial algebra $K[T]$ then yields a unique K-algebra homomorphism $K[T] \to \mathrm{End}\, V_{\mathbb{Z}};\ T \mapsto t$ representing V as a $K[T]$-module. Note that two K-endomorphisms t_1, t_2 of V are similar iff (V, t_1) and (V, t_2) yield isomorphic $K[T]$-modules. If V is a finitely generated K-module, then it may be expressed as a cokernel of $K[T]$-modules.

Theorem 3.4.1. *Let V be a finitely generated free module over the commutative ring K, with basis $\beta : \mathbb{Z}_n \to V$. Let V be a $K[T]$-module via $K[T] \to \mathrm{End}\, V_K$; $T \mapsto t$. Then there is an exact sequence*

$$K[T]^n \xrightarrow{T - R(\beta)tR(\beta)^{-1}} K[T]^n \xrightarrow{R(\beta)} V \longrightarrow 0. \tag{3.4.1}$$

Proof. As a free K-module with basis $\beta : \mathbb{Z}_n \to V$, the module V is isomorphic with K^n via $R(\beta) : K^n \to V$. Thus $t \in \mathrm{End}\, V_K$ yields $R(\beta)tR(\beta)^{-1} \in \mathrm{End}\, K^n = K(K^n, K^n) \leq K[T](K[T]^n, K[T]^n)$. Moreover, the K-isomorphism $R(\beta) : K^n \to V$ yields a surjective $K[T]$-homomorphism $R(\beta) : K[T]^n \to V$. The elements T and $R(\beta)tR(\beta)^{-1}$ of $K[T](K[T]^n, K[T]^n)$ yield the same maps $V \to V$, so $\mathrm{Im}(T - R(\beta)tR(\beta)^{-1}) \leq \mathrm{Ker}(R(\beta) : K[T]^n \to V)$. Conversely, let $X = \mathrm{Ker}(R(\beta) : K[T]^n \to V)/\mathrm{Im}(T - R(\beta)tR(\beta)^{-1})$. Suppose X is non-zero. Let $(p_1(T), \dots, p_n(T)) + \mathrm{Im}(T - R(\beta)tR(\beta)^{-1})$ be a non-zero element of X, represented by an element $(p_1(T), \dots, p_n(T))$ such that $\max\{\deg p_1(T), \dots, \deg p_n(T)\} = m$ is minimal amongst all such representatives. Note that m is positive, since $\mathrm{Ker}(R(\beta) : K^n \to V) = \{0\}$. Set $p_i(T) = \sum_{j=0}^{m} p_{ij}T^j$. Then $(p_1(T), \dots, p_n(T)) + \mathrm{Im}(T - R(\beta)tR(\beta)^{-1}) = (\sum_{j=0}^{m-1} p_{1j}T^j, \dots, \sum_{j=0}^{m-1} p_{nj}T^j) + (p_{1m}, \dots, p_{nm})R(\beta)tR(\beta)^{-1}T^{m-1} + \mathrm{Im}(T - R(\beta)tR(\beta)^{-1})$, contradicting the minimality of m. Thus (3.4.1) is exact. □

Corollary 3.4.2 (The Cayley-Hamilton Theorem). *The determinant* $\det(T - R(\beta)tR(\beta)^{-1})$ *lies in the annihilator* $\mathrm{An}_{K[T]}V$.

Proof. Set $f = T - R(\beta)tR(\beta)^{-1} \in K[T](K[T]^n, K[T]^n) = K[T]^n_n$, so that $V = \mathrm{Coker}(f : K[T]^n \to K[T]^n)$ by Theorem 3.4.1. Now by (2.4.7) and Theorem 2.4.3(b), one has $\hat{f}f = (f^T\widehat{f^T})^T = |f^T| = |f|$. Then for $p \in K[T]^n$, one has $(p + \mathrm{Im}\, f)|f| = p|f| + \mathrm{Im}\, f = p\hat{f}f + \mathrm{Im}\, f = \mathrm{Im}\, f$. □

Note that $\det(T - R(\beta)tR(\beta)^{-1}) = \det[R(\beta)(T - t)R(\beta)^{-1}] = \det(T - t)$ in Corollary 3.4.2. For a square matrix t or endomorphism t of a finitely generated free module, the monic polynomial $\det(T - t)$ is called the *characteristic polynomial* of t.

Example 3.4.3. Consider the complex number

$$t = \begin{bmatrix} x & y \\ -y & x \end{bmatrix} \in \mathbb{R}_2^2,$$

as in Exercise 2.4Q. Then

$$\det\begin{bmatrix} T - x & -y \\ y & T - x \end{bmatrix} = T^2 - 2xT + (x^2 + y^2).$$

Thus t is a root of the real polynomial $T^2 - (t + \bar{t})T + |t|^2 \in \mathbb{R}[T]$. □

For the remainder of this section, let K be a field, so that $K[T]$ is a Euclidean domain. Let (V, t) be a linear dynamical system of finite dimension n. Note that the exact sequence (3.4.1) becomes a free resolution $0 \longrightarrow K[T]^n \xrightarrow{f} K[T]^n \xrightarrow{R(\beta)} V \longrightarrow 0$ of the $K[T]$-module V, an instance of (3.3.5). By Theorem 3.3.4, there is an $(m - 1)$-simplex $d_1K[T] \supseteq d_2K[T] \supseteq \cdots \supseteq d_mK[T]$ in (Prop $K[T]$, $\supseteq$) yielding an isomorphism

$$\rho: \sum_{i=1}^{m} K[T]/d_iK[T] \longrightarrow V \tag{3.4.2}$$

of $K[T]$-modules. No d_i can be zero, since $K[T]$ is infinite-dimensional as a K-module. By (3.3) and (3.2.1), the polynomials $d_1, \dots, d_m$ may be chosen uniquely as monic polynomials. These polynomials are then called the *invariant factors* of t, and the isomorphism ρ of (3.4.2) is called the *rational canonical form* of t. Note that the algorithm of Section 3.3 may be used to determine the monic polynomials $d_1, \dots, d_m$. Now $d_1K[T] = \operatorname{Ker}(K[T] \to \operatorname{End} V_{\mathbb{Z}}; T \mapsto t)$, the annihilator of the $K[T]$-module V. The monic polynomial d_1 is called the *minimal polynomial* of the endomorphism t.

For a monic polynomial $p(T)$ in $K[T]$, the quotient $K[T]/p(T)K[T]$ is a vector space over K with basis $\{T^i | 0 \leq i < \deg p(T)\}$. The $K[T]$-homomorphism $L(T): K[T] \to K[T]$; $q(T) \mapsto Tq(T)$ induces a K-endomorphism $L(T)_p$ of $K[T]/p(T)K[T]$. Note that the annihilator $\operatorname{An}_{K[T]}K[T]/p(T)K[T]$ is $p(T)K[T]$. Now the degree of the monic characteristic polynomial $\det(T - L(T)_p)$ of $L(T)_p$ is $\deg p(T)$, so the Cayley-Hamilton Theorem shows that this polynomial is just $p(T)$. Returning to the rational canonical form (3.4.2), the endomorphism $\Sigma_{i=1}^{m} L(T)_{d_i}$ of the domain may be expressed as

$$\prod_{i=1}^{m} \left\{ \left(\sum_{j<i} 1_{K[T]/d_jK[T]} \right) \oplus L(T)_{d_i} \oplus \left(\sum_{j>i} 1_{K[T]/d_jK[T]} \right) \right\},$$

whose characteristic polynomial is $\Pi_{i=1}^m d_i$. Since V_K has basis

$$(3.4.3) \qquad \rho : \bigcup_{i=1}^{m} \{T^j + d_i K[T] \| 0 \le j < \deg d_i\} \to V,$$

the characteristic polynomial of t is $\det(T - R(\rho)tR(\rho)^{-1}) = \det(T - \Sigma_{i=1}^m L(t)_{d_i}) = \Pi_{i=1}^m d_i$. Summarizing:

Theorem 3.4.4. *Let V be a vector space of finite dimension n over the field K. Then endomorphisms of V are classified up to similarity by their invariant factors $d_1, d_2, \dots, d_m$, with each factor a multiple of its predecessor. The minimal polynomial of an endomorphism is its first invariant factor, while its characteristic polynomial is the product of its invariant factors.* □

Example 3.4.5. For a positive integer n and field K, elements of the group $\mathrm{GL}_n(K)$ are conjugate iff they are similar. Thus conjugacy classes in $\mathrm{GL}_n(K)$ are classified by the invariant factors of their members. Recall (cf. Exercise 2.4M) the isomorphism of $\mathrm{GL}_2(\mathbb{Z}_2)$ with the symmetric group S_3 given by restricting the action $(\mathbb{Z}_2)_1^2 \times (\mathbb{Z}_2)_2^2 \to (\mathbb{Z}_2)_1^2$ of (1.3.4) to $((\mathbb{Z}_2)_1^2 - \{[0,0]\}) \times \mathrm{GL}_2(\mathbb{Z}_2) \to ((\mathbb{Z}^2)_1^2 - \{[0,0]\})$. Recall Proposition I 3.2.2 classifying conjugacy classes in $((\mathbb{Z}_2)_1^2 - \{[0,0]\})!$ by partitions of 3. Then the S_3-class 1^3 corresponds to invariant factors $(T+1), (T+1)$. The S_3-class $1^1 2^1$ corresponds to the single invariant factor $T^2 + 1$. The S_3-class 3^1 corresponds to the single invariant factor $T^2 + T + 1$. □

As a $K[T]$-module, the linear dynamical system (V, t) is a torsion module, and thus is the codomain of a $K[T]$-isomorphism

$$(3.4.4) \qquad \zeta : \bigoplus_{i=1}^{r} \bigoplus_{j=1}^{m} K[T]/p_i^{e_{ij}} K[T] \to V$$

yielding a primary decomposition, as in (3.3.7). Choosing the prime elements p_i to be monic, the isomorphism ζ is called the *Jordan canonical form* of the K-endomorphism t. (Classically, this nomenclature has been reserved for the special case where all the p_i are linear.) The monic polynomials $p_i^{e_{ij}}$ in the Jordan canonical form are called the *elementary divisors* of t, and the characteristic polynomial of t is the product $\Pi_{i=1}^r \Pi_{j=1}^m p_i^{e_{ij}}$ of its elementary divisors. Endomorphisms of V are classified up to similarity by their Jordan canonical forms, just as they are classified up to similarity by their rational canonical forms according to Theorem 3.4.4. Suppose that p_i is a linear polynomial $T - \lambda_i$. Then $\mathrm{Ker}(t - \lambda_i)$ is called the *eigenspace* of t corresponding to the *eigenvalue* λ_i, and non-zero elements of $\mathrm{Ker}(t - \lambda_i)$ are called *eigenvectors* of t belonging to λ_i. (The synonymous adjectives "proper" or "characteristic" were preferred to "eigen-" in English-speaking countries during or immediately after times of war with German-speaking countries. However, these synonyms are so overworked—"proper subspace", "characteristic polynomial", etc.—that their use is best avoided.) Note that the multiset $\langle e_{ij} | 1 \leq j \leq m, e_{ij} > 0 \rangle$ yields a partition of $\sum_{j=1}^{m} e_{ij}$. Its length $\dim \mathrm{Ker}(t - \lambda_i)$ is called the *geometric multiplicity* of the eigenvalue λ_i, while its sum $\sum_{j=1}^{m} e_{ij}$ is called the *algebraic multiplicity* of λ_i. Note that the eigenvalues of t are just the roots of its characteristic polynomial.

Example 3.4.6. Pick a scalar $\lambda \in K$, and consider the polynomial $p(T) = (T - \lambda)^e$ for $e > 0$. The endomorphism $L(T)_p$ of $K[T]/p(T)K[T]$ has a 1-dimensional eigenspace spanned by $(T - \lambda)^{e-1} + p(T)K[T]$. Thus $L(T)_p$ has eigenvalue λ with geometric multiplicity 1 and algebraic multiplicity e. □

Example 3.4.7. Consider the vehicle suspension system of Example 2.2.2, described by the differential equation (2.2.4) with corresponding homogeneous equation $ks + ls' + ms'' = 0$. This equation may be rewritten in the form

$$[s' \, s''] = [s \, s'] \begin{bmatrix} 0 & -k/m \\ 1 & -l/m \end{bmatrix}.$$

Consider the $\mathbb{R}$-endomorphism $\begin{bmatrix} 0 & -k/m \\ 1 & -l/m \end{bmatrix}$ of $\mathbb{R}^2$. The qualitative behavior of the suspension system is described by the number of eigenvalues of this endomorphism. If there are no eigenvalues, then the differential equation has oscillatory solutions, suggesting that the vehicle body will bounce up and down relative to the axle. Subjectively, the ride is "soft" or "wallowing." If there are two eigenvalues, then the suspension system is slow to respond to displacements of the axle. Subjectively, the ride is "hard" or "rough." The ideal behavior occurs when there is a single eigenvalue (with algebraic multiplicity 2). The suspension absorbs irregularities of the surface over which the vehicle is traveling, but without setting up oscillations. □

EXERCISES

3.4A. Let A be a 2×2 real matrix. Show that A^2 is a real linear combination of A and I_2.

3.4B. Show that each quaternion (cf. Exercise 2.4S) is the root of a real quartic polynomial. Factorize the quartic.

3.4C. Let K be a commutative, unital ring, and let f be an $n \times n$-matrix over K. Show that the characteristic polynomial of f is of the form $T^n - T^{n-1} \operatorname{tr} f + \cdots + (-1)^n \det f$.

3.4D. Give a direct proof that the characteristic polynomial of $L(T) \in \operatorname{End}(K[T]/p(T)K(T))_K$ is $p(T)$, for the monic polynomial $p(T)$ over a field K, without using the Cayley-Hamilton Theorem.

3.4E. For a field K, show that the invariant factors of elements of $\mathrm{GL}_n(K)$ all have non-zero constant term.

3.4F. Classify the conjugacy classes in the groups $\mathrm{GL}_3(\mathbb{Z}_2)$ and $\mathrm{GL}_2(\mathbb{Z}_3)$.

3.4G. Show that elements of $\mathrm{SL}_n(K)$ are conjugate in $\mathrm{SL}_n(K)$ iff they are conjugate in $\mathrm{GL}_n(K)$.

3.4H. Classify the conjugacy classes in $\mathrm{SL}_2(\mathbb{Z}_3)$.

3.4I. Fix a positive integer p. Let ω_p denote rotation of the Cartesian plane $\mathbb{R}^2$ counterclockwise by an angle $2\pi/p$.

(a) Show that ω_p is an element of $\operatorname{End}_{\mathbb{R}}\mathbb{R}^2 = \mathbb{R}^2_2$.

(b) Show that ω_p is a complex number (in the sense of Exercise 2.4Q).

(c) Show that ω_p is a root of $T^p - 1$ in $\mathbb{C}[T]$.

(d) Show that $T^p - 1 = \prod_{j=0}^{p-1}(T - \omega_p^j)$ in $\mathbb{C}[T]$.

(e) Show that the subgroup of $\mathrm{GL}_2(\mathbb{R})$ generated by ω_p and $\begin{bmatrix} 1 & 0 \\ 0 & -1 \end{bmatrix}$ is the dihedral group D_p.

3.4J. Let $M = \langle m \rangle$ be a finite cyclic monoid of index i and period p (cf. Theorem O 4.2.1). For a field K, let V be the free K-module with basis M. Define a K-endomorphism t of V by $t : m^j \mapsto m^{j+1}$ for $0 \le j < i + p$.

(a) Show that the characteristic polynomial of t is $T^i(T^p - 1)$.

(b) Determine the rational canonical form (3.4.2) of t. In particular, for $0 \le j < i + p$, identify $m^j\rho^{-1}$ as an element of Dom ρ.

(c) For $K = \mathbb{C}$, determine the Jordan canonical form (3.4.4) of t. In particular, for $0 \le j < i + p$, identify $m^j\zeta^{-1}$ as an element of Dom ζ. (Hint: Use Exercise 3.4I.)

3.4K. Solve the homogeneous differential equation $ks + ls' + ms'' = 0$ of Example 3.4.7, taking the constants k, l, m in intervals representing the suspension system of Example 2.2.2. In particular, identify the three different cases corresponding to the three possible cardinalities of the set of eigenvalues of $\begin{bmatrix} 0 & -k/m \\ 1 & -l/m \end{bmatrix} \in \operatorname{End}(\mathbb{R}^2)_{\mathbb{R}}$.

3.4L. Let $C(\mathbb{R}, \mathbb{C})$ denote the complex vector space of continuous functions $u: \mathbb{R} \to \mathbb{C}$ (with componentwise operations). Let $D(\mathbb{R}, \mathbb{C})$ denote the subspace of $C(\mathbb{R}, \mathbb{C})$ consisting of continuously differentiable functions. Define $\Delta : D(\mathbb{R}, \mathbb{C}) \to C(\mathbb{R}, \mathbb{C})$; $u \mapsto u'$. For a finite subset Λ of $\mathbb{C}$, let $E\Lambda$ denote the subspace of $D(\mathbb{R}, \mathbb{C})$ spanned by $\{e^{\lambda t} | \lambda \in \Lambda\}$. Determine the rational canonical form and the Jordan canonical form of $(E\Lambda, \Delta)$. [Note: With suitable time units, light whose spectrum comprises the frequencies $\omega_1, \ldots, \omega_n$ is represented by elements of $E\Lambda$ for $\Lambda = \{i\omega_1, \ldots, i\omega_n\}$. The elementary divisors of the $\mathbb{C}[T]$-module $E\Lambda$

are the elements $(T - i\omega_1)\mathbb{C}[T], \ldots, (T - i\omega_n)\mathbb{C}[T]$ of Spec $\mathbb{C}[T]$. This may be the origin of the term "spectrum" for the set of prime ideals of a commutative, unital ring.]

3.5. Elementary Field Theory

Throughout this section, let K be a field. If A is a K-algebra, then one may use the injection $K \to A$; $k \mapsto 1k$ (cf. Exercise 3A) to identify K as a subalgebra of A. Given a non-zero element t of A, consider the K-algebra homomorphism $K[T] \to A$; $p(T) \mapsto p(t)$ of (3.4), with image $K[t] := \{p(t) \mid p(T) \in K[T]\}$. If this homomorphism injects, then t is said to be *transcendental over* K, and the homomorphism yields a K-algebra isomorphism of $K[T]$ with $K[t]$. Otherwise, the homomorphism has a non-trivial kernel of the form $p(T)K[T]$ for a uniquely specified monic polynomial $p(T)$. In this case t is said to be *algebraic of degree* deg $p(T)$ *over* K, and $p(T)$ is called the *minimal polynomial* of the algebraic element t. [If $p(T)$ is quadratic, cubic, etc., then t may be described respectively as *quadratic*, *cubic*, etc.] Complex numbers that are algebraic over the rationals are known as *algebraic numbers*. Thus $3\sqrt{2}$ and i are algebraic numbers, while e and π are transcendental. The quaternions i, j, and k are quadratic over $\mathbb{R}$. In general, the K-algebra A is said to be *algebraic over* K if each non-zero element of A is algebraic over K.

Let t be an algebraic element of A over K, with minimal polynomial $p(T)$. One may identify t with the endomorphism $R_K(t) : a \mapsto at$ of the K-vector space $K[t]$ of dimension deg $p(T)$. The First Isomorphism Theorem applied to (3.4) yields the rational canonical form $\rho : K[T]/p(T)K[T] \to K[t]$ of t. Thus $p(T)$ is both the minimal and the characteristic polynomial of the K-endomorphism t. Motivated by Exercises 2.4Q and 3.4C, one defines $(-1)^{\deg p}p(0)$ to be the *norm* det $R_K(t)$ *of* t *over* K and the negative of the coefficient of $T^{(\deg p)-1}$ in $p(T)$ to be the *trace* tr $R_K(t)$ *of* t *over* K. Thus the norms of the complex number or quaternion i over $\mathbb{R}$ or $\mathbb{Q}$ are all 1, while their traces are all 0.

Proposition 3.5.1. *Suppose that the* K*-algebra* A *has finite dimension as a* K*-vector space.*

(*a*) *Each element* t *of* A *is algebraic over* K, *of degree not exceeding* $\dim A_K$.
(*b*) *If* A *has no zero divisors, then it is a skewfield.*

Proof. (a) The $\mathbb{N}$-tuple $\mathbb{N} \to A$; $i \mapsto t^i$ is dependent, so $K[T] \to A$; $q(T) \mapsto q(t)$ has a non-trivial kernel $p(T)K[T]$, and t is algebraic. Also deg $p(T) = \dim(K[T]/p(T)K[T])_K \leq \dim A_K$.

(b) Let t be a non-zero element of A. By Proposition 2.4.8 or Theorem 2.4.9(c), the injective K-homomorphisms $R(t)$: $A \to A$; $a \mapsto at$ and $L(t)$: $A \to A$; $a \mapsto ta$ are surjective. □

If t is an algebraic element of a (possibly infinite-dimensional) K-algebra A without zero divisors, then the commutative, finite-dimensional subalgebra $K[t]$ is an integral domain. Proposition 3.5.1(b) shows that $K[t]$ is a field. If $K[t]$ is a field, it is usually written in the form $K(t)$, by analogy with (3.1.5). The minimal polynomial of t is then irreducible, or equivalently, prime.

Proposition 3.5.2. *Suppose K, L, M are fields, with $K \leq L \leq M$. Then* $\dim M_K = \dim M_L \cdot \dim L_K$.

Proof. If $f : I \to L$ is a basis for L_K, and $g : J \to M$ is a basis for M_L, then $I \times J \to M; (i, j) \mapsto if \cdot jg$ is a basis for M_K. □

Now suppose that the K-algebra A is an integral domain. Suppose that L_K is a finite-dimensional K-subalgebra of A that is a field. If t is an algebraic element of A over K, then t is also algebraic over L. By Proposition 3.5.2, $\dim_K L(t) = \dim L(t)_L \cdot \dim L_K$, so that $L(t)$ is again a finite-dimensional K-subalgebra of A that is a field. In particular, given algebraic elements $t_1, t_2, \dots, t_n$ of A over K, one may define the finite-dimensional K-field $K(t_1, \dots, t_{r-1})(t_r)$. Now the memberships $t_1^{-1} \in K(t_1), t_1 + t_2 \in K(t_1, t_2)$, and $t_1 t_2 \in K(t_1, t_2)$ show that the union of $\{0\}$ and the set of algebraic elements of A form a subfield $\mathrm{Alg}_K A$ of A. The subfield $\mathrm{Alg}_{\mathbb{Q}} \mathbb{C}$ of $\mathbb{C}$ is called the *field of algebraic numbers*.

Consider a monic polynomial $f(T)$ in $K[T]$. For any K-algebra A, one may consider $f(T)$ as an element of $A[T]$, since the coefficients of f lie in the subfield K of A. If $f(T) = \prod_{i=1}^{\deg f}(T - a_i)$ for elements a_i of A, and $A = K(a_1, \dots, a_{\deg f})$, then the field A is called a *splitting field of $f(T)$ in $K[T]$*. For example, the subfield $\mathbb{Q}(\sqrt{2}) = \mathbb{Q}(\sqrt{2}, -\sqrt{2})$ of $\mathbb{R}$ is a splitting field of $T^2 - 2$ in $\mathbb{Q}[T]$, while $\mathbb{R}$ is the splitting field of $T^2 - 2$ in $\mathbb{R}[T]$. The existence of a splitting field for $f(T)$ may be proved by induction on the degree of $f(T)$. Suppose that $f(T)$ is a member of the maximal ideal $p(T)K[T]$ of $K[T]$, with $p(T)$ monic. Consider the field $L = K[T]/p(T)K[T]$, a K-algebra. As an element of $L[T]$, the polynomial $f(T)$ has a root $a = T + p(T)K[T]$, so that $f(T) = (T - a)f_1(T)$ in $L[T]$ for some element $f_1(T)$ of $L[T]$. Moreover, $L = K(\mathrm{a})$. By induction, $f_1(T)$ in $L[T]$ has a splitting field $L(a_1, \dots, a_r)$. Then $f(T)$ in $K[T]$ has the splitting field $K(a, a_1, \dots, a_r)$. The monic polynomial $f(T)$ in $K[T]$ is said to be *separable* if $f(T) = \prod_{i=1}^{\deg f}(T - a_i)$ with $|\{a_1, \dots, a_{\deg f}\}| = \deg f$ in any splitting field $A = K(a_1, \dots, a_{\deg f})$ of $f(T)$ in $K[T]$.

The *prime field* of a field K is the intersection of all the subfields of K. Thus if K has characteristic 0 (cf. Exercise 3A), the prime field is $\mathbb{Q}$. Otherwise, K has prime characteristic p, in which case the prime field is $\mathbb{Z}_p$. Now for a prime p, one has $(T + U)^p = \sum_{i=0}^{p} \binom{p}{i} T^i U^{p-i} = T^p + U^p$ in $\mathbb{Z}_p[T, U]$. Thus the map $\varphi : K \to K; a \mapsto a^p$ is a ring homomorphism of the field K of characteristic p. It is called the *Frobenius homomorphism* of K.

Suppose that K is a finite field, say with prime field $\mathbb{Z}_p$. Then the order $|K|$ of K is the prime power $q = p^n$, where n is the dimension $\dim K_{\mathbb{Z}_p}$ of the $\mathbb{Z}_p$ vector space $K_{\mathbb{Z}_p}$. Note that K^* is an abelian group of order $q - 1$. For each element a of K, one has $a\varphi^n = a^q = a^{q-1} \cdot a = a$, so the Frobenius homomorphism φ becomes an automorphism, the *Frobenius automorphism*, of K. It will be shown that, for each prime power $q = p^n$, there is a field $\mathrm{F}(q)$ or $\mathrm{GF}(q)$ or F_q of order q, the *Galois field of order* q, that is unique up to isomorphism.

Given a prime power $q = p^n$, the first task is to exhibit a field of order q. If such a field K exists, then each element a of K is a root of $T^q - T$ in $K[T]$.

Theorem 3.5.3. *For $q = p^n$ with a prime number p and a positive integer n, there is a splitting field K of $T^q - T$ in $\mathbb{Z}_p[T]$ with $|K| = q$.*

Proof. It will first be shown that $T^q - T$ in $\mathbb{Z}_p[T]$ is separable. Let L be a field of characteristic p, and let a be a root of $T^q - T$ in $L[T]$. Now $T^q - T = (T^q - a^q) - (T - a) = (T - a)^q - (T - a) = (T - a)[(T - a)^{q-1} - 1]$. Then a is not a root of $(T - a)^{q-1} - 1$. Thus $T^q - T$ is separable.

Since $T^q - T$ is separable, one has $T^q - T = \prod_{i=1}^{q}(T - a_i)$ in $K[T]$, for a q-element subset $S = \{a_1, \ldots, a_q\}$ of the splitting field K. Now $S = \{a \in K | a^q = a\}$ [cf. Exercise 3.2E(a)]. Since $(a + b)^q = a^q + b^q$ and $(ab)^q = a^q b^q$, the set S is an integral domain containing $\mathbb{Z}_p$. By Proposition 3.5.1(b), S is a field. Thus $K = \mathbb{Z}_p(a_1, \ldots, a_q) = S$, of order q. □

Proof of the uniqueness (to within isomorphism) of the finite field of order q requires examination of the structure of finite subgroups of the group of non-zero elements of a field.

Proposition 3.5.4. *Let G be a finite subgroup of the group of units of a field K. Then G is cyclic.*

Proof. Suppose $|G| = n$. Let $d_1\mathbb{Z}, \ldots, d_m\mathbb{Z}$ be the invariant factors of the $\mathbb{Z}$-module G, say with each d_i positive. Then $n = d_1 \ldots d_m$ by (3.3.6), and each element a of G satisfies $a^{d_m} = 1$. Thus the polynomial $T^{d_1} - 1$ in $K[T]$ has n roots in K. It follows that $n = d_1$ [cf. Exercise 3.2E(a)], so $m = 1$ and G is cyclic. □

For a finite field K of order q, Proposition 3.5.4 shows that the group K^* of non-zero elements of K is cyclic. Suppose $K^* = \langle e \rangle$. Then e is called a *primitive element* of the field K. There is a group isomorphism

$$\exp_e : \mathbb{Z}_{q-1} \to K^*; \; x \mapsto e^x. \tag{3.5.1}$$

The inverse $\log_e$ of (3.5.1) is called the *discrete logarithm to base e* in K. Note that

$$K = \mathbb{Z}_p(e) \tag{3.5.2}$$

for the primitive element e.

Theorem 3.5.5. *If the fields K and L have the same finite order $q = p^n$, then they are isomorphic.*

Proof. Let K have a primitive element e with (irreducible) minimal polynomial $m(T)$ over $\mathbb{Z}_p$. By (3.5.2), there is an isomorphism $\alpha : K \to \mathbb{Z}_p[T]/m(T)\mathbb{Z}_p[T]$. Now $T^q - T \in m(T)\mathbb{Z}_p[T]$, so $T^q - T = \prod_{b \in L}(T - b) \in m(T)L[T]$. Thus $m(T)$ has a root b in L. Since $m(T)$ is irreducible in $\mathbb{Z}_p[T]$, it is the minimal polynomial of b over $\mathbb{Z}_p[T]$, and there is an injective homomorphism $\beta : \mathbb{Z}_p[T]/m(T)\mathbb{Z}_p[T] \to L$ with image $\mathbb{Z}_p(b)$. Since $\dim [\mathbb{Z}_p(b)]_{\mathbb{Z}_p} = \deg m(T) = n = \dim L_{\mathbb{Z}_p}$, the homomorphism β surjects, and is thus an isomorphism. Then $\alpha\beta : K \to L$ is the required isomorphism between K and L. □

EXERCISES

3.5A. Let t be an algebraic element of a K-algebra A, with minimal polynomial $p(T) = \sum_{i=0}^{\deg p} p_i T^i$. Suppose that $p(T)$ is an irreducible polynomial in $K[T]$.

(a) Show that t is a unit in A.

(b) Show that t^{-1} is algebraic over K.

(c) Determine the minimal polynomial of t^{-1} over K.

3.5B. Determine the minimal polynomials of $\sqrt{2}$, $\sqrt{3}$, and $\sqrt{2} + \sqrt{3}$ over $\mathbb{Q}$.

3.5C. Determine a splitting field for the polynomial $T^2 + T + 1$ in $\mathbb{Z}_2[T]$.

3.5D. In the notation of Exercise 3.4I, show that $\mathbb{Q}(\omega_p)$ is a splitting field for $T^p - 1$ in $\mathbb{Q}[T]$.

3.5E. Suppose that the K-algebra A is an integral domain.

(a) Show that the norm over K yields a monoid homomorphism $\mathrm{Alg}_K A \to K$.

(b) Show that the trace over K yields a K-homomorphism $(\mathrm{Alg}_K A)_K \to K_K$.

3.5F. Show that the Frobenius homomorphism of the field $\mathbb{Z}_p(T)$ of rational functions over $\mathbb{Z}_p$ (cf. Example 3.1.4) is not surjective.

3.5G. Exhibit a non-commutative finite subgroup of the skewfield $\mathbb{H}$ of quaternions.

3.5H. (a) Construct a field K of order 9.

(b) Identify a primitive element e of K.

(c) Construct a table of the discrete logarithm to base e in K.

(d) Decompose K into orbits under the action of the Frobenius automorphism.

3.5I. Repeat Exercise 3.5H for a field K of order 8.

3.5J. For a prime p and positive integer n, show that the group $GL_n(\mathbb{Z}_p)$ has an element g of order $p^n - 1$. Show that g has a single invariant factor $d_1(T) \in \mathbb{Z}_p[T]$, and that $d_1(T)$ is irreducible.

III

CATEGORIES AND LATTICES

1. POSETS, MONOIDS, AND CATEGORIES

There are many interconnections between the concepts of poset and monoid or semigroup. Semilattices as introduced in Section O 4.3 represent an intersection of the two concepts: Semilattices may be regarded simultaneously as posets with greatest lower bounds and as idempotent commutative semigroups. The main topic of this chapter, categories, offers a union of the two concepts.

Definition 1.1. A *small category* C consists of two sets, an *object set* C_0, or Obj C, and a *morphism* set C_1, or Mor C. Elements of C_0 are called *objects* or *points* of C. Elements of C_1 are called *morphisms* or *arrows* of C. There are two functions assigning points to an arrow: the *domain* or *tail function* $d_0: C_1 \to C_0; f \mapsto fd_0$ and the *codomain* or *head function* $d_1 : C_1 \to C_0; f \mapsto fd_1$. For objects x and y of C, the set $d_0^{-1}\{x\} \cap d_1^{-1}\{y\}$ of arrows of C with tail x and head y is denoted by $C(x, y)$. There is an *identity function*

$$e : C_0 \to C_1; x \mapsto 1_x \tag{1.1}$$

with $ed_0 = ed_1 = 1_{C_0}$, so that $\forall x \in C_0$, $1_x \in C(x, x)$. Finally, for each triple x, y, z of objects of C, there is a *composition*

$$C(x, y) \times C(y, z) \to C(x, z); (f, g) \mapsto fg \tag{1.2}$$

such that

$$\forall x,y \in C_0, \forall f \in C(x, y), 1_x f = f = f 1_y \tag{1.3}$$

and

$$(1.4)\quad \forall x, y, z, t \in C_0, \forall f \in C(x, y), \forall g \in C(y, z), \forall h \in C(z, t),$$
$$(fg)h = f(gh).$$

One usually refers to (1.4) as the *associative law* of composition. □

Note that the identity function e of (1.1) injects, since it has the head and tail functions as retractions. One could thus identify C_0 with $C_0 e$, and suppress explicit mention of the objects altogether.

An element f of $C(x, y)$ may be represented pictorially as $x \xrightarrow{f} y$ or $f : x \to y$. In this convention, the composition (1.2) appears as

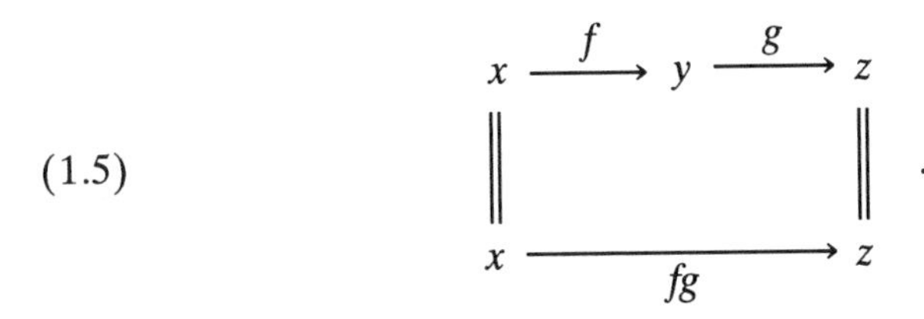

Example 1.2 (Monoids). Let $(M, \cdot, 1)$ be a monoid. Define $M_0 := \{1\}$ and $M_1 := M$. Define the tail function $d_0 : M \to \{1\}$ and the head function $d_1 : M \to \{1\}$. Thus $M = C(1, 1)$. Define the composition $C(1, 1) \times C(1, 1) \to C(1, 1)$; $(m, n) \mapsto mn$ by monoid multiplication. Define $e : \{1\} \to M; 1 \mapsto 1$. Then (1.3) and (1.4) hold by (O 4.3) and (O 4.2). Thus M forms a small category, with a single object. Conversely, composition in a small category with a single object forms a monoid (Exercise 1A). □

Example 1.3 (Posets). Let (X, α) be a poset. Define $\alpha_0 := X$ and $\alpha_1 := \alpha$. Define the tail function $d_0 : \alpha \to X$; $(x, y) \mapsto x$ and the head function $d_1 : \alpha \to X$; $(x, y) \mapsto y$. Using the reflexivity of α, define the identity function to be the diagonal $\Delta : X \to \alpha$; $x \mapsto (x, x)$. Thus $\forall x, y \in X$, $\alpha(x, y) =$ **if** $x \alpha y$ **then** $\{(x, y)\}$ **else** $\varnothing$. A composition (1.2) is then defined (uniquely) by the transitivity of α, and α becomes a small category. □

Example 1.4 (Matrices). Let S be a unital ring. Let S_0 be the set of positive integers, and define $S_1 := \dot\bigcup_{1 \le l,m} S_l^m$ to be the set of all matrices over S. Define d_0 as the disjoint union of the maps $S_l^m \to \{l\}$ and d_1 as the disjoint union of the $S_l^m \to \{m\}$. Define $e : m \mapsto I_m$. Note that $S(l, m) = S_l^m$ for $1 \le l, m$. Define the composition $S_l^m \times S_m^n \to S_l^n$ by matrix multiplication. Then S becomes a small category. □

Example 1.5 (Systems of Sets). Let C_0 be a set of sets. For $X, Y \in C_0$ let $C(X, Y) = Y^X$, the set of all functions from X to Y. Let $C_1 = \dot\bigcup_{X, Y \in C_0} C(X, Y)$. Define $d_0 : C_1 \to C_0$; $f \mapsto \operatorname{Dom} f$ and $d_1 : C_1 \to C_0$; $f \mapsto \operatorname{Cod} f$.

Define $e : C_0 \to C_1;\ X \mapsto \mathrm{id}_X$. Define the composition (1.2) by the usual composition of functions. Then C becomes a category. □

Example 1.6 (Systems of Quasigroups). (a) Let C_0 be a set of quasigroups. For $X, Y \in C_0$, let $C(X,Y)$ be the set of all quasigroup homomorphisms from X to Y. Make the remaining definitions as in Example 1.5. Then C becomes a category, since the composition of quasigroup homomorphisms is a quasigroup homomorphism. This example may be mimicked using other structures on the sets X, Y in place of quasigroup structure, taking $C(X,Y)$ to be the set of homomorphisms of the structure, i.e. the set functions $f : X \to Y$ preserving the structure.

(b) Again taking C_0 to be a set of quasigroups, one may define $C(X,Y)$ to be the set of all quasigroup homotopies from X to Y. Define $((f_1, f_2, f_3) : (X_0, \cdot\,, /, \backslash\,) \to (X_1, \cdot\,, /, \backslash\,))d_i := X_i$ for $i = 0, 1$, and $e : (X, \cdot\,, /, \backslash\,) \mapsto (1_X, 1_X, 1_X)$. Define the composition (1.2) by the composition of homotopies (cf. Exercise I 4A). Then C becomes a category. □

Suppose that a small category C is given as in Definition 1.1. One may then define a new category C^{op} as follows. Take $C_0^{\mathrm{op}} = C_0$ and $C_1^{\mathrm{op}} = C_1$. Take $d_0^{\mathrm{op}} = d_1$ and $d_1^{\mathrm{op}} = d_0$. Thus the arrows of C are reversed (turned head to tail) to yield the arrows of C^{op}. Moreover, $C^{\mathrm{op}}(x, y) = C(y, x)$ for $x, y \in C_0 = C_0^{\mathrm{op}}$. Take $e^{\mathrm{op}} = e$. The composition

$$C^{\mathrm{op}}(x, y) \times C^{\mathrm{op}}(y, z) \to C^{\mathrm{op}}(x, z);\ (f, g) \mapsto f \circ g$$

is defined in terms of the composition in C as

$$C(y, x) \times C(z, y) \to C(z, x);\ (f, g) \mapsto gf.$$

For example, the diagram (1.5) is replaced by

$$(1.6)\qquad \begin{array}{ccccc} x & \xleftarrow{\;f\;} & y & \xleftarrow{\;g\;} & z \\ \Vert & & & & \Vert \\ x & & \xleftarrow[\;f \circ g\;]{} & & z \end{array} \;.$$

The category C^{op} is called the *opposite* or *dual* of the category C. For example, if $M = (M, \cdot\,, 1)$ is a monoid, determining a category M as in Example 1.2, the dual of the category M is the category determined by the opposite monoid $M^{\mathrm{op}} = (M, \breve{\circ}, 1)$ as in (I 1.4). If C is the category determined by the power set poset $(2^B, \subseteq)$ of a set B according to the construction of Example 1.3, then C^{op} is the category determined by the poset $(2^B, \supseteq)$. In general, for a poset (X, α), the poset (X, α') with $x\alpha' y \Leftrightarrow y\alpha x$ is called the *dual* of the poset (X, α). For any category C, the dual $C^{\mathrm{op}\,\mathrm{op}}$ of the dual C^{op} is just the original category C.

A *directed graph* $\Gamma = (\Gamma_0, \Gamma_1, \partial_0, \partial_1)$ is a pair $(\partial_0 : \Gamma_1 \to \Gamma_0, \partial_1 : \Gamma_1 \to \Gamma_0)$ of maps. For example, suppose that α is a simple directed graph on a vertex set X (in the sense of I.3), i.e. an antireflexive binary relation on X. Then $\partial_0 : \alpha \to X;\ (x, y) \mapsto x$ and $\partial_1 : \alpha \to X;\ (x, y) \mapsto y$ yield a directed graph $(X, \alpha, \partial_0, \partial_1)$ in the current sense. A *non-trivial path* in the directed graph Γ is an element $f_1 \ldots f_n$ of the free semigroup Γ_1^+ over Γ_1 with the property $\forall 1 \leq i < n,\ f_i\partial_1 = f_{i+1}\partial_0$. One may then construct a category $\Gamma\Pi$, the *path category* of the graph Γ, as follows. Let $\Gamma\Pi_1$ be the disjoint union of the set Γ_0 and the set $\Gamma\Pi_1^+$ of non-trivial paths in Γ. In this context, elements of Γ_0 are called *trivial paths*, and elements of $\Gamma\Pi_1$ are called *paths* in Γ. Let $\Gamma\Pi_0 = \Gamma_0$. Define $d_0 : \Gamma\Pi_1 \to \Gamma\Pi_0$ to be the disjoint union of the identity $\Gamma_0 \to \Gamma_0$ and the map $\Gamma\Pi_1^+ \to \Gamma_0;\ f_1 \ldots f_n \mapsto f_1\partial_0$. Similarly, define $d_1 : \Gamma\Pi_1 \to \Gamma\Pi_0$ as the disjoint union of the identity $\Gamma_0 \to \Gamma_0$ and $\Gamma\Pi_1^+ \to \Gamma_0;\ f_1 \ldots f_n \mapsto f_n\partial_1$. Define $e : \Gamma\Pi_0 \to \Gamma\Pi_1$ to be the insertion of Γ_0 in the disjoint union $\Gamma_0 \,\dot{\cup}\, \Gamma\Pi_1^+$. Define the composition of non-trivial paths $f_1 \ldots f_m$ and $g_1 \ldots g_n$ with $f_m\partial_1 = g_1\partial_0$ as $f_1 \ldots f_m g_1 \ldots g_n$. Define the composition of a trivial path with any path by requiring (1.3) to hold. Then $\Gamma\Pi$ becomes a category.

Example 1.7 (Hasse Diagrams of Locally Finite Posets). For elements x, y of a poset (X, α), the *closed interval* $[x, y]_\alpha$ is $\{t \in X | x \alpha t \alpha y\}$. The *open interval* $(x, y)_\alpha$ is $[x, y]_\alpha - \{x, y\}$. Note that this notation agrees with the usual notation for intervals in the real line $(\mathbb{R}, \leq)$. The poset (X, α) is said to be *locally finite* if each interval in (X, α) is finite. [Even if X is infinite, the poset $(X, \hat{X})$ is still locally finite.] An element (x, y) of α is a *covering pair* if the closed interval $[x, y]_\alpha$ has 2 elements (necessarily $x \neq y$). In this context, one also says that y *covers* x. The *Hasse diagram* of a locally finite poset (X, α) is the simple directed graph $\Gamma_\alpha := \{(x, y) \in X^2 | 2 = |[x, y]_\alpha|\}$ consisting of the covering pairs. For such a poset, the category α determined as in Example 1.3 is then just the path category $\Gamma_\alpha\Pi$. The Hasse diagram is a convenient way to specify a (locally) finite poset. For example, the Hasse diagram of the power set $(2^{\{0,1\}}, \subseteq)$ is just

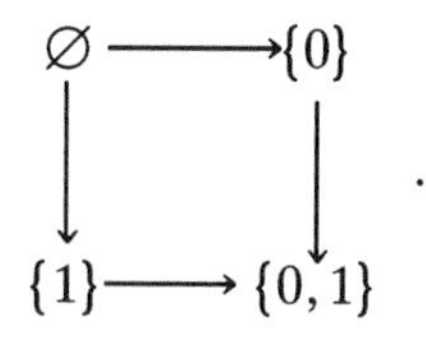

The edges in Hasse diagrams are often drawn without arrowheads, using the convention that the tail of an edge is "lower than" or "south of" its

head. Using this convention, the Hasse diagram of $(2^{\{0,1\}}, \subseteq)$ is

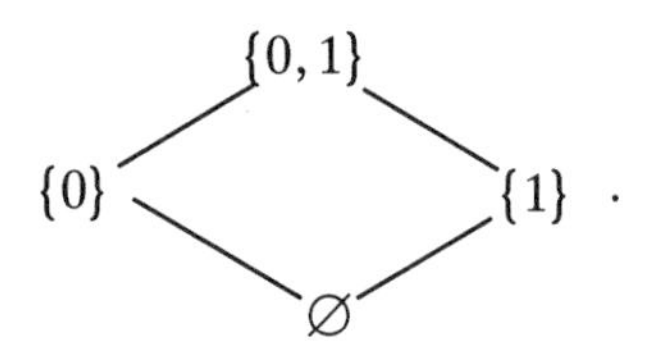

(Advocates of this convention are wont to claim that the Hasse diagram of the dual of a poset is the Australian version of the Hasse diagram of the original poset.) □

Example 1.8 (The Nerve of a Category). A small category C, as in Definition 1.1, yields a directed graph $C = (C_0, C_1, d_0, d_1)$. The path category $C\Pi$ is called the *nerve* of the category C. Elements of the intersection $C_n := C\Pi_1 \cap C_1^n$ of the set of paths with the uniform code of length n over C_1 are known as *n-simplices* in C. For example, the simplices in (the category determined by) the (dual of the) poset of ideals of a principal ideal domain A are just the simplices classifying finitely generated A-modules via the rational canonical form. The successive points along such a path are the invariant factors of the module. □

EXERCISES

1A. Verify that if a small category C has a single object, then C_1 forms a monoid under composition.

1B. (a) Generalize Example 1.3 to show that each pre-ordering yields a small category.

(b) Suppose that for each pair (x, y) of objects of a small category C, the set $C(x, y)$ has at most one element. Show that C_1 is a pre-ordering on C_0.

1C. For the system $C_0 = \{\varnothing, \{0\}, \{0, 1\}\}$ of sets, determine the corresponding category C given as in Example 1.5. [Hint: Represent C by a picture of points and arrows, as in (1.5).]

1D. Construe the dual of the category S of Example 1.4 in terms of transposition of matrices.

1E. Show that the free monoid A^* over an alphabet A is obtained (according to Exercise 1A) as the path category of the directed graph $(\{1\}, A, A \to \{1\}, A \to \{1\})$.

1F. (a) Show that the poset $(\mathrm{Id}\,\mathbb{Z}, \subseteq)$ of ideals of the ring of integers is not locally finite.

(b) Show that the poset $((\mathrm{Id}\,\mathbb{Z}) - \{0\}, \subseteq)$ of non-trivial ideals of the ring of integers is infinite, but locally finite.

1G. Let (X, α) be a locally finite poset. Give a formal proof that α is specified uniquely by its Hasse diagram.

1H. A poset $(X, \leq)$ satisfies the *ascending chain condition* if $x_1 \leq x_2 \leq \cdots \leq x_n \leq \cdots \Rightarrow \exists m.\ x_m = x_{m+1} = x_{m+2} = \cdots$. Show that the ascending chain condition does not imply local finiteness.

1I. Consider the problem of storing a partial order in a computer. Discuss the relative efficiencies of the two following techniques:

(a) Hasse diagrams;

(b) intersections of total orders [cf. Proposition O 3.5.4(b) and Exercise O 3.5E].

1J. Let (X,α) be a locally finite poset. Let K be a unital, commutative ring. Then the set K^α of functions $\alpha \to K$ is a K-algebra $(K^\alpha, \cdot, \zeta)$ with pointwise operations. In particular, the identity ζ is the constant (or *zeta*) function $\zeta : \alpha \to \{1\}$. Now define a new operation of *convolution* on K^α by

$$f * g(x, y) = \sum_{t \in [x, y]_\alpha} f(x, t) g(t, y).$$

(a) Show that $(K^\alpha, *, \delta)$ is a K-algebra with pointwise K-module structure and identity $\delta(x, y) =$ **if** $x = y$ **then** 1 **else** 0.

(b) If K has characteristic 0, prove

$$\zeta * \zeta(x, y) = |X\Pi(x, y) \cap \alpha^2|,$$

where α^2 is the uniform code of length 2 over α (cf. Exercise O 4.1B). Similarly, express $|X\Pi(x, y) \cap \alpha^n|$, for all positive integers n, in terms of the zeta function.

(c) Show that ζ is invertible in $(K^\alpha, *, \zeta)$. Its inverse μ is called the *Möbius function* of (X, α).

(d) Show that δ is invertible in $(K^\alpha, \cdot, \zeta)$ if (X, α) is an antichain, i.e. $\alpha = \hat{X}$.

(e) Let (X, γ) be a chain [cf. (O 3.5.2)]. Prove $\mu(x, y) =$ **if** $x = y$ **then** 1 **else if** $|[x, y]_\alpha| = 2$ **then** -1 **else** 0.

1.1. Diagonalization and Large Categories

Example 1.5 made a category C with a given set of sets as the set C_0 of objects of C, and with $C(x, y)$ as the set of functions $f : x \to y$ for sets x, y in C_0. One would like to be able to consider all sets together in such a category. However, the class of all sets does not itself form a set. To see this, consider the property of sets given by the following *diagonalization argument* due to Cantor.

Proposition 1.1.1. *There is no surjection $f: A \to 2^A$ from a set A to its power set.*

Proof. Suppose that there were such a surjection f. For the subset $B = \{x \in A | x \notin xf\}$, there would then be an element a of A with $B = af$. This would lead to a contradiction, for $a \in B \Rightarrow a \notin af = B$, while $a \notin B \Rightarrow a \in af = B$. □

Corollary 1.1.2. *The class of all sets does not form a set.*

Proof. Suppose that the class of all sets were to form a set A. Then the power set 2^A, as the set of subsets of A, would be a subset of A. Let $j: 2^A \to A;\ X \to X$ be the inclusion. As an injection, the function j would have a right inverse f with $jf = 1_{2^A}$ (Proposition O 3.1.1). Then $f: A \to 2^A$, having j as a left inverse, would surject (Exercise O 3.1C). This would violate Proposition 1.1.1. □

Corollary 1.1.2 shows that the class of all sets is a concept located at a higher hierarchical level than that of sets. Each set is a class, and one may characterize sets as those classes that are members of a class (e.g. the class of all sets). The remaining classes, such as the class of all sets, are called *proper classes*. One may operate with classes and functions between them, just as one operates with sets and functions between them. If the sets appearing in Definition 1.1 are replaced with classes, then one obtains the definition of a *large category*. For example, the category $\underline{\underline{\mathrm{Set}}}$ of sets has the class $\underline{\underline{\mathrm{Set}}}_0$ of all sets as its class of objects. For two sets X, Y, the class $\underline{\underline{\mathrm{Set}}}(X, Y)$ is the set Y^X of functions from X to Y, and the class $\underline{\underline{\mathrm{Set}}}_1$ of arrows of $\underline{\underline{\mathrm{Set}}}$ is the disjoint union $\dot\bigcup_{X, Y \in \underline{\underline{\mathrm{Set}}}_0} \underline{\underline{\mathrm{Set}}}(X, Y)$. Note that this union of sets is a proper class, since it contains the class $\underline{\underline{\mathrm{Set}}}_0$ of all sets as a subclass, namely as the image of the identity function $e: \underline{\underline{\mathrm{Set}}}_0 \to \underline{\underline{\mathrm{Set}}}_1;\ X \mapsto \mathrm{id}_X$. The tail function in $\underline{\underline{\mathrm{Set}}}$ is $d_0: \underline{\underline{\mathrm{Set}}}_1 \to \underline{\underline{\mathrm{Set}}}_0;\ f \mapsto \mathrm{Dom}\, f$, while the head function is $d_1: \underline{\underline{\mathrm{Set}}}_1 \to \underline{\underline{\mathrm{Set}}}_0;\ f \mapsto \mathrm{Cod} f$. The composition in $\underline{\underline{\mathrm{Set}}}$ is the usual composition of set functions. A large category C is said to be *locally small* if $C(x, y)$ is a set for each pair (x, y) of objects of C. Note that $\underline{\underline{\mathrm{Set}}}$ is locally small, since $\underline{\underline{\mathrm{Set}}}(X, Y)$ is the set Y^X.

Example 1.1.3 (Categories of Quasigroups). Just as the category $\underline{\underline{\mathrm{Set}}}$ is a large category corresponding to the small categories of Example 1.5, so there are large categories corresponding to the small categories of Example 1.6.

(a) The *(large) category* $\underline{\underline{\mathrm{Q}}}$ *of quasigroups* has $\underline{\underline{\mathrm{Q}}}_0$ as the class of quasigroups. For quasigroups P and Q, the class $\underline{\underline{\mathrm{Q}}}(P, Q)$ is the set of quasigroup homomorphisms from P to Q. Then $\underline{\underline{\mathrm{Q}}}_1 = \dot\bigcup_{P, Q \in \underline{\underline{\mathrm{Q}}}_0} \underline{\underline{\mathrm{Q}}}(P, Q)$. The tail and head functions are given by the domain and codomain respectively, while the composition is the usual composition of functions.

(b) The *homotopy category* $\underline{\underline{\text{Qtp}}}$ *of quasigroups* again has $\underline{\underline{\text{Qtp}}}_0$ as the class $\underline{\underline{Q}}_0$ of quasigroups. For quasigroups P and Q, the class $\underline{\underline{\text{Qtp}}}(P,Q)$ is the set of quasigroup homotopies from P to Q. Then $\underline{\underline{\text{Qtp}}}_0 = \dot{\bigcup}_{P,Q \in \underline{\underline{Q}}_0} \underline{\underline{\text{Qtp}}}(P,Q)$. Other details are as in Exercise 1.6(b). □

Example 1.1.4 (A Large Semilattice). The large poset or *poclass* $(\underline{\underline{\text{Set}}}_0, \subseteq)$ yields a large category by the construction of Example 1.3. The corresponding large semilattice $(\underline{\underline{\text{Set}}}_0, \cup)$ yields a large monoid $(\underline{\underline{\text{Set}}}_0, \cup, \varnothing)$, which in turn yields a large category by the construction of Example 1.2. Note that this large monoid M has a singleton set $\{1\}$ as M_0, while $M(1,1) = M_1$ is the proper class of sets. In particular, M is not locally small. □

Example 1.1.5 (Partial Functions). The large category $\underline{\underline{\text{Pfn}}}$ has the class $\underline{\underline{\text{Set}}}_0$ of sets as $\underline{\underline{\text{Pfn}}}_0$. Given sets X and Y, a *partial function* f from X to Y is a (*total* or "ordinary") function $\bar{f} : X_f \to Y$ from a subset X_f of X to Y. The subset X_f of X is called the *domain of definition* Dod f of f. Then $\underline{\underline{\text{Pfn}}}(X,Y)$ is the set of partial functions from X to Y, and $\underline{\underline{\text{Pfn}}}_1$ is the class $\dot{\bigcup}_{X,Y \in \underline{\underline{\text{Set}}}_0} \underline{\underline{\text{Pfn}}}(X,Y)$. The composite of two partial functions $f : X \to Y$ and $g : Y \to Z$ is the partial function $fg : X \to Z$ whose domain of definition is $(\bar{f})^{-1}(X_f\bar{f} \cap Y_g)$, and such that $\overline{fg}$ is the total function composite of the restriction of $\bar{f}$ to this domain with the restriction of $\bar{g}$ to $X_f\bar{f} \cap Y_g$. For a set X, the identity 1_X is the total identity function id_X, construed as a partial function with domain of definition X. The category $\underline{\underline{\text{Pfn}}}$ is called the *category of partial functions*. □

EXERCISES

1.1A. Define $\xi : 2^{\mathbb{N}} \to [0,1];\ A \mapsto 2\Sigma_{n=0}^{\infty} 3^{-n-1}\chi_A(n)$.

(a) Show that ξ is a well-defined injection.

(b) A set X is *countable* if there is a surjection $\mathbb{N} \to X$. Using (a) and Proposition 1.1.1, show that the closed real interval $[0,1]$ is uncountable.

1.1B. (a) Show that the abelian group $\coprod_{n \in \mathbb{N}} \mathbb{Z}$ is countable.

(b) Show that the abelian group $\prod_{n \in \mathbb{N}} \mathbb{Z}$ is uncountable.

1.1C. Define a set S of functions $f : \mathbb{N} \to \mathbb{N}$ inductively by the following rules:

(a) $\forall n \in \mathbb{N}$, $(\mathbb{N} \to \{n\}) \in S$;

(b) $1_{\mathbb{N}} \in S$;

(c) $\forall f, g \in S$, $(n \mapsto nf + ng) \in S$;

(d) $\forall f, g \in S$, $(n \mapsto nf.ng) \in S$;

(e) $\forall f, g \in S$, $(n \mapsto (nf)^{ng}) \in S$.

Show that there is a function $f : \mathbb{N} \to \mathbb{N}$ which is not a member of S.

1.1D. Exhibit large categories C with each of the following two properties:
(a) C_0 is a set and C_1 is a proper class;
(b) both C_0 and C_1 are proper classes.

1.1E. Set up a large category $(\underline{\underline{\mathrm{Mon}}};\underline{\underline{\mathrm{Set}}})$ with $(\underline{\underline{\mathrm{Mon}}};\underline{\underline{\mathrm{Set}}})_0$ as the class of monoid actions and with $\overline{(\underline{\underline{\mathrm{Mon}}};\underline{\underline{\mathrm{Set}}})((A,M),(B,N))}$, as the set of action morphisms from the $\overline{M}$-set $\overline{A}$ to the N-set B (cf. Section I 1.2). Define the tail, head, composition, etc.

1.1F. Set up a large *category* $\underline{\underline{\mathrm{Rel}}}$ of *relations*, with $\underline{\underline{\mathrm{Rel}}}_0 = \underline{\underline{\mathrm{Set}}}_0$, as follows. For sets X and Y, one has $\underline{\underline{\mathrm{Rel}}}(X,Y) = 2^{X \times Y}$. For $\alpha \in \underline{\underline{\mathrm{Rel}}}(X,Y)$ and $\beta \in \underline{\underline{\mathrm{Rel}}}(Y,Z)$, define the composition of α and β to be the *relation product* $\alpha \circ \beta = \{(x,z) \in X \times Z \,|\, \exists y \in Y.\ x\alpha y \beta z\}$. Verify that $\underline{\underline{\mathrm{Rel}}}$ is a category.

1.1G. Use the usual constructions of set theory to show that, for sets X and Y, the class $\underline{\underline{\mathrm{Pfn}}}(X,Y)$ is a set.

1.1H. Show that a rational function $p(T)/q(T)$ in the field of fractions $\mathbb{R}(T)$ of $\mathbb{R}[T]$ yields an element $r \mapsto p(r)/q(r)$ of $\underline{\underline{\mathrm{Pfn}}}(\mathbb{R},\mathbb{R})$ with domain of definition $\{x \in \mathbb{R} \,|\, q(x) \neq 0\}$. Can you put an $\mathbb{R}$-algebra structure on $\underline{\underline{\mathrm{Pfn}}}(\mathbb{R},\mathbb{R})$ so that the $\mathbb{R}$-algebra homomorphism $\mathbb{R}[T] \to \underline{\underline{\mathrm{Set}}}(\mathbb{R},\mathbb{R})$ of (II 3.5) extends to an $\mathbb{R}$-algebra homomorphism $\mathbb{R}(T) \to \underline{\underline{\mathrm{Pfn}}}(\mathbb{R},\mathbb{R})$? [Hint: Try $\mathrm{Dod}(f-g) = \mathrm{Dod}\, f \cap \mathrm{Dod}\, g$ in $\underline{\underline{\mathrm{Pfn}}}(\mathbb{R},\mathbb{R})$.]

1.1I. The *graph* of a partial function $f : X \to Y$ is the subset

$$\mathrm{gr} f = \{(x,y) \in (\mathrm{Dod} f) \times Y | y = x\bar{f}\}$$

of $X \times Y$. Define a relation $\leq$ of *extension* on $\underline{\underline{\mathrm{Pfn}}}(X,Y)$ by $f \leq g$ iff $\mathrm{gr}\, f \subseteq \mathrm{gr}\, g$. Show that $(\underline{\underline{\mathrm{Pfn}}}(X,Y), \leq)$ is a poset.

1.1J. Consider the following program to compute a natural number nf for each natural number n:

```
begin i: = 0, 0f = 1
    while i < n do
        (i + 1)f: = if.(i + 1)
        i: = i + 1
    endwhile
output nf
```

(a) Show that each run of the program builds a partial function $f_n : \mathbb{N} \to \mathbb{N}$ whose domain of definition is the closed interval $[0,n]$ in $(\mathbb{N}, \leq)$.

(b) Show that $\{f_n | n \in \mathbb{N}\}$ is a chain in $(\underline{\underline{\mathrm{Pfn}}}(\mathbb{N},\mathbb{N}), \leq)$. What is the upper bound of the chain?

1.2. Functors and Concrete Categories

A monoid homomorphism $f:(M,\cdot,1) \to (N,\cdot,1)$ is a function $f:M \to N$ with $1f = 1$ and with $\forall m_1, m_2 \in M, (m_1 m_2)f = (m_1 f)(m_2 f)$. For example, the exponential function is a monoid homomorphism $\exp:(\mathbb{R},+,0) \to ((0,\infty),\cdot,1)$ from the reals to the positive reals. An *order-preserving* or *monotone* (*increasing*) map $f:(X,\alpha) \to (Y,\beta)$ between posets is a function $f:X \to Y$ such that $(x_1, x_2) \in \alpha \Rightarrow (x_1 f, x_2 f) \in \beta$. For example, the exponential function is a monotone map $\exp:(\mathbb{R},\leq) \to ((0,\infty),\leq)$ from the reals to the positive reals. The concept of a category gives a common generalization of the concepts of monoid and poset. There is a corresponding common generalization of monoid homomorphisms and monotone maps: the concept of a functor.

Definition 1.2.1. Let C and D be categories. Then a (*covariant*) *functor* $F:C \to D$ consists of two functions, an *object part* $F:C_0 \to D_0$ and a *morphism part* $F:C_1 \to D_1$, with the following properties:

(a) $\forall x, y \in C_0, C(x, y)F \subseteq D(xF, yF)$;
(b) $\forall x \in C_0, 1_x F = 1_{xF}$;
(c) $\forall x,y,z \in C_0, \forall f \in C(x, y), \forall g \in C(y, z), (fg)^F = f^F \cdot g^F$. □

Example 1.2.2. Let Y be a fixed set. Given a set Z, define $ZF_Y := Z \times Y$. Given a function $f:Z_1 \to Z_2$, define $fF_Y : Z_1 F_Y \to Z_2 F_Y$ to be the function $(f, 1_Y): Z_1 \times Y \to Z_2 \times Y;\ (z, y) \mapsto (zf, y)$ given by (O 3.2.4). Then $F_Y : \underline{\underline{\text{Set}}} \to \underline{\underline{\text{Set}}}$ is a functor. □

Example 1.2.3. Let (X,α) be a poset, construed as a small category α according to Example 1.3. For $x \in X$, define the *down-set* $xD = \{y \in X | (y, x) \in \alpha\}$. Now for $(x_1, x_2) \in \alpha$, there is a unique arrow $f: x_1 \to x_2$ in the category α. Define $f^D : x_1 D \to x_2 D$ to be the inclusion function of the subset $x_1 D$ in the set $x_2 D$. Then $D: \alpha \to \underline{\underline{\text{Set}}}$ is a functor. □

Example 1.2.4. Let C be a category, and let $C\Pi$ be its nerve (Example 1.8). (If C is large, then so is $C\Pi$.) There is a functor $\varepsilon_C : C\Pi \to C$ whose object part is the identity on $C_0 = C\Pi_0$ and whose morphism part maps a non-trivial path in C to the composite of its constituent arrows. □

Example 1.2.5. Let C be the (category determined by) the poset with Hasse diagram

$$(0,0) \to (1,0) \qquad (1,1) \to (2,1).$$

Let D be the (category determined by) the poset with Hasse diagram $0 \to 1 \to 2$. Let F be the monotone function $F:C \to D; (x, y) \mapsto x$. Then F is a functor $F:C \to D$. Note that the object image set $C_0 F = \{0, 1, 2\}$ and the morphism image set $C_1F = \{1_0, 1_1, 1_2, 0 \to 1, 1 \to 2\}$ do not form an "image category," since there is no composite of $0 \to 1$ and $1 \to 2$ in C_1F. □

Example 1.2.6. A functor $P : \underline{\underline{\mathrm{Set}}} \to \underline{\underline{\mathrm{Set}}}^{\mathrm{op}}$, the *contravariant power set functor*, is defined by $P : \underline{\underline{\mathrm{Set}}}_0 \to \underline{\underline{\mathrm{Set}}}_0^{\mathrm{op}}; X \mapsto 2^X$ and $P : \underline{\underline{\mathrm{Set}}}_1 \to \underline{\underline{\mathrm{Set}}}_1^{\mathrm{op}}; (f : X \to Y) \mapsto (f^{-1} : 2^Y \to 2^X)$, as in Exercise O 3.4E. One may also construe P as a functor $P : \underline{\underline{\mathrm{Set}}}^{\mathrm{op}} \to \underline{\underline{\mathrm{Set}}}$. □

In general, for categories C and D, a functor $F : C \to D^{\mathrm{op}}$, or equivalently $F : C^{\mathrm{op}} \to D$, is sometimes described as a *contravariant functor* $F : C \to D$ from C to D. If C and D are poset categories, then a contravariant functor $F : C \to D$ is called an *antitone* or *order-reversing* map from C to D. If G is a group construed as a monoid category, then inversion $J : G \to G;\ g \mapsto g^{-1}$ yields a contravariant functor from G to itself.

Given functors $F : C \to D$ and $G : D \to E$, the respective composites FG of the object and morphism parts of F and G determine a *composite functor* $FG : C \to E$. Each category C has an *identity functor* $1_C : C \to C$ whose object and morphism parts are the respective identity functions 1_{C_0} and 1_{C_1}. One may thus form a *category* $\underline{\underline{\mathrm{Cat}}}$ *of small categories*, with $\underline{\underline{\mathrm{Cat}}}_0$ as the proper class of small categories. Given small categories C and D, the class $\underline{\underline{\mathrm{Cat}}}(C, D)$ is the set $(D^C)_0$ of functors from C to D. Composition in $\underline{\underline{\mathrm{Cat}}}$ is defined by composition of functors. For arbitrary categories C and D, a functor $F : C \to D$ is an *isomorphism* if there is a functor $G : D \to C$ with $FG = 1_C$ and $GF = 1_D$. In this context, the categories C and D are *isomorphic*, a relation denoted by $C \cong D$. The functor G is the *(two-sided) inverse* F^{-1} of F.

A functor $F : C \to D$ is *full* if, for each pair X, Y of objects of C, the function $C(X, Y) \to D(XF, YF);\ f \mapsto fF$ surjects. For example, let $\underline{\underline{\mathrm{Gp}}}$ be the large category of groups and homomorphisms between them. Then the inclusions $\underline{\underline{\mathrm{Gp}}}_0 \hookrightarrow \underline{\underline{\mathrm{Q}}}_0$ and $\underline{\underline{\mathrm{Gp}}}_1 \hookrightarrow \underline{\underline{\mathrm{Q}}}_1$ yield a full functor from $\underline{\underline{\mathrm{Gp}}}$ to the large category $\underline{\underline{\mathrm{Q}}}$ of quasigroups introduced in Example 1.1.3(a). In general, a category C is said to be a *subcategory* of a category D if there are inclusions $C_0 \hookrightarrow D_0$ and $C_1 \hookrightarrow D_1$ yielding a functor $C \hookrightarrow D$, the *inclusion functor*. For example, $\underline{\underline{\mathrm{Gp}}}$ is a subcategory of $\underline{\underline{\mathrm{Q}}}$ and $\underline{\underline{\mathrm{Q}}}$ is a subcategory of $\underline{\underline{\mathrm{Qtp}}}$ [via $\underline{\underline{\mathrm{Q}}}_1 \to \underline{\underline{\mathrm{Qtp}}}_1;\ (f : P \to Q) \mapsto ((f, f, f) : P \to Q)$]. The subcategory C of D is said to be *full* if the inclusion functor $C \hookrightarrow D$ is full. Thus $\underline{\underline{\mathrm{Gp}}}$ is a full subcategory of $\underline{\underline{\mathrm{Q}}}$, but $\underline{\underline{\mathrm{Q}}}$ is not a full subcategory of $\underline{\underline{\mathrm{Qtp}}}$. Note that, once a category D is given, a full subcategory C of D is specified completely by giving the object subclass C_0 of D_0.

A functor $F : C \to D$ is *faithful* if, for each pair X, Y of objects of C, the function $C(X, Y) \to D(XF, YF);\ f \mapsto fF$ injects. For example, the inclusion functor of any subcategory is faithful. As a second example, consider a module X over a unital ring S. Construe the monoid $(S, \cdot, 1)$ as a category S with $S_0 = \{1\}$ and $S_1 = S$, as in Example 1.2. Construe the representation (II 2.1) as a functor $R : S \to \underline{\underline{\mathrm{Set}}}$, with object part $R : 1 \mapsto X$ and morphism part $R : s \mapsto (x \mapsto xs)$. Then X is a faithful S-module iff the functor R is faithful.

A *concrete category* is a category C supporting a faithful functor $C \to$ Set, the *underlying set functor*. In Section O 3, algebra was characterized as the "study of sets with structure." In studying sets with structure, algebra also studies functions preserving that structure. Since the identity map on a set preserves its structure, and since the composite of structure-preserving maps preserves the structure, algebra is a rich source of concrete categories, such as Q and Gp, of sets with structure. The underlying set functor from such a category then just forgets the structure of an object, "remembering" only the underlying set. Its morphism part forgets the structure-preserving property of a function, "remembering" only the function itself. Functors forgetting (some) structure in this sense are known as *forgetful functors*. For example, the underlying set functor U: Q $\to$ Set has object part $U : (Q, \cdot, /, \backslash) \mapsto Q$ and morphism part $U : (f : (P, \cdot, /, \backslash) \to (Q, \cdot, /, \backslash)) \mapsto (f : P \to Q)$.

Example 1.2.7. Here is a catalogue of concrete categories of "sets-with-structure" and "structure-preserving maps" that have arisen so far.

Q—Quasigroups and quasigroup homomorphisms.

Mon—Monoids and monoid homomorphisms.

Gp—Groups, a full subcategory of Q and of Mon.

Set—Sets and functions (no structure).

Sgp—Semigroups and semigroup homomorphisms.

Sl—Semilattices, a full subcategory of Sgp.

M (For a fixed monoid M)—Msets and M-homomorphisms.

Lp—Loops and loop homomorphisms.

RQ—Right quasigroups and right quasigroup homomorphisms.

RLp—Right loops and right loop homomorphisms.

Rng—Non-unital rings and non-unital ring homomorphisms.

Ring—Unital rings (rings with "i-dentity element") and unital ring homomorphisms.

Mod_S (For a fixed unital ring S)—Unital right S-modules and S-homomorphisms.

${}_S\text{Mod}$ (For a fixed unital ring S)—Unital left S-modules and S-homomorphisms.

Alg_K (For a fixed unital, commutative ring K)—K-algebras and K-algebra homomorphisms. □

The concrete categories of Example 1.2.7 all have obvious underlying set functors, namely the forgetful functors. In other concrete categories, the underlying set functors may not be quite so obvious.

Example 1.2.8. The category Qtp of quasigroups and homotopies is a concrete category. The faithful underlying set functor $U : \mathrm{Qtp} \to \mathrm{Set}$ has object part $U : Q \mapsto Q \dot{\cup} Q \dot{\cup} Q$ and morphism part $U : (f_1, f_2, f_3) \mapsto f_1 \dot{\cup} f_2 \dot{\cup} f_3$. Note that the composite of the inclusion functor $\mathrm{Q} \hookrightarrow \mathrm{Qtp}$ with $U : \mathrm{Qtp} \to \mathrm{Set}$ gives an underlying set functor for the concrete category Q that is different from the forgetful functor. □

Example 1.2.8 shows that a concrete category, such as the concrete category of quasigroups, may have several different underlying set functors. Thus it is sometimes necessary to consider a concrete category C along with a specified underlying set functor $U : C \to \mathrm{Set}$. Following the category theory tradition of flippant definitions that goes right back to the beginning of the subject, one might define such a pair (C, U) as a *reinforced concrete category*.

Example 1.2.9. The category Pfn of partial functions (Example 1.1.5) is a concrete category, with an underlying set functor $G : \mathrm{Pfn} \to \mathrm{Set}$ known as the *(one-point) compactification* functor or *Grimeisen operator*. Fix a singleton $\{*\}$. (In computer science, one may choose to denote $*$ as the element $\perp$, "bottom" or "undefined." In mathematics, one may choose to denote $*$ as the element ∞, "the point at infinity.") Then G has object part $G : X \mapsto X \dot{\cup} \{*\}$ and morphism part $G : (f : X \to Y) \mapsto (\bar{f} \dot{\cup} ((X - X_f) \to \{*\}) \cup (1_{\{*\}})$. □

EXERCISES

1.2A. Verify that $F : \mathrm{Set} \to \mathrm{Set}$ in Example 1.2.2 really is a functor.

1.2B. (a) Show that the composite of monotone maps between posets is again monotone. Conclude that there is a large category Pos with Pos_0 the class of ordered sets and Pos_1 the class of monotone functions between them.

(b) Show that Pos is a full subcategory of the large category Cat, via the construction of Example 1.3.

(c) Construing each set X as the *antichain* or *discrete poset* $(X, \hat{X})$, show that Set is a full subcategory of Pos and Cat. (Cf. Exercise O 4.3A.) Sets regarded in this way are known as *discrete categories*.

1.2C. Let $F : C \to D$ be a functor.

(a) If C is a poset, show that F is faithful.

(b) If D is a poset, is F necessarily full?

1.2D. Give an example of a full and faithful functor that is not an isomorphism.

1.2E. Is $\underline{\underline{\mathrm{Gp}}}$ a full subcategory of $\underline{\underline{\mathrm{Qtp}}}$?

1.2F. Let $F : C \to D$ be a full functor. Show that the image of F, with object class C_0F and morphism class C_1F, forms a category. (Cf. Example 1.2.5.)

1.2G. Show that there is a *covariant power set functor* $P : \underline{\underline{\mathrm{Set}}} \to \underline{\underline{\mathrm{Set}}}$ with $P : \underline{\underline{\mathrm{Set}}}_0 \to \underline{\underline{\mathrm{Set}}}_0;\ X \mapsto 2^X$ and $P : \underline{\underline{\mathrm{Set}}}_1 \to \underline{\underline{\mathrm{Set}}}_1;\ (f : X \to Y) \mapsto (2^X \to 2^Y;\ A \mapsto Af)$.

1.2H. Let G be a fixed abelian group, and let S be a unital ring. Show that duality as in Section II 2.1 yields a contravariant functor $T : \underline{\underline{\mathrm{Mod}}}_S \to {}_S\underline{\underline{\mathrm{Mod}}}$ with object part $X_S \mapsto {}_SX^T = \mathrm{Hom}(X, G)$ and morphism part $f \mapsto f^T = \mathrm{Hom}(f, G)$.

1.2I. (a) Show that

$$(12) \mapsto \begin{bmatrix} 0 & 1 \\ 1 & 0 \end{bmatrix} \quad \text{and} \quad (23) \mapsto \begin{bmatrix} 1 & 0 \\ 1 & 1 \end{bmatrix}$$

extend to (the morphism part of) a functor $S_3 \to \mathbb{Z}_2$ from the symmetric group S_3 (construed as a monoid category according to Example 1.2) to the unital ring $\mathbb{Z}_2$ (construed as a matrix category according to Example 1.4).

(b) Show that

$$(12) \mapsto \begin{bmatrix} 0 & 1 & 0 \\ 1 & 0 & 0 \\ 0 & 0 & 1 \end{bmatrix} \quad \text{and} \quad (23) \mapsto \begin{bmatrix} 1 & 0 & 0 \\ 0 & 0 & 1 \\ 0 & 1 & 0 \end{bmatrix}$$

extend to a functor $S_3 \to \mathbb{Z}_2$.

1.2J. A forgetful functor need not be completely forgetful. For example, there is a forgetful functor $U : \underline{\underline{\mathrm{Mon}}} \to \underline{\underline{\mathrm{Sgp}}}$ with object part $U : (M, \cdot, 1) \mapsto (M, \cdot)$, just forgetting the existence of the identity element. Excluding $\underline{\underline{\mathrm{Set}}}$ from the catalogue of Example 1.2.7, determine all the remaining forgetful functors $U : C \to D$ connecting distinct categories C, D from the catalogue.

1.2K. Verify that U in Example 1.2.8 is a faithful functor.

1.2L. Verify that G in Example 1.2.9 is a faithful functor.

1.2M. Show that the category M of Example 1.1.4 is not concrete.

1.2N. Fix a unital ring S and a positive integer n. Define $U : \underline{\underline{\mathrm{Mod}}}_S \to \underline{\underline{\mathrm{Set}}}$ by $U : \underline{\underline{\mathrm{Mod}}}_S)_0 \to \underline{\underline{\mathrm{Set}}}_0;\ V \mapsto V^n$ and $U : \underline{\underline{\mathrm{Mod}}}_S)_1 \to \underline{\underline{\mathrm{Set}}}_1;\ (f : V \to W) \mapsto ((f, \ldots, f) : V^n \to W^n)$. Show that U is a faithful functor.

1.2O. Show that each small category is concrete.

1.2P. Let n be a natural number, and let K be a commutative unital ring.

(a) Show that there is a contravariant functor $K^n(\ ;K): \underline{\underline{\text{Mod}}}_K \to \underline{\underline{\text{Mod}}}_K$ with object part $V \mapsto K^n(V;K)$ and morphism part $\theta \mapsto K^n(\theta;K)$.

(b) Show that there is a contravariant functor $\tilde{K}^n(\ ;K): \underline{\underline{\text{Mod}}}_K \to \underline{\underline{\text{Mod}}}_K$ with object part $V \to \tilde{K}^n(V;K)$ and morphism part $\theta \mapsto \tilde{K}^n(\theta;K)$.

1.2Q. Show that the function $\underline{\underline{\text{Gp}}}_0 \to \underline{\underline{\text{Set}}}_0$; $G \mapsto Z(G)$ assigning the center to a group cannot be the object part of a functor $\underline{\underline{\text{Gp}}} \to \underline{\underline{\text{Set}}}$.

1.2R. For a set X, let XF be the set of equivalence relations on X, ordered by inclusion and considered as a poset category. For a function $f: X \to Y$, define $fF: YF \to XF$; $\alpha \mapsto \ker(f \text{ nat } \alpha)$. Show that F is a functor $F: \underline{\underline{\text{Set}}}^{\text{op}} \to \underline{\underline{\text{Cat}}}$.

1.3. Commuting Diagrams, Epimorphisms, and Monomorphisms

One of the guiding themes in the study and use of categories is the expression of mathematical statements pictorially, using diagrams of points and arrows (category theory as "comic book mathematics") Such an expression tends to lead to a better understanding of the mathematical statement involved. It also facilitates the transfer of mathematical content from one context to another—specifically, from one category to another. This is a valuable tool for overcoming the progressive fragmentation and specialization that bedevils mathematics (along with other scientific disciplines).

Formally, a *diagram* in a category C is a functor $D: \Gamma\Pi \to C$ from the path category of a directed graph Γ to C. (In a picture of D, one usually just displays $\Gamma_1 D$ or $\Gamma_0 D \cup \Gamma_1 D$.) The diagram is said to *commute* if, for any two non-trivial paths f, g in Γ with common heads and tails, the images fD and gD in C agree. For example, (O 3.2.4)–(O 3.2.6) are commuting diagrams in $\underline{\underline{\text{Set}}}$.

Consider the following simple mathematical statement:

(1.3.1) The function $e: X \to Y$ surjects.

To express (1.3.1) pictorially, one may use the dual of Proposition O 3.1.1 (cf. Exercise O 3.1C) stating that (1.3.1) is equivalent to $\forall f: Y \to Z, \forall g: Y \to Z, ef = eg \Rightarrow f = g$. Thus the commuting of each diagram

(1.3.2)
$$\begin{array}{ccc} X & \xrightarrow{e} & Y \\ {\scriptstyle e}\downarrow & & \downarrow{\scriptstyle g} \\ Y & \xrightarrow[f]{} & Z \end{array}$$

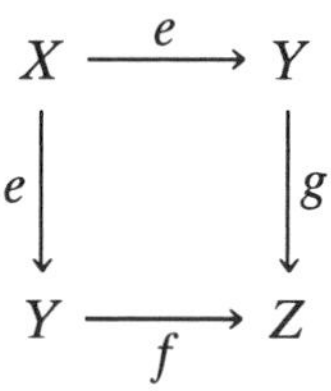

in Set implies $f = g$. If $e : X \to Y$ is a morphism in a category C such that the commuting of each diagram (1.3.2) implies $f = g$, then e is said to be an *epimorphism* in the category C. In particular, epimorphisms in the category Set are just epimorphisms in the sense of (O 3.1.5).

Example 1.3.1. Consider the category Ring of unital rings (cf. Example 1.2.7). If a ring homomorphism $e : R \to S$ is surjective, then e is an epimorphism in Ring. However, the ring homomorphism $e : \mathbb{Z} \to \mathbb{Q}$ embedding the integers in the rationals is an epimorphism in Ring that does not surject. □

A morphism $r : X \to Y$ in a category C is said to be *left invertible* or a *retraction* if there is a morphism $s : Y \to X$ such that

$$sr = 1_Y. \tag{1.3.3}$$

The dual of Proposition O 3.1.1 shows that the concepts of retraction and epimorphism are equivalent in the category Set. On the other hand, each morphism in a poset category (Example 1.3) is an epimorphism, while only the identities are left invertible there. In a general category C, all that survives of the dual of Proposition O 3.1.1 is the observation that retractions are epimorphisms (Exercise 1.3E).

To obtain the dual of a concept in a category C, one may interpret the concept in the dual category C^{op}. The dual of the concept of "epimorphism" is the concept of "monomorphism." Thus a *monomorphism* $m : Y \to X$ in a category C is a morphism such that the commuting of each diagram

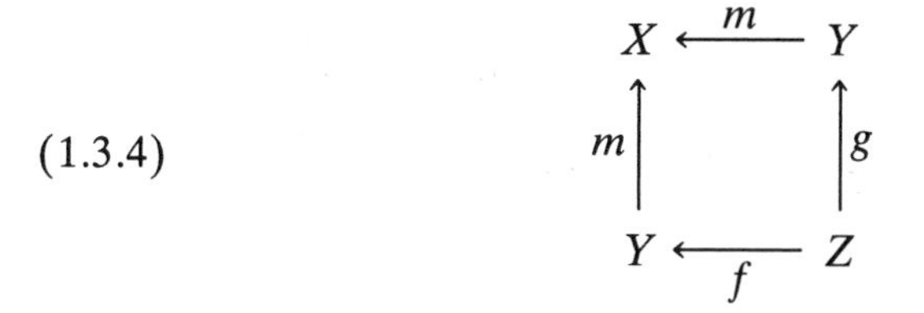

[the dual of (1.3.2)] in C implies $f = g$. Note that monomorphisms in the category Set are just monomorphisms in the sense of (O 3.1.2). A morphism $s: Y \to X$ in a category C is said to be *right invertible* or a *section* if there is a morphism $r : X \to Y$ such that (1.3.3) holds. Applying the result that "retractions are epimorphisms" (Exercise 1.3E) to the category C^{op}, one obtains the result that sections are monomorphisms in any category C. The relationship between these two results is described as duality: thus "retractions are epimorphisms" is dual to "sections are monomorphisms."

In (1.3.3), the morphism s is called a *left inverse* of the left invertible morphism r, and r is called a *right inverse* of the right invertible morphism s. A morphism in a category is *invertible* or an *isomorphism* if it has both a left inverse and a right inverse, i.e. if it is both a retraction and a section.

Proposition 1.3.2. *Let f be a morphism in a category C. Then f is an isomorphism iff it has a unique left inverse and a unique right inverse that coincide.*

Proof. The "if" part is immediate. Conversely, suppose that $f: X \to Y$ is an isomorphism, with a left inverse s and a right inverse r. Then $s = s1_X = sfr = 1_Y r = r$, so that s and r coincide. Moreover, if s' is also a left inverse, then $s' = r = s$. Similarly, the right inverses are unique. □

The coincident right and left inverses of an isomorphism f are denoted by f^{-1}, the *inverse* of the isomorphism f. A pair X, Y of objects of a category C are said to be *isomorphic*, written $X \cong Y$, if they are the domain and codomain of an isomorphism $f: X \to Y$ in C. For the concrete categories of Example 1.2.7, and for the category Cat of small categories, this concept of isomorphism agrees with earlier ones. Note that quasigroups P and Q are isomorphic in Qtp iff they are isotopic. If M is a monoid realized as a category according to Example 1.2, then the isomorphisms in the category M are the units of the monoid M.

Mathematical statements formulated in set-theoretical terms generally involve elements. In order to translate such statements into pictures of points and arrows, one often needs a pictorial way of representing elements. Now an element x of a set X is the unique element of the image of a function $\{\infty\} \to X; \infty \mapsto x$ from a singleton to X. Singleton sets are characterized by the property that there is a unique function from any given set to the singleton. The concept of a terminal object in a category abstracts this property of singletons in Set.

Definition 1.3.3. (a) An object T of a category C is *terminal* if $C(X, T)$ is a singleton for each object X of C.

(b) An object $\perp$ of a category C is *initial* if $C(\perp, X)$ is a singleton for each object X of C.

(c) An object 0 of a category C is a *zero* object if it is both initial and terminal. □

Note that initial objects in a category C are terminal objects in the dual category C^{op}: The concept of initial object is dual to the concept of terminal object. The empty set is initial in the category Set. An upper bound is terminal in a poset category. The zero group $\{0\}$ is a zero object in the category $\text{Mod}_{\mathbb{Z}}$. A unital, commutative ring K is initial in the category Alg_K. A category may have no initial or terminal objects (Exercise 1.3J). In Set, each singleton is a terminal object, and any two singletons are isomorphic. This behavior is typical.

Proposition 1.3.4. *Any two terminal objects of a category C are isomorphic.*

Proof. Suppose that C has terminal objects T_1 and T_2. Consider the diagram

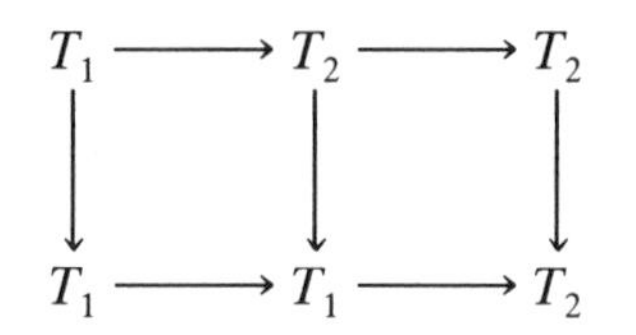

in which each arrow is uniquely specified since its codomain is terminal. The diagram commutes, since each path in the diagram has a terminal codomain. Commutativity of the left square shows that $T_2 \to T_1$ is left invertible. Commutativity of the right square shows that $T_2 \to T_1$ is right invertible. Thus $T_2 \to T_1$ is an isomorphism. □

Corollary 1.3.5. (a) *Any two initial objects of a category C are isomorphic.*
(b) *Any two zero objects of a category C are isomorphic.*

Proof. (a) The dual of Proposition 1.3.4, and
(b) A special case of Proposition 1.3.4. □

The apparent triviality of Corollary 1.3.5(a) is deceptive. In fact, the result is the basis for a major theme of category theory: the specification of mathematical constructs as initial objects in categories. For example, the ring of integers is specified [by Corollary 1.3.5(a), to within isomorphism] as an initial object in the category of unital commutative rings. One may view this as a precise formulation of the vague idea that "you can calculate with commutative rings the same way you calculate with integers."

EXERCISES

1.3A. Show that a small category C is a pre-ordering (cf. Exercise 1B) if the functor ε_C of Example 1.2.4 is a commuting diagram.

1.3B. Interpret each of (O 3.2.4)–(O 3.2.6) formally as a commuting diagram in the category $\underline{\underline{\text{Set}}}$. (For each, exhibit a suitable directed graph Γ and functor $D : \Gamma\Pi \to \underline{\underline{\text{Set}}}$.)

1.3C. Verify the content of Example 1.3.1.

1.3D. Show that a composite of epimorphisms in a category C is again an epimorphism in C.

1.3E. In a category C, prove that left invertible morphisms are epimorphisms.

1.3F. Let $0 \to A \xrightarrow{j} E \xrightarrow{p} Q \to 0$ be an exact sequence of abelian groups.

(a) Show that p is an epimorphism in the category $\underline{\underline{\text{Mod}}}_{\mathbb{Z}}$ (cf. Example 1.2.7).

(b) Show that p is a retraction iff the exact sequence splits (cf. Exercise II 1.1D).

(c) Show that j is a monomorphism in the category $\underline{\underline{\text{Mod}}}_{\mathbb{Z}}$.

(d) Show that j is a section iff the exact sequence splits.

1.3G. Formulate and do the dual of Exercise 1.3D.

1.3H. Show that isomorphism in a category C yields an equivalence relation on C_0.

1.3I. Show that the following three properties of a morphism f of a category C are equivalent:

(a) f is an isomorphism;

(b) f is a monomorphism and a retraction;

(c) f is an epimorphism and a section.

1.3J. Let $\underline{\underline{\text{Twoup}}}$ be the full subcategory of the category $\underline{\underline{\text{Set}}}$ comprising sets with at least two elements. Show that $\underline{\underline{\text{Twoup}}}$ has no terminal object and no initial object.

1.3K. Show that the empty set is a zero object in the category $\underline{\underline{\text{Pfn}}}$ of partial functions.

1.3L. Determine initial objects in each of the concrete categories of Example 1.2.7. In particular, contrast the initial objects of the categories $\underline{\underline{\text{Rng}}}$ and $\underline{\underline{\text{Ring}}}$.

1.3M. Let $\underline{\underline{\text{Fld}}}$ be the full subcategory of $\underline{\underline{\text{Ring}}}$ whose objects are fields. Show that $\underline{\underline{\text{Fld}}}$ has no initial object and no terminal object.

1.4. Natural Transformations and Functor Categories

Category theory studies three levels of transformations. At the first level, categories themselves comprise morphisms between objects. At the second level, functors operate between categories. At the third level, natural transformations operate between functors.

Definition 1.4.1. Let $S : C \to D$ and $T : C \to D$ be functors with common domain and codomain. Then a *natural transformation* $\tau : S \to T$ between S and T is a function

$$\tau : C_0 \to D_1;\ X \mapsto (\tau_X : XS \to XT) \tag{1.4.1}$$

such that, for each morphism $X \xrightarrow{f} Y$ of C, the diagram

$$\begin{array}{ccc} XS & \xrightarrow{\tau_X} & XT \\ {\scriptstyle fS}\big\downarrow & & \big\downarrow{\scriptstyle fT} \\ TS & \xrightarrow[\tau_Y]{} & YT \end{array} \tag{1.4.2}$$

commutes. The D-morphism $\tau_X : XS \to XT$ is called the *component* of τ at the object X of C. The natural transformation is called a *natural isomorphism* if each of its components is an isomorphism of D. □

Note that, to specify a natural transformation, one just has to specify its components, in such a way that the diagrams (1.4.2) commute.

Example 1.4.2. Let M be a monoid, construed as a category M with $M_0 = \{1\}$ and $M_1 = M$ according to Example 1.2. Let (A, M) be an M-set via representation $S : M \to A^A$ and let (B, M) be an M-set via representation $T : M \to B^B$. Construe S as a functor $S : M \to \underline{\underline{\mathrm{Set}}}$ with object part $S : 1 \mapsto A$ and morphism part $S : m \mapsto (A \to A;\ a \mapsto am)$. Construe T similarly. Then an M-homomorphism $f : (A, M) \to (B, M)$ is a natural transformation $f : S \to T$ with component $f : A \to B$ at the unique object 1 of the common domain category M. The diagram (1.4.2) of Definition 1.4.1 becomes

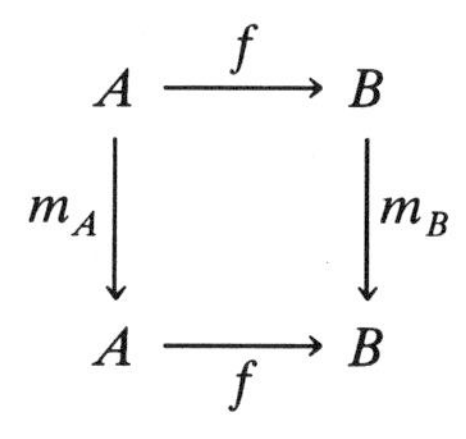

for each element m of M. The natural transformation $f : S \to T$ is a natural isomorphism if and only if $f : (A, M) \to (B, M)$ is an isomorphism of M-sets. □

Example 1.4.3. For sets Y_1 and Y_2, consider the functors $F_{Y_1} : \underline{\underline{\mathrm{Set}}} \to \underline{\underline{\mathrm{Set}}}$ and $F_{Y_2} : \underline{\underline{\mathrm{Set}}} \to \underline{\underline{\mathrm{Set}}}$ of Example 1.2.2, with respective object parts $Z \mapsto Z \times Y_1$ and $Z \mapsto Z \times Y_2$. Let $g : Y_1 \to Y_2$ be a function. Then there is a natural transformation $\gamma : F_{Y_1} \to F_{Y_2}$ with component $\gamma_Z : Z \times Y_1 \to Z \times Y_2;\ (z, y) \mapsto (z, yg)$. The diagram (1.4.2) of Definition 1.4.1, for a function $f : Z_1 \to Z_2$, becomes

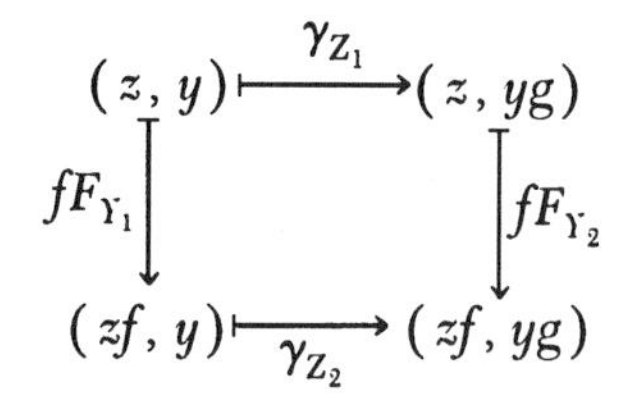

for $z \in Z_1$ and $y \in Y_1$. □

Example 1.4.4. Let K be a commutative, unital ring. Let $S : \underline{\underline{\mathrm{Mod}}}_K \to \underline{\underline{\mathrm{Mod}}}_K$ be the identity functor, and let $T : \underline{\underline{\mathrm{Mod}}}_K \to \underline{\underline{\mathrm{Mod}}}_K$ be the (covariant) composite of the "functional" functors $K^1(\ ;K):\underline{\underline{\text{Mod}}}_K \to \underline{\underline{\text{Mod}}}_K^{\text{op}}$ and $K^1(\ ;K):\underline{\underline{\text{Mod}}}_K^{\text{op}} \to \underline{\underline{\text{Mod}}}_K$ (cf. Exercise 1.2P). Then for each K-module V, the evaluation map of Exercise II 2.1E is the component at V of a natural transformation between S and T. □

Given two natural transformations $\sigma : S \to T$ and $\tau : T \to U$ between functors $S : C \to D$, $T : C \to D$, and $U : C \to D$, a composite $\sigma \cdot \tau$ or $\sigma\tau : S \to U$ is defined by its components $(\sigma\tau)_X = \sigma_X \tau_X$ at objects X of C. For a C-morphism $f : X \to Y$, the diagram

(1.4.3)

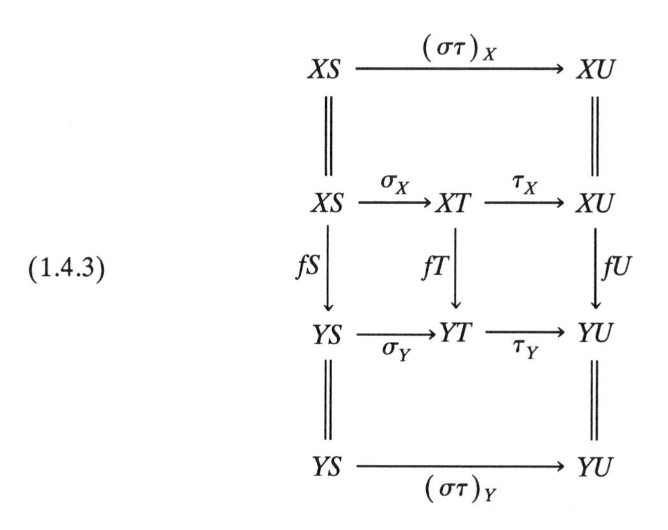

in D commutes, so that $\sigma\tau$ is also a natural transformation. This composition of natural transformations is associative. Given a functor $S : C \to D$, there is a natural isomorphism $1_S : S \to S$ with components $1_{XS} : XS \to XS$ at objects X of C. Such a natural isomorphism 1_S or $1 : S \to S$ is called an *identity transformation*. If $\tau : S \to T$ is a natural isomorphism, then an *inverse transformation* $\tau^{-1} : T \to S$ is defined, with component $(\tau^{-1})_Y = \tau_Y^{-1}$ at an object Y of T. Note that $\tau\tau^{-1} = 1_S$ and $\tau^{-1}\tau = 1_T$.

For small categories C and D, the set $\underline{\underline{\text{Cat}}}(C, D) = (D^C)_0$ of functors from C to D becomes the object set of a category D^C, a so-called *functor category*. For functors $S : C \to D$ and $T : C \to D$, the class $D^C(S, T)$ is the set of natural transformations between S and T. By (1.4.1), $D^C(S, T)$ is a subset of $\underline{\underline{\text{Set}}}(C_0, D_1)$. Composition in D^C is the composition $\sigma\tau$ of natural transformations. If C is small and D is large, one similarly obtains a large category D^C. Note that each functor $S : C \to D$ is represented by a set, the graph $\{(f, fS) \in C_1 \times C_1 S | f \in C_1\}$ of its morphism part. Similarly, each natural transformation $\tau : S \to T$ is represented by a set, the graph $\{(X, \tau_X) \in C_0 \times C_0\tau | X \in C_0\}$ of the function (1.4.1). Thus $(D^C)_0$ is a class and $(D^C)_1$ is a class.

Example 1.4.5. Let M be a monoid. Then Example 1.4.2 shows that the large functor category $\underline{\underline{\mathrm{Set}}}^M$ is (isomorphic to) the category $\underline{\underline{M}}$ of M-sets and M-homomorphisms. □

Now suppose that there are functors $R: B \to C$, $S: C \to D$, $T: C \to D$, and $U: D \to E$, together with a natural transformation $\tau: S \to T$. Define

$$\tau U: C_0 \to E_1;\ X \mapsto (\tau_X U: XSU \to XTU). \tag{1.4.4}$$

A diagram (1.4.2) for τU is obtained by applying the functor U to a diagram (1.4.2) for τ. Thus τU becomes a natural transformation $\tau U: SU \to TU$, with components given by (1.4.4). Dually, define

$$R\tau: B_0 \to D_1;\ W \mapsto (\tau_{WR}: WRS \to WRT). \tag{1.4.5}$$

A diagram (1.4.2) for $R\tau$ at a morphism $f: W \to W'$ of B is obtained as the diagram (1.4.2) for τ at the morphism $fR: WR \to W'R$ of C. Thus $R\tau$ becomes a natural transformation $R\tau: RS \to RT$, with components given by (1.4.5).

EXERCISES

1.4A. Verify the claims in Example 1.4.4.

1.4B. Let K be a field, and let $\underline{\underline{\mathrm{Mod}}}_K^{<\omega}$ be the full subcategory of $\underline{\underline{\mathrm{Mod}}}_K$ consisting of finite-dimensional vector spaces over K. Let $T^{<\omega}$ be the restriction of the functor T of Example 1.4.4 to the subcategory $\underline{\underline{\mathrm{Mod}}}_K^{<\omega}$. Show that the evaluation maps are the components of a natural isomorphism between the identity functor on $\underline{\underline{\mathrm{Mod}}}_K^{<\omega}$ and the functor $T^{<\omega}$.

1.4C. Let C be a category with identity functor 1_C. Show that there is a natural transformation $e_C: 1_C \to 1_C$ given by the function (1.1).

1.4D. (a) For a set (alphabet) A, let A^* be the free monoid on A. For a function $f: A \to B$, define $f^*: A^* \to B^*$ to be the unique monoid homomorphism extending the composite set function $A \xrightarrow{f} B \hookrightarrow B^*$. Show that there is a functor $F: \underline{\underline{\mathrm{Set}}} \to \underline{\underline{\mathrm{Mon}}}$ with object part $A \mapsto A^*$ and morphism part $f \mapsto f^*$.

(b) Let $U: \underline{\underline{\mathrm{Mon}}} \to \underline{\underline{\mathrm{Set}}}$ be the forgetful functor, i.e. with morphism part $(f:(M,\cdot,1) \to (N,\cdot,1)) \mapsto (f: M \to N)$. Show that there are natural transformations $\eta: 1_{\underline{\underline{\mathrm{Set}}}} \to FU$ with components $\eta_A: A \hookrightarrow A^*$ and $\varepsilon: UF \to 1_{\underline{\underline{\mathrm{Mon}}}}$ with each component $\varepsilon_M: M^* \to M$ defined as the unique monoid homomorphism extending the identity $1_M: M \to M$.

(c) Verify the identities

$$(\eta F)_A(F\varepsilon)_A = 1_{AF}$$

for each alphabet A and

$$(U\eta)_M(\varepsilon U)_M = 1_{MU}$$

for each monoid M.

1.4E. Let $(s, f):(A, M, R) \to (B, N, R')$ be an action morphism. Given a monoid homomorphism $e: L \to M$, show that there is an action morphism

$$(es, ef):(A, L, eR) \to (B, N, R').$$

[Hint: Use Example 1.4.2 and (1.4.5) to define es.]

1.4F. Let D be a category with terminal object T. Let C be a small category. Define a functor $T\Delta: C \to D$ with object part $C_0 \to \{T\}$ and with morphism part $C_1 \to \{1_T\}$. Show that $T\Delta$ is a terminal object of the functor category D^C.

1.4G. Let X be a poset category. For a positive integer m, let $\underline{m}$ denote the poset category with Hasse diagram $1 < 2 < \cdots < m$. Identify the objects of the functor category $X^{\underline{m}}$ as the $(m-1)$-simplices in the poset $(X, \leq)$.

2. LIMITS AND LATTICES

Semilattices were introduced in Section O 4.3 as idempotent, commutative semigroups. Given such a semigroup $(S,\cdot)$, an order $\leq$ is defined on S by (O 4.3.3). The partial order $(S, \leq)$ has the special property that each subset $\{a,b\}$ of S has a greatest lower bound glb$\{a,b\}$ given by the product $a \cdot b$ in the semigroup $(S,\cdot)$. Conversely, if $(S, \leq)$ is a partial order in which each subset $\{a, b\}$ has a greatest lower bound glb$\{a, b\}$, then there is a semilattice $(S,\cdot)$ whose product $a \cdot b$ is given by the greatest lower bound glb$\{a, b\}$.

A semilattice S may be realized as a poset category $(S, \leq)$. The defining property of the greatest lower bound glb$\{a, b\}$ or product $a \cdot b$ of two elements (objects) a, b of S may then be expressed diagrammatically in the category S:

$$\begin{array}{ccccc} a & \longleftarrow & a\cdot b & \longrightarrow & b \\ \uparrow & & \uparrow & & \uparrow \\ c & \underset{1_c}{\longleftarrow} & c & \underset{1_c}{\longrightarrow} & c \end{array} \quad . \tag{2.1}$$

Given any object c with arrows $c \to a$ and $c \to b$, there is a unique arrow $c \to a \cdot b$ such that (2.1) commutes. The top row of (2.1) says that $a \cdot b \leq a$ and $a \cdot b \leq b$. The rest of the diagram represents the implication ($c \leq a$ and $c \leq b$) $\Rightarrow c \leq a \cdot b$. Now the diagram (2.1) for the product $a \cdot b$ in the category S is strikingly reminiscent of the diagram (O 3.2.4) for the direct product $A \times B$ of two objects A and B in the category $\underline{\underline{\mathrm{Set}}}$ of sets. One is thus led to the following

Definition 2.1. Let A_1 and A_2 be objects of a category D. Then the *product* of the objects A_1 and A_2 is an object $A_1 \times A_2$ of D together with morphisms $\pi_1 : A_1 \times A_2 \to A_1$ and $\pi_2 : A_1 \times A_2 \to A_2$ of D known as *projections*, having the following *universal property*:

$$\text{(2.2)} \quad \begin{cases} \text{for } B \in D_0 \text{ and for each element } ((f_1 : B \to A_1), (f_2 : B \to A_2)) \\ \text{of } D(B, A_1) \times D(B, A_2), \text{ there is a unique element } (f_1, f_2) \text{ of} \\ D(B, A_1 \times A_2) \text{ such that } (f_1, f_2)\pi_1 = f_1 \text{ and } (f_1, f_2)\pi_2 = f_2. \end{cases}$$

The universality property (2.2) is represented by the diagram

$$\text{(2.3)} \qquad \begin{array}{ccccc} A_1 & \xleftarrow{\pi_1} & A_1 \times A_2 & \xrightarrow{\pi_2} & A_2 \\ {\scriptstyle f_1}\uparrow & & \uparrow{\scriptstyle (f_1, f_2)} & & \uparrow{\scriptstyle f_2} \\ B & \xleftarrow[1_B]{} & B & \xrightarrow[1_B]{} & B \end{array}$$

in D. □

Example 2.2. (a) If X is a poset category, then the product of two elements x, y of X is the greatest lower bound glb$\{x, y\}$. Note that a pair of elements of a poset may or may not have a greatest lower bound (compare the antichains of Exercise O 4.3A). Thus in general, a pair of objects of a category may or may not have a product.

(b) If A_1 and A_2 are sets, then the product $A_1 \times A_2$ of A_1 and A_2 in $\underline{\underline{\mathrm{Set}}}$ is a direct product of A_1 and A_2.

(c) Products in the concrete categories of Example 1.2.7 are given by the componentwise structure on the Cartesian product. For example, (O 4.8) describes products in the category $\underline{\underline{\mathrm{Mon}}}$ of monoids. □

The dual $\underline{\underline{\mathrm{Set}}}^{\mathrm{op}}$ of the category $\underline{\underline{\mathrm{Set}}}$ of sets has a product of each pair A_1, A_2 of objects, namely the disjoint union $A_1 \mathbin{\dot\cup} A_2$ together with the insertions $A_i \xrightarrow{\iota_i} A_1 \mathbin{\dot\cup} A_2$. In general, the product in D^{op} of two elements A_1, A_2 of $(D^{\mathrm{op}})_0 = D_0$ is called the "coproduct" $A_1 + A_2$ in D of the objects A_1, A_2 of D. Thus the coproduct in $\underline{\underline{\mathrm{Set}}}$ is the disjoint union. Formally, one has the following:

Definition 2.3. Let A_1 and A_2 be objects of a category D. Then the *coproduct* of the objects A_1 and A_2 is an object $A_1 + A_2$ of D together with morphisms $\iota_1 : A_1 \to A_1 + A_2$ and $\iota_2 : A_2 \to A_1 + A_2$ of D known as *insertions*, having the following *universal property*:

$$(2.4)\quad \begin{cases} \text{For } b \in D_0 \text{ and for each element } ((f_1 : A_1 \to B), (f_2 : A_2 \to B)) \\ \text{of } D(A_1, B) \times D(A_2, B), \text{ there is a unique element } f_1 + f_2 \\ \text{of } D(A_1 + A_2, B) \text{ such that } \iota_1(f_1 + f_2) = f_1 \text{ and } \iota_2(f_1 + f_2) = f_2. \end{cases}$$

This universality property (2.4) is represented by the diagram

$$(2.5)\quad \begin{array}{ccccc} A_1 & \xrightarrow{\iota_1} & A_1 + A_2 & \xleftarrow{\iota_2} & A_2 \\ {\scriptstyle f_1}\downarrow & & \downarrow{\scriptstyle f_1 + f_2} & & \downarrow{\scriptstyle f_2} \\ B & \xrightarrow[1_B]{} & B & \xleftarrow[1_B]{} & B \end{array}$$

in D. □

Example 2.4. (a) Let S be a unital ring, and let V, W be objects of $\underline{\underline{\text{Mod}}}_S$. Then the direct sum $V \oplus W$, together with the insertions $V \to V \oplus W$ and $W \to V \oplus W$, is the coproduct of V and W in $\underline{\underline{\text{Mod}}}_S$.

(b) Let $(X, \leq)$ be a poset category, with elements x and y. The coproduct of x and y in X is the *least upper bound* $\text{lub}\{x, y\}$, the element of X satisfying $x \leq \text{lub}\{x, y\}$, $y \leq \text{lub}\{x, y\}$, and $((x \leq z$ and $y \leq z) \Rightarrow \text{lub}\{x, y\} \leq z)$.

(c) A coproduct in D^{op} of two elements A_1, A_2 of $(D^{\text{op}})_0 = D_0$ is the product of A_1 and A_2 in D. □

Let $(S, +)$ be a semilattice, written "additively," i.e. with $+$ as the idempotent, commutative, and associative binary operation. Define a relation $\leq_+$ or $\leq$ on S by

$$(2.6)\qquad x \leq y \Leftrightarrow x + y = y.$$

Then $(S, \leq)$ is a partial order in which each subset $\{a, b\}$ has a least upper bound $\text{lub}\{a, b\} = a + b$. Conversely, given such a partial order $(S, \leq)$, one may set $a + b := \text{lub}\{a, b\}$ to obtain a semilattice $(S, +)$. A semilattice $(S, +)$ ordered according to (2.6) is called a *join semilattice* $(S, +, \leq_+)$, while a semilattice $(S, \cdot)$ ordered according to (O 4.3.3) is called a *meet semilattice* $(S, \cdot, \leq_{.})$.

A *lattice* may be defined order-theoretically as a partial order $(L, \leq)$ in which each subset $\{a, b\}$ has both a greatest lower bound $\text{glb}\{a, b\} = a \cdot b$ and a least upper bound $\mathrm{lub}\{a, b\} = a + b$. Thus both $(L, \cdot)$ and $(L, +)$ are semilattices. If L is ordered as a meet semilattice from $(L, \cdot)$, and as a join semilattice from $(L, +)$, then the two orders agree, yielding the original order $(L, \leq)$. Thus a lattice may be characterized as a set $(L, +, \cdot)$ with two semilattice operations $+$ and $\cdot$, such that the join semilattice order $\leq_+$ and meet semilattice order $\leq_.$ agree. It is sometimes convenient to denote the product $x \cdot y$ by the juxtaposition xy. Moreover, in compound expressions, the product or *meet* $x \cdot y$ or xy will bind more strongly than the coproduct or *join* $x + y$. Aside from this notational convention, the concept of a lattice is self-dual. Note that duality on a lattice interchanges the roles of meet and join. Completing the transformation from order-theoretic to algebraic characterizations, one obtains the following:

Proposition 2.5. *Let* $(L, +, \cdot)$ *be a set with two semilattice operations related by the* absorption laws

$$x \cdot y + y = y \tag{2.7}$$

and

$$x \cdot (x + y) = x. \tag{2.8}$$

Then the meet semilattice and join semilattice ordering on L *agree, yielding a lattice* $(L, \leq)$. *Conversely, if* $(L, \leq)$ *is a lattice, then the absorption laws are satisfied.*

Proof. If (2.7) is satisfied, one has $x \leq_. y \Rightarrow x \cdot y = x \Rightarrow x + y = x \cdot y + y = y \Rightarrow x \leq_+ y$. Dually, (2.8) yields $x \leq_+ y \Rightarrow x \leq_. y$. Thus the meet and join semilattice orderings agree if both absorption laws are satisfied. Conversely, in a lattice $(L, \leq)$, one has $y = \mathrm{lub}\{\mathrm{glb}\{x, y\}, y\}$, so (2.7) holds. Dually, (2.8) holds. □

Given the algebraic characterization Proposition 2.5 of lattices, one may apply the usual algebraic constructions to lattices. For example, the *direct product* of two lattices $(L, +, \cdot)$ and $(M, +, \cdot)$ is the set $L \times M$ with componentwise join and meet. A *sublattice* of a lattice $(L, +, \cdot)$ is a subset S of L such that $(S, +)$ is a subsemigroup of $(L, +)$ and $(S, \cdot)$ is a subsemigroup of $(L, \cdot)$. A *congruence* α on a lattice $(L, +, \cdot)$ is an equivalence relation α on L that is a sublattice of $(L^2, +, \cdot)$. A *lattice homomorphism* $f: (L, +, \cdot) \to (M, +, \cdot)$ between lattices $(L, +, \cdot)$ and $(M, +, \cdot)$ is a function $f: L \to M$ that gives semilattice homomorphisms $f: (L, +) \to (M, +)$ and $f: (L, \cdot) \to (M, \cdot)$. The concrete category $\underline{\underline{\mathrm{Lat}}}$ of lattices has the class of all lattices as its object class $\underline{\underline{\mathrm{Lat}}}_0$. Given lattices L and M, the class $\underline{\underline{\mathrm{Lat}}}(L, M)$ is the set of lattice homomorphisms from L to M.

EXERCISES

2A. Let D be a category in which any two objects have a product. Let A_1, A_2, A_3 be objects of D. Exhibit isomorphisms $\tau : A_1 \times A_2 \to A_2 \times A_1$ and $(A_1 \times A_2) \times A_3 \cong A_1 \times (A_2 \times A_3)$.

2B. Let A and B be sets. Show that $(A \,\dot\cup\, B)^*$ is the coproduct in the category $\underline{\underline{\mathrm{Mon}}}$ of the free monoids A^* and B^*.

2C. (a) Let T be a terminal object of the category D. For each object A of D:

(i) show that the product $T \times A$ exists; and

(ii) show that there is an isomorphism $A \cong T \times A$.

(b) Formulate and do the dual of (a).

2D. Does the category $\underline{\underline{\mathrm{Pfn}}}$ (cf. Example 1.1.5) have a product for each pair of objects?

2E. For each of the following partial orders $(L, \leq)$, determine whether L is a lattice:

(a) the intersection of the linear orders $1 < 2 < 3 < 4 < 5 < 6$ and $1 < 3 < 2 < 5 < 4 < 6$;

(b) the intersection of the linear orders $1 < 2 < 4 < 3 < 5 < 6$ and $1 < 3 < 2 < 5 < 4 < 6$ [cf. Proposition O 3.5.4(b) and Exercise O 3.5E].

2F. (a) For a set B, show that the set $\mathscr{P}_{<\omega}(B)$ of finite subsets of B forms a lattice $(\mathscr{P}_{<\omega}(B), \subseteq)$ under the relation of containment.

(b) Define a subset X of the set B to be *cofinite* if its complement $B - X$ is finite. Let $\mathscr{P}^{<\omega}(B)$ denote the set of cofinite subsets of B. Show that $(\mathscr{P}^{<\omega}(B), \subseteq)$ is a lattice.

2G. Let Q be a quasigroup, let $\mathrm{Sb}\, Q$ be the set of subquasigroups of Q, and let $\mathrm{Sb}_{>\varnothing} Q$ be the set of non-empty subquasigroups of Q. Let $\subseteq$ be the containment relation.

(a) Show that $(\mathrm{Sb}\, Q, \subseteq)$ is a lattice.

(b) Show that $(\mathrm{Sb}_{>\varnothing} Q, \subseteq)$ is a join semilattice.

(c) Give an example of a quasigroup Q for which $(\mathrm{Sb}_{>\varnothing} Q, \subseteq)$ is not a lattice.

2H. Let $(L, +, \cdot)$ be a set with two semilattice operations. Let $(L, +)$ be the join semilattice with Hasse diagram $1 \to c \to 0$, and let $(L, \cdot)$ be the meet semilattice with Hasse diagram $c \to 1 \leftarrow 0$. Show that $(L, +, \cdot)$ satisfies the absorption law (2.7), but not the absorption law (2.8). [See J. Gałuszka, "On Bisemilattices with Generalized Absorption Laws, I," *Dem. Math.* **20** (1987), 37–43.]

2I. State and prove the First Isomorphism Theorem for lattices.

2J. Show that any two objects of $\underline{\underline{\mathrm{Lat}}}$ have a product.

2K. Consider the poset C with Hasse diagram

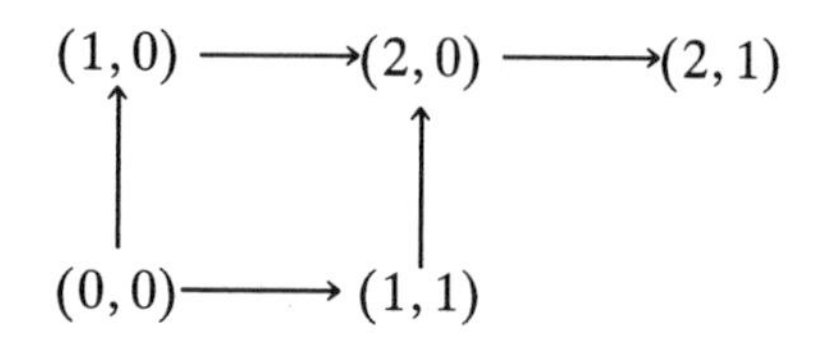

Let D be the poset $0 < 1 < 2$. Show that C and D are lattices, and that $F: C \to D$; $(x, y) \mapsto x$ is an order-preserving map that is not a lattice homomorphism.

2L. Let $F: C \to D$ be a bijection between lattices C and D such that $F:(C, \leq) \to (D, \leq)$ is an isomorphism in $\underline{\underline{\text{Pos}}}$. Show that F is a lattice homomorphism.

2M. (a) Show that the forgetful functor $U: \underline{\underline{\text{Mon}}} \to \underline{\underline{\text{Set}}}$ with object part $(M, \cdot, 1) \to M$ preserves products, in the sense that $((M, \cdot, 1) \times (N, \cdot, 1))U = M \times N$.

(b) Show that U does not preserve coproducts.

2N. Let C be a small category, and let D be a category in which each pair of objects has a product. Show that any two objects S_1, S_2 of the functor category D^C have a product $S_1 \times S_2$ defined *pointwise* by $(f: X \to Y)(S_1 \times S_2) = (\pi_1 f^{S_1}, \pi_2 f^{S_2}): XS_1 \times XS_2 \to YS_1 \times YS_2$ for each C-morphism $f: X \to Y$. [Hint: The projection $\pi^i: S_1 \times S_2 \to S_i$ in D^C has component $\pi^i_X = \pi_i : XS_1 \times XS_2 \to XS_i$ at an object X of C.]

2.1. Products and Coproducts

For a set B, the set $\mathscr{P}_{<\omega}(B)$ of finite subsets of B (cf. Exercise 2F) forms a poset $(\mathscr{P}_{<\omega}(B), \subseteq)$ in which each non-empty multiset $\langle B_j | j \in J \rangle$ of elements has a greatest lower bound, namely $\bigcap_{j \in J} B_j$. If $\langle A_j | j \in J \rangle$ is a multiset of sets, then one may construct the product $\prod_{j \in J} A_j$, along with projections $\pi_i : \prod_{j \in J} A_j \to A_i$ for each i in J, such that, given a multiset $\langle f_j : B \to A_j | j \in J \rangle$ of maps, there is a unique map $f: B \to \prod_{j \in J} A_j$ with $f\pi_i = f_i$ for each i in J (cf. Section O 3.2, including Exercise 3.2Q). For a two-element set J, the greatest lower bound $\bigcap_{j \in J} B_j$ and product $\prod_{j \in J} A_j$ are instances of the two-factor categorical product of Definition 2.1. One is thus led to the following, more general definition:

Definition 2.1.1. Let J be a set, and let D be a category. Let $A: J \to D_0$; $j \mapsto A_j$ be a function. Then the *product* π or π^A of A is an object $\prod_{j \in J} A_j$ of D together with morphisms $\pi_i : \prod_{j \in J} A_j \to A_i$ (known as *projections*) for each i in J, having the following *universal property*:

$$(2.1.1)\quad \begin{cases} \text{For each object } B \text{ of } D \text{ and element } \langle \varphi_j | j \in J \rangle \text{ of } \prod_{j \in J} D(B, A_j), \\ \text{there is a unique element } \prod_{j \in J} \varphi_j \text{ of } D(B, \prod_{j \in J} A_j) \\ \text{such that } \varphi_i = (\prod_{j \in J} \varphi_j)\pi_i \text{ for each } i \text{ in } J. \end{cases}$$

The universality property (2.1.1) is represented by the diagram

$$(2.1.2)\quad \begin{array}{ccc} \prod_{j\in J} A_j & \xleftarrow{\prod_{j\in J}\varphi_j} & B \\ {\scriptstyle \pi_i}\downarrow & & \| \\ A_i & \xleftarrow[\varphi_i]{} & B \end{array}$$

in D. □

By the usual abuse of notation, the object $\prod_{j \in J} A_j$ itself is often called the *product*, leaving tacit the role of the projections.

Example 2.1.2. (a) Let D be one of the concrete categories of Example 1.2.7. Then for any function $A : J \to D_0$ (with domain a set J), the product object $\prod_{i \in J} A_j$ is given by the product set $\prod_{j \in J} A_j U$ with componentwise structure. "Componentwise structure" means precisely that the projections π_i are morphisms of D.

(b) Let $(X, \leq)$ be a poset, construed as a category. Let $A : J \to X$; $j \mapsto x_j$ be a function. Then the product π^A is the greatest lower bound $\prod\{x_j | j \in J\}$ of the set $\{x_j | j \in J\}$, an element $\prod\{x_j | j \in J\}$ of X with the two properties:

(i) $\forall i \in J,\ \prod\{x_j | j \in J\} \leq x_i$;
(ii) $(y \in X$ and $\forall i \in J,\ y \leq x_i) \Rightarrow y \leq \prod\{x_j | j \in J\}$.

Note how (i) corresponds to the existence of the projections π_i, while (ii) corresponds to the existence of the horizontal arrows in (2.1.2). As already seen for the case $|J| = 2$, such a greatest lower bound may or may not exist.

(c) Let D be a category with a terminal object T. Then T is the product of the function $\varnothing : \varnothing \to D_0$. All but the top line of the diagram (2.1.2) disappears, while the existence and uniqueness of the arrow in the top line follows from the Definition 1.3.3(a) of a terminal object.

(d) For an infinite set B, consider the poset $(\mathscr{P}_{<\omega}(B), \subseteq)$. As observed at the beginning of the section, a function $A : J \to \mathscr{P}_{<\omega}(B)$ with non-empty domain has a product, namely $\bigcap_{j \in J} B_j$. On the other hand, the set $\mathscr{P}_{<\omega}(B)$

has no upper bound, i.e. the poset category $(\mathcal{P}_{<\omega}(B), \subseteq)$ has no terminal object. Thus the empty function $\varnothing : \varnothing \to \mathcal{P}_{<\omega}(B)$ has no product. □

Products in the dual D^{op} of a category D are called "coproducts." Formally, one has the following definition, generalizing Definition 2.3.

Definition 2.1.3. Let J be a set, and let D be a category. Let $A : J \to D_0$; $j \mapsto A_j$ be a function. Then the *coproduct* ι or ι^A of A is an object $\Sigma_{j \in J} A_j$ or $\coprod_{j \in J} A_j$ of D together with morphisms $\iota_i : A_i \to \Sigma_{i \in J} A_j$ (known as *insertions*) for each i in J, having the following *universal property*:

$$\text{(2.1.3)} \quad \begin{cases} \text{For each object } B \text{ of } D \text{ and element } \langle \varphi_j | j \in J \rangle \text{ of } \Pi_{j \in J} D(A_j, B), \\ \text{there is a unique element } \Sigma_{j \in J} \varphi_j \text{ of } D(\Sigma_{j \in J} A_j, B) \\ \text{such that } \varphi_i = \iota_i(\Sigma_{j \in J} \varphi_j) \text{ for each } i \text{ in } J. \end{cases}$$

The universality property (2.1.3) is represented by the diagram

$$\text{(2.1.4)} \qquad \begin{array}{ccc} \sum\limits_{j \in J} A_j & \xrightarrow{\sum_{j \in J} \varphi_j} & B \\ {\scriptstyle \iota_i} \uparrow & & \| \\ A_i & \xrightarrow[\varphi_i]{} & B \end{array}$$

in D. □

Again, one sometimes refers to $\Sigma_{i \in J} A_j$ alone as the *coproduct*, not explicitly mentioning the insertions. The notation $\coprod_{j \in J} A_j$ for the coproduct is preferable in categories such as $\underline{\underline{\text{Mod}}}_S$ where one wishes to reserve the notation $\Sigma_{j \in J} A_j$ for other purposes, such as the submodule of a module X generated by the union $\bigcup\{A_j | j \in J\}$ of submodules A_j of X.

Example 2.1.4. (a) In the category $\underline{\underline{\text{Set}}}$, the coproduct $\Sigma_{j \in J} A_j$ of an arbitrary function $A : J \to \underline{\underline{\text{Set}}}_0$ from a (small) set J is given by the disjoint union construction of Exercise O 3.2P.

(b) In the category $\underline{\underline{\text{Mod}}}_{\mathbb{Z}}$ of abelian groups, the coproduct $\coprod_{j \in J} A_j$ is given by the construction of II 1.1 as a subgroup $\Sigma_{i \in J} A_j \iota_j$ of $\Pi_{j \in J} A_j$ [cf. (II 1.1.5)].

(c) For a function $A : J \to \underline{\underline{\text{Set}}}_0$; $j \mapsto A_j$ from a (small) set J, the coproduct in $\underline{\underline{\text{Mon}}}$ of the function $A : J \to \underline{\underline{\text{Mon}}}_0$; $j \mapsto A_j^*$ is the free monoid $(\Sigma_{j \in j} A_j)^*$ on the disjoint union $\Sigma_{j \in J} A_j$ of the alphabets A_j.

(d) Let $(X, \leq)$ be a poset, construed as a category. For a function $A: J \to X;\ j \mapsto x_j$, the coproduct ι^A is the least upper bound $\Sigma\{x_j | j \in J\}$ of the set $\{x_j | j \in J\}$, defined dually to the greatest lower bound.

(e) Let D be a category with an initial object $\perp$. Then $\perp$ is the coproduct of the function $\varnothing : \varnothing \to D_0$. □

For a poset category $(X, \leq)$, the existence of the product (greatest lower bound) of any subset is equivalent to the existence of the coproduct (least upper bound) of any subset.

Proposition 2.1.5. *Let $(X, \leq)$ be a poset, construed as a category. Then the following conditions are equivalent:*

(a) *For each function $A: J \to X;\ j \mapsto x_j$, the greatest lower bound $\Pi\{x_j | j \in J\}$ exists;*

(b) *For each function $A: J \to X;\ j \mapsto x_j$, the least upper bound $\Sigma\{x_j | j \in J\}$ exists.*

Proof. Suppose that (a) holds, and that $A: J \to X;\ j \mapsto x_j$ is a function. Define $U = \{u \in X | \forall j \in J, x_j \leq u\}$, i.e. the set of upper bounds of JA. Consider the inclusion function $I: U \to X;\ u \mapsto u$. By (a), the product ΠU exists. Consider $x_j \in JA$. By Example 2.1.2(b)(ii), $(\forall u \in U, x_j \leq u) \Rightarrow x_j \leq \Pi U$, so that ΠU is an upper bound for JA. Suppose that u is an upper bound for JA. Then by Example 2.1.2(b)(i), one has $\Pi U \leq u$. In other words, ΠU is the least upper bound of JA. Thus (b) holds. The proof (b) $\Rightarrow$ (a) is dual. □

A poset satisfying the equivalent conditions of Proposition 2.1.5 is called a *complete lattice*. The binary product $\cdot$ or meet in a complete lattice $(X, \leq)$ yields a monoid $(X, \cdot, 1)$ whose identity element is the empty product or terminal object [cf. Exercise 2C(a)(ii)]. Dually, the binary coproduct $+$ or join in the complete lattice $(X, \leq)$ yields a monoid $(X, +, 0)$ whose zero element is the empty coproduct or initial object. A lattice $(X, +, \cdot)$ is *bounded above* if the semigroup $(X, \cdot)$ is a monoid $(X, \cdot, 1)$, and *bounded below* if the semigroup $(X, +)$ is a monoid $(X, +, 0)$. A lattice $(X, +, \cdot)$ is *bounded* if it is bounded both above and below. Thus a bounded lattice $(X, +, \cdot, 0, 1)$ comprises monoids $(X, +, 0)$ and $(X, \cdot, 1)$. A *bounded lattice homomorphism* $f: (X, +, \cdot, 0, 1) \to (Y, +, \cdot, 0, 1)$ is a lattice homomorphism with $0f = 0$ and $1f = 1$, i.e. a function $f: X \to Y$ that gives monoid homomorphisms $f: (X, +, 0) \to (Y, +, 0)$ and $f: (X, \cdot, 1) \to (Y, \cdot, 1)$. Note that a finite lattice is complete (Exercise 2.1D).

Example 2.1.6. For a set B, consider the power set $(2^B, \subseteq)$ ordered by containment. The terminal object of 2^B is B. If $A: J \to 2^B;\ j \mapsto A_j$ is a function, then the product π^A or $\prod_{j \in J} A_j$ is the intersection $\bigcap_{j \in J} A_j$. Thus $(2^B, \subseteq)$ is a complete lattice. □

Example 2.1.7. Let Q be a quasigroup, and let $(\mathrm{Cg}Q, \subseteq)$ be the set of congruences on Q ordered by containment. The terminal object of $\mathrm{Cg}Q$ is Q^2. If $A : J \to \mathrm{Cg}Q$; $j \mapsto \alpha_j$ is a function, then the product $\prod_{j \in J} \alpha_j$ is the intersection $\bigcap_{j \in J} \alpha_j$. Thus $(\mathrm{Cg}Q, \subseteq)$ is a complete lattice. □

One of the most striking properties of complete lattices is given by the

Tarski Fixed-Point Theorem 2.1.8. *Let $(X, \leq)$ be a complete lattice, and let T be a covariant functor (i.e. monotone function) $T : X \to X$. Then the induced poset $\{(\mathrm{Fix}(X, \langle T \rangle), \leq\}$ is a complete lattice. In particular, the action $(X, \langle T \rangle)$ does have fixed points.*

Proof. Let A be a subset of $\mathrm{Fix}(X, \langle T \rangle)$. Set $B = \{b \in X \mid \forall a \in A, a \leq bT \leq b\}$. Consider the greatest lower bound g of B in X. Now for b in B, one has $g \leq b$, whence $gT \leq bT \leq b$. Thus $gT \leq g$. Next, consider the least upper bound l of A in X. Then $(\forall b \in B, a \leq bT) \Rightarrow (\forall b \in B, l \leq bT \leq b) \Rightarrow l \leq g$. Thus g is an upper bound for A in X, i.e. $\forall a \in A, a \leq g$. Applying T yields $\forall a \in A, aT = a \leq gT \leq g$, whence $g \in B$. Applying T again yields $\forall a \in A, aT = a \leq gT^2 \leq gT$, whence $gT \in B$. Thus $g \leq gT$, and so $gT = g \in \mathrm{Fix}(X, \langle T \rangle)$. In other words, g is an upper bound for A in $(\mathrm{Fix}(X, \langle T \rangle), \leq)$. If u is any upper bound for A in $(\mathrm{Fix}(X, \langle T \rangle), \leq)$, then $(\forall a \in A, a \leq uT = u) \Rightarrow u \in B \Rightarrow g \leq u$. Thus g, the greatest lower bound for B in X, is the least upper bound for A in $\mathrm{Fix}(X, \langle T \rangle)$. □

Corollary 2.1.9. *Consider functions $f : A \to B$ and $g : B \to A$.*

(a) (**Banach Decomposition Theorem.**) *There are decompositions $A = A_1 \cup A_2$ and $B = B_1 \cup B_2$ with $A_1 f = B_1$ and $B_2 g = A_2$.*

(b) (**Cantor-Schröder-Bernstein Theorem.**) *If f and g inject, then A and B are isomorphic.*

Proof. (a) Consider the following four covariant functors:

$$S_1 : (2^A, \subseteq) \to (2^B, \subseteq); Y \mapsto Yf$$

$$S_2 : (2^B, \subseteq) \to (2^B, \supseteq); Z \mapsto B - Z$$

$$S_3 : (2^B, \supseteq) \to (2^A, \supseteq); Z \mapsto Zg$$

$$S_4 : (2^A, \supseteq) \to (2^A, \subseteq); Y \mapsto A - Y.$$

Their composite $T : (2^A, \subseteq) \to (2^A, \subseteq)$ is a functor on the complete lattice $(2^A, \subseteq)$—cf. Example 2.1.6—and thus has a fixed point A_1, i.e. $A_1 = A_1 T = A_1 S_1 S_2 S_3 S_4$. Now set $A_2 = A_1 S_1 S_2 S_3$, $B_2 = A_1 S_1 S_2$, and $B_1 = A_1 S_1$.

(b) If $f : A \to B$ and $g : B \to A$ inject, then $(f : A_1 \to B_1) \cup (g^{-1} : A_2 \to B_2)$ is a bijection from A to B. □

EXERCISES

2.1A. Verify the claims of Example 2.1.2(a) for the concrete categories $\underline{\underline{\text{Mon}}}$ and $\underline{\underline{\text{Q}}}$.

2.1B. (a) Give a precise definition of the least upper bound $\Sigma\{x_j \mid j \in J\}$ of a subset $\{x_j | j \in J\}$ of a poset $(X, \leq)$.

(b) What is the least upper bound of the empty subset of a poset $(X, \leq)$? When does it exist?

2.1C. Verify the claims of Example 2.1.4.

2.1D. Show that a finite lattice $(X, +, \cdot)$ is complete.

2.1E. Show that the set Eq B of equivalence relations on a set B, ordered by containment, forms a complete lattice (cf. Exercise O 3.3B).

2.1F. Let X and Y be countable, infinite sets [cf. Exercise 1.1A(b)]. Prove $X \cong Y$.

2.1G. Let $X = \{0\} \cup \{n^{-1} | 0 < n \in \mathbb{Z}\}$. Let $Y = \{-1\} \cup X$. Consider X and Y with the orders induced from $(\mathbb{R}, \leq)$.

(a) Show that X and Y are complete lattices.

(b) Define $F: X \to Y$ by $xF =$ **if** $x = 0$ **then** -1 **else** x. Show that F is a bounded lattice homomorphism.

(c) Exhibit a subset $\{x_j \mid j \in J\}$ such that $\Pi\{x_jF \mid j \in J\} \neq [\Pi\{x_j \mid j \in J\}]F$.

2.1H. Let $(X, \leq)$ be a complete lattice. Define $T: (2^X, \subseteq) \to (2^X, \subseteq)$; $Y \mapsto \{\Sigma Y\}D$ (cf. Example 1.2.3). Show that T is a functor, and show that the lattices $(\text{Fix}(X, \langle T \rangle), +, \cdot)$ and $(X, +, \cdot)$ are isomorphic.

2.2. Slice Categories

Let C be a category, and let x be a point of C. One often wishes to consider the category C "relative to" the object x, or to "take a slice of C from the standpoint x." To this end, one constructs a new category C/x, called the *slice category of C-objects over x*, as follows. An object of C/x is a C-morphism $p: a \to x$, i.e. an arrow of C with head x, an element of $d_1^{-1}\{x\}$. A morphism $f: (p: a \to x) \to (q: b \to x)$ of C/x is a C-morphism $f: a \to b$ such that the diagram

$$\begin{array}{ccc} a & \xrightarrow{f} & b \\ {\scriptstyle p}\downarrow & & \downarrow{\scriptstyle q} \\ x & = & x \end{array} \qquad (2.2.1)$$

in C commutes. If $g: (q: b \to x) \to (r: c \to x)$ is a second (C/x)-morphism, then the equation $p = fq = f(gr) = (fg)r$ in C shows that the composite

C-morphism $fg : a \to c$ yields the composite (C/x)-morphism $fg : (p : a \to x) \to (r : c \to x)$. The identity at $p : a \to x$ in (C/x) is the (C/x)-morphism $1_a : (p : a \to x) \to (p : a \to x)$.

Example 2.2.1 (Down-sets). Let $(X, \leq)$ be a poset category. For an element x of X, consider the down-set $xD = \{y \in X | y \leq x\}$ (cf. Example 1.2.3). Then the slice category X/x is isomorphic to the down-set $(xD, \leq)$ with the partial order induced from $(X, \leq)$. The isomorphism is given by the forgetful functor $X/x \to xD$ with object part $(y \to x) \mapsto y$. □

Example 2.2.2 (Augmentations of Rings). An *augmentation* of a unital ring R is a unital ring homomorphism $\varepsilon : R \to \mathbb{Z}$ from R to the ring of integers. For example, the map

$$\varepsilon : \mathbb{Z}[T] \to \mathbb{Z};\ \sum_{i=0}^{n} p_i T^i \mapsto \sum_{i=0}^{n} p_i$$

is an augmentation of the integral polynomial ring $\mathbb{Z}[T]$. Then augmentations of rings are precisely the objects of the slice category $\underline{\underline{\mathrm{Ring}}}/\mathbb{Z}$ of unital rings over $\mathbb{Z}$. □

Example 2.2.3 (Subobjects). Let C be a category. The composite of two monomorphisms in C is again a monomorphism in C (cf. Exercises 1.3D and 1.3G). Moreover, identity morphisms in C are monomorphisms in C. One obtains a subcategory $\mathrm{Mon}(C)$ of C with $\mathrm{Mon}(C)_0 = C_0$ and with $\mathrm{Mon}(C)_1$ the class of monomorphisms in C. Now let x be an object of C. Consider the slice category $\mathrm{Mon}(C)/x$. Objects of $\mathrm{Mon}(C)/x$ are called *subobjects* of x, and an *isomorphism of subobjects* of x is an isomorphism in the category $\mathrm{Mon}(C)/x$. □

For an object x of a category C, the dual slice category $(C^{\mathrm{op}}/x)^{\mathrm{op}}$ is called the *slice category* x/C of C*-objects under* x. Thus an object of x/C is a C-morphism $l : x \to a$, i.e. an arrow of C with tail x, an element of $d_0^{-1}\{x\}$. A morphism $f : (l : x \to a) \to (m : x \to b)$ of x/C is a C-morphism $f : a \to b$ such that the diagram

$$\begin{array}{ccc} x & = & x \\ {\scriptstyle l}\downarrow & & \downarrow{\scriptstyle m} \\ a & \xrightarrow[f]{} & b \end{array} \tag{2.2.2}$$

in C commutes, while the composite of f with the (x/C)-morphism $g : (m : x \to b) \to (n : x \to c)$ is $fg : (l : x \to a) \to (n : x \to c)$ given by $fg : a \to c$ in C.

Example 2.2.4 (Up-sets). If x is an object of a poset category $(X, \leq)$, then the slice category x/X is the *up-set* $\{y \in X | x \leq y\}$ of x with the partial order induced from $(X, \leq)$. (Warning: Elements of the up-set x/X are X-objects under x. Elements of the down-set X/x are X-objects over x. The vertical orientation implicit in the terminology "over x" and "under x" is the reverse of that used in Hasse diagrams and implicit in the terms "down-set" and "up-set".) □

Example 2.2.5 (Commutative Algebras). Let $\underline{\underline{\text{CRing}}}$ be the concrete category of commutative, unital rings and homomorphisms. For an object K of $\underline{\underline{\text{CRing}}}$, (codomains of) objects of the slice category $K/\underline{\underline{\text{CRing}}}$ are commutative K-algebras (cf. Proposition II 3.1). □

Example 2.2.6 (Partial Functions). Fix a singleton $\{*\}$. Then the slice category $\{*\}/\underline{\underline{\text{Set}}}$ embeds into the category $\underline{\underline{\text{Pfn}}}$ of partial functions (cf. Examples 1.1.5 and 1.2.9) via a functor $F: \{*\}/\underline{\underline{\text{Set}}} \to \underline{\underline{\text{Pfn}}}$ whose object part is $(f: \{*\} \to Y) \mapsto Y - \{*\}f$. □

Section 1.3 introduced one of the prime applications of category theory: the specification of mathematical constructions as initial (or dually, terminal) objects in certain categories. By Proposition 1.3.4 and Corollary 1.3.5, such specifications are unique, to within isomorphism. Products and coproducts are typical mathematical constructions. By taking subcategories of slices of certain functor categories, one obtains new categories having coproducts as initial objects, and products as terminal objects.

Consider a category D, a set J, and a function $A: J \to D_0$. The coproduct ι^A will be expressed as an initial object. The first task is to make the function $A: J \to D_0$ into a functor. The set J determines an antichain $(J, \hat{J})$, which may then be construed as a discrete (poset) category J (cf. Exercise 1.2B). The function $A: J \to D_0$ becomes the object part of a uniquely specified functor $A: J \to D$. Thus A is a point of the functor category D^J. Note that each point of D^J is specified completely by its object part $J \to D_0$, since the category J has no non-identity morphisms.

The functor category D^J contains a subcategory $D\Delta$ that (for $J \neq \varnothing$) is an isomorphic copy of the category D, the isomorphism being given by a *diagonal functor* $\Delta: D \to D\Delta$. The object part of Δ sends an object B to the *constant functor* $B\Delta: J \to D$ with object part $B\Delta: J \to D_0$; $j \mapsto B$. The morphism part of Δ sends a D-morphism $f: B' \to B$ to the natural transformation $f\Delta: B'\Delta \to B\Delta$ whose component at each element j of J (i.e. at each object j of the category J) is the same D-morphism $f: B' \to B$. Note that verification of (1.4.2) for $f\Delta$ is trivial.

Now consider the slice category A/D^J of D^J-objects under A. An object of A/D^J is a natural transformation from A to a functor $J \to D$. Let $A/D\Delta$ be the subcategory of A/D^J whose objects are natural transformations $\varphi: A \to B\Delta$ from A to a constant functor $B\Delta$. The component of φ at an object i of J is a D-morphism $\varphi_i : A_i \to B$ as in the bottom row of (2.1.4). The $(A/D\Delta)$-morphisms $f\Delta : (\varphi' : A \to B'\Delta) \to (\varphi : A \to B\Delta)$ are just those natural transformations $f\Delta : B'\Delta \to B\Delta$ for $f : B' \to B$ in D_1 such that, for each i in J, the diagram

$$\begin{array}{ccc} B' & \xrightarrow{f} & B \\ {\scriptstyle \varphi_i'}\uparrow & & \| \\ A_i & \xrightarrow[\varphi_i]{} & B \end{array} \tag{2.2.3}$$

in D commutes. One may readily verify that $A/D\Delta$ is indeed a subcategory of A/D^J (Exercise 2.2I). It is called the *category of constant functors under* A. The universality property (2.1.3) for the coproduct ι^A of A is then expressed precisely by the following:

Proposition 2.2.7. *Let D be a category, and let J be a set or discrete category. Let $A : J \to D$ be a functor. Then the coproduct ι^A of A is characterized as an initial object of the category $A/D\Delta$ of constant functors under A. In particular, the coproduct object $\Sigma_{j \in J} A_j$ of D is unique to within isomorphism.*

Proof. If the coproduct ι^A exists, then (2.1.4) exhibits a unique $(A/D\Delta)$-morphism $\Sigma_{j \in J} \varphi_j \Delta$ from $\iota^A : A \to (\Sigma_{j \in J} A_j)\Delta$ to each $(A/D\Delta)$-object $\varphi : A \to B\Delta$. Conversely, given an initial object $\varphi' : A \to B'\Delta$ of $A/D\Delta$, comparison of (2.2.3) and (2.1.4) shows that f is the unique element of $D(B', B)$ such that $\varphi_i = \varphi_i' f$ for each i in J. If $\varphi' : A \to B'\Delta$ and $\varphi'' : A \to B''\Delta$ are two initial objects of $A/D\Delta$, then there is an $(A/D\Delta)$-isomorphism $f'\Delta : (\varphi' : A \to B'\Delta) \to (\varphi'' : A \to B''\Delta)$, by Corollary 1.3.5. Then $f' : B' \to B''$ is a D-isomorphism between the two coproduct objects B' and B''. □

Dually, one may construct the *category of constant functors over* A, a subcategory $D\Delta/A$ of D^J/A whose objects are natural transformations of the form $f\Delta : B'\Delta \to B\Delta$ for D-morphisms $f : B' \to B$ with $\varphi_i' f = \varphi_i$ for each i in J. Dual to Proposition 2.2.7, one obtains:

Proposition 2.2.8. *Let D be a category, and let J be a set or discrete category. Let $A : J \to D$ be a functor. Then the product π^A of A is characterized as a terminal object of the category $D\Delta/A$ of constant functors over A. In particular, the product object $\Pi_{j \in J} A_i$ of D is unique to within isomorphism.* □

EXERCISES

2.2A. Complete the specification of the category isomorphism between X/x and $(xD, \leq)$ in Example 2.2.1. In particular, determine the morphism part of the forgetful functor, and both parts of the inverse functor.

2.2B. Let T be a terminal object of a category C.

(a) Show that C is isomorphic to the slice category C/T.

(b) Show that C is isomorphic to the slice category $(C/T)/(T \to T)$.

(c) What is the terminal object of the slice category $(C/T)/(T \to T)$?

2.2C. (a) Verify that $\mathbb{Z}[T] \to \mathbb{Z};\ \Sigma_{i=0}^{n} p_i T^i \mapsto \Sigma_{i=0}^{n} p_i$ is a unital ring homomorphism.

(b) Show that the polynomial ring $\mathbb{Z}[T]$ may be augmented by the map $\varepsilon : \mathbb{Z}[T] \to \mathbb{Z};\ \Sigma_{i=0}^{n} p_i T^i \mapsto p_0$.

2.2D. Let $\mathbb{Z}[S_3]$ denote the set $\underline{\underline{\mathrm{Set}}}(S_3, \mathbb{Z})$ of functions from the symmetric group S_3 to the ring $\mathbb{Z}$, with pointwise subtraction. Define a binary operation $*$ of *convolution* on $\mathbb{Z}[S_3]$ by

$$f * g(T) = \sum_{UV=T} f(U)g(V)$$

for each T in S_3. Show that $(\mathbb{Z}[S_3], -, *)$ is a unital ring, and that $\varepsilon : f \mapsto \Sigma_{T \in S_3} f(T)$ is an augmentation of $\mathbb{Z}[S_3]$.

2.2E. Let x be an object of a category C. Define a relation σ on $(\mathrm{Mon}(C)/x)_0$ by $y \sigma z \Leftrightarrow \exists (f : y \to z) \in (\mathrm{Mon}(C)/x)_1$.

(a) Show that σ is a pre-order.

(b) Consider the equivalence relation α on $(\mathrm{Mon}(C)/x)_0$ given by $y \alpha z \Leftrightarrow y \sigma z$ and $z \sigma y$ (cf. Exercise I 1.3H). Show that α is the relation of isomorphism of subobjects defined in Example 2.2.3.

(c) Consider the order relation $\leq$ on $(\mathrm{Mon}(C)/x)_0^{\alpha}$ given by $y^{\alpha} \leq z^{\alpha} \Leftrightarrow y \sigma z$ (cf. Exercise I 1.3H). For a set B, show that there is an isomorphism $(\mathscr{P}(B), \subseteq) \to ((\mathrm{Mon}(\underline{\underline{\mathrm{Set}}})/B)_0^{\alpha}, \leq);\ X \mapsto (X \hookrightarrow B)^{\alpha}$.

2.2F. Let $(L, +, \cdot)$ be a lattice, construed as a poset category $(L, \leq)$. For elements x, y of L, prove

(a) $(L/x)_0 \cap (L/y)_0 = (L/xy)_0$;

(b) $(x/L)_0 \cap (y/L)_0 = ((x+y)/L)_0$.

2.2G. Complete the specification of the functor F of Example 2.2.6, and determine its inverse.

2.2H. Let $\Delta : D \to D^{\varnothing}$ be the diagonal functor from a category D to the functor category $D^{\varnothing}$. Determine the category $D^{\varnothing}$ and its subcategory $D\Delta$.

2.2I. Verify that (2.2.3) leads to a subcategory $A/D\Delta$ of A/D^J.

2.2J. Let A and J be sets. Show that the product of the object $A\Delta$ of $\underline{\underline{\mathrm{Set}}}^J$ is the *evaluation* natural transformation

$$\mathrm{ev} : \underline{\underline{\mathrm{Set}}}(J, A)\Delta \to A\Delta$$

whose component at the element j of J is

$$\mathrm{ev}_j : \underline{\underline{\mathrm{Set}}}(J, A) \to A;\ f \mapsto jf.$$

2.2K. Let x be an object of a category C.

(a) Show that $1_x : x \to x$ is a terminal object of the slice category C/x.

(b) Show that $1_x : x \to x$ is an initial object of the slice category x/C.

2.3. Equalizers and Pullbacks

The Definition 2.1 of the product of two objects in a category was obtained as a generalization of the Definition O 3.2.2 of the direct product of two sets. This section examines the construction (O 3.3.7) of the natural projection of a set A onto its quotient A^α by an equivalence relation α, formulating the construction categorically so that it may be applied readily in other contexts.

The equivalence relation α on the set A is a certain subset of the direct product $A \times A$. The direct product $A \times A$ or A^2 comes equipped with the two projections $\pi_i : A^2 \to A; (a_1, a_2) \mapsto a_i$ for $i = 1,2$, each of which restricts to the subset α of A^2. In the diagram

$$\begin{array}{ccc} \alpha & \xrightarrow{\pi_2} & A \\ {\scriptstyle \pi_1}\downarrow & & \downarrow{\scriptstyle \text{nat } \alpha} \\ A & \xrightarrow[\text{nat } \alpha]{} & A^\alpha, \end{array} \tag{2.3.1}$$

one has $\pi_1(\text{nat } \alpha) = \pi_2(\text{nat } \alpha)$. In other words, the diagram (2.3.1) commutes. Moreover, given a function $f : A \to B$ with $\pi_1 f = \pi_2 f$, there is a unique, well-defined function $\bar{f} : A^\alpha \to B; x^\alpha \mapsto xf$ such that $(\text{nat } \alpha)\bar{f} = f$. One is thus led to the following:

Definition 2.3.1. Let D be a category.

(a) A *parallel pair* $A_1 \overset{f_1}{\underset{f_2}{\rightrightarrows}} A_2$ *of arrows* in D is a pair (f_1, f_2) of arrows of D with common tail A_1 and common head A_2.

(b) Let $A_1 \overset{f_1}{\underset{f_2}{\rightrightarrows}} A_2$ be a parallel pair of arrows in D. Then the *coequalizer* of the pair (f_1, f_2) is (an object Q of D together with) a morphism $q : A_2 \to Q$ of D, satisfying $f_1 q = f_2 q$ and the following *universal property*:

$$\left\{ \begin{array}{l} \text{For each } D\text{-morphism } f : A_2 \to B \text{ with } f_1 f = f_2 f, \text{ there is} \\ \text{a unique element } \bar{f} \text{ of } D(Q, B) \text{ such that } q\bar{f} = f. \end{array} \right. \tag{2.3.3}$$

The universal property (2.3.2) is represented by the commuting diagram

(2.3.3)

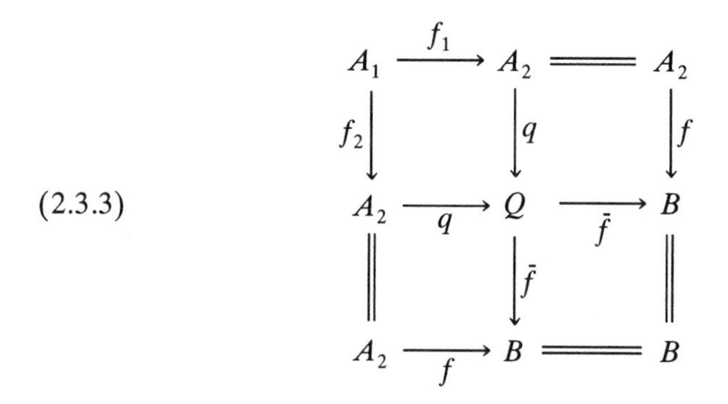

in D. □

As in the case of products, one often refers to the object Q alone as the *coequalizer*, suppressing the (important) role of the morphism q.

Example 2.3.2 (Cokernels). Let $A_1 \xrightarrow{g} A_2 \xrightarrow{q} A_3 \to 0$ be an exact sequence of abelian groups. Consider the pair $(g, 0)$ of homomorphisms from A_1 to A_2. Note that $gq = 0 = 0q$. Suppose there is a homomorphism $f: A_2 \to B$ with $gf = 0f = 0$. Then $A_1 g \leq \operatorname{Ker} f$, so there is a well-defined composite homomorphism $\bar{f}: A_3 \cong A_2/A_1 g \to A_2/\operatorname{Ker} f \cong A_2 f \hookrightarrow B$ with $q\bar{f} = f$. In other words, $q: A_2 \to A_3$ is the coequalizer of the pair $(g, 0)$ in the category $\underline{\underline{\mathrm{Mod}}}_{\mathbb{Z}}$. Conversely, let $A_1 \underset{f_2}{\overset{f_1}{\rightrightarrows}} A_2$ be a parallel pair of arrows in $\underline{\underline{\mathrm{Mod}}}_{\mathbb{Z}}$. Then their coequalizer is the projection $A_2 \to \operatorname{Coker}(f_1 - f_2)$ (Exercise 2.3B). □

Example 2.3.3 (Presentations of Groups). Let $X = \{x_1, x_2\}$ and $Y = \{y_1, y_2, y_3\}$. Define a group homomorphism $f_1^G : YG \to XG$ between free groups, determined via (I 1.4.8) by the set map $f_1 : Y \to XG$; $y_1 \mapsto x_1^2$, $y_2 \mapsto x_2^2$, $y_3 \mapsto x_1 x_2 x_1$. Similarly, define $f_2^G : YG \to XG$ from $f_2 : Y \to XG$; $y_1 \mapsto 1$, $y_2 \mapsto 1$, $y_3 \mapsto x_2 x_1 x_2$. Then the coequalizer in $\underline{\underline{\mathrm{Gp}}}$ of the pair (f_1^G, f_2^G) of group homomorphisms is the group homomorphism $q: XG \to S_3$ onto the symmetric group given by $x_1 \mapsto (12)$, $x_2 \mapsto (23)$. Indeed, $f_1 q = f_2 q$. Suppose $f: XG \to Q$ is a a group homomorphism with $f_1 f = f_2 f$. Set $s = x_1 f \in Q$ and $t = x_2 f \in Q$. Then $s^2 = y_1 f_1 f = y_1 f_2 f = 1$, $t^2 = y_2 f_1 f = y_2 f_2 f = 1$, and $(st)^3 = ststst = (sts)(tst)^{-1} = (y_3 f_1 f)(y_3 f_2 f)^{-1} = 1$, whence there is a group homomorphism $\bar{f}: S_3 \to Q$; $(12) \mapsto s$, $(23) \mapsto t$ (Exercise I 3.2I) satisfying $q\bar{f} = f$. In general, set $X = \{x_1, \ldots, x_n\}$ and $Y = \{y_1, \ldots, y_r\}$. Then the coequalizer of a pair $YG \underset{f_2}{\overset{f_1}{\rightrightarrows}} XG$ of group homomorphisms is called the *group with presentation* $\langle x_1, \ldots, x_n | y_1 f_1 = y_1 f_2, \ldots, y_r f_1 = y_r f_2 \rangle_{\underline{\underline{\mathrm{Gp}}}}$. The relation $\{(y_i f_1, y_i f_2) | 1 \leq i \leq r\}$ on XG is called the set of *relations imposed* by the presentation, while the set $\{x_1, \ldots, x_n\}$ is called the set of *generators* in the presentation. □

Examples 2.3.2 and 2.3.3 illustrate the specification of algebras as coequalizers: modules as cokernels in Example 2.3.2 and groups by presentations in Example 2.3.3. The precision of such specifications depends on the uniqueness of coequalizers (to within isomorphism). This uniqueness is demonstrated by exhibiting coequalizers as initial objects in suitably contrived categories, as was done for coproducts in Proposition 2.2.7. Thus cokernels and group presentations become typical examples of the theme broached in Section 1.3: the specification of mathematical constructs as initial objects.

The first task is to describe parallel pairs $A_1 \overset{f_1}{\underset{f_2}{\rightrightarrows}} A_2$ of arrows in a category D in terms of functors. As a directed graph Γ, a parallel pair of arrows has the form $\cdot \rightrightarrows \cdot$. Technically, one considers the directed graph $\Gamma = (\Gamma_0, \Gamma_1, \partial_0, \partial_1) = (\{1, 2\}, \{1, 2\}, \{1, 2\} \to \{1\}, \{1, 2\} \to \{2\})$, i.e. the pair $(\partial_0 : \{1, 2\} \to \{1\}, \partial_1 : \{1, 2\} \to \{2\})$ of maps. Let J be the path category $\Gamma\Pi$. Then the parallel pair $A_1 \overset{f_1}{\underset{f_2}{\rightrightarrows}} A_2$ in D corresponds to the functor $A : J \to D$ with object part $\{1, 2\} \to D_0$; $1 \mapsto A_1$, $2 \mapsto A_2$ and with morphism part determined by $\{1, 2\} \to D_1$; $1 \mapsto f_1$, $2 \mapsto f_2$. In other words, parallel pairs of arrows in D become objects of the functor category D^J.

Given an object B of D, one has a *constant functor* $B\Delta : J \to D$ with object part $1 \mapsto B$, $2 \mapsto B$ and with morphism part determined by $\{1, 2\} \to D_1$; $1 \mapsto 1_B$, $2 \mapsto 1_B$. Given a D-morphism $f : B' \to B$, there is a natural transformation $f\Delta : B'\Delta \to B\Delta$ whose component at each of the two objects of J is the same D-morphism $f : B' \to B$. Note that (1.4.2) for $f\Delta$ reduces to $1_{B'}f = f1_B$. One obtains a subcategory $D\Delta$ of the functor category D^J as the image of the *diagonal functor* $\Delta : D \to D^J$ with object part $D_0 \to (D^J)_0$; $B \mapsto B\Delta$ and morphism part $D_1 \to (D^J)_1$; $(f : B' \to B) \mapsto (f\Delta : B'\Delta \to B\Delta)$.

Now consider the slice category A/D^J of D^J-objects under the functor $A : J \to D$ corresponding to a parallel pair $A_1 \overset{f_1}{\underset{f_2}{\rightrightarrows}} A_2$ in D. Consider the subcategory $A/D\Delta$ of A/D^J whose objects are natural transformations $\varphi : A \to B\Delta$ for an object B of D, with components $\varphi_1 : A_1 \to B$ and $\varphi_2 : A_2 \to B$. The diagram (1.4.2) in D corresponding to the morphism $1 : 1 \to 2$ in J yields $\varphi_1 1_B = f_1\varphi_2$. The diagram (1.4.2) corresponding to $2 : 1 \to 2$ in J yields $\varphi_1 1_B = f_2\varphi_2$. Thus $f_1\varphi_2 = \varphi_1 = f_2\varphi_2$. In particular, the coequalizer $q : A_2 \to Q$ of the parallel pair (f_1, f_2) yields an object $\iota^A : A \to Q\Delta$ of $A/D\Delta$ with components $\iota_2^A = q : A_2 \to Q$ and $\iota_1^A = f_1 q = f_2 q : A_1 \to Q$. The morphisms $\bar{f}\Delta : (\varphi' : A \to B'\Delta) \to (\varphi : A \to B\Delta)$ of the subcategory $A/D\Delta$ of A/D^J are just those natural transformations $\bar{f}\Delta : B'\Delta \to B\Delta$ for $\bar{f} : B' \to B$ in D_1 such that, for each object i of J, the diagram

$$
\begin{array}{ccc}
A_i & = & A_i \\
\varphi'_i \downarrow & & \downarrow \varphi_i \\
B' & \xrightarrow[\bar{f}]{} & B
\end{array}
\tag{2.3.4}
$$

commutes. In fact, if (2.3.4) commutes for $i = 2$, i.e. $\varphi'_2 \bar{f} = \varphi_2$, then $\varphi'_1 \bar{f} = f_1 \varphi'_2 \bar{f} = f_1 \varphi_2 = \varphi_1$, so that (2.3.4) automatically commutes for $i = 1$. The diagram (2.3.3) then shows that the object $\iota^A : A \to Q\Delta$ of $A/D\Delta$ determined by the coequalizer $q : A_2 \to Q$ is initial in $A/D\Delta$. Conversely, an initial object $\iota : A \to Q\Delta$ of $A/D\Delta$ yields a coequalizer $\iota_2 : A_2 \to Q$ of the parallel pair $(1A, 2A)$ of morphisms in D. Summarizing, one obtains for coequalizers the following analogue of Proposition 2.2.7 for coproducts.

Proposition 2.3.4. *Let D be a category, and let J be the path category of the directed graph* $\cdot \rightrightarrows \cdot$ *or* $(\{1,2\} \to \{1\}, \{1,2\} \to \{2\})$. *Let $A : J \to D$ be a functor. Then the coequalizer $\iota_2^A : A_2 \to Q$ of the parallel pair $(1A, 2A)$ of arrows in D is characterized as an initial object $\iota : A \to Q\Delta$ of the category $A/D\Delta$ of constant functors under A. In particular, the coequalizer object Q of D is unique to within isomorphism.* □

Dual to coequalizers, one has the concept of the *equalizer* E or $e : E \to A_1$ of a parallel pair $A_1 \underset{f_2}{\overset{f_1}{\rightrightarrows}} A_2$ of arrows in a category D. Thus $ef_1 = ef_2$, and if $ff_1 = ff_2$ for a D-morphism $f : B \to A_1$, then there is a D-morphism $\bar{f} : B \to E$ with $\bar{f}e = f$. Using the same category J as for equalizers, one may consider the slice category D^J/A and its subcategory $D\Delta/A$ of natural transformations to A from constant functors, the category of *constant functors* over A. One then obtains the following dual of Proposition 2.3.4.

Proposition 2.3.5. *Let D be a category, and let J be the path category of the directed graph* $(\{1,2\} \to \{1\}, \{1,2\} \to \{2\})$. *Let $A : J \to D$ be a functor. Then the equalizer $\pi_1^A : E \to A_1$ of the parallel pair $(1A, 2A)$ of arrows in D is characterized as a terminal object $\pi : E\Delta \to A$ of the category $D\Delta/A$ of constant functors over A. In particular, the equalizer object E of D is unique to within isomorphism.* □

If $A_1 \underset{f_2}{\overset{f_1}{\rightrightarrows}} A_2$ is a parallel pair of arrows in $\underline{\text{Set}}$, then their equalizer $e : E \to A_1$ certainly exists. One may take e to be the inclusion in A_1 of the subset $E = \{a \in A_1 | af_1 = af_2\}$. Then if $ff_1 = ff_2$ for a map $f : B \to A_1$, one has $Bf \subseteq E$, so the restriction $\bar{f} : B \to E$ of the map $f : B \to A_1$ satisfies $\bar{f}e = f$. In fact, this construction works in any of the concrete categories of Example 1.2.7. The particular case of $\underline{\underline{\mathrm{Mod}}}_{\mathbb{Z}}$ is worth considering in detail, as it offers a dual to Example 2.3.2 (Exercise 2.3K).

Slice categories were used in Propositions 2.3.4 and 2.3.5 to demonstrate the uniqueness of coequalizers and equalizers. Conversely, one may use equalizers and coequalizers to determine properties of slice categories, namely the existence of products in slices C/x of a category C with products, and dually the existence of coproducts in slices x/C of a category C with coproducts.

Let C be a category, and let x be an object of C. Let $p_1 : a_1 \to x$ and $p_2 : a_2 \to x$ be objects of C/x. Let $\pi_i : a_1 \times a_2 \to a_i$ be the product of a_1 and a_2 in C. Let $e : a_1 \times_x a_2 \to a_1 \times a_2$ be the equalizer of the parallel pair $(\pi_1 p_1, \pi_2 p_2)$ of C-morphisms. Set $p = e\pi_1 p_1 = e\pi_2 p_2 : a_1 \times_x a_2 \to x$. Then one may readily verify the following (cf. Exercise 2.3L).

Proposition 2.3.6. *The object $p : a_1 \times_x a_2 \to x$, with projections $e\pi_i : (p : a_1 \times_x a_2 \to x) \to (p_i : a_i \to x)$, is the product of the two objects $p_i : a_i \to x$ $(i = 1,2)$ in the slice category C/x.* □

In the category C, the object $a_1 \times_x a_2$, together with the maps $e\pi_i : a_1 \times_x a_2 \to a_i$, is called the *pullback* of the pair (p_1, p_2) of C-morphisms with common codomain x. Such a pullback is described by the commuting diagram

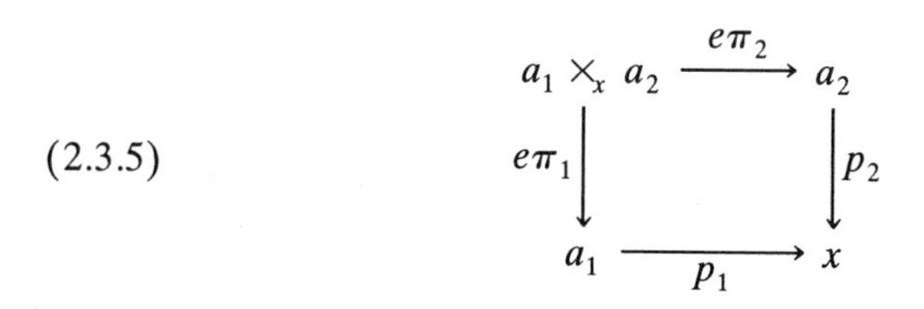

in C. Taking J to be the poset category with Hasse diagram $1 \to 0 \leftarrow 2$, one may characterize the pullback of $a_1 \xrightarrow{p_1} x \xleftarrow{p_2} a_2$ as a terminal object in the category $C\Delta/A$ of constant functors over A, a subcategory of the slice category C^J/A, for the functor $A : J \to C$ with morphism part $((1 \to 0) \mapsto (p_1 : a_1 \to x)), ((2 \to 0) \mapsto (p_2 : a_2 \to x))$ (Exercise 2.3M). Duals of pullbacks are called *pushouts*. They provide coproducts in slice categories $x \setminus C$.

This section began by considering equivalence relations. It concludes the same way.

Example 2.3.7. Let $f: A \to B$ be a set map. Then its kernel is the pullback

$$\begin{array}{ccc} \ker f & \longrightarrow & A \\ \downarrow & & \downarrow f \\ A & \underset{f}{\longrightarrow} & B \end{array} . \qquad \square$$

EXERCISES

2.3A. Let $q: A_2 \to Q$ be the coequalizer in $\underline{\underline{\text{Set}}}$ of a parallel pair $A_1 \overset{f_2}{\underset{f_2}{\rightrightarrows}} A_2$ of functions. Show that Q is isomorphic to the quotient of A_2 by the transitive closure α of the reflexive, symmetric relation $\hat{A}_2 \cup A_1(f_1, f_2) \cup A_1(f_2, f_1)$ on A_2 (cf. Exercise O 3.3D).

2.3B. Verify that the projection $A_2 \to \text{Coker}(f_1 - f_2)$ is the coequalizer of a parallel pair of morphisms $A_1 \overset{f_1}{\underset{f_2}{\rightrightarrows}} A_2$ in the category $\underline{\underline{\text{Mod}_{\mathbb{Z}}}}$ of abelian groups.

2.3C. Let t be an endomorphism of a finitely generated free module V over a commutative ring K with basis $\beta: \mathbb{Z}_n \to V$. Show that the $K[T]$-homomorphism $R(\beta): K[T]^n \to V$ is the coequalizer of the parallel pair $(T, R(\beta)tR(\beta)^{-1})$ of endomorphisms of $K[T]^n$.

2.3D.
(a) Let A be an object of a category D with initial object $\perp$. Show that the coequalizer of the pair $\perp \rightrightarrows A$ is the identity on A.
(b) For a set $X = \{x_1, \ldots, x_n\}$, show that the free group XG on X has the presentation $\langle x_1, \ldots, x_n | \rangle_{\underline{\underline{\text{Gp}}}}$, i.e. is the coequalizer of $\varnothing G \rightrightarrows XG$.

2.3E. Determine the group with presentation $\langle x_1 | x_1^4 = 1 \rangle_{\underline{\underline{\text{Gp}}}}$.

2.3F. Determine the group with presentation $\langle x_1, x_2 | x_1^4 = 1,\ x_1^{-1} x_2 x_1 = x_2^{-1} \rangle_{\underline{\underline{\text{Gp}}}}$.

2.3G. Let J be the path category of the directed graph $\cdot \rightrightarrows \cdot$ or $(\{1,2\} \to \{1\}, \{1,2\} \to \{2\})$. Show that the identity functor on J does not have a coequalizer or equalizer.

2.3H.
(a) Show that the coequalizer of an equal pair of arrows in a category is the identity on their codomain.
(b) Formulate and do the dual of (a).

2.3I. Write out formal definitions of the categories D^J/A and $D\Delta/A$ involved in the formulation of Proposition 2.3.5. Then give a careful proof of the proposition.

2.3J. Let X, Y, f_1 and f_2 be as in Example 2.3.3.

(a) For a given pair (σ, τ) of distinct transpositions in S_3, show that the group homomorphism $q_{(\sigma,\tau)}: XG \to S_3$ given by $x_1 \mapsto \sigma$, $x_2 \mapsto \tau$ is a coequalizer of the pair (f_1^G, f_2^G) of group homomorphisms.

(b) For a pair $((\sigma, \tau), (\sigma', \tau'))$ of pairs of distinct transpositions in S_3, show that there is an automorphism $\Pi_{(\sigma,\tau)}^{(\sigma',\tau')}$ of S_3 such that $q_{(\sigma',\tau')}\Pi_{(\sigma,\tau)}^{(\sigma',\tau')} = q_{(\sigma,\tau)}$.

(c) Determine the automorphism group Aut S_3 of S_3.

(d) Does S_3 have any outer automorphisms?

2.3K. (a) Let $0 \to E \xrightarrow{e} A_1 \xrightarrow{f} A_2$ be an exact sequence of abelian groups. Show that $e: E \to A_1$ is the equalizer of the pair $(f, 0)$ of $\underline{\underline{\text{Mod}}}_{\mathbb{Z}}$-morphisms.

(b) Let $A_1 \underset{f_2}{\overset{f_1}{\rightrightarrows}} A_2$ be a parallel pair of arrows in $\underline{\underline{\text{Mod}}}_{\mathbb{Z}}$. Show that their equalizer is the injection $\text{Ker}(f_1 - f_2) \to A_1$.

2.3L. (a) Verify Proposition 2.3.6.

(b) Formulate and prove the dual of Proposition 2.3.6.

2.3M. Let J be the poset category with Hasse diagram $1 \to 0 \leftarrow 2$. Let D be a category. Let $A: J \to D$ and $B: J^{\text{op}} \to D$ be functors.

(a) Show that the pullback of JA in D is characterized as a terminal object π of the category $D\Delta/A$ of constant functors (from J to D) over A.

(b) Show that the pushout of $J^{\text{op}}B$ in D is characterized as an initial object ι of the category $B/D\Delta$ of constant functors under B.

2.3N. Let C be a category with terminal object T.

(a) Show that the product $A_1 \times A_2$ of two objects A_i of C is the pullback

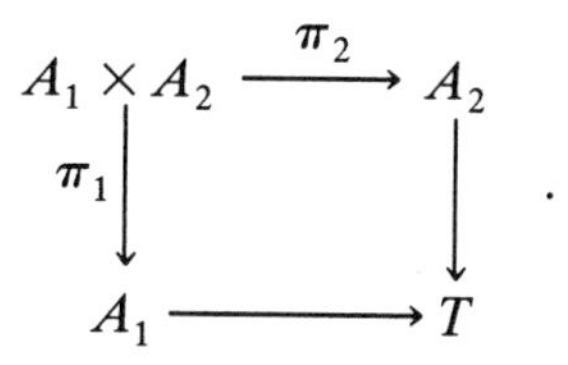

(b) Show that the equalizer $e: E \to A_1$ of a parallel pair $A_1 \underset{f_2}{\overset{f_1}{\rightrightarrows}} A_2$ of arrows in C is the pullback

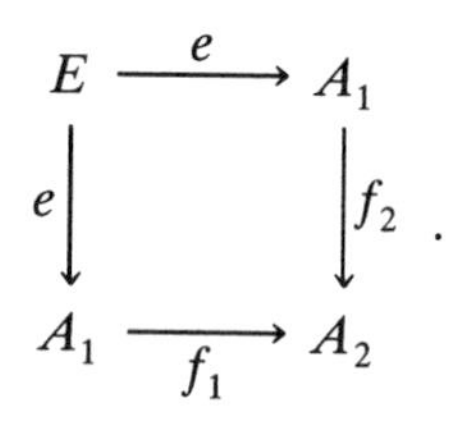

(c) Identify coproducts as pushouts.

(d) Identify coequalizers as pushouts.

2.30. Let C be a small category. Describe the domain of the composition of morphisms in C as a pullback $C_1 \times_{C_0} C_1$.

2.4. Groups in Categories

In I 1.3, a group (S, μ, J, e) was defined as a set S equipped with an associative multiplication $\mu : S^2 \to S;\ (s,t) \mapsto st$, an identity element e satisfying $(\forall s \in S,\ se = s = es)$, and an inversion $J : S \to S;\ s \mapsto s^{-1}$ satisfying $(\forall s \in S,\ ss^{-1} = e = s^{-1}s)$. As discussed in Section 1.3, the element e of S may be described by the map $e : T \to S$ from the terminal object $T = \{\infty\}$ to S with image $\{e\}$. Since an empty product is terminal [Example 2.1.2(c)], one may also describe the element or map e as $e : S^0 \to S$. Given $e : S^0 \to S$, it is convenient to denote the composite $S \to S^0 \xrightarrow{e} S$ by e_1. The associative law or semigroup property of S is then specified by the commuting diagram

$$
\begin{array}{ccc}
S^3 & \xrightarrow{(\mu,1)} & S^2 \\
{\scriptstyle (1,\mu)}\downarrow & & \downarrow{\scriptstyle \mu} \\
S^2 & \xrightarrow[\mu]{} & S
\end{array}
\tag{2.4.1}
$$

in the category $\underline{\underline{\text{Set}}}$. The existence of an identity element, i.e. the monoid property of S, is specified by the commuting diagram

$$
\begin{array}{ccccc}
S^2 & \xleftarrow{(1,e_1)} & S & \xrightarrow{(e_1,1)} & S^2 \\
{\scriptstyle \mu}\downarrow & & \downarrow{\scriptstyle 1} & & \downarrow{\scriptstyle \mu} \\
S & = & S & = & S
\end{array}
\tag{2.4.2}
$$

in the category $\underline{\underline{\text{Set}}}$. Finally, the inversion or group property of S is specified by the commuting diagram

$$
\begin{array}{ccccc}
S & \xrightarrow{(1,J)} & S^2 & \xleftarrow{(J,1)} & S \\
{\scriptstyle e_1}\downarrow & & \downarrow{\scriptstyle \mu} & & \downarrow{\scriptstyle e_1} \\
S & = & S & = & S
\end{array}
\tag{2.4.3}
$$

in $\underline{\underline{\text{Set}}}$. In other words, a group is a set S equipped with maps $\mu : S^2 \to S$, $e : S^0 \to S$ and $J : S \to S$ such that the diagrams (2.4.1)–(2.4.3) commute.

Note that the commutative law for a multiplication $\mu : S^2 \to S$ can be expressed, using the isomorphism $\tau : S^2 \to S^2; (s_1, s_2) \mapsto (s_2, s_1)$ of Exercise 2A, as the commuting of the diagram

$$\begin{array}{ccc} S^2 & \xrightarrow{\tau} & S^2 \\ \mu \downarrow & & \downarrow \mu \\ S & = & S \end{array} \qquad (2.4.4)$$

in Set. Moreover, given groups S_1 and S_2, a function $f : S_1 \to S_2$ is a group homomorphism if the diagram

(2.4.5)

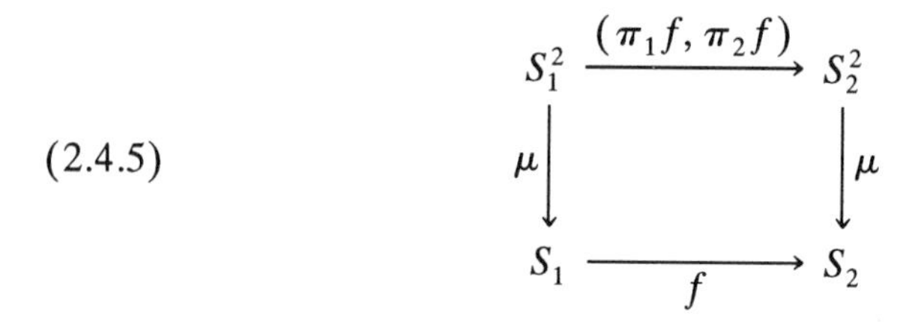

in Set commutes. The only special feature of the category Set of sets used in formulating these properties diagrammatically is the fact that an object S of Set has the powers S^0, S^1, S^2, S^3. The observation leads to the following:

Definition 2.4.1. Let C be a category in which each object S has the powers S^0 (i.e. terminal object of C), $S^1 = S$, $S^2 = S \times S$, and $S^3 = S^2 \times S \cong S \times S^2$.

(a) A *semigroup in the category* C is an object S of C, equipped with a C-morphism $\mu : S^2 \to S$ (called *multiplication*) such that the diagram (2.4.1) in C commutes.

(b) A *monoid in the category* C is a semigroup (S, μ) in C, equipped with a C-morphism $e : S^0 \to S$ and composite $e_1 : S \to S^0 \xrightarrow{e} S$, such that the diagram (2.4.2) in C commutes.

(c) A *group in the category* C is a monoid (S, μ, e) in C, equipped with a C-morphism $J : S \to S$ (called *inversion*) such that the diagram (2.4.3) commutes.

(d) A *morphism of semigroups in the category* C is a C-morphism $f : S_1 \to S_2$, with domain and codomain both semigroups in C, such that the diagram (2.4.5)

commutes.

(e) A semigroup (S, μ) in C is *commutative* if the diagram (2.4.4) commutes. □

Example 2.4.2 (Sequences of Groups). Consider the poset category $(\mathbb{N}, \leq)$, and the functor category $\underline{\underline{\mathrm{Set}}}^{\mathbb{N}}$. Note that $\underline{\underline{\mathrm{Set}}}^{\mathbb{N}}$ has pointwise finite products of objects (cf. Exercises 1.4F and 2N). Then a group in the functor category $\underline{\underline{\mathrm{Set}}}^{\mathbb{N}}$ is a functor $A : \mathbb{N} \to \underline{\underline{\mathrm{Set}}}$ with object part $n \mapsto A_n$ and with morphism part given by $(n \to n+1) \mapsto (A_n \xrightarrow{f_n} A_{n+1})$, such that each A_n is a group and each f_n is a group homomorphism. Such an object is called a *sequence of groups*. Note that a finite sequence (II 1.1) of abelian groups and homomorphisms yields a sequence in the current sense on setting $A_n = 0$ for $n > N$.
□

Example 2.4.3 (Groups in the Category of Groups). If $(A, +, 0)$ is an abelian group, then $+ : A^2 \to A$ is a group homomorphism, so that $(A, +)$ is a group in the category of groups. Conversely, suppose that the group $(G, \cdot, {}^{-1}, 1)$ is a group (G, μ, J, e) in the category of groups. Thus $\forall a, b, c, d \in G$, $(ac, bd)\mu = (a, b)\mu \cdot (c, d)\mu$. Setting $(a, b, c, d) = (a, 1, 1, d)$, one obtains $(a, d)\mu = (a, 1)\mu \cdot (1, d)\mu$. Setting $d = e$, one obtains $a = (a, e)\mu = (a, 1)\mu \cdot (1, e)\mu = (a, 1)\mu \cdot 1 = (a, 1)\mu$. Similarly, $d = (1, d)\mu$. Thus $(a, d)\mu = a \cdot d$, and then $ac \cdot bd = ab \cdot cd$. Setting $a = d = 1$ yields $cb = bc$, i.e. the group G is abelian. Thus groups in the category of groups are precisely the abelian groups. □

Let G be a group, construed as a category G with $G_0 = \{1\}$ and $G_1 = G$ according to Example 1.2. For a unital ring S, an *S-representation* of G is a functor $R : G \to \underline{\underline{\mathrm{Mod}}}_S$. The S-module $1R$, the unique element of the image of the object part of R, is called a *G-module*, or the *space* of the representation R. (Such representations are sometimes described as *linear*, in order to distinguish them from the representations $G \to X!$ or G-sets X.) Let M be a G-module, with a representation (having morphism part) $R : G \to \mathrm{Aut}(M, -)$; $g \mapsto (m \mapsto mg)$. It will be shown that M determines an abelian group $p : E \to G$ in the slice category $\underline{\underline{\mathrm{Gp}}}/G$ of groups over G. The domain E of the group homomorphism p is called the *split extension* $G[M$ or $G[_R M$ *of* M *by* G *via the representation* R. Its underlying set is $G \times M$, with multiplication

$$(2.4.6) \qquad (g_1, m_1)(g_2, m_2) = (g_1 g_2, m_1 g_2 + m_2),$$

identity element $(1, 0)$, and inversion $(g, m) \mapsto (g^{-1}, -mg^{-1})$ (Exercise 2.4D). The group morphism p is the projection $p : E \to G$; $(g, m) \mapsto g$. By Proposition 2.3.6, the direct square of p in the slice category $\underline{\underline{\mathrm{Gp}}}/G$ is $E \times_G E \to G$; $((g, m_1), (g, m_2)) \mapsto g$. Define $+_G : E \times_G E \to E$; $((g, m_1), (g, m_2)) \mapsto (g, m_1 + m_2)$. Define $0_G : G \to E$; $g \mapsto (g, 0)$. Define a negation $-_G : E \to E$;$(g, m) \mapsto (g, -m)$. Then $(p : E \to G, +_G, -_G, 0_G)$ is an abelian group in the slice category $\underline{\underline{\mathrm{Gp}}}/G$ (Exercise 2.4E).

Example 2.4.4 (Dihedral Groups). Let G be the two-element group $\langle J \rangle = \{1, J\}$, and let M be the abelian group $\mathbb{Z}_n$. Consider the representation

$R:\langle J\rangle \to \operatorname{Aut}\mathbb{Z}_n;\ J \mapsto (x \mapsto -x)$. Then the split extension of $\mathbb{Z}_n$ by $\langle J\rangle$ via R is the dihedral group D_n. □

Example 2.4.5 (Direct Products). Let M be an abelian group. Let G be a group, and let $R: G \to \operatorname{Aut} M$ be the trivial representation $g \mapsto 1_M$. Then the split extension of M by G via R is the direct product $G \times M$. □

Example 2.4.6 (Affine Groups). Let K be a commutative, unital ring, and let n be a positive integer. Consider the general linear group $\mathrm{GL}_n(K)$ of dimension n over K (cf. Exercise II 2.4K), the group of invertible elements of the monoid $(K_n^n, \cdot, I_n)$. The matrix product $K_1^n \times K_n^n \to K_1^n$ (II 1.3.4) yields a K-representation $R: \mathrm{GL}_n(K) \to \operatorname{Aut} K_1^n$. The split extension of $(K_1^n, +)$ by $\mathrm{GL}_n(K)$ via R is called the *general affine group* $\mathrm{GA}_n(K)$ of *dimension* n over K. The split extension of $(K_1^n, +)$ by $\mathrm{SL}_n(K)$ via the restriction of R to $\mathrm{SL}_n(K)$ is called the *special affine group* of *dimension* n over K. □

Given the split extension $p: G[M \to G$, one may recover the module M as $\operatorname{Ker} p = \{(1,m) | m \in M\}$. Note that $(\operatorname{Ker} p)^2$ is a subgroup of the pullback $(G[M) \times_G (G[M)$, and the addition on $\operatorname{Ker} p$ is the restriction of $+_G$ to this subgroup. Moreover, $(1,m)T((g,0)) = (g^{-1},0)(1,m)(g,0) = (g^{-1},m)(g,0) = (1,mg)$. In other words, the action of G on M is recovered as conjugation of $G0_G$ on $\operatorname{Ker} p$. More generally, noting that $(1,m)T((g,n)) = (g^{-1}, -ng^{-1})(1,m)(g,n) = (g^{-1}, -ng^{-1})(g, mg+n) = (1, mg+n)$, one obtains an action

$$(g,n): m \mapsto mg + n \tag{2.4.7}$$

of the split extension $G[M$ on M.

EXERCISES

2.4A. (a) Verify the claims of Example 2.4.2.

(b) Let A and B be sequences of groups. Show that a natural transformation $\sigma: A \to B$ is a morphism of groups in $\underline{\underline{\mathrm{Set}}}^{\mathbb{N}}$ iff each component $\sigma_n: A_n \to B_n$ is a group homomorphism.

2.4B. Show that monoids in the category of monoids are commutative.

2.4C. Let an object $(M, \cdot, 1)$ of $\underline{\underline{\mathrm{Mon}}}$ be a monoid (M, μ, e) in the category $\underline{\underline{\mathrm{Mon}}}$. Show that the image of the monoid homomorphism $e_1: M \to M$ is $\{1\}$.

2.4D. Verify that (2.4.6) makes E a group.

2.4E. Verify that $p: E \to G$ is an abelian group $(E, +_G, -_G, 0_G)$ in the slice category $\underline{\underline{\mathrm{Gp}}}/G$.

2.4F. Verify the claim of Example 2.4.4.

2.4G. Verify the claim of Example 2.4.5.

2.4H. Show that the symmetric group S_3 is isomorphic to the affine group $\mathrm{GA}_1(\mathbb{Z}_3)$.

2.4I. Is the symmetric group S_4 isomorphic to the affine group $\mathrm{GA}_2(\mathbb{Z}_2)$? Justify your answer. [Hint: Consider (2.4.7).]

2.4J. Show that $\mathrm{GA}_n(K)$ is isomorphic to the subgroup of $\mathrm{GL}_{n+1}(K)$ consisting of matrices whose first column is $[1,0,\ldots,0]^T$.

2.4K. Let G be a group, and let $(p: E \to G, \mu, J, e)$ be an abelian group in the slice category $\underline{\underline{\mathrm{Gp}}}/G$.

(a) Show that $e: G \to E$ is an injective group homomorphism.

(b) Show that $(\mathrm{Ker}\, p)^2$ is a subgroup of E, and that $\mu: E \times_G E \to E$ restricts to a commutative group operation $+: (\mathrm{Ker}\, p)^2 \to \mathrm{Ker}\, p$.

(c) Let $r: \mathrm{Inn}\, E \to \mathrm{Aut}\,\mathrm{Ker}\, p$ be the restriction of conjugations to automorphisms of the normal subgroup $\mathrm{Ker}\, p$ of E. Show that the composite $R: G \overset{e}{\to} E \overset{T}{\to} \mathrm{Inn}\, E \overset{r}{\to} \mathrm{Aut}\,\mathrm{Ker}\, p$ yields a linear representation of G.

(d) Show that $p: E \to G$ and $G[\mathrm{Ker}\, p \to G;\ (g, n) \mapsto g$ are isomorphic objects of $\underline{\underline{\mathrm{Gp}}}/G$.

2.4L. Let G be a (not necessarily abelian) group, with automorphism group $\mathrm{Aut}\, G$. Define a multiplication on the set $(\mathrm{Aut}\, G) \times G$ by $(\alpha, g)(\beta, h) = (\alpha\beta, g^{\beta}h)$. Show that $(\mathrm{Aut}\, G) \times G$ with this multiplication becomes a group (known as the *holomorph* $\mathrm{Hol}\, G$ of G). If G is abelian, show that $\mathrm{Hol}\, G$ is the split extension $(\mathrm{Aut}\, G)[G$ of G by $\mathrm{Aut}\, G$ via the representation $1: \mathrm{Aut}\, G \to \mathrm{Aut}\, G$.

2.5. Limits

Proposition 2.2.7, Proposition 2.3.4, and Exercise 2.3M(b) reveal a certain similarity amongst the concepts of coproduct, coequalizer, and pushout. Dually, Proposition 2.2.8, Proposition 2.3.5, and Exercise 2.3M(a) connect products, equalizers, and pullbacks. These similarities are formalized by the following:

Definition 2.5.1. Let D be a category, and let J be a small category.

(a) For an object B of D, the *constant functor* $B\Delta: J \to D$ has morphism part $(i \overset{u}{\to} j) \mapsto (B \overset{1_B}{\to} B)$.

(b) For a D-morphism $f: B \to B'$, the *diagonal transformation* $f\Delta: B\Delta \to B'\Delta$ has component $f: B \to B'$ at each object i of J.

(c) The *diagonal functor* $\Delta: D \to D^J$ has object part $\Delta: B \mapsto B\Delta$ and morphism part $\Delta: (f: B \to B') \mapsto (f\Delta: B\Delta \to B'\Delta)$.

(d) For a functor $F: J \to D$, the *category $D\Delta/F$ of constant functors over F* is the subcategory of the slice category D^J/F whose objects are natural transformations $\varphi: B\Delta \to F$ from constant functors, and whose morphisms are of the form $f\Delta : (\varphi : B\Delta \to F) \to (\varphi' : B'\Delta \to F)$ for a diagonal transformation $f\Delta : B\Delta \to B'\Delta$.

(e) For a functor $F : J \to D$, the *limit* is a terminal object π^F or $\pi : (\varprojlim F)\Delta \to F$ of $D\Delta/F$ or, by abuse of notation, the object $\varprojlim F$ of D. For objects i of J, the components $\pi_i : \varprojlim F \to iF$ are known as the *projections* of the limit. Given an object $\varphi : B\Delta \to F$ of $D\Delta/F$, the *limit morphism* $\varprojlim \varphi : B \to \varprojlim F$ is the D-morphism yielding the unique $(D\Delta/F)$-morphism $(\varprojlim \varphi)\Delta : (\varphi : B\Delta \to F) \to (\pi : (\varprojlim F)\Delta \to F)$.

(f) For a functor $F : J \to D$, the *category $F/D\Delta$ of constant functors under F* is the subcategory of the slice category F/D^J whose objects are natural transformations $\varphi : F \to B\Delta$ to constant functors, and whose morphisms are of the form $f\Delta : (\varphi : F \to B\Delta) \to (\varphi' : F \to B'\Delta)$ for a diagonal transformation $f\Delta : B\Delta \to B'\Delta$.

(g) For a functor $F : J \to D$, the *colimit* is an initial object ι^F or $\iota : F \to (\varinjlim F)\Delta$ of $F/D\Delta$ or, by abuse of notation, the object $\varinjlim F$ of D. For objects i of J, the components $\iota_i : iF \to \varinjlim F$ are known as the *insertions* of the colimit. Given an object $\varphi : F \to B\Delta$ of $F/D\Delta$, the *colimit morphism* $\varinjlim \varphi : \varinjlim F \to B$ is the D-morphism yielding the unique $(F/D\Delta)$-morphism $(\varinjlim \varphi)\Delta : (\iota : F \to (\varinjlim F)\Delta) \to (\varphi : F \to B\Delta)$. □

Remarks 2.5.2. (a) As a mnemonic for the direction of the arrow under the symbol "lim," display a typical component with the functor value on the left: $iF \xleftarrow{\pi_i} \varprojlim F$ and $iF \xrightarrow{\iota_i} \varinjlim F$. Note that the direction of the arrow under the symbol "lim" is the same as the direction of the component. "Projections are thrown out from limits: insertions are forced in to colimits."

(b) Some authors use "lim" in place of "$\varprojlim$" and "colim" in place of "$\varinjlim$.

(c) Some authors use "projective limit" or "inverse limit" in place of "limit." Dually, they use "inductive limit" or "direct limit" in place of "colimit." Occasionally, these terms are used only in restricted circumstances, such as the "directed" case in which the small category J is a poset category having upper bounds (not necessarily least upper bounds) of finite subsets. □

Example 2.5.3. (a) Let $F : (\mathbb{N}, \leq) \to (\mathbb{R} \cup \{\infty\}, \leq)$ be a functor between the poset categories. Then the colimit $\varinjlim F$ is the limit $\lim_{n \to \infty}(nF)$ of the sequence $(nF | n \in \mathbb{N})$ of (extended) real numbers. The existence of the insertions $\iota_n : nF \to \varinjlim F$ says that $\varinjlim F$ is an upper bound of the real numbers nF. If B is any such upper bound, the arrow $(F \to (\varinjlim F)\Delta) \to (F \to B\Delta)$ in $F/\mathbb{R}\Delta$ yields the diagonal transformation $(\varinjlim F)\Delta \to B\Delta$ in $\mathbb{R}\Delta$, i.e. the colimit arrow $\varinjlim F \to B$ in $\mathbb{R}$, i.e. the inequality $\varinjlim F \leq B$. Thus $\varinjlim F$ is the least upper bound of the monotonic sequence $(nF | n \in \mathbb{N})$.

(b) Let K be a unital, commutative ring. Let $K^{\mathbb{N}} = \underline{\underline{\mathrm{Set}}}(\mathbb{N}, K)$ have the pointwise K-module structure. Define an operation $*$ on $K^{\mathbb{N}}$ by $f * g(n) = \sum_{l+m=n} f(l)g(m)$. Define δ in $K^{\mathbb{N}}$ by $\delta(n) =$ **if** $n = 1$ **then** 1 **else** 0. Then

$(K^{\mathbb{N}}, *, \delta)$ is a commutative monoid making $K^{\mathbb{N}}$ a K-algebra. Define a functor $F : (\mathbb{N}, \geq) \to \underline{\underline{\mathrm{Alg}}}_K$ by $(n \to n-1) \mapsto (K[T]/T^nK[T] \to K[T]/T^{n-1}K[T];\ p(T) + T^nK[T] \mapsto p(T) + T^{n-1}K[T])$. Then the limit $\varprojlim F$ is the object $K^{\mathbb{N}}$ together with the projections $\pi_n : K^{\mathbb{N}} \to K[T]/T^nK[T];\ f \mapsto [\Sigma_{i=0}^{n-1} f(i)T^i] + T^nK[T]$. The K-algebra $K^{\mathbb{N}}$ is often denoted by $K[[T]]$, and called the *algebra of power series over K in the indeterminate T*. A function $f : \mathbb{N} \to K$ is then written as a *formal power series* $\Sigma_{i=0}^{\infty} f(i)T^i$. The property $\forall n \in \mathbb{N},\ f\pi_n = \Sigma_{i=0}^{n-1} f(i)T^i + T^nK[T]$ is sometimes expressed as $f = \lim_{n \to \infty} \Sigma_{i=0}^{n-1} f(i)T^i$.

(c) Let G be a group, and let $(\mathscr{P}_{<\omega}(G), \subseteq)$ be the set of finite subsets of G, ordered by containment. Define $F : \mathscr{P}_{<\omega}(G) \to \underline{\underline{\mathrm{Gp}}}$ with morphism part sending a containment $X \hookrightarrow Y$ to the inclusion $\langle X \rangle \hookrightarrow \langle Y \rangle$ of the subgroup of G generated by X in the subgroup generated by Y. Then $G = \varinjlim F$, with insertions $\iota_X : \langle X \rangle \hookrightarrow G$. □

As for the special cases discussed at the beginning of the section, the issue of uniqueness of general limit and colimit objects arises.

Proposition 2.5.4. *Let $F : J \to D$ be a functor with small domain.*

(a) *If the limit object $\varprojlim F$ exists in D, then it is unique to within isomorphism in D.*

(b) *If the colimit object $\varinjlim F$ exists in D, then it is unique to within isomorphism in D.*

Proof. (a) Suppose that $\pi : L\Delta \to F$ and $\pi' : L'\Delta \to F$ are both limits of F. Then they are both terminal objects of $D\Delta/F$. By Proposition 1.3.4, there is a $(D\Delta/F)$-isomorphism $f\Delta : (\pi : L\Delta \to F) \to (\pi' : L'\Delta \to F)$. This yields a diagonal isomorphism $f\Delta : L\Delta \to L'\Delta$, and so a D-isomorphism $f : L \to L'$.

(b) Dual to (a). □

At the end of Section 2.3, the pullback $a_1 \times_x a_2$ of $a_1 \xrightarrow{p_1} x \xleftarrow{p_2} a_2$ in a category C was constructed as the equalizer $a_1 \times_x a_2 \xrightarrow{e} a_1 \times a_2$ of the parallel pair $a_1 \times a_2 \overset{\pi_1 p_1}{\underset{\pi_2 p_2}{\rightrightarrows}} x$ of morphisms, i.e. of the parallel pair $(\pi_1 p_1, \pi_2 p_2)$ of morphisms $\prod_{i=1}^{2} a_i \to x$ between products. Thus if the category C has appropriate products and equalizers, it has pullbacks. This is an example of a general property of limits: They may be constructed by equalizers and products.

Theorem 2.5.5. *Let $F : J \to D$ be a functor with small domain. Then $\varprojlim F$ is the equalizer of the parallel pair $(\prod_{u \in J_1} \pi_{ud_0} u^F, \prod_{u \in J_1} \pi_{ud_1})$ of D-morphisms $\prod_{j \in J_0} jF \to \prod_{u \in J_1} ud_1 F$: The limit exists if the products and equalizer exist.*

Proof. For an object $\varphi : B\Delta \to F$ of $D\Delta/F$ and objects i, k of J, consider the commuting diagram in D:

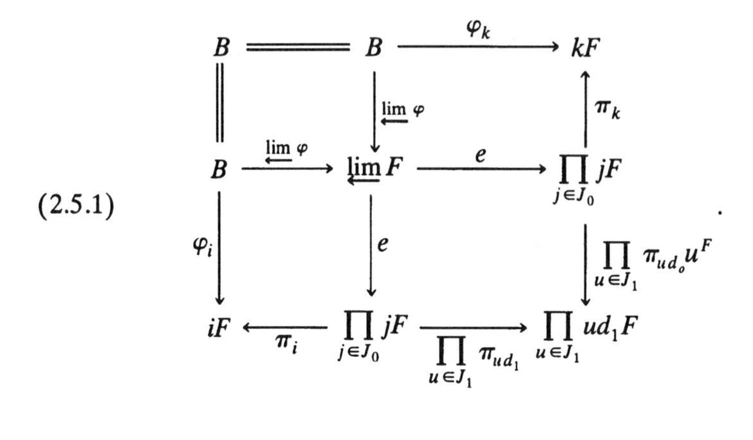

Here the central object, putatively labeled $\varprojlim F$, is the equalizer of the bottom right and right lower morphisms in D. Set $\pi_i^F := e\pi_i : \varprojlim F \to iF$. Consider a J-morphism $v : vd_0 \to vd_1$. Since $e\prod_{u\in J_1}\pi_{ud_0}u^F = e\prod_{u\in J_1}\pi_{ud_1}$ by the equalizer property, one has $e\pi_{vd_0}v^F = e\pi_{vd_1}$. In other words, condition

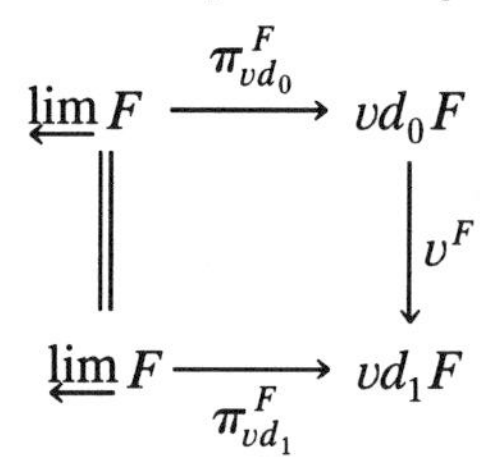

(1.4.2) at v in J_1 is verified, showing that there is a natural transformation $\pi^F : (\varprojlim F)\Delta \to F$. Now in order that the bottom left and top right-hand squares commute independently of the choice of i or k in J_0, the diagonals from B to $\prod_{j\in J_0} jF$ have to be $\prod_{j\in J_0}\varphi_j$. For a morphism $ud_0 \xrightarrow{u} ud_1$ in J, consider condition (1.4.2) on the natural transformation $\varphi : B\Delta \to F$:

$$\begin{array}{ccc} B & \xrightarrow{\varphi_{ud_0}} & ud_0F \\ \| & & \downarrow{\scriptstyle u^F} \\ B & \xrightarrow[\varphi_{ud_1}]{} & ud_1F \end{array} \tag{2.5.2}$$

Thus $\varphi_{ud_0}u^F = \varphi_{ud_1}$. It follows that $\prod_{j\in J_0}\varphi_j\prod_{u\in J_1}\pi_{ud_0}u^F$ and $\prod_{j\in J_0}\varphi_j\prod_{u\in J_1}\pi_{ud_1}$ are equal morphisms from B to $\prod_{u\in J_1}ud_1F$. By the universality property of the equalizer called $\varprojlim F$, one then obtains a unique element $\varprojlim \varphi$ of $D(B, \varprojlim F)$ such that the top right and bottom left-hand squares in (2.5.1) commute. The commuting of these squares shows that there is a unique $(D\Delta/F)$-morphism $(\varprojlim \varphi)\Delta : (\varphi : B\Delta \to F) \to (\pi^F : (\varprojlim F)\Delta \to F)$. Thus $\pi^F : \varprojlim F \to F$ is indeed a terminal object of $D\Delta/F$. □

Corollary 2.5.6. *Let $F : J \to D$ be a functor with small domain. Then $\varinjlim F$ is the coequalizer of the parallel pair $(\Sigma_{u \in J_1} u^F \iota_{ud_1}, \Sigma_{u \in J_1} \iota_{ud_0})$ of D-morphisms $\Sigma_{u \in J_1} ud_0 F \to \Sigma_{j \in J_0} jF$: The colimit exists if the coproducts and coequalizer exist.*

Proof. Dual to Theorem 2.5.5. □

Example 2.5.7. (a) Consider how Theorem 2.5.5 builds the pullback $A_1 \times_X A_2$ of $A_1 \xrightarrow{p_1} X \xleftarrow{p_2} A_2$ in the category $\underline{\underline{\text{Set}}}$. The bottom right-hand square of (2.5.1) produces the equalizer of the functions $(\pi_{A_1} p_1, \pi_{A_2} p_2) : A_1 \times X \times A_2 \to X \times X$ and $(\pi_X, \pi_X) : A_1 \times X \times A_1 \to X \times X$, i.e. the subset $\{(a_1, x, a_2) | a_1 p_1 = x,\ a_2 p_2 = x\}$ of $A_1 \times X \times A_2$.

(b) (**The Ring of p-adic Integers.**) For a prime p, consider the functor $F : (\mathbb{N}, \geq) \to \underline{\underline{\text{Ring}}}$ with morphism part $(n : (n+1) \to n) \mapsto (\mathbb{Z}/p^{n+1}\mathbb{Z} \to \mathbb{Z}/p^n\mathbb{Z};\ x + p^{n+1}\mathbb{Z} \mapsto x + p^n\mathbb{Z})$. The ring $\varprojlim F$ is called the *ring* $\hat{\mathbb{Z}}_{p\mathbb{Z}}$ *of p-adic integers*. Theorem 2.5.5 builds $\hat{\mathbb{Z}}_{p\mathbb{Z}}$ as a subring of $\Pi_{n \in \mathbb{N}}(\mathbb{Z}/p^n\mathbb{Z})$. Writing elements of $\Pi_{n \in \mathbb{N}}(\mathbb{Z}/p^n\mathbb{Z})$ as sequences $(x_0 + \mathbb{Z},\ x_1 + p\mathbb{Z}, x_2 + p^2\mathbb{Z}, \dots)$, the subring is the equalizer of the identity homomorphism and the "shift"$(x_0 + \mathbb{Z},\ x_1 + p\mathbb{Z},\ x_2 + p^2\mathbb{Z}, \dots) \mapsto (x_1 + \mathbb{Z},\ x_2 + p\mathbb{Z},\ x_3 + p^2\mathbb{Z}, \dots)$. In other words, it is the set of sequences $(x_0 + \mathbb{Z},\ x_1 + p\mathbb{Z},\ x_2 + p^2\mathbb{Z}, \dots)$ with $x_{n+1} + p^n\mathbb{Z} = x_n + p^n\mathbb{Z}$ for all n. □

A category D is said to be (*small*) *complete* if each functor $F : J \to D$ with small domain has a limit in D. Dually, D is (*small*) *cocomplete* if each such functor has a colimit in D. Theorem 2.5.5 shows that a category is complete if it has all products and equalizers. Corollary 2.5.6 shows that a category is cocomplete if it has all coproducts and coequalizers. For example, $\underline{\underline{\text{Set}}}$ is complete and cocomplete. In general, a category is said to be (*small*) *bicomplete* if it is both complete and cocomplete.

EXERCISES

2.5A. Let $F : (\mathbb{N}, \geq) \to (\mathbb{R} \cup \{-\infty\}, \leq)$ be a functor. Prove $\varprojlim F = \lim_{n \to \infty} nF$.

2.5B. Verify the claims of Example 2.5.3(b).

2.5C. Prove $(1 - T)^{-1} = \lim_{n \to \infty} \Sigma_{i=0}^{n-1} T^i$ in $K[[T]]$.

2.5D. Verify the claims of Example 2.5.3(c).

2.5E. Define $F:(\mathbb{N}, \leq) \to \underline{\underline{\mathrm{Gp}}}$ with morphism part $(m \to n) \mapsto (S_m \hookrightarrow S_n)$. Prove $S_\infty = \varinjlim F$.

2.5F. Define $F:(\mathscr{P}_{<\omega}(\mathbb{N}), \subseteq) \to \underline{\underline{\mathrm{Gp}}}$ with morphism part $(X \subseteq Y) \mapsto (X! \hookrightarrow Y!)$. Prove $S_\infty = \varinjlim F$.

2.5G. Let G be the group $\mathbb{N}!$, and let F be the corresponding functor of Example 2.5.3(c). How does this functor differ from the functor of Exercise 2.5F?

2.5H. Draw the analogue of (2.5.1) that describes the proof of Corollary 2.5.6.

2.5I. Use Corollary 2.5.6 to construct the pushout of the diagram $A_1 \xleftarrow{p_1} X \xrightarrow{p_2} A_2$ in the category $\underline{\underline{\mathrm{Mod}}}_{\mathbb{Z}}$ of abelian groups.

2.5J. For a prime p, show that the sequence $(1, 1, 1, \ldots)$ generates a subring of $\hat{\mathbb{Z}}_{p\mathbb{Z}}$ isomorphic to $\mathbb{Z}$.

2.5K. Show that the category $\underline{\underline{\mathrm{Mod}}}_{\mathbb{Z}}$ of abelian groups is complete and cocomplete.

2.5L. Let p be a prime. Let $\underline{\underline{\mathrm{Mod}}}_{\mathbb{Z}}^{(p)}$ denote the full subcategory of $\underline{\underline{\mathrm{Mod}}}_{\mathbb{Z}}$ consisting of those abelian groups $(A, +, 0)$ with the property: $\exists n \in \mathbb{N}.\ \forall a \in A,\ p^n a = 0$.

(a) Is each abelian p-group an object of $\underline{\underline{\mathrm{Mod}}}_{\mathbb{Z}}^{(p)}$?

(b) Show that $\underline{\underline{\mathrm{Mod}}}_{\mathbb{Z}}^{(p)}$ has equalizers.

(c) Show that $\underline{\underline{\mathrm{Mod}}}_{\mathbb{Z}}^{(p)}$ has finite products.

(d) Show that $\underline{\underline{\mathrm{Mod}}}_{\mathbb{Z}}^{(p)}$ is not complete.

2.5M. Let p be a prime. Let C_{p^∞} denote the set of complex numbers c with the property: $\exists n \in \mathbb{N}.\ c^{p^n} = 1$.

(a) Show that C_{p^∞} is a subgroup of $\mathbb{C}^*$.

(b) Show that C_{p^∞} is a Sylov p-subgroup of $(\{z \in \mathbb{C} | z\bar{z} = 1\}, \cdot, 1)$.

(c) Show that C_{p^∞} is the colimit of the functor $F:(\mathbb{N}, \leq) \to \underline{\underline{\mathrm{Mod}}}_{\mathbb{Z}}$ with morphism part $(n \to n+1) \mapsto (\mathbb{Z}/p^n\mathbb{Z} \to \mathbb{Z}/p^{n+1}\mathbb{Z};\ x + p^n\mathbb{Z} \mapsto px + p^{n+1}\mathbb{Z})$.

2.5N. Let G be a group. Show that the category $\underline{\underline{G}}$ of G-sets and G-homomorphisms is complete and cocomplete.

2.5O. Let $F: J \to D$ be a functor with small domain.

(a) If J has a terminal object T, show that $\varinjlim F = TF$.

(b) Formulate and do the dual of (a).

2.5P. Show that -1 has a square root in the ring $\hat{\mathbb{Z}}_{5\mathbb{Z}}$ of 5-adic integers. (Hint: Consider roots of $T^2 + 1$ in $\{0\}$, $\mathbb{Z}/5\mathbb{Z}$, $\mathbb{Z}/5^2\mathbb{Z}, \ldots$.)

3. ADJOINT FUNCTORS

Let B be a subgroup of a group A. Consider the category $\underline{\underline{A}}$ of A-sets and A-homomorphisms, and the category $\underline{\underline{B}}$ of B-sets and B-homomorphisms.

The forgetful functor $G:\underline{\underline{A}}\to\underline{\underline{B}}$ has morphism part $(k:(Y,A)\to(Y',A))\mapsto(k:(Y,B)\to(Y',B))$. One may also write $G=\downarrow^A_B$, identifying G with the concept of restriction introduced in Section I 2.3. In the other direction, induction $\uparrow^A_B$ leads to a functor $F:\underline{\underline{B}}\to\underline{\underline{A}}$ with object part $(X,B)\mapsto(X,B)\uparrow^A_B=((X\times A)/B,A)$. Thus for X in $\underline{\underline{B}}_0$, the A-set XF is the free A-set over the B-set X. Definition of the morphism part of F, e.g. acting on a $\underline{\underline{B}}$-morphism $f:X'\to X$ or B-homomorphism $f:(X',B)\to(X,B)$, depends on the B-homomorphisms $\eta_{X'}:(X',B)\to((X'\times A)/B,B)$ and $\eta_X:(X,B)\to((X\times A)/B,B)$ of (I 2.3.2). Note that there is a composite B-homomorphism $f\eta_X:(X',B)\to((X\times A)/B,A)\downarrow^A_B$. By (I 2.3.3), there is then a unique A-homomorphism $\overline{f\eta_X}:(X',B)\uparrow^A_B\to(X,B)\uparrow^A_B$ with $\eta_{X'}\overline{f\eta_X}=f\eta_X$. The image $f^F:X'F\to XF$ of $f:X'\to X$ under the morphism part of F is defined to be this $\underline{\underline{A}}$-morphism or A-homomorphism $\overline{f\eta_X}$. It is straightforward to verify that F does indeed give a functor $F:\underline{\underline{B}}\to\underline{\underline{A}}$ (Exercise 3A).

The restriction functor G and induction functor F are closely related. Their relation is described as *adjunction*. More specifically, G is described as a *right adjoint* of F, and F is described as a *left adjoint* of G. The relationship may be summarized by the isomorphism

$$\underline{\underline{A}}(XF,Y)\cong\underline{\underline{B}}(X,YG) \tag{3.1}$$

obtaining for a given B-set X and A-set Y. More explicitly, the isomorphism is

$$\varphi^X_Y:\underline{\underline{A}}(XF,Y)\to\underline{\underline{B}}(X,YG);\ g\mapsto\eta_X g^G, \tag{3.2}$$

with two-sided inverse

$$\left(\varphi^X_Y\right)^{-1}:\underline{\underline{B}}(X,YG)\to\underline{\underline{A}}(XF,Y);\ h\mapsto\bar{h}, \tag{3.3}$$

the A-homomorphism $\bar{h}$ being given by (I 2.3.3). In order to describe $(\varphi^X_Y)^{-1}$ more directly, consider the $\underline{\underline{A}}$-morphism $\varepsilon_Y=\overline{1_{YG}}:YGF\to Y$, i.e. the A-homomorphism $\varepsilon_Y:((Y\times A)/B,A)\to(Y,A);(y,a)B\mapsto ya$. Then for $h\in\underline{\underline{B}}(X,YG)$, one has $\bar{h}=h^F\varepsilon_Y:(XF\to YGF\to Y)$. One may thus replace (3.3) by

$$\left(\varphi^X_Y\right)^{-1}:\underline{\underline{B}}(X,YG)\to\underline{\underline{A}}(XF,Y);\ h\mapsto h^F\varepsilon_Y. \tag{3.4}$$

Incidentally, if B is the trivial subgroup $\{1\}$ of A, and Y is an A-set (Y,A), then $\varepsilon_Y:Y\times A\to Y;\ (y,a)\mapsto ya$ is just the action (I 1.7) of A on Y.

Remark 3.1. Various mnemonics are available to help one distinguish a right adjoint from a left adjoint. In the isomorphism (3.1), putting the unfunctored objects Y and X next to the isomorphism sign $\cong$, the left adjoint F is the functor appearing on the left-hand side of the isomorphism, while the right adjoint G is the functor appearing on the right-hand side. Another mnemonic, usually reliable when dealing with concrete categories, is that right adjoints like $\downarrow_B^A$ tend to be "trivial," "forgetful," or "destructive," while left adjoints like $\uparrow_B^A$ tend to be "complicated," "free," or "constructive." [Note that Example 3.1.6(b) below presents one case where this mnemonic has to be reversed.] □

The isomorphism (3.1) is natural. More explicitly, for each X in $\underline{\underline{B}}_0$, there is a natural isomorphism φ^X whose component at an object Y of $\underline{\underline{A}}$ is the isomorphism (3.2), while for each Y in $\underline{\underline{A}}_0$, there is a natural isomorphism φ_Y whose component at an object X of $\underline{\underline{B}}$ is the isomorphism (3.2). Yet more explicitly, for fixed X in $\underline{\underline{B}}_0$, consider the functor $\underline{\underline{A}}(XF, \_\_): \underline{\underline{A}} \to \underline{\underline{\text{Set}}}$ with morphism part $(k : Y \to Y') \mapsto (\underline{\underline{A}}(XF, k) : \underline{\underline{A}}(XF, Y) \to \underline{\underline{A}}(XF, Y');\ g \mapsto gk)$. Consider the functor $\underline{\underline{B}}(X, \_\_G) : \underline{\underline{A}} \to \underline{\underline{\text{Set}}}$ with morphism part $(k : Y \to Y') \mapsto (\underline{\underline{B}}(X, kG) : \underline{\underline{B}}(X, YG) \to \underline{\underline{B}}(X, Y'G);\ h \mapsto hk^G)$. Then $\varphi^X : \underline{\underline{A}}(XF, \_\_) \to \underline{\underline{B}}(X, \_\_G)$ is a natural isomorphism. For an $\underline{\underline{A}}$-morphism $k : Y \to Y'$, the diagram (1.4.2) becomes

$$\begin{array}{ccc} \underline{\underline{A}}(XF, Y) & \xrightarrow{\varphi_Y^X} & \underline{\underline{B}}(X, YG) \\ {\scriptstyle \underline{\underline{A}}(XF,k)}\downarrow & & \downarrow{\scriptstyle \underline{\underline{B}}(X,kG)} \\ \underline{\underline{A}}(XF, Y') & \xrightarrow[\varphi_{Y'}^X]{} & \underline{\underline{B}}(X, Y'G) \end{array} \quad . \tag{3.5}$$

This diagram in $\underline{\underline{\text{Set}}}$ commutes. Indeed, for $g \in \underline{\underline{A}}(XF, Y)$, one has

$$\begin{aligned} g\varphi_Y^X \underline{\underline{B}}(X, kG) &= (\eta_X g^G)\underline{\underline{B}}(X, kG) = \eta_X g^G k^G \\ &= \eta_X (gk)^G = (gk)\varphi_{Y'}^X = g\underline{\underline{A}}(XF, k)\varphi_{Y'}^X, \end{aligned}$$

the crucial middle equality holding since G is a functor. Dually, consider the functor $\underline{\underline{A}}(\_\_F, Y) : \underline{\underline{B}}^{\text{op}} \to \underline{\underline{\text{Set}}}$ with morphism part $(f : X' \to X) \mapsto (\underline{\underline{A}}(XF, Y) \to \underline{\underline{A}}(X'F, Y);\ g \mapsto f^F g)$. Consider the functor $\underline{\underline{B}}(\_\_, YG) : \underline{\underline{B}}^{\text{op}} \to \underline{\underline{\text{Set}}}$ with morphism part $(f : X' \to X) \mapsto (\underline{\underline{B}}(X, YG) \to \underline{\underline{B}}(X', YG);\ h \mapsto fh)$. Then φ_Y^X is the component at X of a natural isomorphism $\varphi_Y : \underline{\underline{A}}(\_\_F, Y) \to \underline{\underline{B}}(\_\_, YG)$. Verification of the naturality of φ_Y is dual to the above verification for φ^X (cf. Exercise 3B). The duality between φ^X and φ_Y can be emphasized by writing them in the form

$$\begin{cases} \varphi^X : (k \mapsto (g \mapsto gk)) \to (k \mapsto (h \mapsto hk^G)) \\ \varphi_Y : (f \mapsto (g \mapsto f^F g)) \to (f \mapsto (h \mapsto fh)) \end{cases}, \tag{3.6}$$

where the relevant functors are being described by their morphism parts. Note how the right adjoint G appears on the right-hand side of the composite hk^G in (3.6), while the left adjoint F appears on the left-hand side of the composite $f^F g$ there.

In a similar vein, the B-homomorphism $\eta_X : (X, B) \to (X, B)\uparrow_B^A \downarrow_B^A$ becomes natural. It is the component at the $\underline{\underline{B}}$-object X of a natural transformation $\eta : 1_{\underline{\underline{B}}} \to FG$ from the identity functor on $\underline{\underline{B}}$ to the composite $FG : (\underline{\underline{B}} \to \underline{\underline{A}} \to \underline{\underline{B}})$. To demonstrate the naturality, one needs to verify the commuting of the diagram

$$(3.7) \qquad \begin{array}{ccc} X' & \xrightarrow[\eta_{X'}]{} & X'FG \\ {\scriptstyle f}\downarrow & & \downarrow{\scriptstyle f^{FG}} \\ X & \xrightarrow[\eta_X]{} & XFG \end{array}$$

for a $\underline{\underline{B}}$-morphism $f : X' \to X$. But

$$\begin{aligned} \eta_{X'} f^{FG} &= 1_{X'F}\varphi^{X'}_{X'F}\underline{\underline{B}}(X', f^F G) = 1_{X'F}\underline{\underline{A}}(X'F, f^F)\varphi^{X'}_{XF} = (1_{X'F} f^F)\varphi^{X'}_{XF} \\ &= (f^F 1_{XF})\varphi^{X'}_{XF} = 1_{XF}\underline{\underline{A}}(fF, XF)\varphi^{X'}_{XF} = 1_{XF}\varphi^{X}_{XF}\underline{\underline{B}}(f, XFG) = f\eta_X, \end{aligned}$$

as required. The second equality here is the naturality of $\varphi^{X'}$ [viz. the diagram (1.4.2) applied to the $\underline{\underline{A}}$-morphism $f^F : X'F \to XF$]. The penultimate equality is the naturality of φ_{XF} [viz. the diagram (1.4.2) applied to the $\underline{\underline{B}}$-morphism $f : X' \to X$]. The natural transformation $\eta : 1_{\underline{\underline{B}}} \to FG$ is described as the *unit* of the adjunction (3.1). Dually, the $\underline{\underline{A}}$-morphism $\varepsilon_Y : YGF \to Y$ becomes the component at the $\underline{\underline{A}}$-object Y of a natural transformation $\varepsilon : GF \to 1_{\underline{\underline{A}}}$ (Exercise 3C). This natural transformation is called the *counit* of the adjunction (3.1).

A major issue is the uniqueness of the isomorphism (3.2) or the unit component $\eta_X : X \to XFG$ at an object X of B. The issue is addressed by exhibiting η_X as an initial object of a specially constructed category, and then appealing to Corollary 1.3.5(a). The specially constructed category is a more general version of the category of constant functors under a given functor [Definition 2.5.1(f)] in which the colimit of the functor is an initial object. Indeed, no new construction would be needed, were it not for the fact (exhibited in Example 1.2.5) that the image of a functor need not be a subcategory of its codomain.

Definition 3.2. Let $S : C \to D$ be a functor, and let x be an object of D.

(a) The *comma category* (x, S) *of* x *over* S has objects $f : x \to cS$ that are D-morphisms from x to the image cS of a C-object c under the functor S. An (x, S)-morphism γ or $\gamma S : (f : x \to cS) \to (f' : x \to c'S)$ is a C-morphism $\gamma : c \to c'$ whose image γS under S yields a morphism $\gamma S : (f : x \to cS) \to$

$(f' : x \to c'S)$ of the slice category x/D. The composite of two (x, S)-morphisms $\gamma S : (f : x \to cS) \to (f' : x \to c'S)$ and $\gamma' S : (f' : x \to c'S) \to (f'' : x \to c''S)$ is $(\gamma\gamma')S : (f : x \to cS) \to (f'' : x \to c''S)$.

(b) The *comma category* (S, x) *of* S *over* x has objects $f : cS \to x$ that are D-morphisms to x from the image cS of a C-object c under the functor S. An (S, x)-morphism γ or $\gamma S : (f : cS \to x) \to (f' : c'S \to x)$ is a C-morphism $\gamma : c \to c'$ whose image γS under S yields a morphism $\gamma S : (f : cS \to x) \to (f' : c'S \to x)$ of the slice category D/x. The composite of two (S, x)-morphisms $\gamma S : (f : cS \to x) \to (f' : c'S \to x)$ and $\gamma' S : (f' : c'S \to x) \to (f'' : c''S \to x)$ is $(\gamma\gamma')S : (f : cS \to x) \to (f'' : c''S \to x)$. □

Example 3.3 (Slice Categories). Let $1_C : C \to C$ be the identity functor on a category C. Let x be an object of C. Then the slice category C/x is (isomorphic with) the comma category $(1_C, x)$, while x/C is (isomorphic with) the comma category $(x, 1_C)$. □

Example 3.4 (Constant Functors Under a Given Functor). For functor $F : J \to D$ with small domain, the category $F/D\Delta$ is (isomorphic with) the comma category (F, Δ). □

Using comma categories, one may now recognize components of the unit and counit of (3.1) as initial and terminal objects.

Proposition 3.5. *Consider the adjunction* (3.1).

(a) *Let* X *be a* $\underline{\underline{B}}$*-object. Then* $\eta_X : X \to XFG$ *is an initial object of the comma category* (X, G).

(b) *Let* Y *be an* $\underline{\underline{A}}$*-object. Then* $\varepsilon_Y : YGF \to Y$ *is a terminal object of the comma category* (F, Y).

Proof. (a) Let $h : X \to YG$ be an object of (X, G). Then for $h(\varphi_Y^X)^{-1} : XF \to Y$ in $\underline{\underline{A}}$, one has $\eta_X \cdot h(\varphi_Y^X)^{-1}G = h$, so there is an (X, G)-morphism $h(\varphi_Y^X)^{-1}G : (\eta_X : X \to XFG) \to (h : X \to YG)$. For any such morphism $\tilde{h}G : (\eta_X : X \to XFG) \to (h : X \to YG)$, one has $h = \eta_X \cdot \tilde{h}G = \tilde{h}\varphi_Y^X$, whence $\tilde{h} = h(\varphi_Y^X)^{-1}$ and $\tilde{h}G = h(\varphi_Y^X)^{-1}G$. In other words, the (X, G)-morphism from η_X to h is unique. Thus the object η_X of (X, G) is initial.

(b) Dual to (a). □

Corollary 3.6. (*a*) *The unit* $\eta : 1_{\underline{\underline{B}}} \to FG$ *of the adjunction* (3.1) *is uniquely determined by the functors* F *and* G.

(*b*) *The counit* $\varepsilon : GF \to 1_{\underline{\underline{A}}}$ *of the adjunction* (3.1) *is uniquely determined by the functors* F *and* G.

Proof. (a) Apply Corollary 1.3.5(a) and Proposition 3.5(a).

(b) Apply Proposition 1.3.4 and Propostion 3.5(b). □

Corollary 3.7. *One has the following identities:*
(a) $(\eta F)(F\varepsilon) = 1$;
(b) $(G\eta)(\varepsilon G) = 1$.

Proof. (a) Consider an object X of $\underline{\underline{B}}$. Then at X, (a) reduces to the commuting of the diagram

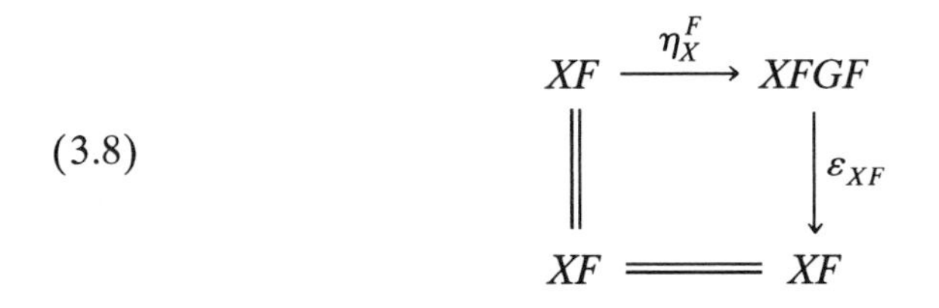

in $\underline{\underline{A}}$ [cf. (1.4.4) and (1.4.5)]. Since ε_{XF} is terminal in (F, XF), there is a unique $\underline{\underline{B}}$-morphism $h: X \to XFG$ such that $1_{XF} = h^F \cdot \varepsilon_{XF} = h(\varphi^X_{XF})^{-1}$. Thus $h = 1_{XF}\varphi^X_{XF} = \eta_X 1^G_{XF} = \eta_X 1_{XFG} = \eta_X$, and $h^F = \eta^F_X$, whence (3.8) commutes.
(b) Dual to (a). □

EXERCISES

3A. If B is a subgroup of a group A, verify that induction $\uparrow^A_B$ does give a functor $F: \underline{\underline{B}} \to \underline{\underline{A}}$.

3B. Verify that $\varphi_Y: \underline{\underline{A}}(\_\_F, Y) \to \underline{\underline{B}}(\_\_, YG)$ is a natural isomorphism.

3C. Verify that $\varepsilon: GF \to 1_{\underline{\underline{A}}}$ is a natural transformation.

3D. For what groups A and B does the counit ε become a natural isomorphism?

3E. Let B be the trivial subgroup $\{1\}$ of A. Show that the multiplication in the group A leads to a natural transformation $\mu: FGFG \to FG$.

3F. Let $F: J \to D$ be a functor with small domain. Show that the category $D\Delta/F$ of constant functors over F is isomorphic with the comma category (Δ, F).

3G. Let $F: C \to D$ be the functor of Example 1.2.5. Show that the comma category $(0, F)$ is not a subcategory of the slice category $0/D$.

3H. Write out a detailed proof of Proposition 3.5(b).

3I. Write out a detailed proof of Corollary 3.7(b).

3J. Let B be a subgroup of a monoid A that is not necessarily a group. Show that the forgetful functor $G: \underline{\underline{A}} \to \underline{\underline{B}}$ with morphism part $(k: (Y, A) \to (Y', A)) \mapsto (k: (Y, B) \to (Y', B))$ has a left adjoint (cf. Exercise I 2.3E).

3.1. Adjunctions

The adjunction relationship (3.1) between the restriction and induction functors is the prototype for a phenomenon that is extremely common, throughout mathematics and particularly in algebra.

Definition 3.1.1. Let A and B be categories. Let $F: B \to A$ and $G: A \to B$ be functors. Then F is a *left adjoint* of G, and G is a *right adjoint* of F, if there is a natural isomorphism

$$A(XF, Y) \cong B(X, YG) \tag{3.1.1}$$

for each object X of B and each object Y of A. In other words, the isomorphism (3.1.1) or

$$\varphi_Y^X : A(XF, Y) \to B(X, YG) \tag{3.1.2}$$

is the component at Y of a natural isomorphism $\varphi^X : A(XF, \_\_) \to B(X, \_\_G)$ and the component at X of a natural isomorphism $\varphi_Y : A(\_\_F, Y) \to B(\_\_, YG)$. The relationship between F and G is described as *adjunction.* □

Example 3.1.2 (Currying). Let Y be a fixed set. Consider the "law of exponents" (O 3.4.1): $X^{Z \times Y} \cong (X^Y)^Z$. In the form (O 3.4.4), it becomes $\underline{\underline{\mathrm{Set}}}(Z \times Y, X) \cong \underline{\underline{\mathrm{Set}}}(Z, X^Y)$. Consider the functor $F_Y : \underline{\underline{\mathrm{Set}}} \to \underline{\underline{\mathrm{Set}}};\ Z \mapsto Z \times Y$ of Example 1.2.2, along with the functor $G_Y : \underline{\underline{\mathrm{Set}}} \to \underline{\underline{\mathrm{Set}}}$ whose morphism part is $(f : X \to X') \mapsto (X^Y \to X'^Y;\ g \mapsto gf)$. Then (O 3.4.4) becomes a bijection $\varphi_X^Z : \underline{\underline{\mathrm{Set}}}(ZF_Y, X) \to \underline{\underline{\mathrm{Set}}}(Z, XG_Y)$. This isomorphism is natural [Exercise 3.1A(b),(c)]. Thus F_Y is a left adjoint of G_Y. Comparing with Remark 3.1, note how the left adjoint F_Y incorporates the construction of the direct product of a set Z with Y, while the right adjoint G_Y merely involves the concatenation of a function $g : Y \to X$ with a function $f : X \to X'$. □

The adjunction (3.1) had a natural unit η and a natural counit ε. This is completely typical.

Proposition 3.1.3. *Consider the adjunction* (3.1.1).

(a) *There is a natural transformation* $\eta : 1_B \to FG$ *whose component at a* B*-object* X *is* $1_{XF}\varphi_{XF}^X$.

(b) *There is a natural transformation* $\varepsilon : GF \to 1_A$ *whose component at an* A*-object* Y *is* $1_{YG}(\varphi_Y^{YG})^{-1}$.

Proof. (a) Cf. (3.7). (b) Dual to (a). □

The natural transformation η is called the *unit* of the adjunction (3.1.1); the natural transformation ε is called its *counit*. Note that in the codomain FG

of the unit, the left adjoint F comes on the left and the right adjoint G comes on the right.

Spotting adjunctions is greatly facilitated by the following result, a key illustration of the theme broached in Section 1.3 that mathematical constructs are specified as initial objects.

Theorem 3.1.4. *Let $G : A \to B$ be a functor. Then G has a left adjoint if and only if, for each object X of B, the comma category (X, G) has an initial object. Dually, G has a right adjoint iff (G, X) has a terminal object for each X in B_0.*

Proof. The "only if" direction follows as in Proposition 3.5(a), via Proposition 3.1.3(a). The "if" direction builds the left adjoint $F : B \to A$ of $G : A \to B$ in the way that induction was construed as a functor starting from (I 2.3.2) and (I 2.3.3). Suppose that, for each object X of B, there is an initial object η_X of (X, G). Now η_X is a B-morphism whose codomain is the image XFG of an A-object XF. One thus defines a function $F : B_0 \to A_0$; $X \mapsto XF$ which will become the object part of the budding left adjoint $F : B \to A$. Given a B-morphism $f : X' \to X$, the A-morphism $fF : X'F \to XF$ is defined as the A-morphism yielding the unique (X', G)-morphism $fFG: (\eta_{X'} : X' \to X'FG) \to (f\eta_X : X' \to XFG)$. As in Exercise 3A, one may readily verify that $F : B \to A$ is a functor. Moreover, the definition of the morphism part of F yields $\eta_{X'} f^{FG} = f\eta_X$, so the initial object $\eta_X : X \to XFG$ of (X, G) becomes the component at X of a natural transformation $\eta : 1_B \to FG$ [cf. (3.7)].

Now for X in B_0 and Y in A_0, define

$$\text{(3.1.3)} \qquad \varphi_Y^X : A(XF, Y) \to B(X, YG);\ g \mapsto \eta_X g^G.$$

Fix Y, and consider the B-morphism $f : X' \to X$. For an element g of $A(XF, Y)$, one has $gA(fF, Y)\varphi_X^Y = (f^F g)\varphi_{X'}^Y = \eta_{X'}(f^F g)^G = \eta_{X'} f^{FG} g^G = f\eta_X g^G = (\eta_X g^G)B(f, YG) = g\varphi_X^Y B(f, YG)$. Thus $A(fF, Y)\varphi_{X'}^Y = \varphi_X^Y B(f, YG)$, i.e. there is a natural transformation $\varphi_Y : A(\_\_F, Y) \to B(\_\_, YG)$. For fixed X, the existence of a natural transformation $\varphi^X : A(XF, \_\_) \to B(X, \_\_G)$ follows as in (3.5). Finally, for an element h of $B(X, YG)$, define $h\psi_X^Y$ to be the unique element of $A(XF, Y)$ yielding an (X, G)-morphism $(h\psi_X^Y)^G : (\eta_X : X \to XFG) \to (h : X \to YG)$. Then φ_Y^X and ψ_X^Y are mutually inverse, so that (3.1.3) is an isomorphism. □

Example 3.1.5 (Free Monoids). Let $U : \underline{\underline{\text{Mon}}} \to \underline{\underline{\text{Set}}}$ be the forgetful functor whose object part sends a monoid $(M, \cdot, 1)$ to its underlying set M. Consider an object X of $\underline{\underline{\text{Set}}}$ as an alphabet. The universality property (O 4.1.2) shows that the inclusion $\eta_X : X \hookrightarrow X^*U$ of the alphabet as the set of words of length 1 in X^*U is an initial object of the comma category (X, U). Theorem 3.1.4 then shows that $X \mapsto X^*$ is the object part of a functor $F : \underline{\underline{\text{Set}}} \to \underline{\underline{\text{Mon}}}$, the *free monoid functor*, that is a left adjoint to the underlying set functor $U : \underline{\underline{\text{Mon}}} \to \underline{\underline{\text{Set}}}$. □

Example 3.1.6 (Colimits and Limits). Let $F : J \to D$ be a functor with small codomain.

(a) By definition, the colimit $\iota : F \to (\varinjlim F)\Delta$ is an initial object of the comma category (F, Δ) (cf. Example 3.4). Thus the diagonal functor $\Delta : D \to D^J$ is the right adjoint of a functor $\varinjlim : D^J \to D$ with object part $F \mapsto \varinjlim F$.

(b) Dually, the limit $\pi : (\varprojlim F)\Delta \to F$ is a terminal object of the comma category (Δ, F) (cf. Exercise 3.3F). Thus the diagonal functor $\Delta : D \to D^J$ is the left adjoint of a functor $\varprojlim : D^J \to D$ with object part $F \mapsto \varprojlim F$. In this latter case, the heuristic of Remark 3.1 needs to be reversed: The left adjoint Δ here is trivial (or at least repetitive), while it is the right adjoint that incorporates the construction of the limit object. □

Theorem 3.1.4 characterizes an adjunction between functors $F : B \to A$ and $G : A \to B$ in terms of initial objects for (X, G) or terminal objects for (F, Y). One may also characterize adjunctions without explicitly mentioning objects or morphisms of the categories A or B.

Proposition 3.1.7. *Let $F : B \to A$ and $G : A \to B$ be functors. Then F is a left adjoint of G iff there are natural transformations $\eta : 1_B \to FG$ and $\varepsilon : GF \to 1_A$ satisfying $(\eta F)(F\varepsilon) = 1$ and $(G\eta)(\varepsilon G) = 1$.*

Proof. The "only if" direction follows as in Corollary 3.7. For the "if" direction, suppose that such natural transformations exist. For an object X of B and an object Y of A, define $\varphi_Y^X : A(XF, Y) \to B(X, YG);\ g \mapsto \eta_X g^G$ and $\psi_X^Y : B(X, YG) \to A(XF, Y);\ h \mapsto h^F \varepsilon_Y$ [cf. (3.1.3), (3.2), (3.4)]. Verification of the naturality of $\varphi^X : A(XF, \_\_) \to B(X, \_\_G)$ and $\varphi_Y : A(\_\_F, Y) \to B(\_\_, YG)$ follows as in (3.5) and its dual (Exercise 3B). Now for g in $A(XF, Y)$, one has $g\varphi_Y^X \psi_X^Y = (\eta_X g^G)\psi_X^Y = (\eta_X g^G)^F \varepsilon_Y = \eta_X^F g^{GF} \varepsilon_Y = \eta_X^F \varepsilon_{XF} g$ $= (\eta F)_X (F\varepsilon)_X g = 1_{XF} g = g$. The middle equality holds by (1.4.2) for ε applied to $g : XF \to Y$. The next equality uses (1.4.4) and (1.4.5), while the penultimate equality follows from the hypothesis. Dually, one has $\psi_X^Y \varphi_Y^X = 1_{B(X, YG)}$ (cf. Exercise 3.1G). Thus φ_Y^X is an isomorphism, as required. □

In view of Proposition 3.1.7, one may specify the adjunction (3.1.1) by the quadruple $(F, G, \eta, \varepsilon)$ of left adjoint, right adjoint, unit, and counit.

EXERCISES

3.1A. In the "Currying" Example 3.1.2:

(a) verify that G_Y is a functor;

(b) verify that φ^Z is natural for fixed Z;

(c) verify that φ_X is natural for fixed X;

(d) determine the unit $\eta_Z : Z \to (Z \times Y)^Y$;

(e) determine the counit $\varepsilon_X : X^Y \times Y \to X$.

3.1B. In the proof of Theorem 3.1.4, verify that $F : B \to A$ is indeed a functor.

3.1C. Determine the counit of the adjunction in Example 3.1.5.

3.1D. Determine the unit of the adjunction in Example 3.1.6(a), and the counit of the adjunction in Example 3.1.6(b).

3.1E. By analogy with Example 3.1.5, show that the construction (I 1.4.3) yields a *free group functor* G : Set → Gp that is a left adjoint to the forgetful functor U : Gp → Set.

3.1F. Let F : Gp → Mon be the forgetful functor that forgets inversion. Show that F has a right adjoint G : Mon → Gp whose object part $M \mapsto M^*$ sends a monoid to its group of units (cf. Exercise I 1.3M). What is the counit of this adjunction?

3.1G. Give a detailed verification of $\psi_X^Y \varphi_Y^X = 1_{B(X,YG)}$ in the proof of Proposition 3.1.7.

3.1H. (a) In the adjunction (3.1.1), suppose that A has a terminal object T. Prove that TG is a terminal object of B.

(b) Formulate and do the dual of (a).

3.1I. Suppose that $G : A \to B$ has left adjoints $F : B \to A$ and $F' : B \to A$. Show that there is a natural isomorphism $\iota : F \to F'$. [Hint: Use Theorem 3.1.4 and Corollary 1.3.5(a).]

3.1J. Let CSgp be the category of commutative semigroups and homomorphisms. Let G : CSgp → Set be the forgetful functor.

(a) Show that the construction of free commutative semigroups over sets (Exercise I 1.5E) yields a left adjoint F : Set → CSgp to G : CSgp → Set.

(b) Show that CSgp has a terminal object T.

(c) Identity TGF as the semigroup of positive integers.

(d) Identify $TGFGF$ as the semigroup of integer partitions.

(e) Identify $TGFGFGF$ as the semigroup of Segre characteristics.

Now let ε be the counit of the adjunction.

(f) Identify the component $(GF\varepsilon)_T$ of the natural transformation $GF\varepsilon$ as the map determining the sum of an integer partition.

(g) Identify the component $(\varepsilon GF)_T$ of the natural transformation εGF as the map determining the length of an integer partition.

(h) Describe the map determining the sum of a Segre characteristic in terms of T, G, F, and ε.

3.1K. Let $F_1 : A_0 \to A_1$ be a left adjoint of $G_1 : A_1 \to A_0$. Let $F_2 : A_1 \to A_2$ be a left adjoint of $G_2 : A_2 \to A_1$. Use Proposition 3.1.7 to show that $F_1F_2 : A_0 \to A_2$ is a left adjoint of $G_2G_1 : A_2 \to A_0$.

3.2. Equivalence and Boolean Algebras

Let $F : \underline{B} \to \underline{A}$ be an isomorphism of categories, with two-sided inverse $G : \underline{A} \to \underline{B}$. Since $1_{\underline{B}} = FG$ and $GF = 1_{\underline{A}}$, there is an adjunction between F and G whose unit is the identity transformation $1 : 1_{\underline{B}} \to FG$ and whose counit is the identity transformation $1 : GF \to 1_{\underline{A}}$. In this adjunction, F is a left adjoint of G. But there is also an adjunction between G and F whose unit is the identity transformation $1 : 1_{\underline{A}} \to GF$ and whose counit is the identity transformation $1 : FG \to 1_{\underline{B}}$. In this adjunction, F is a right adjoint of G.

It is rare for the unit and counit of an adjunction to be identity transformations. On the other hand, it often happens that the unit and counit are natural isomorphisms. An adjunction $(F, G, \eta, \varepsilon)$ is an *equivalence* if the unit and counit are natural isomorphisms.

Example 3.2.1 (Finite Cardinals). Let $\underline{\underline{\text{Set}}}^{<\omega}$ be the full subcategory of $\underline{\underline{\text{Set}}}$ consisting of finite sets. Let $\underline{\underline{\text{Card}}}^{<\omega}$ be the small full subcategory of $\underline{\underline{\text{Set}}}$ or $\underline{\underline{\text{Set}}}^{<\omega}$ whose set of objects is the set $\{[0, n) | n \in \mathbb{N}\}$ of finite down-sets of the poset $(\mathbb{N}, \leq)$. Let $G : \underline{\underline{\text{Card}}}^{<\omega} \to \underline{\underline{\text{Set}}}^{<\omega}$ be the inclusion functor. For each finite set X, pick a bijection $\eta_X : X \to [0, |X|)$. Since η_X bijects, it is an initial object of the comma category (X, G). By Theorem 3.1.4, there is then an adjunction $(F, G, \eta, \varepsilon)$. Note that, for $n \in \mathbb{N}$, the counit has component $\varepsilon_{[0,n)} = 1_{[0,n)}$. Thus the adjunction $(F, G, \eta, \varepsilon)$ is an equivalence. □

If $(F, G, 1, 1)$ is an equivalence, it was noted that $(G, F, 1, 1)$ is also an equivalence. This symmetry holds for general equivalences:

Proposition 3.2.2. *Let $(F, G, \eta, \varepsilon)$ be an equivalence. Then $(G, F, \varepsilon^{-1}, \eta^{-1})$ is also an equivalence.*

Proof. By Proposition 3.1.7, one has $1_{XF} = (\eta F)_X (F\varepsilon)_X = \eta_X^F \varepsilon_{XF}$ and $1_{YG} = (G\eta)_Y (\varepsilon G)_Y = \eta_{YG} \varepsilon_Y^G$ for X an object of the domain of F and Y an object of the domain of G. Thus $1_{YG} = (\varepsilon_Y^G)^{-1} \eta_{YG}^{-1} = (\varepsilon_Y^{-1})^G \eta_{YG}^{-1} = (\varepsilon^{-1} G)_Y (G\eta^{-1})_Y$ and $1_{XF} = \varepsilon_{XF}^{-1} (\eta_X^F)^{-1} = \varepsilon_{XF}^{-1} (\eta_X^F)^{-1} = (F\varepsilon^{-1})_X (\eta^{-1} F)_X$, so that $(\varepsilon^{-1} G)(G\eta^{-1}) = 1$ and $(F\varepsilon^{-1})(\eta^{-1} F) = 1$. The other direction of Proposition 3.1.7 then shows that $(G, F, \varepsilon^{-1}, \eta^{-1})$ is an adjunction, and thus an equivalence. □

Two categories $\underline{A}$ and $\underline{B}$ are said to be *equivalent* if there are functors $F : \underline{B} \to \underline{A}$ and $G : \underline{A} \to \underline{B}$ such that there is an equivalence $(F, G, \eta, \varepsilon)$. Equivalence is an equivalence relation on any set of categories (Exercise 3.2A). A functor $F : \underline{B} \to \underline{A}$ is said to be an *equivalence* iff it appears as the left adjoint in an equivalence $(F, G, \eta, \varepsilon)$. By Proposition 3.2.2, a functor $G : \underline{A} \to \underline{B}$ is an equivalence iff it appears as the right adjoint in an equivalence $(F, G, \eta, \varepsilon)$. Now a functor $F : \underline{B} \to \underline{A}$ is said to be *dense* if each object of $\underline{A}$ is isomorphic in $\underline{A}$ to the image of an object of $\underline{B}$ under F. Note that the functor G of Example 3.2.1 is dense: Each finite set X is isomorphic (via η_X) to the image of the object $[0, |X|)$ of $\underline{\underline{\mathrm{Card}}}^{<\omega}$ under the inclusion functor G. By the definition of $\underline{\underline{\mathrm{Card}}}^{<\omega}$, the functor G is full. Moreover, the functor G is faithful, since it is the inclusion of a subcategory. These three properties of the equivalence G are entirely typical:

Theorem 3.2.3. *Let $F : B \to A$ be a functor. Then the following three conditions are equivalent:*

(a) *F is an equivalence;*

(b) *There is a functor $G : A \to B$ with natural isomorphisms $\eta : 1_B \to FG$ and $\varepsilon : GF \to 1_A$;*

(c) *F is full, faithful, and dense.*

Proof. (a) $\Rightarrow$ (b) is trivial.

(b) $\Rightarrow$ (c): Suppose that (b) holds. Note that (b) is symmetrical in F and G. The properties for F listed under (c) will be verified in reverse order.

(i) F dense: For an object Y of A, there is an A-isomorphism $\varepsilon_Y : YG^F \to Y$ of Y with the image of the object YG of B under F.

(ii) F faithful: Consider B-objects X' and X. The naturality (1.4.2) of η at an element f of $B(X', X)$ yields $f = \eta_{X'} f^{FG} \eta_X^{-1}$. Thus $r : A(X'F, XF) \to B(X', X);\ k \mapsto \eta_{X'} k^G \eta_X^{-1}$ is a retraction for $F : B(X', X) \to A(X'F, XF)$, whence F is faithful. Similarly, G is faithful.

(iii) F full: Consider B-objects X' and X. Given $k : X'F \to XF$, the naturality (1.4.2) of η at $k^r : X' \to X$ yields $k^{rFG} = \eta_{X'}^{-1} k^r \eta_X = k^G$. Since G is faithful, one then has $k^{rF} = k$, so that $F : B(X', X) \to A(X'F, XF)$ surjects.

(c) $\Rightarrow$ (a): Suppose that (c) holds. It will be shown that F appears in an equivalence $(G, F, \varepsilon^{-1}, \eta^{-1})$. Given an object Y of A, use the density of F to choose an object YG of B with an A-isomorphism $\varepsilon_Y^{-1} : Y \to YGF$. This A-isomorphism becomes an initial object of the comma category (Y, F). Indeed, given an object $h : Y \to XF$ of this comma category, the fullness and faithfulness of F yield a unique B-morphism $g : YG \to X$ such that $g^F = \varepsilon_Y h : YGF \to XF$. Theorem 3.1.4 now shows that there is an adjunction $(G, F, \varepsilon^{-1}, \theta)$. By Proposition 3.1.7, one has $1_{XF} = (F\varepsilon)_X^{-1}(\theta F)_X = \varepsilon_{XF}^{-1}\theta_X^F$ at each object X of B, i.e. $\theta_X^F = \varepsilon_{XF} : XFGF \to XF$. Since F is full, there is a B-morphism $f : X \to XFG$ with $f^F = \varepsilon_{XF}^{-1} : XF \to XFGF$. Since F is faithful, the equations $(\theta_X f)^F = \theta_X^F f^F = \varepsilon_{XF}\varepsilon_{XF}^{-1} = 1_{XFGF}$ and $(f\theta_X)^F = f^F\theta_X^F = \varepsilon_{XF}^{-1}\varepsilon_{XF} = 1_{XF}$ imply $\theta_X f = 1_{XFG}$ and $f\theta_X = 1_X$. Thus θ is a natural isomorphism, and the adjunction $(G, F, \varepsilon^{-1}, \theta)$ becomes an equivalence. □

The conditions of Theorem 3.2.3 are very useful for determining the equivalence of categories. In the following example, condition (c) of Theorem 3.2.3 is used.

Example 3.2.4 (Elementary Linear Algebra). Let K be a field construed as the small category of matrices over K (Example 1.4). Consider the category $\underline{\underline{\mathrm{Mod}}}_K^{<\omega}$ of finite-dimensional vector spaces over K (cf. Exercise 1.4B). Define a functor $F : K \to \underline{\underline{\mathrm{Mod}}}_K^{<\omega}$ with object part $m \mapsto \bigoplus_{i=1}^m K$ and with morphism part given by II (1.3.5). By definition of $\underline{\underline{\mathrm{Mod}}}_K^{<\omega}$, the functor F is dense. By the isomorphism $K(\bigoplus_{i=1}^l K, \bigoplus_{j=1}^m K) = K(\coprod_{i=1}^l K, \prod_{j=1}^m K) \cong K_l^m$, the functor F is full and faithful. Thus F satisfies condition (c) of Theorem 3.2.3. It follows that the category of matrices over the field K is equivalent to the category of finite-dimensional vector spaces over K. □

A second, major application of Theorem 3.2.3, this time using condition (b), is to Boolean rings. A unital ring $(R, +, \cdot, 1)$ is *Boolean* if $(R, \cdot)$ is a semilattice (cf. Example II 1.3.4).

Proposition 3.2.5. (a) *A unital ring R is Boolean iff* $\forall x \in R, x^2 = x$.

(b) *A Boolean ring has characteristic* 2.

(c) *The only Boolean integral domain is the Galois field* $\mathbb{Z}_2$.

Proof. The "only if" part of (a) is trivial. Conversely, suppose that the multiplication in a unital ring R is idempotent. Then $\forall x, y \in R, x + y = (x + y)^2 = x^2 + xy + yx + y^2 = x + y + xy + yx$, so $\forall x, y \in R, xy + yx = 0$. Taking $x = y$, one obtains $2x = x + x = x^2 + x^2 = 0$. In particular, (b) holds. In the ring R, one has $xy = -yx = yx$, so the multiplication in R is also commutative, and R becomes Boolean. Towards (c), let K be a Boolean integral domain. Each element of K is a root of the quadratic polynomial $T^2 - T \in K[T]$. Thus $K = \{0, 1\} \cong \mathrm{GF}(2)$. □

In a Boolean ring $(R, +, \cdot, 1)$, define $x \cup y := x + y + xy$ [cf. Exercise II 1.3A(a)] and $x \cap y := xy$. Then $(R, \cup, \cap)$ is a lattice: The absorption laws (2.7) and (2.8) are readily verified. In fact, $(R, \cup, \cap, 0, 1)$ is a bounded lattice (Exercise 3.2G). Now the distributive laws in the ring $(R, +, \cdot, 1)$ yield the *meet distributive law*

$$x \cap (y \cup z) = (x \cap y) \cup (x \cap z) \tag{3.2.1}$$

and the *join distributive law*

$$x \cup (y \cap z) = (x \cup y) \cap (x \cup z) \tag{3.2.2}$$

in the lattice $(R, \cup, \cap)$ (Exercise 3.2H). In general, a lattice $(L, \cup, \cap)$ satisfying (3.2.1) and (3.2.2) is said to be *distributive*. Each of (3.2.1) and (3.2.2) alone is actually equivalent to the distributivity of a lattice $(L, \cup, \cap)$ (Exercise 3.2I).

Let $(L, \cup, \cap, 0, 1)$ be a bounded lattice. Let x be an element of L. Then an element x' of L is a *complement* to x in L if

$$\begin{cases} (a) & x \cap x' = 0 \\ (b) & x \cup x' = 1 \end{cases} \tag{3.2.3}$$

Note that an element of a bounded lattice may have several complements. For example, the subgroup $\langle(12)(34)\rangle$ of the Vierergruppe V (Exercise I 3.2E) has complements $\langle(13)(24)\rangle$ and $\langle(14)(23)\rangle$ in the bounded lattice $\mathrm{Sb}\,V$ of subgroups of V (cf. Exercise 2G). However, an element x of a bounded distributive lattice has at most one complement (Exercise 3.2J). If x is an element of a Boolean ring $(R, +, \cdot, 1)$, then $x' = 1 - x$ is a complement to x in $(R, \cup, \cap, 0, 1)$ (Exercise 3.2K). A bounded distributive lattice in which each element has a complement is called a *Boolean algebra*. Note that the power sets of sets are Boolean algebras.

Proposition 3.2.6. *Let $f: R \to S$ be a lattice homomorphism between Boolean algebras. Then f is a bounded lattice homomorphism iff $\forall x \in R$, $x'f = xf'$.*

Proof. If f is a bounded lattice homomorphism, and x is an element of R, then $xf \cup x'f = (x \cup x')f = 1f = 1$ and dually $xf \cap x'f = 0$. Thus $x'f$ equals the unique complement xf' to xf in S. Conversely, $x'f = xf'$ implies $0f = (x \cap x')f = xf \cap x'f = xf \cap xf' = 0$, and dually $1f = 1$, so that f is a bounded lattice homomorphism. □

A lattice homomorphism between Boolean algebras satisfying the equivalent conditions of Proposition 3.2.6 is called a *Boolean algebra homomorphism*. Let $\underline{\underline{\mathrm{BAlg}}}$ be the category of Boolean algebras and homomorphisms. Let $\underline{\underline{\mathrm{BRing}}}$ be the category of Boolean rings and homomorphisms. Given a Boolean ring $(R, +, 0, \cdot, 1)$, let $(R, +, 0, \cdot\, 1)F$ be the Boolean algebra $(R, \cup, \cap, ')$. Given a Boolean algebra $(R, \cup, \cap, ')$, let $(R, \cup, \cap, ')G = (R, +, 0, \cdot, 1)$ with $x + y = (x \cap y') \cup (y \cap x')$ (cf. Example II 1.3.4), with $xy = x \cap y$, with $0 = x \cap x'$, and with $1 = x \cup x'$ (the latter two equations for any x in R). Then RG is a Boolean ring [Exercise 3.2L(a)]. Given a Boolean ring homomorphism $f: R \to S$, define $f^F: RF \to SF$ to be $f: RF \to SF$. Observe that f^F is a Boolean algebra homomorphism [Exercise 3.2L(b)]. Given a Boolean algebra homomorphism $g: R \to S$, define $g^G: RG \to RS$ to be $g: RG \to SG$. Observe that g^G is a Boolean ring homomorphism [Exercise 3.2L(c)]. One readily obtains:

Proposition 3.2.7. *The assignments $F: \underline{\underline{\mathrm{BRing}}} \to \underline{\underline{\mathrm{BAlg}}}$ and $G: \underline{\underline{\mathrm{BAlg}}} \to \underline{\underline{\mathrm{BRing}}}$ yield mutually inverse isomorphisms between the categories of Boolean rings and of Boolean algebras.* □

Now let $\underline{\underline{\mathrm{BRing}}}^{<\omega}$ be the full subcategory of $\underline{\underline{\mathrm{BRing}}}$ consisting of finite Boolean rings. Let $\underline{\underline{\mathrm{Set}}}^{<\omega}$ be as in Example 3.2.1. Define a functor Spec: $\underline{\underline{\mathrm{BRing}}}^{<\omega} \to (\underline{\underline{\mathrm{Set}}}^{<\omega})^{\mathrm{op}}$ with object part $R \mapsto$ Spec R and with morphism part $(f: R \to S) \mapsto (f^{-1}:$ Spec $S \to$ Spec $R; P \mapsto f^{-1}(P))$ (cf. Exercise II 1.4S). Given a set X, let $\mathbb{Z}_2^X = \underline{\underline{\mathrm{Set}}}(X, \mathbb{Z}_2)$ be the Boolean ring of functions to $\mathbb{Z}_2 = \{0, 1\}$, with componentwise ring operations. Given a function $f: X' \to X$, define $\underline{\underline{\mathrm{Set}}}(f, \mathbb{Z}_2): \underline{\underline{\mathrm{Set}}}(X, \mathbb{Z}_2) \to \underline{\underline{\mathrm{Set}}}(X', \mathbb{Z}_2)$; $h \mapsto fh$. Note that $\underline{\underline{\mathrm{Set}}}(f, \mathbb{Z}_2)$ is a unital ring homomorphism. One obtains a functor $\underline{\underline{\mathrm{Set}}}(\_\_\_, \mathbb{Z}_2) : (\underline{\underline{\mathrm{Set}}}^{<\omega})^{\mathrm{op}} \to \underline{\underline{\mathrm{BRing}}}$. There is then a natural isomorphism η whose component at a finite set X is the function $\eta_X : X \to \operatorname{Spec} \mathbb{Z}_2^X;\ x \mapsto \{f : X \to \mathbb{Z}_2 | xf = 0\}$. Note that η_X clearly injects. To show that it surjects, consider a prime ideal P of the finite Boolean ring $\mathbb{Z}_2^X$. Let b be the upper bound ΣP of the sublattice PF of $\mathbb{Z}_2^X F$. Note $b \in P \subset \mathbb{Z}_2^X$ implies $\exists x \in X.\ xb = 0$, for otherwise $b = 1$. Then $P \subseteq \{f : X \to \mathbb{Z}_2 | xf = 0\}$. By Proposition 3.2.5(c), the prime ideal P is maximal. Thus $P = \{f : X \to \mathbb{Z}_2 | xf = 0\}$, so that η_X also surjects.

In the other direction, there is a natural isomorphism ε whose component at a finite Boolean ring S is $\varepsilon_S : \mathbb{Z}_2^{\operatorname{Spec} S} \to S;\ \chi_Y \mapsto \Sigma \bigcap(\operatorname{Spec} S - Y)$. Here elements of $\mathbb{Z}_2^{\operatorname{Spec} S}$ are construed as characteristic functions χ_Y of subsets Y of Spec S. The intersection $\bigcap(\operatorname{Spec} S - Y)$ of the complement of such a subset is then an ideal of S. Now each ideal I of S can be expressed as the intersection

$$I = \bigcap\{P \in \operatorname{Spec} S | I \leq P\}. \tag{3.2.4}$$

Certainly I is contained in the right-hand side of (3.2.4). Conversely, let x be an element of the right-hand side. Suppose $x \notin I$. Now the ideal $(x' + I)(S/I)$ of S/I is contained in a maximal ideal of S/I. This maximal ideal is of the form Q/I for $I \leq Q \in \operatorname{Spec} S$ (cf. Exercise II 1.4J). Now $x \notin Q$, for otherwise $1 = x \cup x' = x + x' + xx' \in Q$, a contradiction. But $x \notin Q \geq I$ in turn contradicts the assumption that x was an element of the right-hand side of (3.2.4). Given (3.2.4), one obtains $x \mapsto \chi_{\{P \in \operatorname{Spec} S | x S \nleq P\}}$ as a two-sided inverse to ε_S. Thus ε_S bijects. Since ε_S is order-preserving, it follows that ε_S is a lattice isomorphism (cf. Exercise 2L). Finally, since ε_S is a bounded lattice isomorphism, Proposition 3.2.6 shows that it is a morphism in $\underline{\underline{\mathrm{BAlg}}}$ or $\underline{\underline{\mathrm{BRing}}}$.

Theorem 3.2.8. *The categories of finite Boolean rings and of finite Boolean algebras are each equivalent to the opposite of the category of finite sets.*

Proof. The functors Spec: $\underline{\underline{\mathrm{BRing}}}^{<\omega} \to (\underline{\underline{\mathrm{Set}}}^{<\omega})^{\mathrm{op}}$ and $\underline{\underline{\mathrm{Set}}}(\_\_, \mathbb{Z}_2) : (\underline{\underline{\mathrm{Set}}}^{<\omega})^{\mathrm{op}} \to \underline{\underline{\mathrm{BRing}}}^{<\omega}$ satisfy condition (b) of Theorem 3.2.3, so that $\underline{\underline{\mathrm{BRing}}}^{<\omega}$ and $(\underline{\underline{\mathrm{Set}}}^{<\omega})^{\mathrm{op}}$ are equivalent. Proposition 3.2.7 then shows that $(\underline{\underline{\mathrm{Set}}}^{<\omega})^{\mathrm{op}}$ is equivalent to the category $\underline{\underline{\mathrm{BAlg}}}^{<\omega}$ of finite Boolean algebras. □

EXERCISES

3.2A. Show that equivalence is an equivalence relation on any set of categories. (Hint: Use Exercise 3.1K.)

3.2B. Let $G : \underline{\underline{\mathrm{Set}}} \to \underline{\underline{\mathrm{Gp}}}$ be the free group functor of Exercise 3.1E. Determine precisely which of the three properties of Theorem 3.2.3(c) are possessed by G.

3.2C. Let $U : \underline{\underline{\mathrm{Gp}}} \to \underline{\underline{\mathrm{Set}}}$ be the forgetful functor. Determine precisely which of the three properties of Theorem 3.2.3(c) are possessed by U.

3.2D. Give an example of a full, dense functor that is not faithful.

3.2E. A functor $G : A \to B$ is a *reflection* if it is part of an adjunction $(F, G, \eta, \varepsilon)$—also known as a reflection—in which the counit ε is a natural isomorphism. Prove that G is a reflection iff it is full and faithful.

3.2F. In Example 3.2.4, replace the field K by an arbitrary commutative, unital ring S. Which of the properties (c) of Theorem 3.2.3 does the corresponding functor F still possess?

3.2G. Let $(R, +, \cdot, 1)$ be a Boolean ring. Verify that $(R, \cup, \cap, 0, 1)$ is a bounded lattice.

3.2H. Let $(R, +, \cdot, 1)$ be a Boolean ring. Verify that $(R, \cup, \cap)$ is a distributive lattice.

3.2I. Let $(L, \cup, \cap)$ be a lattice. Show that the following three conditions are equivalent:

(a) L satisfies (3.2.1);

(b) L satisfies (3.2.2);

(c) L is distributive.

3.2J. Prove that an element x of a bounded distributive lattice has at most one complement.

3.2K. Let x be an element of a Boolean ring R. Show that $x' = 1 - x$ is a complement to x in the lattice $(R, \cup, \cap, 0, 1)$.

3.2L. (a) Given a Boolean algebra $(R, \cup, \cap, ')$, verify that RG is a Boolean ring.

(b) Given a Boolean ring homomorphism f, verify that f^F is a Boolean algebra homomorphism.

(c) Given a Boolean algebra homomorphism g, verify that g^G is a Boolean ring homomorphism.

(d) Complete the proof of Proposition 3.2.7.

3.2M. Given a Boolean ring $(S, +, \cdot, 1)$, define $(S, +, \cdot, 1)F$ to be the corresponding ring $(S, \oplus, \circ, 0)$ of Exercise II 1.3A. Given a Boolean ring homomorphism $f : R \to S$, define $f^F : RF \to SF$ to be $f : RF \to SF$. Show that F forms a functor $F : \underline{\underline{\mathrm{BRing}}} \to \underline{\underline{\mathrm{BRing}}}$ that is an equivalence.

3.2N. Let R be a finite, non-trivial Boolean ring.

(a) Show that $\log_2 |R|$ is a positive integer n.

(b) Show that the automorphism group of R is the symmetric group S_n.

3.2O. Suppose that a unital ring R has the property $\forall x \in R, x^3 = x$. Prove that R is commutative.

3.2P. **(Atomic Boolean algebras.)** An *atom* of a bounded lattice $(L, +, \cdot, 0, 1)$ is an element covering 0. The bounded lattice L is *atomic* if, for each non-zero element x of L, the interval $[0, x]_{\leq}$ contains an atom.

(a) For any set B, show that the power set $(2^B, \cup, \cap, ')$ is an atomic Boolean algebra.

(b) Let R be the intersection of all the Boolean subalgebras of the power set $(2^{\mathbb{R}}, \cup, \cap, ')$ that contain $\{(-\infty, r), [r, \infty) \mid r \in \mathbb{R}\}$. Show that R has no atoms.

(c) Exhibit a Boolean algebra that possesses an atom, but is not atomic.

(d) Let x be an element of a Boolean ring S. Prove: $(x'S \in \text{Spec } S) \Leftarrow x$ is an atom of S.

(e) Show that the converse of (d) is true if S is finite.

(f) Show that the converse of (d) is false if S is infinite.

(g) For any Boolean algebra R, define $\text{At } R := \{a \in R \mid 2 = |[0, a]|\}$, the set of atoms of R. Show that

$$\varepsilon_R : R \to 2^{\text{At} R}; x \mapsto \{a \in \text{At} R \mid a \leq x\}$$

is a Boolean algebra homomorphism.

(h) Show that ε_R injects if R is atomic.

(i) Show that ε_R surjects if the lattice $(R, \cup, \cap)$ is complete.

3.2R. For a set X, define $\mathscr{P}_{<\omega}^{<\omega}(X) = \mathscr{P}_{<\omega}(X) \cup \mathscr{P}^{<\omega}(X)$, the set of finite or cofinite subsets of X (cf. Exercise 2F). Show that the Boolean algebra $(\mathscr{P}_{<\omega}^{<\omega}(\mathbb{Z}), \cup, \cap, ')$ is not isomorphic to the power set $(2^B, \cup, \cap, ')$ of any set. [Hint: B finite $\Rightarrow 2^B$ finite, while B infinite $\Rightarrow 2^B$ is uncountable, by Proposition 1.1.1. Show $\mathscr{P}_{<\omega}^{<\omega}(\mathbb{Z})$ is infinite and countable.]

3.3. Galois Connections and Galois Theory

Let $(A, \leq)$ and $(B, \leq)$ be poset categories. An adjunction $(S, R, \eta, \varepsilon)$ between functors $S : B \to A$ and $R : A \to B$ is known as a *Galois connection.* The relationship (3.1.1) reduces to:

$$\forall x \in B, \forall y \in A, x^S \leq y \Leftrightarrow x \leq y^R. \tag{3.3.1}$$

The existence of the unit η and counit ε reduces to:

$$\begin{cases} (a) & \forall x \in B, x \leq x^{SR}; \\ (b) & \forall y \in A, y^{RS} \leq y. \end{cases} \tag{3.3.2}$$

On the other hand, given functors $S : B \to A$ and $R : A \to B$ such that (3.3.2) holds, applying S to the B-morphism (3.3.2)(a) and R to the A-morphism

(3.3.2)(b) yield $x^S \leq x^{SRS}$ and $y^{RSR} \leq y^R$. Meanwhile, setting $y = x^S$ in (3.3.2)(b) yields $x^{SRS} \leq x^S$; setting $x = y^R$ in (3.3.2)(a) yields $y^R \leq y^{RSR}$. Thus:

$$(3.3.3) \qquad \begin{cases} (a) & \forall x \in B, x^S = x^{SRS}; \\ (b) & \forall y \in A, y^R = y^{RSR}. \end{cases}$$

In particular, Proposition 3.1.7 then shows that S is left adjoint to R. Summarizing:

Proposition 3.3.1. *Let $(A, \leq)$ and $(B, \leq)$ be poset categories. Then a functor $S: B \to A$ is left adjoint to a functor $R: A \to B$ iff (3.3.2) holds.* □

Example 3.3.2. (a) (**Annihilators.**) Let M be a module over a commutative ring K. Let $B = (2^M, \subseteq)$ and let $A = (\mathrm{Id}(K), \supseteq)$. Define $S: B \to A$; $X \mapsto \{k \in K | \forall m \in X, mk = 0\}$ and $R: A \to B; I \mapsto \{m \in M | \forall k \in I, mk = 0\}$. Since $\forall X \subseteq M, X \subseteq X^{SR}$ and $\forall I \lhd K, I \leq I^{RS}$, Proposition 3.3.1 shows that the functors S and R yield a Galois connection.

(b) (**Images and inverse images.**) Let $f: D \to C$ be a function. Define $S: (2^D, \subseteq) \to (2^C, \subseteq); X \mapsto Xf$ and $R: (2^C, \subseteq) \to (2^D, \subseteq); Y \mapsto f^{-1}Y$. Since $Xf \subseteq Y \Leftrightarrow X \subseteq f^{-1}Y$, the functors R and S yield a Galois connection.

(c) (**Polarities.**) A *relation* between sets I and J is a subset α of $I \times J$ (cf. Exercise 1.1F). As in the special case of binary relations on a single set, it is convenient to use the notation $x \alpha y$ for the membership $(x, y) \in \alpha$. Given a relation α between I and J, define $S: (2^I, \subseteq) \to (2^J, \supseteq); X \mapsto \{y \in J | \forall x \in X, x \alpha y\}$ and $R: (2^J, \supseteq) \to (2^I, \subseteq); Y \mapsto \{x \in I | \forall y \in Y, x \alpha y\}$. Now for $X \subseteq I$ and $Y \subseteq J$, one has $X^S \supseteq Y \Leftrightarrow \forall x \in X, \forall y \in Y, x \alpha y \Leftrightarrow X \subseteq Y^R$. Thus R and S yield a Galois connection, known as the *polarity* determined by the relation α. □

In a Galois connection (3.3.1), elements of the subsets BS of A and AR of B are described as *closed*. One obtains induced posets $(BS, \leq)$ and $(AR, \leq)$. By (3.3.3), the restricted functors $S: AR \to BS$ and $R: BS \to AR$ then yield category isomorphisms. This isomorphism between the posets of closed elements is called the *Galois correspondence* for the Galois connection (3.3.1). For an element x of B, the closed element x^{SR} of B is called the *closure* of x. By Theorem 3.1.4 [cf. Proposition 3.5(a)], the closure of x is the product of the set of closed elements above x in B, i.e.

$$(3.3.4) \qquad x^{SR} = \prod \{y^R | x \leq y^R\}.$$

Dually, one has

$$(3.3.5) \qquad y^{RS} = \sum \{x^S | x^S \leq y\}$$

for y in A (Exercise 3.3C). The closed element y^{RS} of A is called the *kernel* of the element y of A. Given a Galois connection, the determination of the closed elements on each side is often a primary concern.

Example 3.3.3. (a) Let G be a group. Consider the polarity determined by the binary relation $\{(x, y) \in G^2 | xy = yx\}$. Then the closed elements are subgroups of G.

(b) Let X be a finite set, and let $\mathbb{Z}_2^X$ be the Boolean ring of functions $f : X \to \mathbb{Z}_2$. Define a relation α on $X \times \mathbb{Z}_2^X$ by $x \alpha f \Leftrightarrow xf = 0$. Consider the polarity determined by α. Then the closed subsets of $\mathbb{Z}_2^X$ are precisely the ideals of $\mathbb{Z}_2^X$.

(c) If B is a complete lattice in the Galois connection (3.3.1), the Tarski Fixed-Point Theorem 2.1.8 shows that the set of fixed points of the composite functor $SR : B \to B$ forms a complete lattice. But this set of fixed points is precisely the poset of closed elements of B. Similarly, if A is a complete lattice in a Galois connection (3.3.1), then its closed elements form a complete lattice. If both A and B are complete lattices, Exercise 2L then shows that the Galois correspondence yields lattice isomorphisms. □

Let G be a group, and let (X, G) be a G-set. The subset $\{(x, T) \in X \times G | xT = x\}$ of $X \times G$ used in the proof of Burnside's Lemma (Theorem I 3.1.2) is a relation between X and G that yields a polarity. The right adjoint is the *fixed-point functor* $F : (2^G, \supseteq) \to (2^X, \subseteq); H \mapsto \mathrm{Fix}(X, \langle H \rangle)$. The left adjoint is the *stabilizer functor* $S : (2^X, \subseteq) \to (2^G, \supseteq)$; $Y \mapsto \{T \in G | \forall y \in Y, yT = y\}$. Note that, for a singleton subset $\{x\}$ of X, one has $\{x\}^S = G_x$. There is an interesting correspondence between G-subsets of X and normal subgroups of G.

Proposition 3.3.4. (a) *If Y is a G-subset of X, then $Y^S \lhd G$.*

(b) *If $N \lhd G$, then N^F is a G-subset of X.*

Proof. (a) $\forall U \in G, \quad \forall T \in Y^S, \quad \forall y \in Y, \quad yU^{-1}TU = yU^{-1}U = y \Rightarrow U^{-1}TU \in Y^S$.

(b) $\forall U \in G, \quad \forall y \in N^F, \quad \forall T \in N, yUT = yUTU^{-1}U = yU \Rightarrow yU \in N^F$. □

Galois connections and Galois correspondences arose initially in "Galois Theory." a particular case of the adjointness between fixed-point and stabilizer functors. Suppose that a field A contains a subfield K, making A a K-algebra A_K via the inclusion $K \hookrightarrow A$. Let G be the *Galois group* of A_K, the group $\underline{\mathrm{Alg}}_K(A, A)^*$ of K-algebra automorphisms of A_K. Consider the G-set (A, G). One wishes to find conditions under which each subgroup of G is closed and each field B with $K \leq B \leq A$ is closed. An obvious first requirement is that K itself be a closed subfield of A_K. To see that this need not necessarily happen, note that any $\mathbb{Q}$-algebra automorphism U of $\mathbb{R}_{\mathbb{Q}}$ has the set $[0, \infty)$ of squares as a $\langle U \rangle$-subset. Then $U:(\mathbb{R}, \leq) \to (\mathbb{R}, \leq)$ is a functor, since $x \leq y \Rightarrow y - x \in [0, \infty) \Rightarrow yU - xU = (y - x)U \in [0, \infty) \Rightarrow xU \leq yU$. But $\forall x \in \mathbb{R}, \forall r \in (-\infty, x) \cap \mathbb{Q}, \forall s \in (x, \infty) \cap \mathbb{Q}, r \leq x \leq s \Rightarrow r \leq xU \leq s$. Thus $x = xU$, and U is the identity. This shows that $\mathbb{R}$ is the only closed subfield of $\mathbb{R}_{\mathbb{Q}}$, and in particular $\mathbb{Q}$ is not a closed subfield. For K to be a closed subfield of A_K with Galois group G, it is necessary and sufficient that $K = G^F$. Indeed, $K = Y^F$ for $G \supseteq Y$ implies $K \subseteq G^F \subseteq Y^F = K$. Note that the faithfulness of (A, G) implies $A^S = \{1\}$, while $K^S = G$ by definition, and of course $\{1\}^F = A$. Identification of further closed elements is facilitated by a dual pair of inequalities.

Proposition 3.3.5. *If $K \leq B \leq C \leq A$ for subfields B and C of A with* $\dim C_B$ *finite, then* $|C^S \setminus B^S| \leq \dim C_B$.

Proof. By induction on $\dim C_B$. If $B = C$, the result is trivial. If there is a field D with $B \leq D \leq C$, then $|C^S \setminus B^S| = |C^S \setminus D^S| \cdot |D^S \setminus B^S| \leq \dim C_D \cdot \dim D_B = \dim C_B$, by induction and Proposition II 3.5.2. One may thus assume that $C = B(t)$ for an element t of C. Let $p(T)$ be the minimal polynomial of t in C_B. Since $C^S \setminus B^S \to \{x \in C | p(x) = 0\}; C^S U \mapsto t^U$ is a well-defined injection, $|C^S \setminus B^S| \leq \deg p(T) = \dim C_B$. $\square$

Proposition 3.3.6. *If* $\underline{\underline{\mathrm{Alg}}}_K\,(A, A)^* = G \geq H \geq J$ *for subgroups H and J of G with $|J \setminus H|$ finite, then* $\dim J^F_{HF} \leq |J \setminus H|$.

Proof. Let $\{1 = U_1, U_2, \ldots, U_n\}$ be a normalized right transversal to J in H. Suppose that there were an independent subset $\{a_1, \ldots, a_{n+1}\}$ of the HF-space J^F. Consider the element $[a_i U_j]$ of A^n_{n+1} as the coefficient matrix of a homogeneous system of n equations in $n + 1$ unknowns. By Gaussian elimination, $[a_i U_j]$ is reduced to a column-reduced matrix of column rank $r \leq n$. Thus the homogeneous system has a solution $(x_1, \ldots, x_r, 1, 0, \ldots, 0)$ in A^{n+1}_1. Now if $(\forall 1 \leq i \leq r, x_i \in HF)$ were true, the first equation of the homogeneous system would yield the contradiction $x_1 a_1 + \cdots + x_r a_r + a_{r+1} = 0$ to the supposed independence of the subset $\{a_1, \ldots, a_{n+1}\}$ of the HF-space J^F. Thus $\exists 1 \leq i \leq r.\ x_i \notin HF$, whence

$$\exists V \in H. \quad \exists 1 \leq i \leq r. \quad x_i - x_i^V \neq 0. \tag{3.3.6}$$

Applying the field automorphism V to the original homogeneous system, one obtains a new homogeneous system with solution $(x_1^V, \ldots, x_r^V, 1, 0, \ldots, 0)$ in A^{n+1}_1. Consider the element T of S_n with $U_j V^{-1} \in JU_{jT}$ or $U_{jT} V \in JU_j$ for $1 \leq j \leq n$. Then the j-th equation of the old system becomes the jT-th equation of the new system. In other words, the coefficient matrix of the new system is obtained from the old coefficient matrix, postmultiplying by the permutation matrix ρ_T. Thus $(x_1^V, \ldots, x_r^V, 1, 0, \ldots, 0)$ is also a solution to the original system, whence the difference $(x_1 - x_1^V, \ldots, x_r - x_r^V, 0, 0, \ldots, 0)$ of the two solutions is again a solution. But by (3.3.6), the i-th equation of the column-reduced system yields the contradiction $x_i - x_i^V = 0$. $\square$

Propositions 3.3.5 and 3.3.6 combine to yield the

Fundamental Theorem of Galois Theory 3.3.7. *Let K be a field. Let A be a field containing K, with* $\dim A_K$ *finite. Let G be the Galois group of A_K. Suppose that K is a closed subfield of A_K. Then $|G| = \dim A_K$. Moreover, each subgroup of G is closed, and each field C with $K \leq C \leq A$ is closed. In particular, the Galois correspondence yields a bijection between the set of subgroups H of G and the set of fields C between K and A, with* $\dim C_K = |C^S \setminus G|$ *and* $|H| = \dim A_{HF}$.

Proof. For each intermediate field C, one has $\dim C_K^{SF} = \dim C_{KSF}^{SF} \leq |C^S \setminus K^S| \leq \dim C_K \leq \dim C_K^{SF}$, and also $C \leq C^{SF}$ by (3.3.2)(a). Thus $C = C^{SF}$ is closed. Also $|G| = |A^S \setminus K^S| = \dim A_K$. Then for each subgroup H of G, one has $|H^{FS}| = |1^{FS} \setminus H^{FS}| \leq \dim 1_{HF}^F \leq |1 \setminus H| = |H| \leq |H^{FS}|$, and also $H^{FS} \supseteq H$ by (3.3.2)(b). Thus $H = H^{FS}$ is closed. □

EXERCISES

3.3A. In the annihilator Example 3.3.2(a):

(a) Verify the functoriality of R and S;

(b) Show that the closed elements of B are submodules of M.

3.3B. In the image Example 3.3.2(b):

(a) Show that each element of 2^C is closed;

(b) Exhibit a function f such that 2^D contains an element that is not closed.

3.3C. Verify (3.3.4) and (3.3.5).

3.3D. (a) Verify the claim of Example 3.3.3(a).

(b) Setting $G = S_4$ in Example 3.3.3(a), determine which subgroups are closed.

3.3E. Verify the claim of Example 3.3.3(b).

3.3F. Show that the only Galois connections between antichains A and B are pairs of mutually inverse isomorphisms $S : B \to A$ and $R : A \to B$.

3.3G. **(Concepts.)** Let I be a set of *items*, classified according to whether or not they possess various *properties* from a set J. Let α be the subset of $I \times J$ consisting of those pairs (i, j) such that item i has property j. Consider the polarity determined by α. A *concept* is an element (X, Y) of $2^I \times 2^J$ with $X^S = Y$ and $X = Y^R$, i.e. a pair of elements that correspond in the Galois correspondence. The set X of items is called the *extent* of the concept (X, Y), while the set Y of properties is called its *intent*. By Example 3.3.3(c), the concepts form a complete lattice.

Consider the following 1987 classification of major west European countries. ("Major" means having at least 2^{18} inhabitants, while "west

European" means anywhere in Europe between the longitudes $\pm 10°$.) Countries are identified as "small" (having less than 2^{24} inhabitants) or otherwise "large," as "coastal" or otherwise "inland," and as "monarchies" or otherwise "republics." (For Luxembourg, the term "monarchy" is considered to include the term "duchy.")

	Small?	Large?	Coastal?	Inland?	Monarchy	Republic?
Austria	1	0	0	1	0	1
Belgium	1	0	1	0	1	0
Denmark	1	0	1	0	1	0
Eire	1	0	1	0	0	1
France	0	1	1	0	0	1
Germany	0	1	1	0	0	1
Italy	0	1	1	0	0	1
Luxembourg	1	0	0	1	1	0
Netherlands	1	0	1	0	1	0
Norway	1	0	1	0	1	0
Portugal	1	0	1	0	0	1
Spain	0	1	1	0	1	0
Switzerland	1	0	0	1	0	1
UK	0	1	1	0	1	0

(a) Show that {France, Germany, Italy} forms the extent of a concept whose intent is {large, coastal, republic}.

(b) Determine the concept whose extent is the closure of {Belgium, Denmark}.

(c) Determine the concept whose intent is the kernel of {inland}.

3.3H. (a) Show that $\mathbb{Q}$ is a closed subfield of $\mathbb{Q}(2^{\frac{1}{2}})_{\mathbb{Q}}$.

(b) Show that $\mathbb{Q}$ is not a closed subfield of $\mathbb{Q}(2^{\frac{1}{3}})_{\mathbb{Q}}$.

3.3I. Let $A = K(t)$ be such that t has a separable minimal polynomial $f(T)$ in $K[T]$. Prove that K is a closed subfield of A_K. (Hint: Mimic the proof of Proposition 3.3.6.)

3.3J. Let q be a power of a prime number p. Prove that $\mathrm{GF}(p)$ is a closed subfield of the $\mathrm{GF}(p)$-algebra $\mathrm{GF}(q)$. (Hint: Consider the Frobenius automorphism.)

3.3K. Determine all the automorphisms and the subfields of the field $\mathrm{GF}(2^8)$.

3.3L. **(Symmetric functions.)** Let K be a field.

(a) For a positive integer n, show that the commutative K-algebra $K[T_1, \dots, T_n]$ of Theorem II 3.7 is an integral domain.

(b) Let $A = K(T_1, \dots, T_n)$ be the field of fractions of $K[T_1, \dots, T_n]$. Show that there is an action of the symmetric group S_n on $K(T_1, \dots, T_n)$ given by $\forall U \in S_n, \forall 1 \le i \le n, T_i\, U = T_{iU}$.

(c) Show that S_n is a subgroup of the Galois group of A_K.

(d) Let $S = \mathrm{Fix}(A, S_n)$. Show that S is a subfield of A properly containing K.

(e) Show that S_n is the Galois group of the S-algebra A_S.

(f) Set $\prod_{i=1}^{n}(T - T_i) = \sum_{j=0}^{n}(-1)^j \sigma_j T^{n-j} \in S[T]$. The elements $\sigma_1, \ldots, \sigma_n$ of S are called the *elementary symmetric functions* of $T_1, \ldots, T_n$. Derive *Viète's formulas* $\sigma_j = \sum_{1 \le i_1 < \cdots < i_j \le n} T^{i_1} \ldots T^{i_j}$ for $1 \le j \le n$.

(g) Set $E = K(\sigma_1, \ldots, \sigma_n)$. Prove $\dim K(T_1, \ldots, T_n)_E \le n!$.

(h) Conclude $K(\sigma_1, \ldots, \sigma_n) = \mathrm{Fix}(K(T_1, \ldots, T_n), S_n)$.

3.3M. Show that, for any finite group G, there is a closed subfield K of a K-algebra A_K such that G is the Galois group of A_K. [Hint: Consider Exercise 3.3L(e) and the right regular representation of the group G.]

3.4. Continuity and Topology

Let $R : A \to B$ be a functor, and let $F : J \to A$ be a functor with small domain. Let Δ or $\Delta_A : A \to A^J$ and Δ or $\Delta_B : B \to B^J$ be diagonal functors. The functor R is said to *preserve the limit of* F if the existence of the limit $\pi : (\varprojlim F)\Delta_A \to F$ of F implies that $\pi R : (\varprojlim F)\, R\Delta_B \to FR$ is the limit of $FR : J \to B$. Dually, R *preserves the colimit of* F if the existence of the colimit $\iota : F \to (\varinjlim F)\Delta_A$ of F implies that $\iota R : FR \to (\varinjlim F)R\Delta_B$ is the colimit of $FR : J \to B$. The functor R is *continuous* if it preserves all limits (of functors $F : J \to A$ with small domain), and *cocontinuous* if it preserves all colimits. Continuity of functors follows from adjointness:

Theorem 3.4.1. (a) *Right adjoints are continuous.*

(b) *Left adjoints are cocontinuous.*

Proof. (a) Let $R : A \to B$ be right adjoint to $S : B \to A$ in the adjunction $(S, R, \eta, \varepsilon)$. Let $F : J \to A$ have small domain, and let $\pi : (\varprojlim F)\Delta_A \to F$ be the limit of F. Given an object $\varphi : b\Delta_B \to FR$ of $B\Delta_B/FR$, there is then a unique $(B\Delta_B/FR)$-morphism to $\pi R : (\varprojlim F)R\Delta_B \to FR$, namely $\{\eta_b[\varprojlim (\varphi X)(F\varepsilon)]^R\}\Delta_B$. This follows, via the adjointness, from the existence of the unique $(A\Delta_A/F)$-morphism $[\varprojlim (\varphi S)(F\varepsilon)]\Delta_A : ((\varphi S)(F\varepsilon) : bS\Delta_A \to F) \to (\pi : (\varprojlim F)\Delta_A \to F)$. To see that $\{\eta_b[\varprojlim (\varphi S)(F\varepsilon)]^R\}\Delta_B$ really is a $(B\Delta_B/FR)$-morphism, consider an object j of J. Then $[\{\eta_b[\varprojlim (\varphi S)(F\varepsilon)]^R\}\Delta_B]_j(\pi R)_j = \eta_b[\varprojlim (\varphi S)(F\varepsilon)]^R(\pi_j)^R = \eta_b[(\varphi S)(F\varepsilon)]_j^R = \eta_b \varphi_j^{SF} \varepsilon_{jF}^R = \varphi_j \eta_{jFR} \varepsilon_{jF}^R = \varphi_j[(R\eta)(\varepsilon R)]_{jF} = \varphi_j$, as required. The second equality holds since F has a limit, while the fourth equality is the naturality of η. The final equality holds by Proposition 3.1.7.

(b) Is dual to (a). $\square$

Example 3.4.2. (a) Consider the adjointness of Example 3.1.5 between the free monoid functor $F : \underline{\underline{\mathrm{Set}}} \to \underline{\underline{\mathrm{Mon}}}$ and the underlying set functor $U : \underline{\underline{\mathrm{Mon}}} \to \underline{\underline{\mathrm{Set}}}$. The fact observed in Exercise 2B, that the free monoid over a disjoint union is the coproduct of the free monoids over the uniands, becomes an instance of Theorem 3.4.1(b). The fact observed in Exercise 2M(a), that the underlying set of a product monoid is the product of the underlying sets of the factors, becomes an instance of Theorem 3.4.1(a). Exercise 2M(b) shows that right adjoints need not be cocontinuous. Dually, left adjoints need not be continuous.

(b) Consider the currying adjointness of Example 3.1.2. The cocontinuity of the left adjoint F_Y, resulting by Theorem 3.4.1(b), yields the distributive law $(Z_1 \cup Z_2) \times Y \cong (Z_1 \times Y) \cup (Z_2 \times Y)$, and indeed the infinite version $[\Sigma\{Z_i \mid i \in I\}] \times Y \cong \Sigma\{Z_i \times Y | i \in I\}$ for arbitrary index sets I.

(c) For complete lattices $(A, \leq)$ and $(B, \leq)$, suppose that a functor $R : A \to B$ is right adjoint to a functor $S : B \to A$. As observed in Example 3.3.3(c), the induced posets $(BS, \leq)$ and $(AR, \leq)$ of closed elements are themselves complete lattices, and the Galois correspondence yields a lattice isomorphism $R : BS \to AR$. Theorem 3.4.1(a) now shows that R preserves all products. (Contrast with Exercise 2.1G.) □

Let $(L, +, \cdot, 0, 1)$ be a bounded lattice. For each element a of L, consider the left multiplication $S(a) : L \to L; x \mapsto ax$ in the commutative, idempotent monoid $(L, \cdot, 1)$. Now $x \leq y \Rightarrow xy = x \Rightarrow ax \cdot ay = ax \Rightarrow ax \leq ay$, so $S(a) : L \to L$ is a functor. The bounded lattice L is a *Heyting algebra* if each of these functors $S(a) : L \to L$ (for any a in L) has a right adjoint $R(a) : L \to L; y \mapsto a \setminus y$. In other words,

$$\forall a, x, y \in L, \quad a \cdot x \leq y \Leftrightarrow x \leq a \setminus y \tag{3.4.1}$$

by (3.3.1). If L is a Heyting algebra, then the left multiplications $S(a)$, being left adjoints, preserve the coproduct $+$ in the category $(L, \leq)$ by Theorem 3.4.1(b), so that (3.2.1) holds in L. By Exercise 3.2I, it follows that Heyting algebras are distributive lattices. Similarly, since the functors $R(a)$ are right adjoints, they preserve the product $\cdot$ in the category $(L, \leq)$ by Theorem 3.4.1(a), so that $a \setminus (b \cdot c) = (a \setminus b) \cdot (a \setminus c)$ for elements a, b, c of L. Note also $a \setminus a = 1$ for each element a of L.

Example 3.4.3. (a) **(Boolean Algebras.)** Let $(L, \cup, \cap, ')$ be a Boolean algebra. Define $a \setminus y := a' \cup y$. Then $a \cap x \leq y \Rightarrow x = (a' \cup a) \cap x = (a' \cap x) \cup (a \cap x) \leq a' \cup (a \cap x) \leq a' \cup y = a \setminus y$. Conversely, $x \leq a \setminus y \Rightarrow a \cap x \leq a \cap (a' \cup y) = (a \cap a') \cup (a \cap y) = a \cap y \leq y$. Thus Boolean algebras are Heyting algebras.

(b) **(Bounded Chains.)** Suppose that the bounded lattice $(L, +, \cdot, 0, 1)$ has a chain as its partial order $(L, \leq)$. Define $a \setminus y :=$ **if** $a \leq y$ **then** 1 **else** y. Since $(L, \leq)$ is a chain, $a \cdot x \leq y \Leftrightarrow (a \leq y \text{ or } x \leq y) \Leftrightarrow x \leq a \setminus y$. Thus bounded chains are Heyting algebras. □

In the Boolean algebra Example 3.4.3(a), one has $a' = a' \cup 0 = a \setminus 0$. In a general Heyting algebra $(L, +, \cdot, \setminus, 0, 1)$, one may thus define the *pseudo-*

complement

$$a' = a \setminus 0 \tag{3.4.2}$$

for $a \in L$. By (3.4.1), $a' \leq a \setminus 0 \Rightarrow a \cdot a' \leq 0$, so that $a \cdot a' = 0$ [cf. (3.2.3)(a)]. On the other hand, in the bounded chain Example 3.4.3(b), one has $0' = 0 \setminus 0 = 1$, but $a' = a \setminus 0 = 0$ for $0 < a$. Thus (3.2.3)(b) is not satisfied in general —hence the "pseudo" in (3.4.2). Now for elements x, y of any Heyting algebra $(L, +, \cdot, \setminus, 0, 1)$, one has

$$y \leq x \setminus 0 = x' \Leftrightarrow xy = 0 \Leftrightarrow x \leq y \setminus 0 = y'. \tag{3.4.3}$$

Thus there is a Galois connection with $' : (L, \leq) \to (L, \geq)$ left adjoint to $' : (L, \geq) \to (L, \leq)$. Each part of Theorem 3.4.1 then yields

$$(a + b)' = a' \cdot b' \tag{3.4.4}$$

for any a, b in L.

Proposition 3.4.4. (a) *In any Heyting algebra* $(L, \leq)$, *the induced poset* $(L', \leq)$ *forms a Boolean algebra.*

(b) *A Heyting algebra L is a Boolean algebra iff the Galois connection* (3.4.3) *is a category equivalence.*

Proof. (a) Since $'' : (L, \leq) \to (L, \leq)$ is a functor, one has $a \cdot b \leq a \Rightarrow (a \cdot b)'' \leq a''$, and similarly $(a \cdot b)'' \leq b''$. Thus $(a \cdot b)'' \leq a'' \cdot b''$. Conversely, $(a \cdot b) \cdot (a \cdot b)' = 0 \Rightarrow b \cdot [a \cdot (a \cdot b)'] \leq 0 \Rightarrow a \cdot (a \cdot b)' \leq b \setminus 0 = b' = b''' = b'' \setminus 0 \Rightarrow b'' \cdot a \cdot (a \cdot b)' \leq 0 \Rightarrow b'' \cdot (a \cdot b)' \leq a \setminus 0 = a' = a''' = a'' \setminus 0 \Rightarrow a'' \cdot b'' \cdot (a \cdot b)' \leq 0 \Rightarrow a'' \cdot b'' \leq (a \cdot b)' \setminus 0 = (a \cdot b)''$. Thus $(a \cdot b)'' = a'' \cdot b''$, i.e. $'' : L \to L$ is a meet semilattice homomorphism, and its image L' is a meet subsemilattice of L. For a, b in L, one has $a \leq a'' \leq (a + b)''$ and $b \leq b'' \leq (a + b)''$, so that $(a + b)''$ is an upper bound of $\{a, b\}$ in L'. For any such upper bound c' in L', one has $a + b \leq c' \Rightarrow (a + b)'' \leq c''' = c'$. Thus $(a + b)''$ is the least upper bound of $\{a, b\}$ in L'. In particular, $(L', \leq)$ is a join semilattice $(L, \cup)$ with $a' \cup b' = (a' + b')''$ for a, b in L. Since $1 = 0' \in L'$ and $0 = 1' \in L'$, the poset $(L', \leq)$ becomes a bounded lattice $(L', \cup, \cdot, 0, 1)$. Now for a, b, c in L', one has $c \cdot (a \cup b) = c'' \cdot (a + b)'' = [c \cdot (a + b)]'' = [(c \cdot a) + (c \cdot b)]'' = (c \cdot a) \cup (c \cdot b)$, so that $(L', \cup, \cdot, 0, 1)$ is distributive. Finally, for c in L', one has $c \cup c' = (c + c')'' = (c' \cdot c'')' = 0' = 1$, the second equality holding by (3.4.4). Thus the pseudocomplement c' of c is a genuine complement, and $(L', \cup, \cdot, 0, 1)$ becomes a Boolean algebra.

(b) Note that (3.4.3) is an equivalence iff $x'' = x$ for all x in L. Then if L is a Boolean algebra, one has $\forall x \in L, x = x''$. Conversely, $(\forall x \in L, x = x'')$ implies $(L, +) = (L', \cup)$, so that L is a Boolean algebra, by (a). □

Since left adjoints tend to embody useful and significant mathematical constructions, it is helpful to have indications of their presence. In other words, when does a functor $R : A \to B$ possess a left adjoint $S : B \to A$? By

Theorem 3.4.1(a), the functor R has to be continuous. For this necessary condition to have much force as part of a collection of sufficient conditions, the category A should be richly endowed with limits, and preferably be complete. One is led to the question: When does a continuous functor $R: A \to B$ with (small-) complete domain have a left adjoint? If A and B are poset categories, the answer ("always") is given by the following:

Theorem 3.4.5. (a) *Let $R: A \to B$ be a continuous functor from a complete lattice $(A, \leq)$ to a poset $(B, \leq)$. Then R has a left adjoint $S: B \to A$.*

(b) *Let $S: B \to A$ be a cocontinuous functor from a complete lattice $(B, \leq)$ to a poset $(A, \leq)$. Then S has a right adjoint $R: A \to B$.*

Proof. (a) Define $S: B \to A; x \mapsto \prod\{z \in A \mid x \leq z^R\}$. Note $x_1 \leq x_2 \Rightarrow \{z \in A \mid x_1 \leq z^R\} \supseteq \{z \subset A \mid x_2 \leq z^R\} \Rightarrow \prod\{z \in A | x_1 \leq z^R\} \leq \prod\{z \in A | x_2 \leq z^R\} \Rightarrow x_1^S \leq x_2^S$, so that S is a functor. To verify its left adjointness to R, Proposition 3.3.1 will be used. Indeed, $\forall x \in B, x \leq \prod\{z^r | : z \in A, x \leq z^R\} = [\prod\{z \mid z \in A, x \leq z^R\}]^R = x^{SR}$, the first equality holding by the continuity of R. Also, $\forall y \in A, y^{RS} = \prod\{z \in A \mid y^R \leq z^R\} \leq y$. Thus R and S satisfy (3.3.2), as required.

(b) Is dual to (a). □

A *completely (meet) distributive lattice* is a complete lattice $(L, \cup, \cap, 0, 1)$ such that

$$(3.4.5) \quad \forall x \in L, \forall J \subseteq L, x \cap \sum\{y \mid y \in J\} = \sum\{x \cap y \mid y \in J\}.$$

For example, any finite bounded distributive lattice is completely distributive. Each left multiplication functor $S(x): L \to L; y \mapsto x \cap y$ of a completely distributive lattice is cocontinuous. Theorem 3.4.5(b) then shows that such lattices are actually Heyting algebras, known as *complete Heyting algebras*. Now if the underlying lattice of a Heyting algebra is complete, Theorem 3.4.1(b) shows that (3.4.5) is satisfied, so that the Heyting algebra is a complete Heyting algebra. For example, complete Boolean algebras [cf. Exercise 3.2P(i)], such as the power set of any set (cf. Example 2.1.6), are complete Heyting algebras. More generally, any complete sublattice T of a power set 2^X, for which the inclusion is a cocontinuous bounded lattice homomorphism, is a complete Heyting algebra: (3.4.5) in $(T, \cup, \cap, \varnothing, X)$ is inherited from (3.4.5) in $(2^X, \cup, \cap, \varnothing, X)$. Such a sublattice T of 2^X is called a *topology* on the set X, and the pair (X, T) is called a *topological space*. Elements of T are called *open subsets of X in the topology T*.

Example 3.4.6. (a) For any set X, the set $\{\varnothing, X\}$ forms a topology, the *indiscrete topology* on X.

(b) For any set X, the full power set 2^X forms a topology, the *discrete topology* on X.

(c) If $(X, \leq)$ is a poset, then the set XD of down-sets of S (cf. Example 1.2.3) forms a topology, the *Alexandrov topology* on X. In particular, the poset $\{0 < 1\}$ with the Alexandrov topology is called the *Sierpiński space*.

(d) For any set X, the union of $\{\emptyset\}$ and the set $\mathscr{P}^{<\omega}(X)$ of cofinite subsets of X [cf. Exercise 2F(b)] forms a topology, the *cofinite topology* on X.

(e) Let (X, M) be an M-set for a monoid M. Then the set T of M-subsets of X is a topology on X.

(f) Let $(B, \cap, X)$ be a submonoid of $(2^X, \cap, X)$. Then $\{Y \subseteq X \mid \forall y \in Y, \exists Z \in B.\ y \in Z \subseteq Y\}$ is a topology on X, the *topology spanned* or *generated by the base* B.

(g) Let S be a subset of 2^X. Let $(\langle S \rangle, \cap, X)$ be the submonoid of $(2^X, \cap, X)$ generated by S. Then the topology spanned by the base $\langle S \rangle$ is called the *topology spanned by the subbase* S. For example, if V is a finite-dimensional real vector space, then the set $\{f^{-1}(r, \infty) \mid r \in \mathbb{R}, f \in \mathbb{R}(V, \mathbb{R})\}$ of *open half-spaces* is a subbase for the *Euclidean topology* E on V.

(h) Let Y be a subset of a topological space (X, T). Then $U = \{Z \subseteq Y \mid \exists G \in T.\ Z = G \cap Y\}$ is a topology on Y, the *subspace topology* on Y. □

If T is a topology on a set X, consider $T^C = \{X - G \mid G \in T\}$, the set of complements of open sets. Such complements are called *closed subsets of* X *in the topology* T. Note that $(T^C, \cap, \cup, X, \emptyset)$ is then a complete Heyting algebra. For example, the union of $\{X\}$ with the set $\mathscr{P}_{<\omega}(X)$ of finite subsets of a set X [cf. Exercise 2F(a)] is the set of closed sets of the cofinite topology on X.

If (X, T) and (Y, U) are topological spaces, consider a function $f: X \to Y$ and the right adjoint $f^{-1}: (2^Y, \subseteq) \to (2^X, \subseteq);\ Z \mapsto f^{-1}Z$ to the functor $(2^X, \subseteq) \to (2^Y, \subseteq);\ W \mapsto Wf$ [cf. Example 3.3.2(b)]. The function f is said to be *continuous* if f^{-1} restricts to a functor $f^{-1}: (U, \subseteq) \to (T, \subseteq)$. There is a large concrete category $\underline{\underline{\mathrm{Top}}}$ whose objects are topological spaces and whose morphisms are continuous functions. It is the domain of the forgetful functor $G: \underline{\underline{\mathrm{Top}}} \to \underline{\underline{\mathrm{Set}}}$ with morphism part $(f: (X, T) \to (Y, U)) \mapsto (f: X \to Y)$. This forgetful functor G is left adjoint to the *indiscrete topology functor* $I: \underline{\underline{\mathrm{Set}}} \to \underline{\underline{\mathrm{Top}}}$ with morphism part $(f: X \to Y) \mapsto (f: (X, \{\emptyset, X\}) \to (Y, \{\emptyset, Y\}))$. The forgetful functor G is right adjoint to the *discrete topology functor* D: $\underline{\underline{\mathrm{Set}}} \to \underline{\underline{\mathrm{Top}}}$ with morphism part $(f: X \to Y) \mapsto (f: (X, 2^X) \to (Y, 2^T))$.

EXERCISES

3.4A. Write out a proof of Theorem 3.4.1(b).

3.4B. **(Creation of limits.)** Let $R: A \to B$ be a functor, and let $F: J \to A$ be a functor with small domain. The functor R is said to *create the limit of* F if the existence of the limit $\omega: \underleftarrow{\lim}\,(FR)\Delta_B \to FR$ of FR implies the existence of the limit $\pi: (\underleftarrow{\lim}\,F)\Delta_A \to F$ of F such that $\pi R = \omega$. For a unital ring K, show that the forgetful functor $U : \underline{\underline{\text{Mod}}}_K \to \underline{\underline{\text{Set}}}$ creates the limit of each functor $F : J \to \underline{\underline{\text{Mod}}}_K$. (Hint: By Theorem 2.5.5, it suffices to show that U creates products and equalizers.)

3.4C. Let $R : A \to B$ be a functor, and let $F : J \to A$ be a functor with small domain. If FR has a limit, and R creates the limit of F, show that R preserves the limit of F.

3.4D. (a) Use Theorem 3.4.1(a) and Exercise 3.1F to show that the group of units of an arbitrary product of monoids is the product of the groups of units of the factors.

(b) Determine the primary decomposition of the abelian group $(\mathbb{Z}_{255}, \cdot, 1)^*$. (Hint: Use Proposition II 3.5.4.)

3.4E. Let $f : D \to C$ be a function, and let $\{Y_i \mid i \in I\}$ be a set of subsets of the codomain C. Prove $f^{-1} \bigcap_{i \in I} Y = \bigcap_{i \in I} f^{-1} Y$.

3.4F. Give a direct proof [i.e. independent of Theorem 3.4.1(a)] of the continuity of the right adjoint $R : A \to B$ in a Galois connection (3.3.1). [Hint: You only have to prove that R preserves products, i.e. greatest lower bounds of (possibly empty) subsets of A.]

3.4G. Let A be a cofinite subset of an infinite set X. Determine the pseudocomplement A' of A in the cofinite topology on X [cf. Example 3.4.6(d)].

3.4H. Let α be a relation between sets I and J. Consider the polarity determined by α [cf. Example 3.3.2(c)].

(a) Show that the closed subsets of I, under the order induced from $(2^I, \subseteq)$, form a complete lattice L.

(b) Show that the closed subsets of J, under the order induced from $(2^J, \subseteq)$, form a complete lattice M.

(c) Show that the inclusion functor $(L, \subseteq) \hookrightarrow (2^I, \subseteq)$ is continuous.

(d) Show that the inclusion functor $(M, \subseteq) \hookrightarrow (2^J, \subseteq)$ is continuous.

3.4I. Let $\backslash : L \times L \to L; (a, b) \mapsto a \backslash b$ be a binary operation on a bounded lattice $(L, +, \cdot, 0, 1)$. Show that $(L, +, \cdot, \backslash, 0, 1)$ is a Heyting algebra iff the following four identities hold for all a, b, c in L:

(a) $a \backslash a = 1$; (b) $a \cdot (a \backslash b) = a \cdot b$;

(c) $b \cdot (a \backslash b) = b$; (d) $a \backslash (b \cdot c) = (a \backslash b) \cdot (a \backslash c)$.

3.4J. Verify the claims of Example 3.4.6(e)–(h).

3.4K. For posets $(X, \leq)$ and $(Y, \leq)$, let $f : (X, \leq) \to (Y, \leq)$ be a functor. Show that $f : (X, XD) \to (Y, YD)$ is a continuous map between Alexandrov spaces [cf. Example 3.4.6(c)].

3.4L. **(The Zariski topology.)** Let n be a positive integer, and let K be a field. Define a relation α between K^n and $K[T_1, \ldots, T_n]$ by $(x_1, \ldots, x_n) \alpha p(T_1, \ldots, T_n) \Leftrightarrow p(x_1, \ldots, x_n) = 0$ [cf. II(3.8)]. Subsets of K^n closed under the polarity determined by α are called *varieties*.

(a) For varieties $V_1 = Y_1^R$ and $V_2 = Y_2^R$ with $Y_1, Y_2 \subseteq K[T_1, \ldots, T_n]$, show that $V_1 \cup V_2 = (Y_1 Y_2)^R$, where $Y_1 Y_2 = \{p_1 p_2 \mid p_i \in Y_i\}$. [Hint:

$(x_1, \ldots, x_n) \notin V_1 \cup V_2 \Rightarrow \forall i \in \{1, 2\}, \exists p_i \in Y_i.\ p_i(x_1, \ldots, x_n) \neq 0.$ Then $(p_1 p_2)(x_1, \ldots, x_n) \neq 0.$]

(b) Show that arbitrary intersections of varieties are varieties. [Hint: Use Exercise 3.4H(c).]

(c) Conclude that the set of varieties is the set of closed sets of a topology on K^n. This topology is called the *Zariski topology*.

3.4M. Identifying $\mathbb{R}_2^2$ with $\mathbb{R}^4$, show that $\mathrm{SL}_2(\mathbb{R})$ is a variety.

3.4N. Let n be a positive integer, and let $1 : \mathbb{R}^n \to \mathbb{R}^n$ be the identity mapping. Show that $1 : (\mathbb{R}^n, E) \to (\mathbb{R}^n, Z)$ is a continuous function from $\mathbb{R}^n$ with the Euclidean topology to $\mathbb{R}^n$ with the Zariski topology.

3.4O. Verify that G : Top → Set is right adjoint to the discrete topology functor D : Set → Top.

3.4P. Verify that G : Top → Set is left adjoint to the indiscrete topology functor I : Set → Top.

3.4Q. **(Completeness of Top.)** (a) Show that the forgetful functor G : Top → Set creates limits.

(b) Show that the forgetful functor G : Top → Set preserves limits.

(c) Show that Top is complete.

3.4R. **(Cocompleteness of** Top.) (a) Dually to Exercise 3.4B, formulate a concept of "creation of colimits." Then show that the forgetful functor G : Top → Set creates colimits.

(b) Show that the forgetful functor G : Top → Set preserves colimits.

(c) Show that Top is cocomplete.

3.4S. Consider $\mathrm{SL}_2(\mathbb{R})$ as a subspace of the space $\mathbb{R}^4$ with the Euclidean topology, so that $\mathrm{SL}_2(\mathbb{R})$ has the subspace topology. Show that $\mathrm{SL}_2(\mathbb{R})$ is a *topological group*, i.e. a group in the category Top.

3.5. Existence of Adjoints

Theorem 3.4.5 gave a partial answer to the question of the existence of a left adjoint to a continuous functor $R : A \to B$ with (small-)complete domain. Dealing with the case where A is a poset category, i.e. A_0 a set and each morphism class $A(x, y)$ having at most one element, it showed that a left adjoint always existed. The main result of this section, the Freyd Adjoint Functor Theorem 3.5.2, answers the question for the more general case where A is locally small, i.e. each morphism ciass $A(x, y)$ is a set. In such cases, a left adjoint need not necessarily exist. Take A to be the poclass $(\text{Set}_0, \supseteq)$ (cf. Example 1.1.4). Note that A is (small-)complete, having products given by unions and equalizers given by domains [cf. Exercise 2.3H(b)]. Consider the continuous functor $R : A \to$ Set whose object part sends each set to the terminal object T of Set (i.e. a singleton). Suppose that R had a left adjoint S in an adjunction $(S, R, \eta, \varepsilon)$. For each set Y, the counit component $(\varepsilon_Y : YRS \to Y) = (\varepsilon_Y : TS \to Y) = (TS \supseteq Y)$ in A would then exhibit Y as a subset of TS. No set TS is that big!

Theorem 3.1.4 reduced the existence of a left adjoint to the existence of certain initial objects (in comma categories). The first step towards the Adjoint Functor Theorem addresses the existence of initial objects in complete categories. Let D be a category. A set K_0 of objects of D is said *to dominate* D if each object of D is the codomain of a D-morphism whose domain lies in K_0. For example, if S is a non-trivial ring, then Max S dominates the poset category (Prop S, $\supseteq$).

Proposition 3.5.1. *Let D be a (small-)complete, locally small category. Then D has an initial object iff it has a dominating set.*

Proof. If the category D has an initial object $\perp$, then D is dominated by the singleton $\{\perp\}$. Conversely, suppose that D has a dominating set K_0. Let K be the full subcategory of D whose object class is the set K_0. Since D is locally small, the category K is small. Let $F : K \to D$ be the inclusion of K in D. Since D is (small-)complete, the functor F has a limit $\pi : l\Delta \to F$ with limit object l. It will be shown that l is initial in D.

Each object k of K is the codomain of the identity D-morphism $\kappa_k : k \to k$ whose domain lies in K_0. Extend the function $\kappa : K_0 \to D_1;\ k \mapsto \kappa_k$ to a function $\kappa : D_0 \to D_1;\ x \mapsto \kappa_x$ such that dom $\kappa_x \in K_0$ and cod $\kappa_x = x$ for each object x of D. Then define $\lambda : D_0 \to D_1;\ x \mapsto \lambda_x$ such that $\lambda_x = \pi_j \kappa_x$ for $\kappa_x : j \to x$. Let $L : D \to D$ be the constant functor with morphism part $(f : x \to y) \mapsto (1_l : l \to l)$. It will first be shown that λ is a natural transformation $\lambda : L \to 1_D$ from L to the identity functor on D. Consider the following diagram:

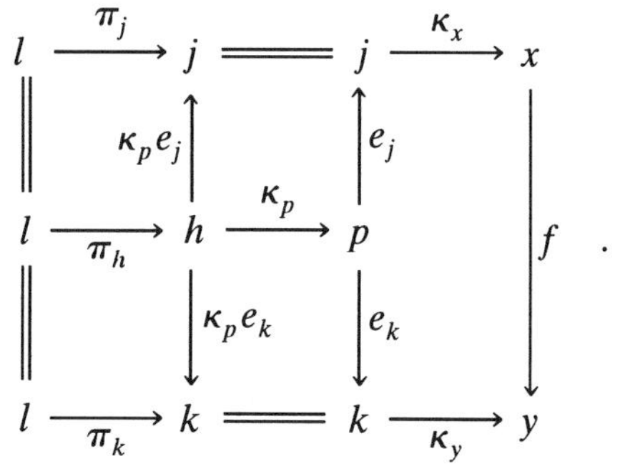

$$
\begin{array}{ccccccc}
l & \xrightarrow{\pi_j} & j & = & j & \xrightarrow{\kappa_x} & x \\
\| & & \uparrow{\scriptstyle \kappa_p e_j} & & \uparrow{\scriptstyle e_j} & & \\
l & \xrightarrow[\pi_h]{} & h & \xrightarrow{\kappa_p} & p & & \Big\downarrow{\scriptstyle f} \\
\| & & \downarrow{\scriptstyle \kappa_p e_k} & & \downarrow{\scriptstyle e_k} & & \\
l & \xrightarrow[\pi_k]{} & k & = & k & \xrightarrow[\kappa_y]{} & y
\end{array}
.
$$

The outer rectangle is the naturality diagram (1.4.2) for λ at the D-morphism $f : x \to y$. Thus $\lambda_x f = \pi_j \kappa_x f = \pi_k \kappa_y = \lambda_y$ must be verified. The right hand rectangle is the pullback diagram (2.3.5) for the pair $(\kappa_y, \kappa_x f)$ of D-morphisms. The pullback exists, since D is (small-)complete. Then $\pi_j \kappa_x f = \pi_h \kappa_p e_j \kappa_x f = \pi_h \kappa_p e_k \kappa_y = \pi_k \kappa_y$, as required. The second equality is the commuting of the pullback diagram (2.3.5), while the other two equalities hold by the naturality of π.

Given that $\lambda : L \to 1_D$ is natural, it will now be shown that l is initial in D. For k in K_0, the naturality (1.4.2) of λ at the D-morphism $\pi_k : l \to k$ yields the commuting of

$$\begin{array}{ccc} l & \xrightarrow{\lambda_l} & l \\ \| & & \downarrow{\scriptstyle \pi_k} \\ l & \xrightarrow[\lambda_k]{} & k \end{array} \tag{3.5.1}$$

Since $\kappa : D_0 \to D_1$ extends the identity function (1.1) of the category K, the function $\lambda : D_0 \to D_1$ extends the function $\pi : K_0 \to D_1$. The commuting diagram (3.5.1) may thus be rewritten as

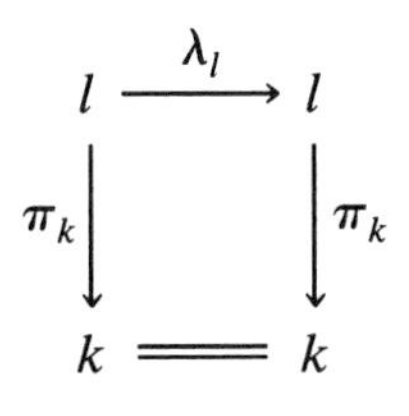

showing that $\lambda_l \Delta : (\pi : l\Delta \to F) \to (\pi : l\Delta \to F)$ is a $(K\Delta/F)$-morphism. The terminality of $\pi : l\Delta \to F$ in $K\Delta/F$ then yields $\lambda_l = 1_l$. Now consider an object x of D. The set $D(l, x)$ is certainly non-empty, since it contains the component λ_x of λ. To complete the proof that l is initial in D, it will be shown that λ_x is the unique element of $D(l, x)$. Indeed, given any element f of $D(l, x)$, the naturality (1.4.2) of λ at $f : l \to x$ yields the commuting diagram

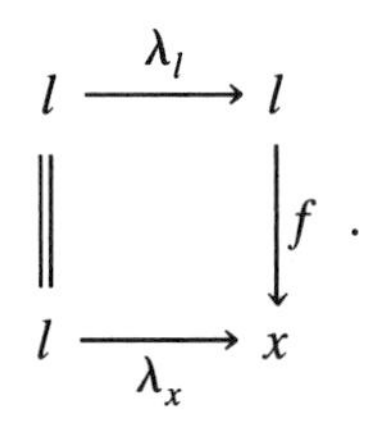

Thus $f = 1_l f = \lambda_l f = \lambda_x$, as required. $\square$

A functor $R : A \to B$ is said *to satisfy the solution set condition* if, for each object x of B, the comma category (x, R) possesses a dominating set. (This dominating set is sometimes called a *solution set for x with respect to R*.) One may then formulate the

Freyd Adjunct Functor Theorem 3.5.2. *Let A be a (small-)complete, locally small category. If a continuous functor $R : A \to B$ satisfies the solution set condition, then it is a right adjoint.*

Proof. Let x be an object of B. By Theorem 3.1.4, it suffices to exhibit an initial object of the comma category (x, R). By the solution set condition, the comma category (x, R) has a dominating set. Let $G:(x, R) \to A$ be the faithful forgetful functor with morphism part $(fR:(x \to y_1 R) \to (x \to y_2 R)) \mapsto (f: y_1 \to y_2)$. Since G is faithful and A is locally small, it follows that (x, R) is locally small. By Proposition 3.5.1, it thus remains to show that (x, R) is (small-)complete.

Let $F: J \to (x, R)$ be a functor with small domain. Since A is complete, the composite functor $FG: J \to A$ has a limit $\omega: y\Delta_A \to FG$ with limit object y. Since $R: A \to B$ is continuous, it preserves the limit of FG. Thus $\omega R: yR\Delta_B \to FGR$ is the limit of $FGR: J \to B$. Now the functor $F: J \to (x, R)$ may be construed as a natural transformation $\varphi: x\Delta_B \to FGR$ whose component φ_j at an object j of J is the B-morphism $jF: x \to jFGR$. It follows that there is a unique B-morphism $\varprojlim \varphi: x \to yR$ yielding a $(B\Delta_B/FGR)$-morphism $(\varprojlim \varphi)\Delta_B : (\varphi : x\Delta_B \to FGR) \to (\omega R : yR\Delta_B \to FGR)$. Let $\Omega:(\varprojlim \varphi: x \to yR)\Delta_{(x, R)} \to F$ be the natural transformation whose component at the object j of J is the (x, R)-morphism $\omega_j R:(\varprojlim \varphi: x \to yR) \to (jF: x \to jFGR)$. Note that $(\varprojlim \varphi)\omega_j^R = \varphi_j = jF$ (since $(\varprojlim \varphi)\Delta_B$ is a $(B\Delta_B/FGR)$-morphism), so that $\omega_j R$ is indeed an (x, R)-morphism. It will be shown that $\Omega:(\varprojlim \varphi: x \to yR)\Delta_{(x, R)} \to F$ is the limit of F.

Let $\psi:(f: x \to zR)\Delta_{(x, R)} \to F$ be an object of $(x, R)\Delta/F$. Then $\psi GR: zR\Delta_B \to FGR$ is an object of $B\Delta_B/FGR$. Since ωR is terminal in $B\Delta_B/FGR$, there is a unique B-morphism $\varprojlim(\psi GR) = (\varprojlim \psi G)R$ yielding a $(B\Delta_B/FGR)$-morphism $\varprojlim(\psi GR)\Delta_B : (\psi GR : zR\Delta_B \to FGR) \to (\omega R: yR\Delta_B \to FGR)$. Now for each object j of J, one has $f(\varprojlim \psi GR)(\omega R)_j = f[(\varprojlim \psi G)\omega_j]^R = f(\psi_j G)^R = jF = \varphi_j$. The second equality holds since ω is the limit of FG, while the third holds since ψ is an object of $(x, R)\Delta/F$. It follows that $[f(\varprojlim \psi GR)]\Delta_b$ is the unique element $(\varprojlim \varphi)\Delta_B$ of $(B\Delta_B/FGR)(\varphi : x\Delta_B \to FGR, \omega R \;\; : yR\Delta_B \to FGR)$. Thus $f(\varprojlim \psi GR) = \varprojlim \varphi$, and one obtains the unique element $\varprojlim \psi$ of $[(x, R)\Delta/F](\psi, \Omega)$: The component of $\varprojlim \psi$ at each object j of J is the (x, R)-morphism $(\varprojlim \psi G)R:(f: x \to zR) \to (\varprojlim \varphi: x \to yR)$. □

Example 3.5.3. Let N be a submonoid of a monoid M. Let $R:\underline{\underline{M}} \to \underline{\underline{N}}$ be the forgetful functor from the category $\underline{\underline{M}}$ of M-sets to the category $\underline{\underline{N}}$ of N-sets with morphism part $(f:(A_1, M) \to (A_2, M)) \mapsto (f:(A_1, N) \to (A_2, N))$. For example, if N and M are groups, then R is the restriction functor $\downarrow_N^M$ of Section I 2.3. Let X be an N-set. Recall the construction (Exercise I 1U) of the free M-set $X \times M$ over the set X, together with the insertion $\eta: X \to X \times M; x \mapsto (x, 1)$. Let K_0 be the set of composite maps $k: X \xrightarrow{\eta} X \times M \xrightarrow{\text{nat } \alpha} (X \times M)^\alpha R$ for M-congruences α on $(X \times M, M)$ such that $\forall x \in X, \forall n \in N, (xn, 1)\alpha(x, n)$. Note that $k: X \to (X \times M)^\alpha R$ is an $\underline{\underline{N}}$-morphism, since $xnk = (xn, 1)^\alpha = (x, n)^\alpha = (x, 1)n^\alpha = (x, 1)^\alpha n = xkn$ for x in X and n in N. Now K_0 is a dominating set for the comma category (X, R). Consider an object $f: X \to YR$ of this category. As in Exercise I 1U, there is a unique $\underline{\underline{M}}$-morphism $\bar{f}: X \times M \to Y$ with $\eta\bar{f} = f$. The First Isomorphism Theorem for M-actions yields a commuting diagram

$$
\begin{array}{ccc}
X \times M & \xrightarrow{\ \bar{f}\ } & Y \\
{\scriptstyle \text{nat ker } \bar{f}} \downarrow & & \uparrow \\
(X \times M)^{\ker \bar{f}} & \xrightarrow[\theta]{} & XfM
\end{array}
\tag{3.5.2}
$$

in $\underline{\underline{M}}$. The right-hand arrow is the inclusion of the M-subset $XfM = \{xfm \mid x \in X, m \in M\}$ of Y, while θ is an isomorphism. One then obtains a commuting diagram

(3.5.3)

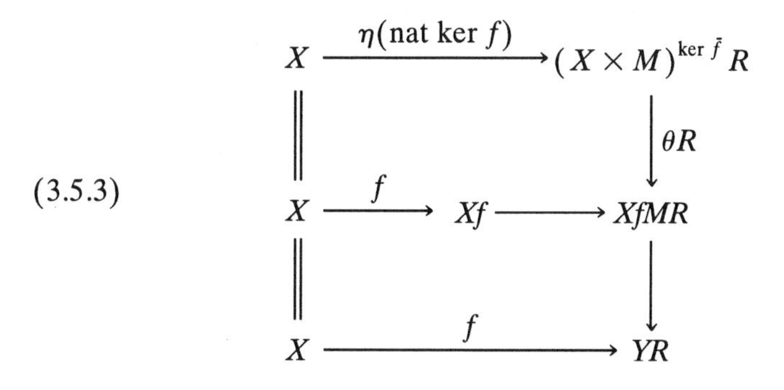

in $\underline{\underline{N}}$, the unlabeled arrows being inclusions. For x in X and n in N, note $(x, n)^{\ker \bar{f}} = (x, 1)n^{\ker \bar{f}} = (x, 1)^{\ker \bar{f}}n = xf(\theta^{-1})^R n = xfn(\theta^{-1})^R = xnf(\theta^{-1})^R = (xn, 1)^{\ker \bar{f}}$, so that the top arrow in the diagram is an element of K_0 (and thus of $\underline{\underline{N}}_1$). The outer rectangle of (3.5.3) now shows that K_0 dominates (X, R). Thus R satisfies the solution set condition. Moreover, the category $\underline{\underline{M}}$ is complete and $R : \underline{\underline{M}} \to \underline{\underline{N}}$ is continuous (Exercise 3.5D). The Freyd Adjoint Functor Theorem 3.5.2 then shows that R has a left adjoint $S : \underline{\underline{N}} \to \underline{\underline{M}}$. If N and M are groups, then S is the induction functor $\uparrow_N^M$ of Section I 2.3. □

EXERCISES

3.5A. (a) Let C be a small category. Show that the limit of the identity functor $C \to C$ is the colimit of the empty functor $\varnothing \to C$. (Hint: Mimic the last paragraph of the proof of Proposition 3.5.1.)

(b) Let D be a small, complete category. Show that D is equivalent to a complete lattice. [Hint: For each x in D_0, show that x is an initial object of the full subcategory of D with object set $(d_0^{-1}\{x\})d_1$.]

3.5B. Verify the naturality of Ω in the proof of the Freyd Adjunct Functor Theorem 3.5.2.

3.5C. Let $R: A \to B$ be a functor with small domain and codomain.

(a) Verify that R satisfies the solution set condition.

(b) Show that Theorem 3.4.5(a) is a special case of the Freyd Adjoint Functor Theorem 3.5.2.

(c) Dualize the Freyd Adjoint Functor Theorem 3.5.2 to obtain a generalization of Theorem 3.4.5(b).

3.5D. Let N be a submonoid of a monoid M.

(a) Show that the category $\underline{\underline{M}}$ is complete. (Hint: Show that the forgetful functor $U: \underline{\underline{M}} \to \underline{\underline{\mathrm{Set}}}$ creates limits, in the sense of Exercise 3.4B.)

(b) Show that the forgetful functor $R: \underline{\underline{M}} \to \underline{\underline{N}}$ of Example 3.5.3 is continuous.

3.6. Tensor Products of Modules

Throughout this section, let K be a commutative unital ring. For K-modules Y and X, the set $K(Y, X) = \underline{\underline{\mathrm{Mod}}}_K(Y, X)$ of K-module homomorphisms $g: Y \to X$ itself forms a K-module (cf. Exercise II 2.1D). For a fixed module Y, one thus has a functor $R_Y: \underline{\underline{\mathrm{Mod}}}_K \to \underline{\underline{\mathrm{Mod}}}_K$ with morphism part $(f: X \to X') \mapsto (K(Y, X) \to K(Y, X'); g \mapsto gf)$. If $U: \underline{\underline{\mathrm{Mod}}}_K \to \underline{\underline{\mathrm{Set}}}; (X, +, K) \mapsto X$ is the forgetful functor, and $G_Y: \underline{\underline{\mathrm{Set}}} \to \underline{\underline{\mathrm{Set}}}; X \mapsto X^Y$ is the right adjoint in the Currying adjunction of Example 3.1.2, then one has the commuting diagram

$$(3.6.1) \qquad \begin{array}{ccc} \underline{\underline{\mathrm{Mod}}}_K & \xrightarrow{U} & \underline{\underline{\mathrm{Set}}} \\ {\scriptstyle R_Y}\downarrow & & \downarrow{\scriptstyle G_{YU}} \\ \underline{\underline{\mathrm{Mod}}}_K & \xrightarrow[U]{} & \underline{\underline{\mathrm{Set}}} \end{array}$$

of functors. Since U is faithful and creates limits (cf. Exercise 3.4B), while the right adjoint G_{YU} is continuous, it follows that $\underline{\underline{\mathrm{Mod}}}_K$ is locally small and complete, and that the functor R_Y is continuous. (For a direct verification of the continuity, cf. Exercise 3.6A.) The Freyd Adjoint Functor Theorem 3.5.2 will be used to produce a left adjoint S_Y to R_Y. The adjunction between S_Y and R_Y is the linear algebraic analogue of the Currying adjunction of Example 3.1.2.

For the Freyd Adjoint Functor Theorem to apply, the functor $R_Y: \underline{\underline{\mathrm{Mod}}}_K \to \underline{\underline{\mathrm{Mod}}}_K$ must satisfy the solution set condition. Fix a K-module Z. Now an object $g: Z \to K(Y, X)$ of the comma category (Z, R_Y) is an element of the subset $K(Z, K(Y, X))$ of $\underline{\underline{\mathrm{Set}}}(Z, X^Y) = \underline{\underline{\mathrm{Set}}}(Z, XG_Y)$. The Currying adjunction provides a bijection $\psi_Z^X: \underline{\underline{\mathrm{Set}}}(Z, XG_Y) \to \underline{\underline{\mathrm{Set}}}(ZF_Y, X)$ with inverse φ_X^Z. Under this bijection, the image $g\psi_Z^X$ of the object of the comma category is a bilinear function $g\psi_Z^X : Z \times Y \to X$, an element of the K-module $K(Z, Y; X)$ in the notation of Section II 2.4. Speaking loosely, the comma category (Z, R_Y) "is" the category of bilinear functions on $Z \times Y$.

Let $\iota : Z \times Y \to M$ be the insertion II(2.3.4) of the set $Z \times Y$ into the free K-module $M = \coprod_{Z \times Y} K$ on the generating set $Z \times Y$. It will be shown that the union $\bigcup_{N \leq M} K(Z, K(Y, M/N))$, over all submodules N of M, dominates (Z, R_Y). Consider an element $\gamma\varphi_X^Z : Z \to K(Y, X)$ of (Z, R_Y). Consider the following diagram:

(3.6.2)
$$\begin{array}{ccccc} Z \times Y & \xrightarrow{\iota} & M & \xrightarrow{p} & M/N \\ \downarrow{\scriptstyle\gamma} & & \downarrow{\scriptstyle\overline{\gamma}} & & \downarrow{\scriptstyle i} \\ X & = & X & \xleftarrow[j]{} & \operatorname{Im} \overline{\gamma} \end{array}.$$

The left-hand square is an instance of II(2.3.4), while the right hand square results from the First Isomorphism Theorem. Thus $N = \operatorname{Ker} \overline{\gamma}$, while p is the projection, i is an isomorphism, and j is the inclusion. One then has the following (Z, R_Y)-morphism:

(3.6.3)
$$\begin{array}{ccc} Z & \xrightarrow{(\iota p)\varphi_{M/N}^Z} & K(Y, M/N) \\ \| & & \downarrow{\scriptstyle (ij)R_Y} \\ Z & \xrightarrow[\gamma\varphi_X^Z]{} & K(Y, X) \end{array}.$$

Note that $\gamma = \iota\overline{\gamma} = \iota pij$ bilinear in (3.6.2) implies that ιp is bilinear, and thus that the top row of (3.6.3) is a well-defined K-homomorphism. The solution set condition is verified, so the Freyd Adjoint Functor Theorem 3.5.2 yields an adjunction $(S_Y, R_Y, \eta^Y, \varepsilon^Y)$ for each K-module Y.

Definition 3.6.1. Let K be a commutative, unital ring, and let Z and Y be K-modules. Then the *tensor product* $Z \otimes_K Y$ or $Z \otimes Y$ of Z and Y is the image ZS_Y of the module Z under the left adjoint S_Y to the functor $R_Y : \underline{\underline{\text{Mod}}}_K \to \underline{\underline{\text{Mod}}}_K; X \mapsto K(Y, X)$. □

Combining the tensor product adjunction with the restriction of the Currying adjunction, one obtains isomorphisms

(3.6.4)
$$K(Z \otimes Y, X) \cong K(Z, K(Y, X)) \cong K(Z, Y; X)$$

for K-modules X, Y, Z. In particular, consider the identity $1_{Z \otimes Y}$ as an element of $K(Z \otimes Y, Z \otimes Y)$. Its image under (3.6.4) yields a bilinear function

$$\otimes : Z \times Y \to Z \otimes Y; (z, y) \mapsto z \otimes y. \tag{3.6.5}$$

The isomorphism $Z \times Y \cong Y \times Z$ yields the commutative law

$$Z \otimes Y \cong Y \otimes Z. \tag{3.6.6}$$

The cocontinuity of the left adjoint S_Y [Theorem 3.4.1(b)] yields the distributive law

$$Y \otimes \coprod_{i \in I} Z_i \cong \coprod_{i \in I} Y \otimes Z_i. \tag{3.6.7}$$

Example 3.6.2. For any K-module Z, one has $K(K \otimes_K Z, X) \cong K(K, K(Z, X)) \cong K(Z, X)$. The uniqueness of adjoints yields an isomorphism $K \otimes_K Z \cong Z$. Explicitly, the isomorphism is given as $K \otimes_K Z \to Z$; $k \otimes z \mapsto kz$. □

Example 3.6.3. Tensor products of free modules are free. Indeed, using (3.6.7) and Example 3.6.2, one obtains $(\coprod_{i \in I} K) \otimes (\coprod_{j \in J} K) \cong \coprod_{I \times J} K \otimes_K K \cong \coprod_{I \times J} K$. □

Example 3.6.4. Let L be a K-algebra. Then the bilinear multiplication $L \times L \to L$; $(l_1, l_2) \mapsto l_1 l_2$ yields a linear $\mu : L \otimes L \to L$; $l_1 \otimes l_2 \mapsto l_1 l_2$. □

Example 3.6.5. Let J be an ideal of the ring K. Let X be a K-module. Then there are mutually inverse isomorphisms $X \otimes K/J \to X/XJ$; $x \otimes (k + J) \mapsto xk + XJ$ and $X/XJ \to X \otimes K/J$; $x + XJ \mapsto x \otimes 1$. In particular, if J' is a second ideal of K, one has $K/J' \otimes K/J \cong K/(J' + J)$. □

EXERCISES

3.6A. Verify directly that the functor $R_Y : \underline{\underline{\mathrm{Mod}}}_K \to \underline{\underline{\mathrm{Mod}}}_K$ preserves both products and equalizers. Conclude that R_Y is continuous.

3.6B. Let $X_i (1 \leq i \leq 3)$ be K-modules. Verify the associative law

$$(X_1 \otimes X_2) \otimes X_3 \cong X_1 \otimes (X_2 \otimes X_3).$$

3.6C. Let V be a K-module. Define $\otimes_{i=1}^n V$ inductively by $\otimes_{i=1}^1 V = V$ and $\otimes_{i=1}^n V = V \otimes [\otimes_{i=1}^{n-1} V]$. Prove $K^n(V; K) \cong K(\otimes_{i=1}^n V, K)$.

3.6D. Verify the commuting of (3.6.3).

3.6E. Describe the inverse $Z \to K \otimes Z$ of the isomorphism $K \otimes Z \to Z$; $k \otimes z \mapsto kz$ of Example 3.6.2.

3.6F. Let Z and Y be K-modules. Elements of $Z \otimes Y$ in the image of (3.6.5) are called *primitive*. Exhibit a ring K and K-modules Z and Y such that $Z \otimes Y$ has elements that are not primitive. (Hint: Take Z and Y to be vector spaces over a finite field K. Use Example 3.6.3 to compare $|Z \times Y|$ and $|Z \otimes Y|$.]

3.6G. Let L be a K-algebra. Determine K-homomorphisms l and r such that the associative law in the monoid $(L, \cdot, 1)$ becomes the commuting of a diagram

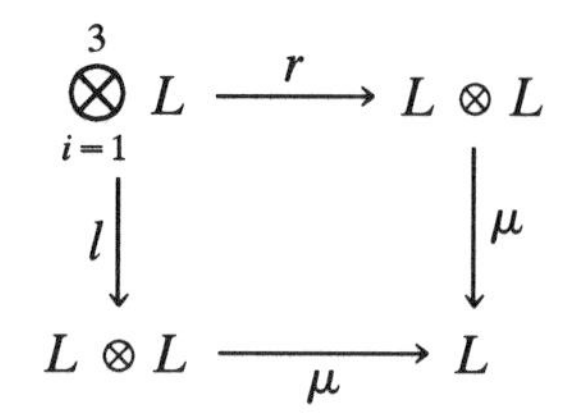

in $\underline{\underline{\text{Mod}}}_K$, using the notation of Exercise 3.6C.

3.6H. Prove (O 3.7).

3.6I. Verify the commutative law (3.6.6).

3.6J. Verify the claims of Example 3.6.5.

3.6K. Compute the tensor product of the abelian groups $\mathbb{Z}_6$ and $\mathbb{Z}_{14}$. (Hint: Use Example 3.6.5.)

3.6L. Let L be a commutative K-algebra. Show that the forgetful functor $\underline{\underline{\text{Mod}}}_L \to \underline{\underline{\text{Mod}}}_K$ has a left adjoint whose object part is of the form $X \mapsto X \otimes_K L$. (If X is a real vector space, then $X \otimes_{\mathbb{R}} \mathbb{C}$ is known as the *complexification* of X.)

3.6M. Let A be a splitting field of a separable polynomial $f(T)$ in $K[T]$, for a field K. Prove $A \otimes_K A \cong \coprod_{i=1}^{\deg f} A$.

3.6N. (a) For positive integers m and n, prove $K_m^m \otimes K_n^n \cong K_{mn}^{mn}$.

(b) For $[x_{ij}]_{m \times m}$ in K_m^m and $[y_{kl}]_{n \times n}$ in K_n^n, prove that $[x_{ij}]_{m \times m} \otimes [y_{kl}]_{n \times n}$ in $K_m^m \otimes K_n^n$ corresponds to $[x_{ij} y_{kl}]_{mn \times mn}$ in K_{mn}^{mn} under the isomorphism of (a).

3.6O. Let $0 \to A \xrightarrow{f} B \xrightarrow{g} C \to 0$ be an exact sequence of abelian groups. Let G be an abelian group. Prove that the sequence $A \otimes G \to B \otimes G \to C \otimes G \to 0$ is exact [cf. Exercise II 2.1B(a).]

IV

UNIVERSAL ALGEBRA

1. SETS WITH OPERATIONS

By this stage, readers should be acquainted with a wide range of algebraic structures: monoids, semilattices, monoid actions, groups, quasigroups, loops, right quasigroups, right loops, rings, fields, integral domains, modules, algebras, lattices, Boolean algebras, Heyting algebras, etc. In the study of these structures, certain concepts keep recurring: homomorphisms, products, substructures, congruences, the First Isomorphism Theorem, freeness properties, etc. The mathematician's response to recurring patterns is to seek an explanation at a more basic level. This is the origin of universal algebra. Thus the first task of universal algebra is to provide a framework for the study of the properties shared by the great majority of the kinds of algebra so far encountered. Once this framework is established, it may then be used to facilitate an approach to the various new kinds of algebra one may meet in the future.

Universal algebra is a broad and powerful branch of mathematics. Such breadth and power have their price. However, there is at least a choice of modes of payment: a direct and concrete but cumbersome and formidable formalism on the one hand, or a clean and sophisticated but apparently rather abstract approach on the other. Choices like this are not uncommon in mathematics. For example, differential geometry may be studied either by classical index-ridden tensor analysis, or with a more modern, sheaf-theoretic approach. Although readers may have their own tastes and preferences amongst these alternative modes of study, it is best if one is able to master them both.

Monoid actions provide a prototype for much of universal algebra. The approach taken in the first part of this chapter—sets with operations—generalizes the concept of an automaton presented in Section I 1.1 as a state space X together with a set A of elementary events. The construction of word algebras or absolutely free algebras in Section 1.3 corresponds to the generation of the free monoid A^* of words in the alphabet A. Thus this first part restricts itself to the analogue of actions of free monoids. The next part, introducing the concept of a variety of algebras, presents the analogue of M-sets (X, M) for general monoids M, i.e. for quotients of free monoids. These two parts both take the direct, formalistic approach to universal algebra. The more sophisticated approach adopted in the third part, with the concept of an algebraic theory, corresponds to taking an M-set as a functor $M \to \underline{\underline{\text{Set}}}$, regarding the monoid as a category in the sense of Example III 1.2. Finally, the "monadic" approach of the fourth part corresponds to taking the action $X \times M \to X$ of an M-set X as a component ε_X of the counit in the adjunction between the forgetful functor $\underline{\underline{M}} \to \underline{\underline{\text{Set}}}$; $(X, M) \mapsto X$ and its left adjoint. This approach makes precise the extent to which universal algebra is modeled on monoid actions.

1.1. Operations and Types

Let A be a set, and let n be a natural number. Then an *n-ary operation on the set A* is a function $\omega : A^n \to A$ from the n-th direct power of A to A. For $n = 0, 1, 2, 3, 4$ respectively, the term "n-ary" becomes: *nullary*, *unary*, *binary*, *ternary*, *quaternary* (cf. Section I 3). A unary operation is just an element of A^A, a function from the set to itself. Now the zeroth direct power A^0 of any set A is a terminal object T of the category $\underline{\underline{\text{Set}}}$, i.e. a singleton set $\{\infty\}$. A nullary operation ω or *constant* on a (necessarily non-empty) set A is thus a function $\omega : \{\infty\} \to A$, corresponding to the element $\infty\omega$ of A (cf. Section III 1.3). One often identifies an element of a set A with the nullary operation $A^0 \to A$ selecting that element. For example, the identity element e of a group Q is selected by the nullary operation $e : Q^0 \to Q$; $\infty \mapsto e$.

Given an n-ary operation $\omega : A^n \to A$ on a set A, various notations are used for elements of A^n and their images under ω. Recalling that the direct power A^n may be realized as the uniform code A^n of length n in the alphabet A (cf. Exercise O 4.1B), one often writes a typical element $(a_1, a_2, \dots, a_n)$ of the direct power A^n as a word $a_1 a_2 \dots a_n$. The n-ary operation ω may then be described as $\omega : A^n \to A$; $(a_1, a_2, \dots, a_n) \mapsto a_1 a_2 \dots a_n \omega$. The form $a_1 a_2 \dots a_n \omega$ for the image element is described as *postfix notation*, or occasionally as "reverse Polish notation." This notation is very convenient for the construction of absolutely free algebras in Section 1.3. On the other hand, familiar binary operations such as addition in an abelian group A are usually written in the form $+ : A^2 \to A$; $(a, b) \mapsto a + b$. The form $a + b$ for the image element is described as *infix notation*. Occasionally infix notation is used for ternary or other operations, such as the operation $P : A^3 \to A$; $(a, b, c) \mapsto a - b + c$ of (I 2.2.5) in an abelian group A. Notation with the operation preceding its arguments is called *prefix notation*. For example, the unary operation of negation in an abelian group A is usually written as $- : A \to A$; $a \mapsto -a$. (Of course, one has to be careful not to confuse this with infix notation for the binary operation of subtraction.) Finally, the juxtaposition of two or more factors to denote their product (e.g. under a semigroup multiplication) might be called an instance of *nofix* *notation*. Nofix notation is also used occasionally to tacitly denote (or rather: not to denote!) the nullary operation selecting the empty word in the free monoid $(A^*, \cdot, 1)$ over an alphabet A.

Example 1.1.1 ("Black Boxes"). A useful intuition about operations on sets may be gained by considering them as components in electrical circuits. Let the set A represent the set of potentials or voltages that may be present in any of the connecting wires of a circuit. Then an n-ary operation $\omega : A^n \to A$ represents a circuit component or "black box" which has n input wires and one output wire. If the voltages in the n input wires are $a_1, a_2, \ldots, a_n$ respectively, then the voltage of the output wire is the element $a_1\ a_2 \ldots a_n\, \omega$ of A.

(a) (**xor** gates.) In logical circuitry, one assumes a voltage set $A = \{0, 1\}$. The addition operation $+$ in the abelian group $(\mathbb{Z}_2, +, 0)$ then corresponds to the circuit component $\omega : A^2 \to A$ sometimes known as an "exclusive-or" gate. Thus $a_1 a_2\, \omega = 1$ iff (a_1 is 1 or a_2 is 1, but not both simultaneously). (Cf. Example II 1.3.4.)

(b) (**Circuit Breakers.**) Consider a DC power circuit with voltages from the real interval $A = [0, N]$. (Here the large number N might represent the voltage at which the insulation of the wires begins to break down.) For a value $r \in (0, N]$, a circuit breaker measures the voltage in input a_1. It passes current directly from input a_2 to the output unless the voltage a_1 exceeds r, at which point it breaks the connection between a_2 and the output. Thus the circuit breaker is represented by the binary operation $\omega : A^2 \to A$ with $a_1 a_2\, \omega =$ **if** $a_1 \leq r$ **then** a_2 **else** 0.

(c) (**Batteries.**) Consider a DC circuit with voltages in the range $A = [0, N]$. Then (the ungrounded terminal of) a battery of voltage $V \in [0, N]$ is represented by the nullary operation $A^0 \to A;\ \omega \mapsto V$.

(d) (**Threshold Gates.**) The voltages present in a neural network lie in the real interval $A = [0, N]$, with $1 < N$. A neuron firing when sufficiently stimulated is modeled in the neural network by a "threshold gate." This circuit component has a positive number n of inputs, and one output. Its "threshold" is a value $r \in [0, N]$. The threshold gate corresponds to the n-ary operation $\omega : A^n \to A$ with $a_1 \ldots a_n\, \omega =$ **if** $a_1 + \ldots + a_n < r$ **then** 0 **else** 1. □

A group A is described as a structure $(A, \cdot, {}^{-1}, 1)$ with three operations, namely with a binary operation $A^2 \to A;\ (a, b) \mapsto a \cdot b$ of multiplication, a unary operation $A \to A;\ a \mapsto a^{-1}$ of inversion, and a nullary operation $A^0 \to A; \infty \mapsto 1$ selecting the identity element. A right loop A is also described as a structure $(A, \cdot, /, 1)$ with three operations, namely a binary operation of multiplication, a binary operation $A^2 \to A;\ (a, b) \mapsto a/b$ of right division, and a nullary operation selecting the identity element. In these terms, the key difference between the group and right-loop structures is that the second group operation is unary, while the second right-loop operation is binary. Given an n-ary operation $\omega : A^n \to A$, the natural number n is called the *arity* $\omega\tau$ of the operation ω. One thus makes the following:

Definition 1.1.2. A (*finitary*) *type* is a function $\tau : \Omega \to \mathbb{N}$. The domain Ω of the type τ is called its *operator domain*, and the elements of Ω are called *operators* (or sometimes: *operation symbols*). The image multiset $\langle \omega\tau | \omega \in \Omega \rangle$ of τ is called the *signature* of τ. For $\omega \in \Omega$, the natural number $\omega\tau$ is called the *arity* of the operator ω. □

Given a finitary type $\tau : \Omega \to \mathbb{N}$, a *$\tau$-algebra* or an *$\Omega$-algebra* (A, τ) or (A, Ω) *of type* τ or *of signature* $\langle \omega\tau | \omega \in \Omega \rangle$ is defined to be a set A equipped with an operation $\omega : A^{\omega\tau} \to A$ corresponding to each operator or element ω of the domain Ω of τ. The type τ is often described by (or identified with) its graph $\{(\omega, \omega\tau) \in \Omega \times \mathbb{N} | \omega \in \Omega\}$. Thus the group $(A, \cdot, {}^{-1}, 1)$ is an algebra of type $\{(\cdot, 2), ({}^{-1}, 1), (1, 0)\}$, while the right loop $(A, \cdot, /, \backslash)$ is an algebra of type $\{(\cdot, 2), (/, 2), (1, 0)\}$ and signature $\langle 2, 2, 0 \rangle$. Note that an abelian group $(A, +, -, 0)$ written additively may be considered as an algebra of type $\{(+, 2), (-, 1), (0, 0)\}$ using the unary operation $-$ of negation as an inversion, or as an algebra of type $\{(+, 2), (-, 2), (0, 0)\}$ using the binary operation $-$ of subtraction as a right division. An algebra of signature $\langle 2 \rangle$, i.e. with a single binary operation, is called a *magma*. (The term "groupoid" is also used in this context, but is better reserved to denote a category in which each morphism is an isomorphism.)

Example 1.1.3. (a) A dynamical system (X, T) is an algebra of type $\{(T, 1)\}$.

(b) For a monoid M, an M-set (X, M) is an algebra of type $M \times \{1\}$.

(c) For a unital ring S, an S-module $(X, +, S)$ is an algebra of type $\{(+, 2)\} \cup (S \times \{1\})$. □

Example 1.1.4. Let $(M, \cdot, 1)$ be a monoid. For elements $(x_1, \ldots, x_n)$ of M^n, define $\prod_{i=1}^n x_i$ inductively by $\prod_{i=1}^0 x_i = 1$ and $\prod_{i=1}^r x_i = (\prod_{i=1}^{r-1} x_i) x_r$ for $r \leq n$. Given a finitary type $\tau : \Omega \to \mathbb{N}$, one then obtains a τ-algebra (M, Ω) with

$$(1.1.1) \qquad \omega : M^{\omega\tau} \to M; \ (x_1, \ldots, x_{\omega\tau}) \mapsto \prod_{i=1}^{\omega\tau} x_i$$

for each ω in Ω. □

Consider a fixed set A. For a natural number n, the full set of all n-ary operations on A is $\underline{\underline{\mathrm{Set}}}(A^n, A)$, described by the type function $\underline{\underline{\mathrm{Set}}}(A^n, A) \to \{n\}$. Then the full set of all finitary operations on A is the disjoint union $\sum_{n \in \mathbb{N}} \underline{\underline{\mathrm{Set}}}(A^n, A)$, described by the coproduct type function T or

$$(1.1.2) \quad T_A = \left(\sum_{n \in \mathbb{N}} (\underline{\underline{\mathrm{Set}}}(A^n, A) \to \{n\}) : \sum_{n \in \mathbb{N}} \underline{\underline{\mathrm{Set}}}(A^n, A) \to \mathbb{N} \right).$$

For a monoid M, an M-set structure on A may be described as (A, M), or more precisely by means of the representation $R: M \to \underline{\underline{\mathrm{Set}}}\ (A^1, A)$ as in (I 1.1). In fact, this representation may be construed as a morphism $R: (M \to \{1\}) \to (\underline{\underline{\mathrm{Set}}}(A^1, A) \to \{1\})$ of the slice category $\underline{\underline{\mathrm{Set}}}/\{1\}$. Then for a finitary type $\tau: \Omega \to \mathbb{N}$, a τ-algebra structure (A, τ) or (A, Ω) on A may be described more precisely by a *representation of* τ *on* A, i.e. by a morphism $R: \tau \to T_A$ of the slice category $\underline{\underline{\mathrm{Set}}}/\mathbb{N}$. This added precision is desirable when a set carries different structures of the same type. For example, consider the type $\tau = \{(\mu, 2)\}$ and a lattice $(L, +, \cdot)$. Then the join semilattice structure (L, τ) of L is defined by the representation $J: \tau \to T_L$ with $\mu^J: L^2 \to L; (x, y) \mapsto x + y$, while the meet semilattice structure (L, τ) has representation $M: \tau \to T_L$ with $\mu^M: L^2 \to L; (x, y) \mapsto x \cdot y$.

Example 1.1.5. Consider the poset $(\mathbb{Z}^+, \leq)$ of positive natural numbers with the usual ordering as a category C. Then the category's codomain function $d_1: C_1 \to C_0; (i \leq n) \mapsto n$ is a finitary type $\perp: C_1 \to \mathbb{N}; (i \leq n) \mapsto n$. Given a set A, one has a representation $\Pi_A: \perp \to T_A$ with $(i \leq n)\Pi_A$ defined as the *n-ary projection onto the i-th factor* $\pi_i^n: A^n \to A; (a_1, \ldots, a_i, \ldots, a_n) \mapsto a_i$. The representation Π_A is called the *projection representation.* The image of Π_A is denoted by $\perp_A$, or more explicitly as the type $\perp_A: \overline{\varnothing} \to \mathbb{N}$ [cf. (1.3.7)(c)]. □

EXERCISES

1.1A. Construe the digital stopwatch of Example I 1.1.2 as an algebra of type $\{R, S, T\} \times \{1\}$.

1.1B. Determine a type for each of the following kinds of algebraic structure: semigroup, semilattice, quasigroup, non-unital ring, unital ring, K-algebra (over a unital commutative ring K), lattice, bounded lattice, Boolean algebra, and Heyting algebra.

1.1C. Let A be a set. Show that (A, π_1^2, π_1^2) is an associative right quasigroup and that (A, π_2^2, π_2^2) is an associative left quasigroup.

1.1D. Let K be a commutative, unital ring, and let n be a positive integer. Define a type $\tau: K[T_1, \ldots, T_n] \to \{n\}$. Show that II (3.8) is a representation yielding a τ-algebra (K, τ).

1.2. The Isomorphism Theorems

Throughout this section, let $\tau: \Omega \to \mathbb{N}$ be a finitary type. Given two algebras (A, Ω) and (B, Ω) of type τ, a function $f: A \to B$ is said to be an *Ω-homomorphism* or a *τ-homomorphism*, or just a *homomorphism*, if

$$a_1 f \ldots a_{\omega\tau} f \omega = a_1 \ldots a_{\omega\tau}\, \omega f \tag{1.2.1}$$

for all ω in Ω and a_i in A. Note that the identity 1_A is an Ω-homomorphism, and that the composite of Ω-homomorphisms is again an Ω-homomorphism. One obtains a large category $\underline{\underline{\tau}}$ such that $\underline{\underline{\tau}}_0$ is the class of all algebras of type τ, and $\underline{\underline{\tau}}_1$ is the class of all Ω-homomorphisms. Note that the forgetful functor $G: \underline{\underline{\tau}} \to \underline{\underline{\text{Set}}}$, with morphism part $(f:(A, \Omega) \to (B, \Omega)) \mapsto (f: A \to B)$, is faithful.

Example 1.2.1. Let τ be the type $\{(\cdot, 2),(1, 0)\}$. Then the concrete category $\underline{\underline{\text{Mon}}}$ of monoids is a full subcategory of $\underline{\underline{\tau}}$. In particular, (1.2.1) reduces to $1f = 1$ for the nullary operation 1. In general, (1.2.1) reduces to $\omega f = \omega$ for a nullary operation $\omega: A^0 \to A;\ \infty \mapsto \omega$ on A and $\omega: B^0 \to B;\ \infty \mapsto \omega$ on B. □

Let B be a subset of an algebra (A, Ω) of type τ.

Definition 1.2.2. (a) Suppose $\forall \omega \in \Omega$,

$$(\forall 1 \leq i \leq \omega\tau,\ b_i \in B) \Rightarrow b_1 \dots b_{\omega\tau}\,\omega \in B.$$

Then the subset B is said to form a *subalgebra* (B, Ω) of (A, Ω)—notation $(B, \Omega) \leq (A, \Omega)$ or $B \leq A$—with $\omega: B^{\omega\tau} \to B;\ (b_1, \dots, b_{\omega\tau}) \mapsto b_1 \dots b_{\omega\tau}\,\omega$ for each ω in Ω. One also says that (A, Ω) is a *superalgebra* of (B, Ω).

(b) Suppose $\forall \omega \in \Omega$,

$$(\forall 1 \leq i \leq \omega\tau,\ b_i \in B) \Leftrightarrow b_1 \dots b_{\omega\tau}\,\omega \in B.$$

Then the subalgebra (B, Ω) of (A, Ω) is said to be a *wall* of (A, Ω).

(c) Suppose $\forall \omega \in \Omega, \forall a_1, \dots, a_{\omega\tau} \in A$,

$$(\exists 1 \leq i \leq \omega\tau.\ a_i \in B) \Rightarrow a_1 \dots a_{\omega\tau}\,\omega \in B.$$

Then B is said to be a *sink* of (A, Ω). Note that a sink is a subalgebra. In particular, a sink contains each (element selected by a) nullary operation of Ω. □

Example 1.2.3. (a) Consider the "arithmetic mean" quasigroup $(\mathbb{R}, \circ)$ of Example I 2.3 as an algebra of type $\{(\circ, 2)\}$. Then the closed unit interval $[0, 1]$ forms a subalgebra $([0, 1], \circ)$. The singletons $\{0\}$ and $\{1\}$ form walls of $([0, 1], \circ)$.

(b) Consider a ring S as an algebra $(S, \cdot\,)$ of type $\{(\cdot, 2)\}$. Then an ideal of S is a sink of $(S, \cdot\,)$. □

Definition 1.2.4. Let (A, Ω) be an algebra of type τ.

(a) The *subalgebra poset* $\mathrm{Sb}(A, \Omega)$ or $\mathrm{Sb}\,A$ of (A, Ω) is the subset of the power set 2^A comprising the subalgebras of (A, Ω), with the partial order induced from $(2^A, \subseteq)$.

(b) The *wall poset* $\mathrm{Wl}(A, \Omega)$ or $\mathrm{Wl}\,A$ of (A, Ω) is the subset of the power set 2^A comprising the walls of (A, Ω), with the partial order induced from $(2^A, \subseteq)$.

(c) The *sink poset* $\mathrm{Sk}(A, \Omega)$ or $\mathrm{Sk}\,A$ of (A, Ω) is the subset of the power set 2^A comprising the sinks of (A, Ω), with the partial order induced from $(2^A, \subseteq)$. □

If A is a set with an empty set $\varnothing$ of operations, then $\mathrm{Sb}(A, \varnothing) = \mathrm{Wl}(A, \varnothing) = \mathrm{Sk}(A, \varnothing) = 2^A$.

In the other direction, let Ψ be a subset of Ω, and let $\sigma : \Psi \to \mathbb{N}$ be the corresponding restriction of τ. Given an algebra (A, Ω) of type τ, one then obtains an algebra (A, Ψ) of type σ. The algebra (A, Ψ) is said to be a (*basic*) *reduct* or *impoverishment* of the algebra (A, Ω). One also says that (A, Ω) is an *augment* or *enrichment* of the algebra (A, Ψ). Note that there is a *restriction* functor $R^\tau_\sigma : \underline{\underline{\tau}} \to \underline{\underline{\sigma}}$, a faithful forgetful functor with morphism part $(f : (A, \Omega) \to (B, \Omega)) \mapsto (f : (A, \Psi) \to (B, \Psi))$. For the empty type $\varnothing : \varnothing \to \mathbb{N}$, the category $\underline{\underline{\varnothing}}$ is just the category $\underline{\underline{\mathrm{Set}}}$. Thus the underlying set functor $U : \underline{\underline{\tau}} \to \underline{\underline{\mathrm{Set}}}$ is an example of impoverishment. Note that reducts of a given algebra may have many more subalgebras than the original algebra. For example, $\mathbb{N}$ is a subalgebra of the reduct $(\mathbb{Z}, +)$ of the group $(\mathbb{Z}, +, -, 0)$, but not a subgroup.

Let (A, Ω) and (B, Ω) be algebras of type τ. The direct product $A \times B$ then becomes an algebra $(A \times B, \Omega)$ of type τ via

$$(1.2.2) \qquad \omega : (A \times B)^{\omega\tau} \to A \times B;\ ((a_1, b_1), \ldots, (a_{\omega\tau}, b_{\omega\tau})) \mapsto (a_1 \ldots a_{\omega\tau}\,\omega, b_1 \ldots b_{\omega\tau}\,\omega)$$

for each ω in Ω. In particular, direct powers A^r of A become algebras of type τ for all positive integers r. Furthermore, $A^0 = T = \{\infty\}$ is also an algebra of type τ, with $\omega = 1_T$ for all ω in Ω. Now consider τ-homomorphisms $f : (C, \Omega) \to (A, \Omega)$ and $g : (C, \Omega) \to (B, \Omega)$. The product map $(f, g) : C \to A \times B$ (cf. Proposition O 3.2.1) then becomes a τ-homomorphism. Indeed, the diagram (O 3.2.4) becomes an instance of the product diagram III (2.3) in the category $\underline{\underline{\tau}}$. For any algebra (A, Ω) of type τ, the diagonal $\hat{A}$ [cf. (O 3.3.8)] forms a subalgebra of A^2. In general, if (A, Ω) and (B, Ω) are τ-algebras, then a relation α between A and B, i.e. a subset α of $A \times B$ [cf. Example III 3.3.2(c)], is called a τ- *relation* (*between A and B*) if it is a subalgebra of $A \times B$. An equivalence relation α on A is a *congruence* (*relation*) or *τ-congruence* if it is a τ-relation. In particular, a $\varnothing$-congruence is just an equivalence relation. A τ-algebra (A, Ω) always has the *improper* τ-congruence A^2 and the *trivial* congruence $\hat{A}$. If it has no proper, non-trivial congruences, then it is said to be *simple*. A simple algebra whose only improper subalgebras are empty or singletons is described as *plain* or *strictly simple*.

Definition 1.2.5. Let (A, Ω) be an algebra of type τ. The *congruence poset* $\mathrm{Cg}(A, \Omega)$ or $\mathrm{Cg}\ A$ of (A, Ω) is the subset of the power set $2^{A \times A}$ comprising the congruence relations of (A, Ω), with the partial order induced from $(2^{A \times A}, \subseteq)$. □

Recall (from Section I 3) that an n-ary relation α on a set A is a subset α of A^n. Such a relation is again called a *τ-relation* (*on* A) if it is a subalgebra of the direct power algebra (A^n, Ω). In particular, the subalgebras of the τ-algebra A are precisely the unary τ-relations.

Proposition 1.2.6. *Let (A, Ω) be an algebra of type τ. Then the posets* $\mathrm{Sb}(A, \Omega)$, $\mathrm{Wl}(A, \Omega)$, $\mathrm{Sk}(A, \Omega)$, *and* $\mathrm{Cg}(A, \Omega)$ *form complete lattices.*

Proof. Let $(L, \subseteq)$ denote any of the four posets under consideration. Then if $J \to L;\ j \mapsto l_j$ is a function, the product $\prod_{j \in J} l_j$ is the intersection $\bigcap_{j \in J} l_j$. □

If X is a subset of a τ-algebra (A, Ω), then the *subalgebra* $\langle X \rangle_{\mathrm{Sb}(A, \Omega)}$, *wall* $\langle X \rangle_{\mathrm{Wl}(A, \Omega)}$ or *sink* $\langle X \rangle_{\mathrm{Sk}(A, \Omega)}$ *generated by* X is the intersection of the respective set of all subalgebras, walls, or sinks of (A, Ω) containing X. For a binary relation β on A, one similarly defines the *congruence* $\langle \beta \rangle_{\mathrm{Cg}(A, \Omega)}$ *generated by* β as the intersection of the set of all congruences of (A, Ω) containing β.

If α is a congruence relation on (A, Ω), then A^α is a τ-algebra with $a_1^\alpha \dots a_{\omega\tau}^\alpha \omega = a_1 \dots a_{\omega\tau} \omega^\alpha$ for $\omega \in \Omega$ and $a_i \in A$. Algebras of the form (A^α, Ω) are called *quotients* of (A, Ω). Note that the natural projection yields a τ-homomorphism

(1.2.3) $$\mathrm{nat}\ \ker \alpha : (A, \Omega) \to (A^\alpha, \Omega);\ a \mapsto a^\alpha.$$

Conversely, one has the following:

First Isomorphism Theorem 1.2.7. *Let $f : (A, \Omega) \to (B, \Omega)$ be a τ-homomorphism. Then*

(1.2.4) $$\ker f \in \mathrm{Cg}(A, \Omega) \text{ and } Af \in \mathrm{Sb}(B, \Omega).$$

One obtains the following commuting diagram in $\underline{\underline{\tau}}$:

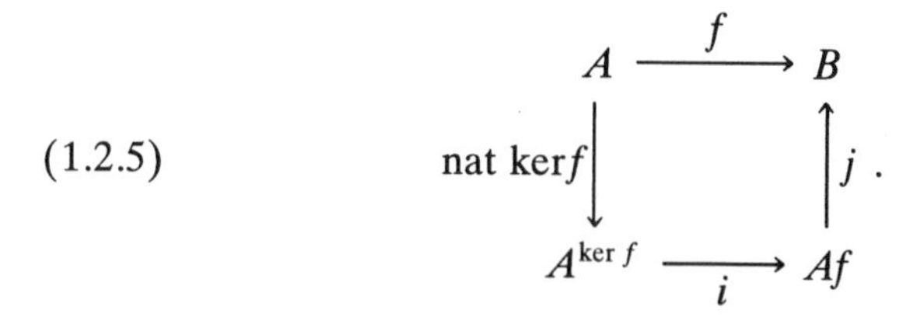

Here i is an isomorphism, and $j : Af \to B$ is the inclusion of the image of f.

Proof. By O 3.3.1, one already has (1.2.5) as a commuting diagram in $\underline{\underline{\text{Set}}}$. Now (1.2.4) is readily verified. The rest of the theorem follows by (1.2.3) and the definition of $(A^{\ker f}, \Omega)$. □

There are two more classical isomorphism theorems to accompany the first. (Although the theorems are classical, their ordering varies.) Their proofs are left as Exercises 1.2E and 1.2F.

Second Isomorphism Theorem 1.2.8. *Let* (B, Ω) *be a subalgebra of a* τ*-algebra* (A, Ω). *Then for* $\alpha \in \mathrm{Cg}(A, \Omega)$, *one has*:

(a) $\cup B^{\alpha} = \{a \in A \mid \exists b \in B.\ a\alpha b\} \in \mathrm{Sb}(A, \Omega)$;

(b) $(\cup B^{\alpha})^{\alpha} \cong B^{(\alpha \cap B^2)}$. □

Let $f:(A, \Omega) \to (B, \Omega)$ be a τ-homomorphism. Define

$$f^* : \mathrm{Cg}\, B \to \mathrm{Cg}\, A;\ \beta \mapsto \ker(f \text{ nat } \beta). \tag{1.2.6}$$

Note that $f^* : (\mathrm{Cg}\, B, \subseteq) \to (\mathrm{Cg}\, A, \subseteq)$ is functorial.

Third Isomorphism Theorem 1.2.9. *Let* α *be a congruence on a* τ*-algebra* (A, Ω). *Then* $(\text{nat } \alpha)^* : \mathrm{Cg}(A^{\alpha}) \to \mathrm{Cg}\, A$ *induces an isomorphism*

$$(\text{nat } \alpha)^* : \mathrm{Cg}(A^{\alpha}) \to [\alpha, A^2];\ \beta \mapsto (\text{nat } \alpha)^*(\beta) \tag{1.2.7}$$

between $\mathrm{Cg}(A^{\alpha})$ *and the interval* $[\alpha, A^2]$ *in the poset* $(\mathrm{Cg}\, A, \subseteq)$. *Moreover, for* $\beta \in \mathrm{Cg}(A^{\alpha})$, *one has*

$$(A^{\alpha})^{\beta} \cong A^{(\text{nat } \alpha)^*(\beta)}. \tag{1.2.8}$$

EXERCISES

1.2A. Let (A, Ω) and (B, Ω) be τ-algebras. Show that a function $f: A \to B$ is a τ-homomorphism iff its graph is a τ-relation between A and B.

1.2B. Let $\alpha \in \mathrm{Cg}(A, \Omega)$.

(a) Show that $a_1^{\alpha} \ldots a_{\omega\tau}^{\alpha}\, \omega = a_1 \ldots a_{\omega\tau}\, \omega^{\alpha}$ for $\omega \in \Omega$ and $a_i \in A$ yields a well-defined τ-algebra (A^{α}, Ω).

(b) Verify that (1.2.3) is a τ-homomorphism.

1.2C. Let (A, Ω) be a τ-algebra. Explain the distinction between id_A and nat $\hat{A}$.

1.2D. Verify (1.2.4).

1.2E. Prove the Second Isomorphism Theorem.

1.2F. Prove the Third Isomorphism Theorem.

1.2G. Let $j:(B,\Omega) \to (A,\Omega)$ be the inclusion of a subalgebra B in A. For $\alpha \in \mathrm{Cg}(A,\Omega)$, prove $j^*(\alpha) = \alpha \cap B^2$.

1.2H. Realize Exercise II 1.4J as a special case of the Third Isomorphism Theorem.

1.2I. (a) Show that a quasigroup $(Q,\cdot,/,\backslash)$ has no proper walls.

(b) Criticize the following argument: Consider the "arithmetic mean" quasigroup of Examples I 2.3 and 1.2.3. The closed unit interval forms a subalgebra. Since this subalgebra is a quasigroup, it has no proper walls.

1.2J. In the notation of Example 1.2.3, determine all the walls of the product $([0,1] \times [0,1], \circ)$. (Hint: Draw a picture!)

1.2K. Determine all the sinks of the semigroup $(\mathbb{Z}_4, \cdot)$.

1.2L. Show that there is a functor $\underline{\underline{\tau}} \to \underline{\underline{\mathrm{Pos}}}^{\mathrm{op}}$ with morphism part $(f: A \to B) \mapsto (f^*: \mathrm{Cg}\, B \to \mathrm{Cg}\, A)$.

1.2M. Let A be a set. Define $t: A^2 \to A^2; (a,b) \mapsto (b,a)$ and $k:(A^2)^2 \to A^2$; $((a,b),\ (c,d)) \mapsto$ **if** $b = c$ **then** (a,d) **else** (c,d). For each element (a,a) of the diagonal $\hat{A}$, consider the nullary operation $(a,a): T \to A^2$; $\infty \mapsto (a,a)$ selecting (a,a). Show that the subalgebra poset $\mathrm{Sb}(A^2,k,t,\hat{A})$ is the poset $\mathrm{Cg}(A,\varnothing)$ of equivalence relations on A.

1.2N. (a) Exhibit a plain quasigroup.

(b) Exhibit a simple quasigroup that is not plain.

1.2O. Let $\tau: \Omega \to \mathbb{N}$ be a type. Given a monoid $(M,\cdot,1)$, let (M,Ω) be the τ-algebra of Example 1.1.4.

(a) Show that there is a functor $\mathrm{Mon}^\tau: \underline{\underline{\mathrm{Mon}}} \to \underline{\underline{\tau}}$ with morphism part $(f:(M,\cdot,1) \to (N,\cdot,1)) \mapsto (f:(M,\Omega) \to (N,\Omega))$.

(b) If $\tau = \{(\mu,2),(\iota,0)\}$, show that Mon^τ is the inclusion functor on $\underline{\underline{\mathrm{Mon}}}$.

(c) If Ω is empty, show that Mon^τ is the forgetful functor $\underline{\underline{\mathrm{Mon}}} \to \underline{\underline{\mathrm{Set}}}$.

1.2P. Let n be a positive integer, and let K be a field. The quotient $\mathrm{GL}_n(K)/K^*$ is called the *projective general linear group* $\mathrm{PGL}_n(K)$ *of dimension* n over K. The subgroup $K^*\mathrm{SL}_n(K)/K^*$ of $\mathrm{PGL}_n(K)$ is called the *projective special linear group* or linear fractional group $\mathrm{PSL}_n(K)$.

(a) Prove $\mathrm{PSL}_n(K) \triangleleft \mathrm{PGL}_n(K)$.

(b) Prove $\mathrm{PSL}_n(K) \cong \mathrm{SL}_n(K)/\{k \in K | k^n = 1\}$.

(c) For a finite field K, prove $|\mathrm{PGL}_n(K)| = |\mathrm{SL}_n(K)|$.

For a prime power q, set $\mathrm{GL}_n(q) = \mathrm{GL}_n(\mathrm{GF}(q))$, etc.

(d) Prove $|\mathrm{GL}_n(q)| = \prod_{i=0}^{n-1}(q^n - q^i)$.

(e) Prove $|\mathrm{PGL}_n(q)| = q^{n-1}\prod_{i=0}^{n-2}(q^n - q^i)$.

(f) Prove $|\mathrm{PSL}_n(q)| = (q^{n-1}/d)\prod_{i=0}^{n-2}(q^n - q^i)$, where d is the greatest common divisor of n and $q - 1$.

1.3. Word Algebras

Throughout this section, let $\tau : \Omega \to \mathbb{N}$ be a finitary type. An automaton with state space S and set E of elementary events is an algebra (S, E) of type $E \times \{1\}$. It is also an E^*-set (S, E^*) for the free monoid E^*, the monoid of words in the alphabet E. The "word algebras" or "absolutely free algebras" constructed in this section are the analogues of E^* for the general finitary type $\tau : \Omega \to \mathbb{N}$.

Let X be a set. Form the free monoid $(X \uplus \Omega)^*$ of words in the alphabet $X \uplus \Omega$. The free monoid becomes a τ-algebra $((X \uplus \Omega)^*, \Omega)$ on defining

$$(1.3.1) \qquad \omega : [(X \uplus \Omega)^*]^{\omega\tau} \to (X \uplus \Omega)^*; \, (a_1, \ldots, a_{\omega\tau}) \mapsto a_1 \ldots a_{\omega\tau}\, \omega$$

for each ω in Ω. The (τ-)*word algebra* or *absolutely free* (τ-)*algebra* $X\Omega$ or $(X\Omega, \Omega)$ *over* X is then defined to be the subalgebra of $((X \uplus \Omega)^*, \Omega)$ generated by X. The elements of $X\Omega$ are called Ω-*words* or τ-*words in* X, or sometimes just *words*. This justifies the term "word algebra."

Example 1.3.1. For the type $\mathrm{E} \times \{1\}$ of the automaton (S, E), the word algebra $XE = \{xw | x \in X, \, w \in E^*\}$. Thus if $T = \{\infty\}$ is a terminal object of $\underline{\underline{\mathrm{Set}}}$, one has the bijection $E^* \to TE; \, w \mapsto \infty w$. $\square$

The name "absolutely free τ-algebra" arises because the injective function

$$(1.3.2) \qquad \eta_X : X \to X\Omega U; \, x \mapsto x,$$

defined identifying X with its image under the insertion into the disjoint union $X \uplus \Omega$, will become the component at X of the unit η in an adjunction $(\Omega, U, \eta, \varepsilon)$ between a functor $\Omega : \underline{\underline{\mathrm{Set}}} \to \underline{\underline{\tau}}$ with object part $X \mapsto X\Omega$ and the underlying set functor $U : \underline{\underline{\tau}} \to \underline{\underline{\mathrm{Set}}}$. Indeed, by Theorem III 3.1.4, it suffices to show that (1.3.2) is an initial object in the comma category (X, U). Consider an object $h : X \to AU$ of this comma category. Identifying functions with their graphs, one has a subset h of the product algebra $X\Omega \times A$. Let f be the subalgebra of $X\Omega \times A$ generated by h. Then the τ-relation f is a τ-homomorphism $f : X\Omega \to A$ (cf. Exercise 1.2A), and one has the commuting diagram

$$(1.3.3) \qquad \begin{array}{ccc} X & \xrightarrow{\eta_X} & X\Omega U \\ \| & & \downarrow{\scriptstyle fU} \\ X & \xrightarrow[h]{} & AU \end{array}$$

in Set. Now f is the unique τ-homomorphism yielding (1.3.3), so that η_X really is initial in (X, U), and one obtains the desired adjunction $(\Omega, U, \eta, \varepsilon)$. Note that the homomorphism $f: X\Omega \to A$ may then be identified as the composite $h^{\Omega}\varepsilon_A$.

Consider the power set monoid $(\mathscr{P}(X), \cup, \varnothing)$. As in Example 1.1.4, it may be realized as a τ-algebra $(\mathscr{P}(X), \Omega)$ with $Y_1 \dots Y_{\omega\tau}\omega = \bigcup_{i=1}^{\omega\tau} Y_i$ for $\omega \in \Omega$ and $Y_i \subseteq X$. There is a function $s: X \to \mathscr{P}(X);\ x \mapsto \{x\}$ embedding X in $\mathscr{P}(X)$ as the set of singletons. The corresponding homomorphism $s^{\Omega}\varepsilon_{\mathscr{P}(X)}: X\Omega \to \mathscr{P}(X)$ is called the *argument map* $\arg: X\Omega \to \mathscr{P}(X);\ w \mapsto \arg(w)$. Elements of the set $\arg(w)$ are called *arguments* of the word w. For example, one has $\arg(x_1 \dots x_{\omega\tau}\omega) = \{x_1, \dots, x_{\omega\tau}\}$, and of course $\arg(x) = \{x\}$ for x in X. Note that the image of the argument map is contained in the subset $\mathscr{P}_{<\omega}(X)$ of $\mathscr{P}(X)$.

From now on, it will often be convenient to write

$$P = \{x_1 < x_2 < \cdots\} \tag{1.3.4}$$

for the ordered set of positive integers. (There are typographic advantages to writing x_n instead of n.) Consider the function $\max \circ \arg: P\Omega \to \mathbb{N}$ selecting the maximum of the (finite) argument set of an Ω-word in P. (Note that one reverts to the usual symbols $0, 1, 2, \dots$ for the elements of $\mathbb{N}$.) Since the maximum of the empty set of natural numbers is 0, one has $\max \circ \arg\ \omega = 0$ for a nullary operation ω. Now the codomain $\mathbb{N}$ of the function $\max \circ \arg$ carries the usual order relation $\leq$. In general, for a function $f: A \to B$ whose codomain is an ordered set $(B, \leq)$, the *epigraph* of f is the set

$$\operatorname{epi} f = \{(a, b) \in A \times B \mid af \leq b\}. \tag{1.3.5}$$

Define

$$\overline{\Omega} := \operatorname{epi} \max \circ \arg \tag{1.3.6}$$

and $\overline{\tau}: \overline{\Omega} \to \mathbb{N};\ (w, n) \mapsto n$. Note that there is a (Set/$\mathbb{N}$)-morphism $j: (\tau: \Omega \to \mathbb{N}) \to (\overline{\tau}: \overline{\Omega} \to \mathbb{N})$ with $\omega j = (x_1 \dots x_{\omega\tau}\omega, \omega\tau)$ for $\omega \in \Omega$. Since j injects, it is often convenient to identify Ω with its image Ωj in $\overline{\Omega}$. One also identifies $P\Omega$ with the graph of $\max \circ \arg$, a subset of $\overline{\Omega}$, and denotes the restriction $\max \circ \arg$ of $\overline{\tau}$ to $P\Omega$ by τ'. The original type $\tau: \Omega \to \mathbb{N}$ thus yields three types:

$$\begin{cases} \text{(a)} & \tau: \Omega \to \mathbb{N};\ \omega \mapsto \omega\tau; \\ \text{(b)} & \tau': P\Omega \to \mathbb{N};\ w \mapsto \max \circ \arg w; \\ \text{(c)} & \overline{\tau}: \overline{\Omega} \to \mathbb{N};\ (w, n) \mapsto n. \end{cases} \tag{1.3.7}$$

The type $\tau': P\Omega \to \mathbb{N}$ is called the *type derived* from $\tau: \Omega \to \mathbb{N}$. The type $\overline{\tau}: \overline{\Omega} \to \mathbb{N}$ is called the *closure* of the type τ. The words in $P\Omega$ are called *derived operators* or *terms*, while the elements of Ω are called *basic operators*. The set $\overline{\Omega}$ will be described as a *closed operator domain*, and specifically as the *closure* of the operator domain Ω.

Example 1.3.2. For each operator domain Ω, the closure $\overline{\Omega}$ contains the elements (x_i, n) for $1 \leq i \leq n$. In parallel with the notation of Example 1.1.5, the element (x_i, n) is called the *n-ary projection operator* π_i^n *onto the* i-th *factor*. □

The following simple result is worth noting.

Proposition 1.3.3. *There are nullary operators in* $\overline{\Omega}$ *iff there are nullary operators in* Ω. *Indeed,* $\tau^{-1}\{0\} = \bar{\tau}^{-1}\{0\}$. *In particular,* $\tau^{-1}\{0\} = \tau'^{-1}\{0\}$.

Proof. One has $(w, 0) \in \bar{\tau}^{-1}\{0\} \Leftrightarrow (w, 0) \in \overline{\Omega} \Leftrightarrow \max \circ \arg w \leq 0 \Leftrightarrow \max \circ \arg w = 0 \Leftrightarrow \arg w = \varnothing \Leftrightarrow w \in \tau^{-1}\{0\}$. □

Now consider an algebra (A, Ω) of type τ. Consider an element (w, n) of $\overline{\Omega}$. Set $X = \{x_1, \ldots, x_n\}$. Note $\arg(w) \subseteq X$, so $w \in X\Omega U$. Realize the direct power A^n as $\underline{\underline{\mathrm{Set}}}(X, A)$: The element a of $\underline{\underline{\mathrm{Set}}}\ (X, A)$ corresponds to $(x_1^a, \ldots, x_n^a)$ in the Cartesian power A^n. Now the derived operator w is an element of $X\Omega U$. Then using the counit from the adjunction $(\Omega, U, \eta, \varepsilon)$, one defines the n-ary operation $(w, n)_A$ or (w, n) on A by

$$(1.3.8) \qquad x_1^a \ldots x_n^a(w, n) = wa^{\Omega U}\varepsilon_{AU}.$$

This makes the algebra (A, Ω) of type τ into an algebra $(A, \overline{\Omega})$ of type $\bar{\tau}$ or an algebra $(A, P\Omega)$ of type τ'. Now $(A, P\Omega)R_\tau^{\tau'} = (A, \Omega)$, while $(A, \overline{\Omega})R_{\tau'}^{\bar{\tau}} = (A, P\Omega)$ and $(A, \overline{\Omega})R_\tau^{\bar{\tau}} = (A, \Omega)$. The operations of $P\Omega$ on A are called the *derived operations* or *term operations* of the τ-algebra (A, Ω). A basic reduct of $(A, \overline{\Omega})$ is called a *reduct* of (A, Ω). The set $\overline{\Omega}$ of operations on A is called the *closed set of operations* or *the clone* of the τ-algebra (A, Ω), and the *closure* of the set Ω of operations on A. The operations from $\overline{\Omega}$ are called the *clone operations* of the τ-algebra (A, Ω). The set Ω of operations on A is said to be *closed* or *a clone* if $\Omega = \overline{\Omega}$.

Example 1.3.4. (a) **(Opposite Monoids.)** Consider a monoid $(M, \cdot, 1)$ as an algebra (M, Ω) of type $(\tau : \Omega \to \mathbb{N}) = \{(\mu, 2), (\iota, 0)\}$, with $\iota : T \to M; \{\infty\} \mapsto 1$ and $\mu : M^2 \to M;\ (m, n) \mapsto mn$. Then the binary derived operation $x_2 x_1 \mu_M$ on M is the multiplication $x_2 x_1 \mu : M^2 \to M;\ (m, n) \mapsto nm = m \,\breve{\circ}\, n$ of the opposite of the monoid $(M, \cdot, 1)$ {cf. I (1.4)}.

(b) Consider a quasigroup $(Q, \cdot, /, \backslash)$ as an algebra of type $\tau = \{\beta, \delta, \gamma\} \times \{2\}$ with $x_1 x_2 \beta = x_1 x_2$, $x_1 x_2 \delta = x_1 / x_2$, and $x_1 x_2 \gamma = x_1 \backslash x_2$. Then the ternary operation P of (I 2.2.5) is the derived operation $x_1 x_2 x_2 \gamma \delta x_2 x_3 \delta \beta$.

(c) **(Convex Sets.)** A real vector space V may be construed as an algebra $(V, +, \mathbb{R})$ of type $(\tau : \Omega \to \mathbb{N}) = \{(+, 2)\} \cup (\mathbb{R} \times \{1\})$. Using infix notation for addition and postfix notation for scalars, denote the subset $\{x_1(1-r) + x_2 r | r \in (0,1)\}$ of $P\Omega$ by I°. [The notation originates on writing I for the closed unit interval $[0,1]$ and I° for its interior $(0,1)$, the open unit interval.] Consider the reduct (V, I°) of $(V, +, \mathbb{R})$. It is convenient to write the derived operations from I° in the form $\underline{r} : V^2 \to V;\ (x, y) \mapsto xy\underline{r} = x(1-r) + yr$. (Note that the quasigroup multiplication $\circ$ on $\mathbb{R}$ in Example 1.2.3 is the derived operation $\underline{1/2}$.) Then the convex subsets of the vector space V are precisely the subalgebras of the reduct (V, I°) of $(V, +, \mathbb{R})$. □

One readily verifies the following (cf. Exercise 1.3J):

Proposition 1.3.5. *Let* (A, Ω) *be an algebra of type* $\tau : \Omega \to \mathbb{N}$. *Then* $\mathrm{Sb}(A, \Omega) = \mathrm{Sb}(A, \overline{\Omega})$. □

On the other hand, consider the type $(\tau : \Omega \to \mathbb{N}) = \{(1/2, 2)\}$, and the subalgebra $([0,1], \Omega)$ of the algebra $(\mathbb{R}, \circ)$ of Examples I 2.3, 1.2.3(a), and 1.3.4(c). Then $\{0\}$ is a wall of $([0,1], \Omega)$, but not of $([0,1], \overline{\Omega})$, since $0 = 01\pi_1^2 \in \{0\}$, but $1 \notin \{0\}$. Thus $\mathrm{Wl}\,([0,1], \Omega) \neq \mathrm{Wl}\,([0,1], \overline{\Omega})$.

One of the themes of this chapter is that monoid actions provide a prototype for universal algebra. In the other direction, each algebra of a given type supports two monoid actions.

Proposition 1.3.6. *Let* (A, Ω) *be an algebra of type* τ. *Then the set* $\overline{\tau}^{-1}\{1\}$ *of unary clone operations and* $\tau'^{-1}\{1\}$ *of unary derived operations form submonoids of* $(A^A, \cdot, 1_A)$.

Proof. The identity 1_A is the unary derived operation $(x_1, 1)$. Now the unary clone operations are either of the form $(a, 1)$ for a nullary derived operation a selecting the element a of A, or else of the form $(x_1 u, 1)$ for a unary derived operation $x_1 u$. Taking a further nullary derived operation b and a further unary derived operation v, one has the following composition table:

$\cdot$	$(b,1)$	$(x_1 v, 1)$
$(a,1)$	$(b,1)$	$(av, 1)$
$(x_1 u, 1)$	$(b,1)$	$(x_1 uv, 1)$

Since av is again a nullary derived operation, and $x_1 uv$ a unary derived operation, the unary clone operations form a subsemigroup of A^A, and in particular the unary derived operations form a subsemigroup of A^A. □

Example 1.3.7. Consider a finite quasigroup Q as an algebra (Q, Ω) of type $(\tau : \Omega \to \mathbb{N}) = \{(\beta, 2)\} \cup (Q \times \{0\})$ with multiplication β. Then $\tau'^{-1}\{1\}$ is the multiplication group $\mathrm{Mlt}(Q, \beta)$. For example, let q_1 and q_2 be elements of Q. Then $R(q_1) = x_1 q_1 \beta$, $R(q_1)R(q_2) = x_1 q_1 \beta q_2 \beta$, $R(q_1)L(q_2) = q_2 x_1 q_1 \beta\beta$, etc. If $R(q_1)^{-1} = R(q_1)^r$, then $R(q_1)^{-1} = x_1 q_1 \beta q_1 \beta \dots q_1 \beta$, a word of length $2r+1$ in $P\Omega$. □

The device used in Example 1.3.7—adjoining a basic nullary operation $a : A^0 \to A; \infty \mapsto a$ corresponding to each element of the underlying set A of an algebra—is often useful. Given an algebra (A, Ω) of type $\tau : \Omega \to \mathbb{N}$, consider the corresponding algebra $(A, A \cup \Omega)$ of type

$$(1.3.9) \qquad (\tau_A : A \cup \Omega \to \mathbb{N}) = (A \to \{0\}) \cup (\tau : \Omega \to \mathbb{N}).$$

Then the operations of the clone $\overline{A \cup \Omega}$ are called *polynomial operations* of the algebra (A, Ω). For example, realize a commutative, unital ring K as an algebra $(K, +, \cdot)$ of signature $\langle 2, 2 \rangle$. Then for a polynomial $p(T) \in K[T]$, the K-algebra homomorphism II (3.5) yields a unary polynomial operation $K \to K;\ k \mapsto p(k)$ of $(K, +, \cdot)$.

Definition 1.3.8. Let (A, Ω) and (A, Ψ) be two algebra structures on a set A.

(a) Then (A, Ω) and (A, Ψ) are *(clonally) equivalent* if $(A, \overline{\Omega}) = (A, \overline{\Psi})$.

(b) Then (A, Ω) and (A, Ψ) are *(polynomially) equivalent* if $(A, \overline{A \cup \Omega}) = (A, \overline{A \cup \Psi})$. □

Example 1.3.9. (a) Let R be a Boolean ring, and let RF be the corresponding Boolean algebra (cf. Proposition III 3.2.7). Construe R as an algebra $(R, +, 0, \cdot, 1)$ of type $(\tau : \Omega \to \mathbb{N}) = (\{0,1\} \times \{0\}) \cup (\{+, \cdot\} \times 2)$. Construe RF as an algebra $(R, \cup, \cap, ', 0, 1)$ of type $(\sigma : \Psi \to \mathbb{N}) = (\{0,1\} \times \{0\}) \cup \{(', 1)\} \cup (\{\cup, \cap\} \times \{2\})$. Then (R, Ω) and (R, Ψ) are clonally equivalent.

(b) Let Q be a group. Construe Q as an algebra $(Q, \cdot, {}^{-1}, 1)$ of type $(\tau : \Omega \to \mathbb{N}) = \{(\cdot, 2), ({}^{-1}, 1), (1, 0)\}$. Again, construe Q as a quasigroup $(Q, \beta, \delta, \gamma)$ of type $(\sigma : \Psi \to \mathbb{N}) = \{\beta, \gamma, \delta\} \times \{2\}$ [cf. Example 1.3.4(b)]. Then (Q, Ω) and (Q, Ψ) are polynomially equivalent, but not clonally equivalent, since $\bar{\tau}^{-1}\{0\} \neq \emptyset = \bar{\sigma}^{-1}\{0\}$. □

EXERCISES

1.3A. (a) If X and Ω are finite, show that $X\Omega$ is countable.

(b) If X and Ω are countable, show that $X\Omega$ is countable.

1.3B. Let C be a convex subset of a real vector space V. Show that a real-valued function $f : C \to \mathbb{R}$ is convex iff its epigraph is a convex subset of $C \times \mathbb{R}$.

1.3C. Let $f : L \to M$ be a function between lattices. Show that f is functorial iff the epigraph of f is a meet subsemilattice of $L \times M$.

1.3D. Consider a quasigroup Q as an algebra (Q, Ω) of type $(\tau' : \Omega \to \mathbb{N}) = (\{\beta, \delta, \gamma\} \times \{2\}) \cup (Q \times \{0\})$, extending the type from Example 1.3.4(b).

(a) Exhibit a quasigroup Q with $\text{Mlt}\, Q = \tau'^{-1}\{1\}$.

(b) Exhibit a quasigroup Q for which $\text{Mlt}\, Q$ is a proper subgroup of $\tau'^{-1}\{1\}$.

1.3E. Let (A, Ω) be a τ-algebra with both A and Ω countable. Prove that the number of polynomial operations of (A, Ω) is countable.

1.3F. Construe the set S of Exercise III 1.1C as the set of polynomial operations of an algebra $(\mathbb{N}, \Omega)$ with a finite operator domain Ω. Then complete Exercise III 1.1C if you have not already done so.

1.3G. Let (A, Ω) be a τ-algebra. Show that the clone operation on A determined by the n-ary projection operator π_i^n of Example 1.3.2 is the n-ary projection operation π_i^n of Example 1.1.5.

1.3H. Let $(A, \varnothing)$ be a set equipped with the empty set of operations. Prove that the clone of $(A, \varnothing)$ is the set $\overline{\varnothing} = \{\pi_i^n : A^n \to A | 1 \leq i \leq n \in \mathbb{N}\}$ of projections of Example 1.1.5.

1.3I. Let $\tau : \Omega \to \mathbb{N}$ be a type.

(a) Show that the poset category $(\mathbb{N}, \leq)$ has finite coproducts.

(b) Realize $\mathbb{N}$ as a bounded join semilattice $(\mathbb{N}, \max, 0)$.

(c) Realize the monoid $(\mathbb{N}, \max, 0)$ as a τ-algebra $(\mathbb{N}, \Omega)$ according to Example 1.1.4.

(d) Show that the (graph of the) type $\bar{\tau}$, the closure of τ, is a subalgebra $(\bar{\tau}, \Omega)$ of $((P \cup \Omega) \times \mathbb{N}, \Omega)$.

(e) Show that the (graph of the) type τ', the type derived from $\bar{\tau}$, is a subalgebra (τ', Ω) of $(\bar{\tau}, \Omega)$.

1.3J. Prove Proposition 1.3.5.

1.3K. Following Example 1.1.1, show how to interpret derived operations as circuits composed of components represented by basic operations.

1.3L. Let $(R, +, 0, \cdot, 1)$ be a Boolean ring, with corresponding Boolean algebra $(R, \cup, \cap, ', 0, 1)$.

(a) Show that $(R, +, \cdot, 1)$ and $(R, \cup, \cap, ', 1)$ are clonally equivalent.

(b) Show that $(R, +, \cdot, 1)$ and $(R, \cup, \cap, 1)$ are not clonally equivalent.

(c) Show that $(R, +, \cdot, 1)$ and $(R, \cup, \cap, ')$ are polynomially equivalent.

1.4. Universal Geometry

Let A be a set. Recall that the full set $\Sigma_{m \in \mathbb{N}} \underline{\underline{\text{Set}}}(A^m, A)$ of all finitary operations on A is the domain of the type function T or T_A of (1.1.2). It is convenient to identify this function with its graph. Now for a natural number

l, consider the power set $\mathscr{P}(A^l)$, the set of all l-ary relations on A (cf. Section I 3). In order to distinguish from the concept of arity for operations, these relations are sometimes described as having *length* l. One thus has the length function $\mathscr{P}(A^l) \to \{l\}$. Then the full set of all relations on A is the disjoint union $\Sigma_{l \in \mathbb{N}} \mathscr{P}(A^l)$, described by the coproduct length function L or

$$(1.4.1) \qquad L_A = \left(\sum_{l \in \mathbb{N}} (\mathscr{P}(A^l) \to \{l\}) : \sum_{l \in \mathbb{N}} \mathscr{P}(A^l) \to \mathbb{N} \right).$$

It is also convenient to identify this function with its graph. Thus an element of the set L_A is a pair (X, l), where $X \subseteq A^l$. A relation ρ of *preservation* is defined between T_A and L_A, namely

$$(1.4.2) \qquad (\omega, m) \rho (X, l) \Leftrightarrow X \leq (A^l, \omega).$$

Thus the operation ω *preserves* the relation X, or equivalently X is a relation *invariant* under ω, iff the subset X of A^l is actually a subalgebra of the direct power algebra (A^l, ω). In the terminology of Section 1.2, the relation X is preserved by ω iff X is an $\{(\omega, m)\}$-relation.

As in Example III 3.3.2(c), the preservation relation between T_A and L_A yields a polarity $(S, R, \eta, \varepsilon)$ between the poset categories $(\mathscr{P}(T_A), \subseteq)$ and $(\mathscr{P}(L_A), \supseteq)$. Thus for a type $\tau : \Omega \to \mathbb{N}$ of operations on A, the domain of τS is the set of relations preserved by each operation from Ω. For a restriction $\lambda : \Gamma \to \mathbb{N}$ of the length function L_A, the type λR is the largest type for which each element of Γ is a (λR)-relation. Consider an element (X, l) of (the graph of the restriction) λ. Then $X \in \mathrm{Sb}(A^l, \lambda R) = \mathrm{Sb}\,(A^l, \overline{\lambda R})$, by Proposition 1.3.5. Thus the types λR of operations on A that are closed in the polarity $(S, R, \eta, \varepsilon)$ actually form clones or closed types in the sense of Section 1.3. Now the poset categories $(\mathscr{P}(T_A), \subseteq)$ and $(\mathscr{P}(L_A), \supseteq)$ are complete lattices. As in Example III 3.3.3(c), the induced posets $(\mathscr{P}(L_A)^R, \subseteq)$ and $(\mathscr{P}(T_A)^S, \supseteq)$ are complete lattices, isomorphic via the Galois correspondence. Note that for the type $\perp_A$ of Example 1.1.5, one has $\perp_A = L_A^R$. Thus the type $\perp_A$ is the lower bound of the complete lattice $(\mathscr{P}(L_A)^R, \subseteq)$.

Example 1.4.1 (The Euclidean Plane). Let $A = \mathbb{R}_1^2$. Let $\mathrm{O}_2(\mathbb{R}) = \{f \in \mathrm{GL}_2(\mathbb{R}) \mid ff^T = 1\}$. Let $R : \mathrm{O}_2(\mathbb{R}) \to \mathrm{Aut}\,\mathbb{R}_1^2$ be the representation making $\mathbb{R}_1^2$ an $\mathrm{O}_2(\mathbb{R})$-module, with action given by the matrix product $\mathbb{R}_1^2 \times \mathrm{O}_2(\mathbb{R}) \to \mathbb{R}_1^2$ (II 1.3.4). Define the *group* E_2 *of Euclidean motions* as the split extension of $(\mathbb{R}_1^2, +)$ by $\mathrm{O}_2(\mathbb{R})$ via R. Thus E_2 is a subgroup of the general affine group $\mathrm{GA}_2(\mathbb{R})$ (cf. Example III 2.4.6). Define

$$(1.4.3) \quad \Gamma = \left\{ \{(a, b) \in A^2 \mid (a - b)(a - b)^T = d\} \;\middle|\; 0 \leq d \in \mathbb{R} \right\}.$$

Thus the elements of Γ are the binary relations on A consisting of all pairs

(a, b) of points a fixed distance d apart. Define $\lambda = \Gamma \times \{2\}$. The unary operations in λR are precisely the elements of E_2. □

Example 1.4.2 (Convex Sets). Let $A = \mathbb{R}$. Define

$$\Omega = \{A^2 \to A;\ (a, b) \mapsto (1 - r)a + rb \mid 0 < r < 1\}.$$

Thus the elements of Ω are the binary operations of weighted mean (e.g. arithmetic mean for $r = 1/2$). Define $\tau = \Omega \times \{2\}$. Then the domain of τS is precisely the set of convex subsets of the finite-dimensional real vector spaces $\mathbb{R}^m$. □

Klein's famous 1872 "Erlanger Programm" set out to characterize geometry as the study of the relations on a set A invariant under the action of a group G on A. Example 1.4.1 illustrates how the most classical of all geometries, the Euclidean plane, fits in to this programme. Euclidean motions are precisely the *rigid motions*—those preserving the distance relations (1.4.3)—and Euclidean geometry concerns itself with those concepts, such as cosines of angles and areas of triangles, that are invariant under the group E_2 of Euclidean motions. From the current perspective, it now appears that Klein's view of geometry is excessively limited by its restriction to groups, i.e. to bijective unary operations on the set A. The polarity $(S, R, \eta, \varepsilon)$ determined by the preservation relation (1.4.2) offers a broader view of geometry: the study of the relations on a set A invariant under an arbitrary type $\tau : \Omega \to \mathbb{N}$ of operations on A, not necessarily unary and not necessarily bijective. Geometry characterized in this way as part of universal algebra may be described as *universal geometry*. Example 1.4.2 illustrates how convex geometry fits in to this framework of universal geometry.

Certain relations on the set A are invariant under each type $\tau : \Omega \to \mathbb{N}$ of operations on A. To describe them, it is convenient to realize each direct power A^l as $\underline{\underline{\mathrm{Set}}}(\mathbb{Z}_l, A)$, the element a of $\underline{\underline{\mathrm{Set}}}(\mathbb{Z}_l, A)$ corresponding to $(1a, \dots, la)$ in the Cartesian power A^l. A function $f : \mathbb{Z}_k \to \mathbb{Z}_l$ induces a function $f^* : \underline{\underline{\mathrm{Set}}}(\mathbb{Z}_l, A) \to \underline{\underline{\mathrm{Set}}}(\mathbb{Z}_k, A);\ a \mapsto fa$, and the corresponding direct image function

$$f^* : \mathcal{P}(A^l) \to \mathcal{P}(A^k);\ \gamma \mapsto f^*\gamma = \{fa \mid a \in \gamma\}.$$

An l-ary relation δ is said to be *diagonal* if it is of the form f^*A^l for some function $f : \mathbb{Z}_k \to \mathbb{Z}_l$. For example, the binary equality relation $\hat{A} = \{(a, a) \mid a \in A\}$ is diagonal in this sense, being f^*A^1 for $f : \mathbb{Z}_2 \to \mathbb{Z}_1;\ 1 \mapsto 1, 2 \mapsto 1$. The quaternary relation $\hat{A} \times \hat{A}$ is diagonal, being f^*A^2 for $f : \mathbb{Z}_4 \to \mathbb{Z}_2;\ 1 \mapsto 1, 2 \mapsto 1, 3 \mapsto 2, 4 \mapsto 2$. Let Δ or Δ_A denote the full set of all diagonal relations on A, of all possible lengths. Let $D_A : \Delta_A \to \mathbb{N}$ be the restriction of L_A to Δ_A.

Note that, if an operation ω preserves an l-ary relation X, then it also preserves each relation f^*X for $f : \mathbb{Z}_k \to \mathbb{Z}_l$. Since each operation preserves each power A^l, it follows that the terminal object of the complete lattice $(\mathscr{P}(T_A)^S, \subseteq)$ is D_A. A type $\tau : \Omega \to \mathbb{N}$ of operations on A, and the algebra (A, Ω), are said to be *functionally complete* if $\overline{\tau} = T_A$, or equivalently if $\tau S = D_A$ (Exercise 1.4B).

Example 1.4.3. Let $(A, +, \cdot, 0, 1)$ be a finite bounded lattice such that $(A, \leq)$ is a chain. Consider the type $(\tau : \Omega \to \mathbb{N}) = (\{+, \cdot\} \times \{2\}) \cup (A \times \{0\}) \cup (A \times \{1\})$. Construe A as a τ-algebra, in which the nullary operation $(a, 0)_A$ selects the element a of A, and where the unary operation $(a, 1)_A$ acts as

$$(1.4.4) \qquad (a,1)_A : A \to A;\ x \mapsto x^a = (\textbf{if } x = a \textbf{ then } 1 \textbf{ else } 0),$$

the "truth value" $[x = a]$. An arbitrary n-ary operation $f : (x_1, \dots, x_{n-1}, x_n) \mapsto x_1 \dots x_{n-1} x_n f$ on A may then be expressed via

$$(1.4.5) \qquad x_1 \dots x_{n-1} x_n f = \sum_{a \in A} x_1 \dots x_{n-1} a f \cdot x_n^a$$

in terms of Ω and $(n-1)$-ary operations. Since the full set $\underline{\underline{\mathrm{Set}}}(A^0, A)$ of nullary operations on A is contained in Ω, one thus has $f \in \overline{\Omega}$ by induction on n, so that the algebra (A, Ω) is functionally complete. □

An algebra (A, Ω) is said to be *polynomially complete* if A with the clone $\overline{A \cup \Omega}$ of polynomial operations is functionally complete. In other words, for each natural number m, each element f of $\underline{\underline{\mathrm{Set}}}(A^m, A)$ can be realized as a polynomial operation of the algebra (A, Ω).

Example 1.4.4. Let $(K, +, \cdot, 0, 1)$ be a finite field. For each element $(x_1, \dots, x_m)$ of K^m, define the *delta polynomial*

$$(1.4.6) \qquad \delta_{(x_1, \dots, x_m)}(T_1, \dots, T_m) = \prod_{j=1}^{m} \prod_{x_j \neq k \in K} \frac{T_j - k}{x_j - k}$$

in $K[T_1, \dots, T_m]$ (cf. II (3.6)). For an element f of $\underline{\underline{\mathrm{Set}}}(K^m, K)$, define the *Lagrange interpolant polynomial*

$$(1.4.7) \qquad \sum_{(x_1, \dots, x_m) \in K^m} (x_1 \dots x_m\, f)\, \delta_{(x_1, \dots, x_m)}(T_1, \dots, T_m).$$

Since the image of (1.4.7) under the K-algebra homomorphism II (3.8) is f, it follows that the finite field K is polynomially complete. □

EXERCISES

1.4A. For a type τ, prove $\tau S = \bar{\tau} S$.

1.4B. For a type τ, prove $\bar{\tau} = \mathrm{T_A} \Leftrightarrow \tau\mathrm{S} = \mathrm{D_A}$.

1.4C. Let G be a permutation group on a finite set A. Let τ be the type $G \times \{1\}$. Determine the set of unary operations of the type τSR.

1.4D. Show that a 2-element Boolean algebra is functionally complete.

1.4E. Let e be a primitive element of the field $\mathbb{Z}_p$ of prime order p. Express the exponential function $\exp : \mathbb{Z}_p \to \mathbb{Z}_p;\ x \mapsto e^x$ [cf. (II 3.5.1)] as a unary polynomial operation of the ring $\mathbb{Z}_p$.

1.5. Clones and Relations

The universal geometry described in the preceding section studies the polarity between finitary relations and finitary operations on a set A. Although each set of operations on A that is closed in the polarity is also a closed set of operations or clone on A, the converse need not hold. There are clones on infinite sets A that are not describable as the exact set of finitary operations on A preserving a given set of finitary relations on A.

Example 1.5.1. Let $(K, +, \cdot, 0, 1)$ be an infinite field. Let $\tau : \Omega \to \mathbb{N}$ be the type of the polynomial operations on K. Thus Ω is a clone on K. Let γ be a non-diagonal, l-ary relation on K. It will be shown that γ is not preserved by Ω. Since $(\gamma, l) \notin D_K$, there is an m-ary operation $f : K^m \to K$ on K that does not preserve γ. Thus there are elements $(a_{1i}, \dots, a_{li})$ of γ, for $1 \le i \le m$, with $(a_{11} \dots a_{1m} f, \dots, a_{l1} \dots a_{lm} f) \notin \gamma$. For $1 \le i \le m$, define the *delta polynomial*

$$\delta_{(a_{i1}, \dots, a_{im})}(T_1, \dots, T_m) = \prod_{j=1}^{m} \prod_{a_{ij} \neq k \in \{a_{hj} \mid 1 \le h \le l\}} \frac{T_j - k}{a_{ij} - k}$$

[cf. (1.4.6), II (3.6)]. Note that $(x_1, \dots, x_m) \in \{(a_{h1}, \dots, a_{hm}) \mid 1 \le h \le l\}$ implies $\delta_{(a_{i1}, \dots, a_{im})}(x_1, \dots, x_m) = \textbf{if } (x_1, \dots, x_m) = (a_{i1}, \dots, a_{im}) \textbf{ then } 1 \textbf{ else } 0$. Now consider the polynomial

$$p(T_1, \dots, T_m) = \sum_{(x_1, \dots, x_m) \in \{(a_{h1}, \dots, a_{hm}) \mid 1 \le h \le l\}} (x_1 \dots x_m f)\, \delta_{(x_1, \dots, x_m)}(T_1, \dots, T_m).$$

Note that $p(a_{il}, \dots, a_{im}) = a_{il} \dots a_{im}\, f$ for $1 \le i \le l$. Thus γ is not preserved by the element p of Ω. It follows that $\tau S = D_K$ and $\tau\, SR = T_K \neq \tau$. In other words, the clone of polynomial operations on K is not closed in the polarity of Section 1.4 between finitary relations and finitary operations. □

The main goal of this section, Theorem 1.5.2 below, is to show that each clone on a set A is realized as the set of operations preserving a single relation. However, two adjustments to the geometrical framework of the preceding section are needed. Firstly, consider an n-ary operation $f: A^n \to A$ on A. For an element i of $\{1, \ldots, n\}$, the i-th argument of f is said to be *fictitious* if $\forall (a_1, \ldots, a_{i-1}, a_{i+1}, \ldots, a_n) \in A^{n-1}$, $\forall a \in A$, $\forall b \in A$, $a_1 \ldots a_{i-1} a a_{i+1} \ldots a_n f = a_1 \ldots a_{i-1} b a_{i+1} \ldots a_n f$. One then identifies two operations f, g on A that can be obtained from one another by the addition or removal of fictitious arguments. For instance, if a_0 is a fixed element of A, then one does not distinguish between the nullary operation $A^0 \to A$; $\infty \mapsto a_0$ and the unary operation $A \to A; a \mapsto a_0$. The second adjustment is to admit "infinitary" relations, i.e. subsets of an infinite power of A. Note that if I is a possibly infinite index set, and A is an algebra of type $\tau : \Omega \to \mathbb{N}$, then the power A^I, realized as the set $\underline{\underline{\mathrm{Set}}}(I, A)$ of functions, inherits the *componentwise* operations

(1.5.1)
$$\omega : \underline{\underline{\mathrm{Set}}}(I, A)^{\omega\tau} \to \underline{\underline{\mathrm{Set}}}(I, A); (f_1, \ldots, f_{\omega\tau}) \mapsto (i \mapsto i^{f_1} \ldots i^{f_{\omega\tau}} \omega)$$

making it an algebra (A^I, Ω) of type τ.

An index set I or I_A will be associated with each set A, such that there are mutually inverse bijections

$$I \to A \times I; x \mapsto (x\lambda, x\rho) \tag{1.5.2}$$

and

$$A \times I \to I; (a, x) \mapsto a \circ x. \tag{1.5.3}$$

The inverse relationship between these bijections reduces to the following "identities":

$$\begin{cases} \text{(a)} & \forall x \in I, x = x\lambda \circ x\rho; \\ \text{(b)} & \forall a \in A, \forall x \in I, (a \circ x)\lambda = a; \\ \text{(c)} & \forall a \in A, \forall x \in I, (a \circ x)\rho = x. \end{cases} \tag{1.5.4}$$

Note that a non-empty set A is infinite iff there is a bijection $A \to A \times A$. In this case, one takes the bijection to be (1.5.2), so that $I = A$ for infinite A. If A is empty, one also takes $I = A$. Otherwise, A has a positive number d of elements, so there is a bijection $\beta : \{0, 1, \ldots, d-1\} \to A$. In this case, one takes $I = \mathbb{N}$, and then (1.5.3) is given by $a \circ x = a\beta^{-1} + dx$. The Division Algorithm (cf. Section II 3.2) yields $n\lambda\beta^{-1}$ as the remainder and $n\rho$ as the quotient on dividing the natural number n by d. Now for each natural number r, there is a surjection $I \to A^r$. If A is empty, it is the identity map on A. Otherwise, define the map

$$I \to A^r; x \mapsto (x\lambda, x\rho\lambda, \ldots, x\rho^{r-1}\lambda). \tag{1.5.5}$$

For a in A, define $L_\circ(a): I \to I; x \mapsto a \circ x$. Fix an element x_0 in I. Then (1.5.5) has the section

$$(1.5.6) \qquad s_r : A^r \to I; (a_1, \dots, a_r) \mapsto a_1 \circ (a_2 \circ \cdots \circ (a_r \circ x_0) \dots).$$

In other words, the image of $(a_1, \dots, a_r)$ under (1.5.6) is $x_0\, L_\circ(a_r) \dots L_\circ(a_1)$. The fact that (1.5.6) is a section of (1.5.5) then follows by the identities (1.5.4).

Extending (1.4.2) to the (potentially) infinite case, one says that a subset R of A^I is *preserved* by an operation $f : A^n \to A$ on A if R is a subalgebra of the power $(A^I, \{f\})$.

Theorem 1.5.2 [I. Rosenberg, "A Classification of Universal Algebras by Infinitary Relations", *Alg. Univ.* **1** (1972), 350–353.] *Let A be a set, with index set I_A or I such that $A \times I \cong I$. Let Ω be a clone of operations on A, of type $\tau : \Omega \to \mathbb{N}$. Then there is a subset $R_{(A,\Omega)}$ or R of A^I such that an operation f on A preserves R iff it belongs to the clone Ω.*

Proof. For empty A, take $R = \{1_\varnothing\}$. The result follows. Otherwise, set

$$(1.5.7) \qquad R_{(A,\Omega)} = \langle \rho^r \lambda \mid r \in \mathbb{N} \rangle_{\mathrm{Sb}(A^I,\Omega)},$$

the subalgebra of (A^I, Ω) generated by $\{\rho^r \lambda \mid r \in \mathbb{N}\}$. Note that R is preserved by the members of the clone Ω. Conversely, an operation $f: A^n \to A$ that preserves R will be shown to lie in the clone Ω. It is convenient to use prefix notation for operations. Since f preserves R, one obtains $f(\lambda, \rho\lambda, \dots, \rho^{n-1}\lambda)$ as an element of the subalgebra R of (A^I, Ω) generated by $\{\rho^r\lambda \mid r \in \mathbb{N}\}$. In other words, there is an element ω of Ω such that $f(\lambda, \rho\lambda, \dots, \rho^{n-1}\lambda) = \omega(\lambda, \rho\lambda, \dots, \rho^{\omega\tau-1}\lambda)$. Since Ω is a clone, one may assume $n \leq \omega\tau$. Adding fictitious arguments to f if necessary, one may assume $n = \omega\tau$. For an element $(a_1, \dots, a_n)$ of A^n, let x be its image in I under the section s_n of (1.5.6). Then $f(a_1, \dots, a_n) = f(x\lambda, x\rho\lambda, \dots, x\rho^{n-1}\lambda)$ $= xf(\lambda, \rho\lambda, \dots, \rho^{n-1}\lambda) = x\omega(\lambda, \rho\lambda, \dots, \rho^{n-1}\lambda) = \omega(a_1, \dots, a_n)$. Thus $f = \omega$ $\in \Omega$, as required. □

EXERCISES

1.5A. Let Ω be the set of continuous operations on the set $\mathbb{R}$ with the Euclidean topology [cf. Example III 3.4.6(g)].

(a) Show that Ω is a clone on $\mathbb{R}$.

(b) Show that Ω is not closed in the polarity $(S, R, \eta, \varepsilon)$ of Section 1.4.

1.5B. Suppose that the set A in Theorem 1.5.2 is finite and non-empty, so that $I_A = \mathbb{N}$.

(a) For each clone Ω on A, show that each element $\alpha : \mathbb{N} \to A$ of $R_{(A,\Omega)}$ is *periodic* in the following sense: $\exists p > 0.\ \forall n \in \mathbb{N},\ n\alpha = (n+p)\alpha$.

(b) Suppose that each element of a subset R of A^I is periodic. Show that an operation $f: A^n \to A$ on A preserves R iff it preserves a subset of (the graph of) the function L_A of (1.4.1).

1.5C. (Continued Fractions.) Let A be the set of positive integers. Let I be the set of irrational numbers greater than 1. Show that the specifications

$$a \circ x = a + x^{-1},$$
$$x\lambda = \lfloor x \rfloor,$$
$$x\rho = (x - \lfloor x \rfloor)^{-1}$$

(so that $x\lambda$ is the largest integer less than or equal to x) yield mutually inverse bijections (1.5.2) and (1.5.3). [Cf. G. H. Hardy and E. M. Wright, *An Introduction to the Theory of Numbers*, Clarendon Press, Oxford, 1979, Chapter X.]

2. VARIETIES

A quasigroup Q may be construed as an algebra $(Q, \cdot, /, \backslash)$ of signature $\langle 2, 2, 2 \rangle$, i.e. equipped with three binary operations. Amongst all the algebras of such signature, quasigroups are distinguished by the requirement that the identities (I 2.2.3) have to hold. The class of all quasigroups is then seen to have special properties. For example, the direct product of two quasigroups is again a quasigroup, subalgebras of quasigroups are quasigroups, and homomorphic images of quasigroups are quasigroups. The universal algebraic concept of a "variety" (modeled somewhat on the concept of Exercise III 3.4L) abstracts these properties of the class of quasigroups. On the one hand, a variety will be the class of all algebras of a certain type that satisfy a certain set of identities. On the other hand, a variety will be closed under the formation of products, subalgebras, and homomorphic images.

In parallel with consideration of varieties, one often has to take account of other classes of algebras. Recall that a quasigroup A was originally defined (Section I 2) as an algebra $(Q, \cdot)$ of signature $\langle 2 \rangle$ in which all the right and left multiplications biject. Example I 2.2.1 showed that the class of all such algebras is not closed under the formation of homomorphic images. The study of varieties begins in Section 2.1 with the study of "prevarieties": classes required to be closed only under the formation of products and subalgebras.

EXERCISES

2A. Construe quasigroups as algebras of signature $\langle 2, 2, 2 \rangle$ satisfying (I 2.2.3). Show that homomorphic images of quasigroups are quasigroups.

2B. Construe quasigroups as algebras of signature $\langle 2 \rangle$ in which all the right and left multiplications biject.

(a) Are products of quasigroups also quasigroups?

(b) Are subalgebras of quasigroups also quasigroups?

2C. Construe unital rings as algebras $(R, -, 0, \cdot, 1)$ of type $(\{-, \cdot\} \times \{2\}) \cup (\{0, 1\} \times \{0\})$.

(a) Are subalgebras of unital rings also rings?

(b) Are subalgebras of fields also fields?

2.1. Replication and Prevarieties

Throughout this section, let $\tau : \Omega \to \mathbb{N}$ be a finitary type. A class of τ-algebras is a (non-empty) subclass $\underline{\underline{K}}_0$ of the object class $\underline{\underline{\tau}}_0$ of the category $\underline{\underline{\tau}}$ of τ-algebras and τ-homomorphisms. For such a subclass $\underline{\underline{K}}_0$, denote $\underline{\underline{K}}$ to be the full subcategory of $\underline{\underline{\tau}}$ whose object class is $\underline{\underline{K}}_0$. Thus $\underline{\underline{K}}_1$ is the class of all τ-homomorphisms between algebras of the class $\underline{\underline{K}}_0$. Let $G_{\underline{\underline{K}}}$ or $G : \underline{\underline{K}} \to \underline{\underline{\tau}}$ denote the inclusion functor from $\underline{\underline{K}}$ to $\underline{\underline{\tau}}$. This is a "forgetful functor". For example, its object part $G : \underline{\underline{K}}_0 \to \underline{\underline{\tau}}_0;\ A \mapsto A$ "forgets" that its argument A belonged to the subclass $\underline{\underline{K}}_0$. The main goal of the current section, Theorem 2.1.3 below, is to find conditions on $\underline{\underline{K}}_0$ and $\underline{\underline{K}}$ guaranteeing that there is an adjunction $(R_{\underline{\underline{K}}}, G_{\underline{\underline{K}}}, \eta, \varepsilon)$ in which the components of the unit surject. Such an adjunction is called a *replication*. The left adjoint $R_{\underline{\underline{K}}}$ or R is called the *replica functor*.

The conditions to be imposed on the classes $\underline{\underline{K}}_0$ involve the formation of subalgebras, products, homomorphic images, etc. It is convenient to set up a calculus for such constructions.

Definition 2.1.1. Let $\underline{\underline{K}}_0$ be a class of τ-algebras.

(h) A τ-algebra is a member of the class $H\underline{\underline{K}}_0$ iff it is isomorphic to the image of a τ-homomorphism whose domain is a member of $\underline{\underline{K}}_0$.

(i) A τ-algebra is a member of the class $I\underline{\underline{K}}_0$ iff it is isomorphic to a member of $\underline{\underline{K}}_0$. If $\underline{\underline{K}}_0 = I\underline{\underline{K}}_0$, one also says that $\underline{\underline{K}}_0$ is *abstract*.

(p) A τ-algebra is a member of the class $P\underline{\underline{K}}_0$ iff it is isomorphic to the product in $\underline{\underline{\tau}}$ of a multiset of members of $\underline{\underline{K}}_0$. (Warning: Note carefully that the products are to be taken in $\underline{\underline{\tau}}$, and not in the category $\underline{\underline{K}}$. Cf. Exercise 2.1J.)

(s) A τ-algebra is a member of the class $S\underline{\underline{K}}_0$ iff it is isomorphic to a subalgebra of a member of $\underline{\underline{K}}_0$.

The notation is extended to the corresponding full subcategories of $\underline{\underline{\tau}}$. For example, $H\underline{\underline{K}}$ denotes the full subcategory with object class $H\underline{\underline{K}}_0$. □

Towards (p), it is helpful to recall the explicit construction used in Exercise O 3.2Q to realize the product $\prod_{i \in I} A_i$ of a multiset $\langle A_i \mid i \in I \rangle$ of sets. Taking

$p : \Sigma_{i \in I} A_i \to I$ as the disjoint union (coproduct in $\underline{\underline{\text{Set}}}$) of the maps $A_i \to \{i\}$, the product is realized as the set

$$\prod_{i \in I} A_i = \left\{ s : I \to \sum_{i \in I} A_i \mid sp = 1_I \right\} \tag{2.1.1}$$

of sections of $p : \Sigma_{i \in I} A_i \to I$. One may then define a τ-algebra structure $(\Pi_{i \in I} A_i, \Omega)$ on (2.1.1) by

$$i(s_1 \ldots s_{\omega\tau}\, \omega) = (is_1) \ldots (is_{\omega\tau})\, \omega \tag{2.1.2}$$

for operators Ω, sections s_j, and indices $i \in I$. The definition (2.1.2) ensures that the projections $\pi_j : \Pi_{i \in I} A_i \to A_j;\ s \mapsto js$ are τ-homomorphisms (Exercise 2.1A). Thus $(\Pi_{i \in I} A_i, \Omega)$ is the product in $\underline{\underline{\tau}}$ of the multiset $\langle (A_i, \Omega) \mid i \in I \rangle$ of τ-algebras.

The operators of Definition 2.1.1 are usually written acting from the left. They obey the following commutation relations.

Proposition 2.1.2. *Let $\underline{\underline{K}}_0$ be a class of τ-algebras.*
(*a*) *Then $PS\underline{\underline{K}}_0 \subseteq SP\underline{\underline{K}}_0$.*
(*b*) *Then $PH\underline{\underline{K}}_0 \subseteq HP\underline{\underline{K}}_0$.*
(*c*) *Then $SH\underline{\underline{K}}_0 \subseteq HS\underline{\underline{K}}_0$.*

Proof. (a) Suppose $\forall i \in I,\ B_i \le A_i \in \underline{\underline{K}}_0$. Then $\Pi_{i \in I} B_i \le \Pi_{i \in I} A_i \in P\underline{\underline{K}}_0$.

(b) Suppose that for each i in I, the algebra B_i is the image of a surjective homomorphism f_i whose domain A_i lies in $\underline{\underline{K}}_0$. Since $f_i : A_i \to B_i$ surjects, it has a section $s_i : B_i \to A_i$ (not necessarily a homomorphism) with $s_i f_i = 1_{B_i}$. Consider the following diagram in $\underline{\underline{\text{Set}}}$:

(2.1.3)

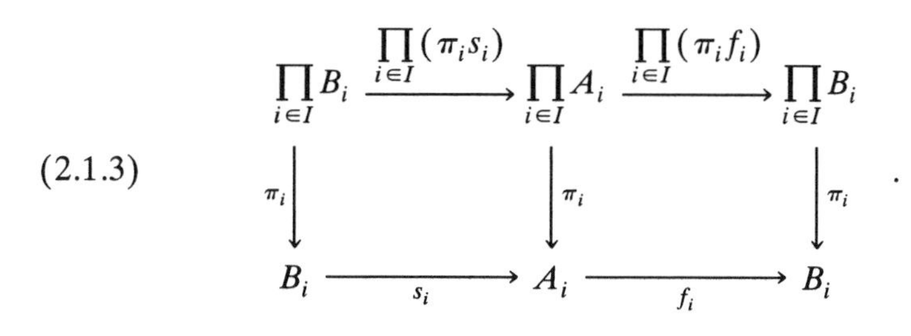

For each i, the composite across the bottom row is 1_{B_i}. Thus the composite across the top row is $\Pi_{i \in I} 1_{B_i} = 1_{\Pi B_i}$. In other words, the homomorphism $\Pi_{i \in I}(\pi_i f_i)$ surjects, since it has $\Pi_{i \in I}(\pi_i s_i)$ as a section in $\underline{\underline{\text{Set}}}$. Thus $\Pi_{i \in I} B_i \in HP\underline{\underline{K}}_0$.

(c) By the First Isomorphism Theorem 1.2.7, it suffices to consider a subalgebra S of a quotient A^α of an algebra A in $\underline{\underline{K}}_0$. Define $B = (\text{nat } \alpha)^{-1}(S)$. Then B is a subalgebra of A. Moreover, $S = (\cup B^\alpha)^\alpha \cong B^{(\alpha \cap B^2)}$

by the Second Isomorphism Theorem 1.2.8(b). Thus $S = B\,\mathrm{nat}\,(\alpha \cap B^2) \in HS\,\underline{\underline{K}}_0$, as required. □

One may now examine necessary and sufficient conditions for the existence of replications.

Theorem 2.1.3. *Let $\tau : \Omega \to \mathbb{N}$ be a finitary type. Let $\underline{\underline{K}}_0$ be an abstract class of τ-algebras, and let $G : \underline{\underline{K}} \to \underline{\underline{\tau}}$ be the inclusion in $\underline{\underline{\tau}}$ of the full subcategory with object class $\underline{\underline{K}}_0$. Then the following conditions are equivalent:*

(a) $\underline{\underline{K}}_0 = SP\underline{\underline{K}}_0$;

(b) *$\underline{\underline{K}}$ is complete, $G : \underline{\underline{K}} \to \underline{\underline{\tau}}$ is continuous, and $\underline{\underline{K}}_0 = S\underline{\underline{K}}_0$;*

(c) *There is a replication $(R, G, \eta, \varepsilon)$.*

Proof. (a) ⇒ (b). By Proposition 2.1.2(a), one has $P\underline{\underline{K}}_0 = PSP\underline{\underline{K}}_0 \subseteq SPP\underline{\underline{K}}_0 = SP\underline{\underline{K}}_0 = \underline{\underline{K}}_0$. Thus $\underline{\underline{\tau}}$-products of $\underline{\underline{K}}_0$-algebras are $\underline{\underline{K}}_0$-products. By Theorem III 2.5.5, it then suffices to show that $\underline{\underline{K}}$ has equalizers given by equalizers in $\underline{\underline{\tau}}$. Consider a parallel pair $A_1 \overset{f_1}{\underset{f_2}{\rightrightarrows}} A_2$ of arrows in $\underline{\underline{K}}$. Then their set equalizer $E = \{a \in A_1 \mid af_1 = af_2\}$ is a subalgebra of A_1, and thus lies in $S\underline{\underline{K}}_0 = SSP\underline{\underline{K}}_0 = SP\underline{\underline{K}}_0$, as required.

(b) ⇒ (c). Considering the first two conditions of (b), one observes that for the Freyd Adjoint Functor Theorem 3.5.2 to apply, it suffices to verify the solution set condition. Well, for a τ-algebra A, the set

$$\{\mathrm{nat}\ \alpha : \mathrm{A} \to \mathrm{A}^\alpha \mid \mathrm{A}^\alpha \in \underline{\underline{K}}_0\} \tag{2.1.4}$$

dominates the comma category (A, G). Indeed, for an object $h : A \to BG$ of this category, the subalgebra Ah of B lies in $S\underline{\underline{K}}_0 = \underline{\underline{K}}_0$. The First Isomorphism Theorem 1.2.7 yields the commutative diagram

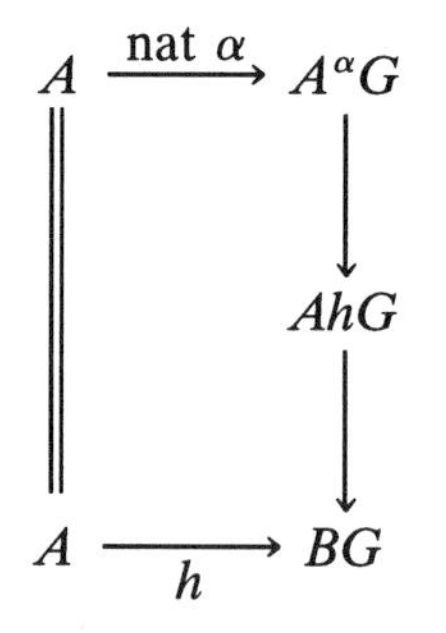

in $\underline{\underline{\tau}}$. Since the upper arrow of the right hand side is an isomorphism, one has $A^\alpha \in \underline{\underline{K}}_0$. Thus nat α belongs to the dominating set. Applying the Freyd Adjoint Functor Theorem, one obtains an adjunction $(R, G, \eta, \varepsilon)$. For a τ-algebra A, consider the initial object $\eta_A : A \to ARG$ of the comma category (A, G). Now $A\eta_A$, as a subalgebra of the algebra AR in $\underline{\underline{K}}_0$, itself lies in

$\underline{\underline{K}}_0$. Then $\eta_A : A \to A\eta_A G$ is also an object of (A, G). Moreover, for an object $h : A \to BG$ of (A, G), consider the unique (A, G)-morphism f or $fG : (\eta_A : A \to ARG) \to (h : A \to BG)$ given by the initiality of $\eta_A : A \to ARG$. One then obtains a unique (A, G)-morphism as the composite $A\eta_A \hookrightarrow AR \xrightarrow{f} B$. Thus $\eta_A : A \to A\eta_A$ is also an initial object of (A, G), so that the components of the unit in $(R, G, \eta, \varepsilon)$ may be taken as surjective.

(c) $\Rightarrow$ (a). Let A be a subalgebra of a member B of $\underline{\underline{K}}_0$, with inclusion $j : A \to B$. Consider the (A, G)-morphism given by the diagram

$$\begin{array}{ccc} A & \xrightarrow{\eta_A} & ARG \\ \| & & \downarrow{\scriptstyle fG} \\ A & \xrightarrow[j]{} & BG \end{array} .$$

Since $j = \eta_A f^G$ injects, it follows that the surjective map η_A also injects, yielding an isomorphism of A with the member AR of $\underline{\underline{K}}_0$. Since $\underline{\underline{K}}_0$ is abstract, A also lies in $\underline{\underline{K}}_0$. Thus $S\underline{\underline{K}}_0 = \underline{\underline{K}}_0$.

Now let A be the product $\prod_{i \in I} A_i$ in $\underline{\underline{\tau}}$ of members A_i of $\underline{\underline{K}}_0$. Consider the diagram

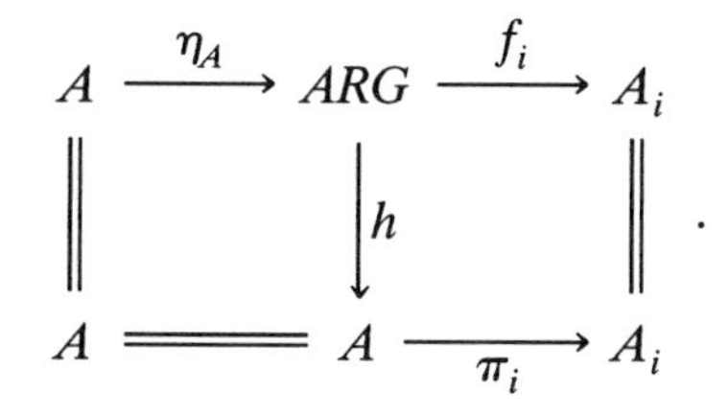

Here $f_i : (\eta_A : A \to ARG) \to (\pi_i : A \to A_i)$ is given by the initiality of η_A in (A, G), while $h : ARG \to A$ is given by $A = \prod_{i \in I} A_i$. Then $1_A \pi_i = \eta_A f_i = \eta_A h \pi_i$ for each i, whence $\eta_A\, h = 1_A$. Thus the surjective map η_A also injects, yielding an isomorphism of A with the member AR of $\underline{\underline{K}}_0$. As before, one obtains $A \in \underline{\underline{K}}_0$ and thus $P\underline{\underline{K}}_0 = \underline{\underline{K}}_0$. □

For a finitary type $\tau : \Omega \to \mathbb{N}$, an abstract class $\underline{\underline{K}}_0$ of τ-algebras satisfying the equivalent conditions of Theorem 2.1.3, or the corresponding full subcategory $\underline{\underline{K}}$, is called a *prevariety*. For a τ-algebra A or (A, Ω), the congruences α with A^α in a prevariety $\underline{\underline{K}}_0$ are called *$\underline{\underline{K}}$-congruences* on A [cf. (2.1.4)]. The subposet of $\mathrm{Cg}(A, \Omega)$ induced on the set of $\underline{\underline{K}}$-congruences is denoted by $\mathrm{Cg}_{\underline{\underline{K}}}(A, \Omega)$ or $\mathrm{Cg}_{\underline{\underline{K}}}\, A$. Note that the singleton τ-algebra, as the product of an empty subset of a prevariety $\underline{\underline{K}}_0$, lies in $\underline{\underline{K}}_0$. Thus $A^2 \in \mathrm{Cg}_{\underline{\underline{K}}}\, A$ for each prevariety $\underline{\underline{K}}$.

Proposition 2.1.4. *Let $\underline{\underline{K}}$ be a prevariety of τ-algebras. Then for each τ-algebra A, the poset $\mathrm{Cg}_{\underline{\underline{K}}}\, A$ is a complete lattice, and a meet subsemilattice of $\mathrm{Cg}\, A$ and $\mathscr{P}(A^2)$.*

Proof. Let $\{\alpha_i \mid i \in I\}$ be a set of $\underline{\underline{K}}$-congruences on A. Then the kernel of $\prod_{i \in I} \mathrm{nat}\ \alpha_i$ is $\bigcap_{i \in I} \alpha_i$. By the First Isomorphism Theorem 1.2.7, one has $A\ \mathrm{nat} \bigcap_{i \in I} \alpha_i \cong \prod_{i \in I} A\ \mathrm{nat}\ \alpha_i \in P\underline{\underline{K}}_0 = \underline{\underline{K}}_0$. Thus $\bigcap_{i \in I} \alpha_i \in \mathrm{Cg}_{\underline{\underline{K}}}\, A$. □

The construction of the absolutely free τ-algebra or word algebra $X\Omega$ over a set X yielded an adjunction $(\Omega, U, \eta, \varepsilon)$ in Section 1.3 for the underlying set functor $U : \underline{\underline{\tau}} \to \underline{\underline{\mathrm{Set}}}$. Let $\underline{\underline{K}}$ be a prevariety of τ-algebras. Consider the restricted underlying set functor $U : \underline{\underline{K}} \to \underline{\underline{\mathrm{Set}}}$. Composing the adjunction $(\Omega, U, \eta, \varepsilon)$ with the replication $(R_{\underline{\underline{K}}}, G_{\underline{\underline{K}}}, \eta^{\underline{\underline{I}}}, \varepsilon^{\underline{\underline{K}}})$, one obtains a left adjoint $\Omega R_{\underline{\underline{K}}} : \underline{\underline{\mathrm{Set}}} \to \underline{\underline{K}}$ to $U : \underline{\underline{K}} \to \underline{\underline{\mathrm{Set}}}$. For a set X, the $\underline{\underline{K}}$-algebra $X\Omega R_{\underline{\underline{K}}}$ is called the *free $\underline{\underline{K}}$-algebra over the set X*. Note that components of the unit of the adjunction $(\Omega R_{\underline{\underline{K}}}, U, \eta, \varepsilon)$ do not necessarily inject. For example, consider a type $\tau = \{\omega\} \times \{0\}$ and the prevariety $\underline{\underline{K}}_0$ of singleton subalgebras. Then for any set X, the free $\underline{\underline{K}}$-algebra $X\Omega R_{\underline{\underline{K}}}$, as a member of $\underline{\underline{K}}_0$, is a singleton.

Example 2.1.5. (a) **(Convex Sets.)** For the type $\tau : \mathrm{I}^\circ \to \{2\}$ of Example 1.3.4(c), one obtains the class $\underline{\underline{K}}_0$ of convex sets as a prevariety. This is readily verified using condition (a) of Theorem 2.1.3. For a finite set $X = \{0, 1, 2, \ldots, n\}$, the free $\underline{\underline{K}}$-algebra over X is called an *n-simplex*. (Cf. Exercise 2.1D).

(b) **(Free Groups.)** For the type $(\tau : \Omega \to \mathbb{N}) = \{(\cdot, 2), (^{-1}, 1), (1, 0)\}$, consider the prevariety $\underline{\underline{\mathrm{Gp}}}_0$ of groups. Then for a set X, the $\underline{\underline{\mathrm{Gp}}}$-algebra $X\Omega\ R_{\underline{\underline{\mathrm{Gp}}}}$ is isomorphic to the free group XG constructed in Section I 1.4. □

A crucial property of free $\underline{\underline{K}}$-algebras, their *projectivity* in $\underline{\underline{K}}$, is described by the following proposition (cf. Exercise II 2.3J).

Proposition 2.1.6. *Let $\underline{\underline{K}}$ be a prevariety of τ-algebras. Consider the free $\underline{\underline{K}}$-algebra $X\Omega R_{\underline{\underline{K}}}$ over a set X, and a surjective $\underline{\underline{K}}$-morphism $g : A \to B$. Then for each τ-homomorphism $h : X\Omega R_{\underline{\underline{K}}} \to B$, there is at least one τ-homomorphism $\overline{h} : X\Omega R_{\underline{\underline{K}}} \to B$ such that the diagram*

$$
\begin{array}{ccc}
X\Omega R_{\underline{\underline{K}}} & \xrightarrow{\ \overline{h}\ } & A \\
\Big\| & & \Big\downarrow g \\
X\Omega R_{\underline{\underline{K}}} & \xrightarrow[\ h\ ]{} & B
\end{array}
\tag{2.1.5}
$$

commutes. (One says that h *lifts* to $\overline{h}$.)

Proof. Consider the adjunction $(\Omega R_{\underline{\underline{K}}}, U, \eta, \varepsilon)$, and the diagram

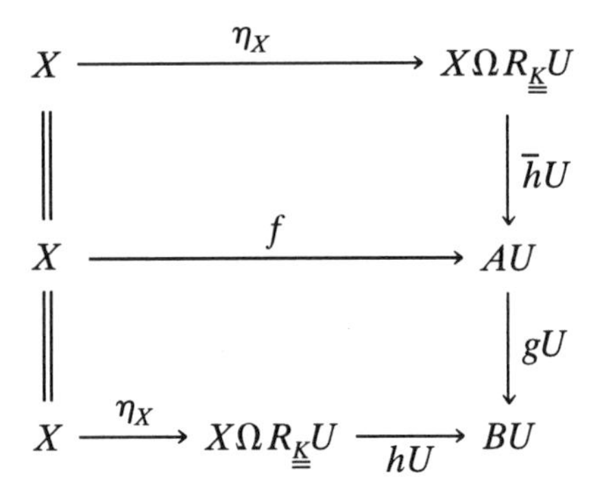

in $\underline{\underline{\text{Set}}}$. Since gU surjects, there is an element f of $\underline{\underline{\text{Set}}}(X, AU)$ making the bottom rectangle commute. The initiality of η_X in $\overline{(X, U)}$ then yields a τ-homomorphism $\overline{h}: X\Omega\ R_{\underline{\underline{K}}} \to A$ making the top rectangle commute. Using the bijection $\varphi_B^X : \underline{\underline{K}}(X\Omega R_{\underline{\underline{K}}}, B) \to \underline{\underline{\text{Set}}}(X, BU)$ of (III 3.1.3), one then has $(\overline{h}g)\varphi_B^X = \eta_X \overline{h}^U g^U = \eta_X h^U = h\varphi_B^X$, so that (2.1.5) commutes. □

EXERCISES

2.1A. Verify that (2.1.2) yields $\prod_{i \in I} A_i$ as a product in $\underline{\underline{\tau}}$.

2.1B. Show that I acts as an identity for the operators of Definition 2.1.1. In other words, for each class $\underline{\underline{K}}_0$ of τ-algebras, verify $IS\underline{\underline{K}}_0 = S\underline{\underline{K}}_0$, $SI\underline{\underline{K}}_0 = S\ \underline{\underline{K}}_0$, etc.

2.1C. Give an alternative proof of (b) $\Rightarrow$ (c) in Theorem 2.1.3 by showing $\eta_A = \text{nat} \bigcap \text{Cg}_{\underline{\underline{K}}} A$ for a τ-algebra A.

2.1D. Show that the $(n-1)$-simplex [in the sense of Example 2.1.5(a)] is isomorphic to the subalgebra of $(\mathbb{R}^n, I^\circ)$ generated by the subset $\{(1,0,0,\ldots,0,0),(0,1,0,\ldots,0,0),\ldots,(0,0,0,\ldots,0,1)\}$.

2.1E. Verify the claim of Example 2.1.5(b).

2.1.F. Let $\tau : \Omega \to \mathbb{N}$ be the type $\{(-,2),(0,0)\}$. Consider the following classes $\underline{\underline{K}}_0$ of abelian groups G. For each, determine whether or not $\underline{\underline{K}}_0$ is a prevariety. If $\underline{\underline{K}}_0$ is a prevariety, describe $\eta : A \to AR_{\underline{\underline{K}}}$ for an abelian group A.

(a) $\underline{\underline{K}}_0 = \{G \mid \text{Tor } G = \{0\}\}$, the class of *torsion-free* abelian groups (cf. Exercise II 3.3.D).

(b) $\underline{\underline{K}}_0 = \{G \mid \text{Tor } G = G\}$, the class of *torsion* abelian groups.

(c) Fix an integer n. Then $\underline{\underline{K}}_0 = \{G \mid Gn = \{0\}\}$.

(d) $\underline{\underline{K}}_0 = \{G \mid \exists n \in \mathbb{Z}.\ Gn = \{0\}\}$.

2.1G. For each prevariety $\underline{\underline{K}}_0$ in Exercise 2.1F, describe the free $\underline{\underline{K}}$-algebra on a set X.

2.1H. Prove that $\underline{\underline{K}}_0$ is a prevariety iff $\underline{\underline{K}}_0 = S\underline{\underline{K}}_0$ and $\underline{\underline{K}}_0 = P\underline{\underline{K}}_0$.

2.1I. Let $(\tau : \Omega \to \mathbb{N}) = \{(\cdot, 2)\}$. Let $\underline{\underline{K}}_0$ denote the class of commutative semigroups.

(a) Show that $\underline{\underline{K}}_0$ is a prevariety.

(b) Describe the free $\underline{\underline{K}}$-algebra $X\Omega R_{\underline{\underline{K}}}$ on a set X.

(c) Show that $X\Omega R_{\underline{\underline{K}}}$ is isomorphic with the semigroup $X^{*\kappa} - \{1\}$ of Exercise I 1.5E.

2.1J. Let A be a principal ideal domain. Let $(\tau : \Omega \to \mathbb{N})$ be the type $\{(-, 2)\} \cup (A \times \{1\}) \cup \{(0, 0)\}$. Let $\underline{\underline{K}}_0$ be the class of *torsion modules*: A-modules W with $W = \operatorname{Tor} W$ (cf. Exercise II 3.3D). Show that $\underline{\underline{K}}$ is not a prevariety in $\underline{\underline{\tau}}$, although $\underline{\underline{K}}$ is complete and $\underline{\underline{K}}_0 = S\underline{\underline{K}}_0$.

2.1K. Set $n = 6$ in Exercise 2.1F(c). Show that $\mathbb{Z}_3$ is projective in $\underline{\underline{K}}$, but not free in $\underline{\underline{K}}$. (Hint: The order of a finite free $\underline{\underline{K}}$-algebra is a power of 6.)

2.1L. Let K be a unital, commutative ring.

(a) Construe the category of K-algebras as a prevariety $\underline{\underline{\mathrm{Alg}}}_K$.

(b) Show that the polynomial algebra $K[T]$ is the free $\underline{\underline{\mathrm{Alg}}}_K$-algebra over the singleton $\{T\}$. [Hint: (II 3.4).]

(c) Is $K[T_1, T_2]$ the free $\underline{\underline{\mathrm{Alg}}}_K$-algebra over the doubleton $\{T_1, T_2\}$?

2.2. Bicompleteness

Throughout this section, let $\tau : \Omega \to \mathbb{N}$ be a finitary type. Theorem 2.1.3 showed that any prevariety $\underline{\underline{K}}$ of τ-algebras, including $\underline{\underline{\tau}}$ itself, is (small-) complete. The main result of this section, Theorem 2.2.3 below, shows that $\underline{\underline{K}}$ is also (small-) cocomplete.

The key task is to show that $\underline{\underline{\tau}}$ is cocomplete. By Corollary III 2.5.6, it suffices to exhibit coequalizers and coproducts in $\underline{\underline{\tau}}$. The construction of coequalizers is immediate. Given a parallel pair (f, g) of τ-homomorphisms $A \to B$, let κ be the congruence on B generated by the binary relation $\{(af, ag) \mid a \in A\}$. Then the coequalizer of (f, g) is the natural projection $\operatorname{nat} \kappa$:

$$A \underset{g}{\overset{f}{\rightrightarrows}} B \xrightarrow{\operatorname{nat} \kappa} B^{\kappa}. \tag{2.2.1}$$

Indeed, if $fh = gh$ for a $\underline{\underline{\tau}}$-morphism $h : B \to C$, the kernel $\ker h$ lies in the interval $[\kappa, B^2]$ of $\operatorname{Cg} B$. By the Third Isomorphism Theorem 1.2.9, there is a unique congruence β on B^{κ} with $(\operatorname{nat} \kappa)^*(\beta) = \ker h$. Consider the commutative diagram

$$\begin{array}{ccccc}
B^{\kappa} & \xleftarrow{\operatorname{nat} \kappa} & B & \xrightarrow{h} & C \\
{\scriptstyle \operatorname{nat} \beta}\downarrow & & \downarrow{\scriptstyle \operatorname{nat} \ker h} & & \uparrow{\scriptstyle j} \\
(B^{\kappa})^{\beta} & \xrightarrow[i_1]{} & B^{\ker h} & \xrightarrow[i_2]{} & Bh
\end{array} \tag{2.2.2}$$

in $\underline{\underline{T}}$, where the right-hand square is an instance of (1.2.5), and i_1 is the isomorphism of (1.2.8). Then $(\text{nat } \kappa)[(\text{nat } \beta)i_1 i_2 j] = h$, verifying that (2.2.1) is a coequalizer in $\underline{\underline{T}}$.

Example 2.2.1 (Free Presentations). Let A be a τ-algebra lying in a prevariety $\underline{\underline{K}}$. Let $\varepsilon_A : A_0 \to A$ be the component at A of the counit ε in the adjunction $(\Omega R_{\underline{\underline{K}}}, U, \eta, \varepsilon)$. Thus $A_0 = AU\Omega R_{\underline{\underline{K}}}$ is the free $\underline{\underline{K}}$-algebra over the underlying set AU of A. Set $A_1 = \ker \varepsilon_A$. Thus $A_1 = \{(a,b) \in A_0^2 \mid a\varepsilon_A = b\varepsilon_A\}$. Note $A_1 \in SP\underline{\underline{K}}_0 = \underline{\underline{K}}_0$, using condition (a) of Theorem 2.1.3. Define $d_0 : A_1 \to A_0; (a,b) \mapsto a$ and $d_1 : A_1 \to A_0; (a,b) \mapsto b$. One obtains a parallel pair (d_0, d_1) of $\underline{\underline{K}}$-morphisms $A_1 \to A_0$. Then the congruence κ of the coequalizer construction in $\underline{\underline{T}}$ is (the congruence on A_0 generated by) the binary relation $\{(a_1 d_0, a_1 d_1) \mid a_1 \in A_1\} = \{(a,b) \in A_0^2 \mid (a,b) \in \ker \varepsilon_A\} = \ker \varepsilon_A$. Thus

$$(2.2.3) \qquad A_1 \underset{d_0}{\overset{d_1}{\rightrightarrows}} A_0 \xrightarrow{\varepsilon_A} A$$

is a coequalizer diagram, both in $\underline{\underline{T}}$ and $\underline{\underline{K}}$. It is called the *free presentation of the $\underline{\underline{K}}$-algebra A*. As an instance of the uniqueness of coequalizers to within isomorphism, the algebra A is specified by the parallel pair (d_0, d_1) to within isomorphism. The elements (a,b) of A_1, sometimes written in the form "$a = b$", are called the *relations of A (relative to $\underline{\underline{K}}$)*. □

Now let $F : J \to \underline{\underline{T}}_0$ be a function. For each element j of J, consider the free presentation $jF_1 \rightrightarrows jF_0 \to jF$ of the $\underline{\underline{T}}$-algebra jF. Let $e_j : jFU \to S$ be the insertion of the underlying set jFU of jF into the disjoint union $S = \dot\bigcup_{j \in J} jFU$. Apply the functor Ω to get the τ-homomorphism $e_j\Omega : jF_0 \to S\Omega$. Define κ to be the congruence on $S\Omega$ generated by

$$(2.2.4) \qquad \bigcup_{j \in J} \left\{ \left(a_1 d_0 e_j^{\Omega}, a_1 d_1 e_j^{\Omega}\right) \middle| a_1 \in jF_1 \right\}.$$

Now for each j in J, one has $d_0 e_j^{\Omega} \text{nat } \kappa = d_1 e_j^{\Omega} \text{nat } \kappa$. Since the free presentation of jF is the coequalizer of the pair (d_0, d_1), one obtains a unique τ-homomorphism $\iota_j : jF \to S\Omega^{\kappa}$ with $\varepsilon_{jF}\iota_j = e_j^{\Omega} \text{ nat } \kappa$. Then the τ-homomorphism $\iota_j : jF \to S\Omega^{\kappa}$ is the insertion of jF into the coproduct $S\Omega^{\kappa} = \sum_{j \in J} jF$ of the jF in $\underline{\underline{T}}$. Informally, the construction of $\sum_{j \in J} jF$ takes the free $\underline{\underline{T}}$-algebra $S\Omega$ on the disjoint union S of the underlying sets jFU of the jF, and then "imposes the relations of each jF relative to $\underline{\underline{T}}$."

With coequalizers and coproducts now constructed in $\underline{\underline{T}}$, Corollary III 2.5.6 yields the following:

Proposition 2.2.2. *The category $\underline{\underline{T}}$ is cocomplete.* □

The main cocompleteness result is then immediate.

Theorem 2.2.3. *Let $\tau : \Omega \to \mathbb{N}$ be a finitary type, and let $\underline{\underline{K}}$ be a prevariety in $\underline{\underline{\tau}}$. Then the category $\underline{\underline{K}}$ is cocomplete.*

Proof. Let $F : J \to \underline{\underline{K}}$ be a functor with small domain. Since $\underline{\underline{\tau}}$ is cocomplete, the composite FG of $F : J \to \underline{\underline{K}}$ with the inclusion $G : \underline{\underline{K}} \to \underline{\underline{\tau}}$ has a colimit $\iota^{FG} : FG \to (\varinjlim FG)\Delta$. Consider the replication $(R, G, \eta, \varepsilon)$. Define $\varinjlim F = (\varinjlim FG)R$. Define a natural transformation $\iota^F : F \to (\varinjlim F)\Delta$ with component $\iota_j^F = \iota_j^{FG}\eta_{\varinjlim FG}$ at the object j of J. Then $\iota^F : F \to (\varinjlim F)\Delta$ is the colimit of the functor $F : J \to \underline{\underline{K}}$. □

EXERCISES

2.2A. Determine the free presentation of the abelian group $\mathbb{Z}_2$ in the prevariety of abelian groups A with $4A = 0$.

2.2B. Let $F : J \to \underline{\underline{\tau}}_0$ be a function. Let A be a τ-algebra. Given τ-homomorphisms $f_j : jF \to A$ for each j in J, show how to construct the unique τ-homomorphism $\Sigma_{j\in J} f_j : \Sigma_{j\in J} jF \to A$ with $\iota_i \Sigma_{j \in J} f_j = f_i$ for each i in J.

2.2C. Let X and Y be sets, and let $\underline{\underline{K}}$ be a prevariety. Prove that the free $\underline{\underline{K}}$-algebra over $X \cup Y$ is the coproduct of the free $\underline{\underline{K}}$-algebras over X and Y.

2.2D. In the proof of Theorem 2.2.3, verify that $\iota^T : F \to (\varinjlim F)\Delta$ is indeed an initial object of $F/\underline{\underline{K}}\Delta$.

2.2E. Let $\tau : \Omega \to \mathbb{N}$ be a finitary type, and let (A, Ω) be a τ-algebra.

(a) Show that $SP\{A\}$ is a prevariety.

(b) Show that $SP\{A\}$ is contained in each prevariety $\underline{\underline{K}}_0$ of which A is a member. (Thus $SP\{A\}$ is called the *prevariety generated by* A).

(c) The *logarithmetic* $\mathscr{L}A$ of A is the subalgebra of A^{AU} generated by $1_{AU} : AU \to AU$. Prove that $\mathscr{L}A$ is the free $SP\{A\}$-algebra over the singleton $\{1_{AU}\}$.

(d) Show that the free $SP\{A\}$-algebra over a set X is the coproduct $\Sigma_X \mathscr{L}A$.

2.2F. Determine the logarithmetics of the groups $\mathbb{Z}_4$ and S_3.

2.2G. Determine the logarithmetic of the quasigroup $(Q, \circ)$ of Exercise I 2.1F.

2.2H. **(Tensor Products of *K*-algebras.)** Let K be a unital commutative ring. Let $\tau = (K \times \{1\}) \cup (\{-, m\} \times \{2\}) \cup (\{0,1\} \times \{0\})$ be the type of K-algebras.

(a) Interpret the category $\underline{\underline{\mathrm{CAlg}_K}}$ of commutative K-algebras as a prevariety of τ-algebras.

(b) For K-algebras A and B, show that the tensor product $A_K \otimes_K B_K$ of the K-modules A_K and B_K becomes a K-algebra $A \otimes B$ under the multiplication given on primitive elements by $a_1 a_2 \otimes b_1 b_2 = (a_1 \otimes b_1)(a_2 \otimes b_2)$.

(c) If A and B are commutative, show that $A \otimes B$ with the insertions $A \to A \otimes B; a \mapsto a \otimes 1$ and $B \to A \otimes B;\ b \mapsto 1 \otimes b$ is the coproduct of A and B in the prevariety $\underline{\underline{\mathrm{CAlg}}}_K$.

(d) Prove $K[T_1, T_2] \cong K[T_1] \otimes K[T_2]$ (cf. Exercise 2.1L).

2.3. Satisfaction and Varieties

Consider the type $\tau = \{\beta, \delta, \gamma\} \times \{2\}$ of signature $\langle 2, 2, 2\rangle$ used for quasigroups in Example 1.3.4(b). Amongst all the algebras of type τ, quasigroups are distinguished by satisfaction of the identities (I 2.2.3). For a quasigroup $(Q, \beta, \delta, \gamma)$, (I 2.2.3) (IR) takes the form

$$x_1 x_2\, \beta x_2\, \delta = x_1. \tag{2.3.1}$$

In the terminology of Section 1.3, the left-hand side of (2.3.1) is the derived operator $u = x_1 x_2\, \beta x_2\, \delta$. The right-hand side is the derived operator $v = x_1$. Satisfaction of (I 2.2.3) (IR) or (2.3.1) by a quasigroup $(Q, \beta, \gamma, \delta)$ means precisely that the binary clone operations $(u, 2)_Q$ or $u_Q : Q^2 \to Q; (p, q) \mapsto pq\, \beta q \delta$ and $(v, 2)_Q$ or $v_Q : Q^2 \to Q; (p, q) \mapsto p$ coincide.

For the rest of this section, fix a finitary type $\tau : \Omega \to \mathbb{N}$. An *identity* (u, v) or $u = v$ of type τ is a pair (u, v) of derived operators, i.e. an element of the direct square $P\Omega^2$ of the word algebra $P\Omega$ on the set P (1.3.4) of positive integers. A τ-algebra A is said *to satisfy the identity* $u = v$ if the n-ary clone operations $(u, n)_A$ and $(v, n)_A$ on A coincide, where $n = \max((\arg u) \cup (\arg v))$. One also says that the words u and v are *synonymous* in A. For example, the magma words $x_1 x_2\, \mu x_3\, \mu$ and $x_1 x_2 x_3\, \mu \mu$ are synonymous in each semigroup.

Proposition 2.3.1. *Let A be a τ-algebra, and let $u = v$ be an identity. The algebra A satisfies $u = v$ iff* $\forall f \in \underline{\underline{\tau}}(P\Omega, A),\ \ uf = vf.$

Proof. Set $n = \max((\arg u) \cup (\arg v))$, and $X = \{x_1, \ldots, x_n\}$. If A satisfies $u = v$ and $f \in \underline{\underline{\tau}}(P\Omega, A)$, then $uf = x_1^f \ldots x_n^f (u, n)_A = x_1^f \ldots x_n^f (v, n)_A = vf$. Conversely, suppose $\forall f \in \underline{\underline{\tau}}(P\Omega, A),\ uf = vf$. If A is empty, then Ω contains no nullary operations. By Proposition 1.3.3, there are then no nullary derived operations. Thus $n > 0$, and $(u, n)_A = 1_\varnothing = (v, n)_A$. Otherwise, let a_0 be an element of A. Given $a \in \underline{\underline{\mathrm{Set}}}(X, A)$, define $\bar{a} = (a : X \to A) \cup ((P - X) \to \{a_0\})$. For $f = \bar{a}^\Omega \varepsilon_A \in \underline{\underline{\tau}}(P\Omega, A)$, one then has $x_1^a \ldots x_n^a (u, n)_A = u a^\Omega \varepsilon_A = uf = vf = x_1^a \ldots\ x_n^a (v, n)_A$. $\square$

Satisfaction of identities is preserved by the constructions of Definition 2.1.1:

Proposition 2.3.2. *Let $\underline{\underline{K}}_0$ be the class of τ-algebras satisfying an identity $u = v$. Then:*

(h) $H\underline{\underline{K}}_0 = \underline{\underline{K}}_0$;
(p) $P\underline{\underline{K}}_0 = \underline{\underline{K}}_0$;
(s) $S\underline{\underline{K}}_0 = \underline{\underline{K}}_0$.

Proof. (h) Suppose A satisfies $u = v$, and that $g : A \to B$ is a surjective τ-homomorphism. By Proposition 2.1.6 applied to the improper prevariety $\underline{\underline{\tau}}$, the free algebra $P\Omega$ is projective in $\underline{\underline{\tau}}$. Thus each element h of $\underline{\underline{\tau}}(P\Omega, B)$ lifts to $\bar{h}$ in $\underline{\underline{\tau}}(P\Omega, A)$ with $h = \bar{h}g$. Then $uh = u\bar{h}g = v\bar{h}g = vh$, so that B satisfies $u = v$.

(p) Suppose each element A_i of a multiset $\langle A_i \mid i \in I \rangle$ of τ-algebras satisfies $u = v$. Then for $f \in \underline{\underline{\tau}}(P\Omega, \prod_{j \in I} A_j)$, one has $uf\pi_i = vf\pi_i$ for each i in I, whence $uf = vf$ and $\prod_{j \in I} A_j$ satisfies $u = v$.

(s) Suppose that A satisfies $u = v$. Let $j : B \to A$ be injective. Then for $f \in \underline{\underline{\tau}}(P\Omega, B)$, the equality $ufj = vfj$ implies $uf = vf$, so that B satisfies $u = v$. □

Proposition 2.3.2 shows that the class of τ-algebras satisfying a given identity is a prevariety. More generally, the class $\underline{\underline{K}}_0$ of τ-algebras satisfying a given set of identities is a prevariety. In addition, $\underline{\underline{K}}_0 = H\underline{\underline{K}}_0$. A *variety* is defined to be a prevariety $\underline{\underline{K}}$ or $\underline{\underline{K}}_0$ satisfying this additional hypothesis $\underline{\underline{K}}_0 = H\underline{\underline{K}}_0$. By Proposition 2.1.2, $\underline{\underline{K}}_0$ is a variety iff $\underline{\underline{K}}_0 = HSP\underline{\underline{K}}_0$ (Exercise 2.3A). The crucial property of varieties is given by a classical converse to Proposition 2.3.2:

Birkhoff's Variety Theorem 2.3.3. *Let $\tau : \Omega \to \mathbb{N}$ be a finitary type. Let $\underline{\underline{K}}_0$ be a class of τ-algebras. Then $\underline{\underline{K}}_0$ is a variety iff it is the class of τ-algebras satisfying a certain set of identities.*

Proof. The "if" direction follows by Proposition 2.3.2. Conversely, suppose that $\underline{\underline{K}}_0$ is a variety. As a prevariety, $\underline{\underline{K}}$ possesses a replication ($R_{\underline{\underline{K}}}$, $G_{\underline{\underline{K}}}$, $\eta^{\underline{\underline{K}}}$, $\varepsilon^{\underline{\underline{K}}}$). It will be shown that $\underline{\underline{K}}_0$ is the class of τ-algebras satisfying the set $\ker(\eta^{\underline{\underline{K}}}_{P\Omega} : P\Omega \to P\Omega RG)$ of identities.

Suppose first that A is a member of $\underline{\underline{K}}_0$. Each τ-homomorphism $f : P\Omega \to AG$ factorizes as $f = \eta^{\underline{\underline{K}}}_{P\Omega}(f^R\varepsilon^{\underline{\underline{K}}}_A)^G$, so that $uf = vf$ for each identity (u, v) in $\ker \eta^{\underline{\underline{K}}}_{P\Omega}$. Conversely, suppose that A is a τ-algebra satisfying $\ker\ \eta^{\underline{\underline{K}}}_{P\Omega}$. Consider the adjunction $(\Omega, U, \eta, \varepsilon)$. It will be shown that $\ker(\eta^{\underline{\underline{K}}}_{AU\Omega} : AU\Omega \to AU\Omega RG) \leq \ker(\varepsilon_A : AU\Omega \to A)$. Then A, as a homomorphic image of the $\underline{\underline{K}}_0$-algebra $AU\Omega RG$, will itself be seen to lie in $\underline{\underline{K}}_0$.

For an element (w_1, w_2) of $\ker\ \eta^{\underline{\underline{K}}}_{AU\Omega}$, let $(\arg w_1) \cup (\arg\ w_2) = \{a_1, \dots, a_n\} = B$ with $|B| = n$. Define $j : B \to P;\ a_i \mapsto x_i$. By the naturality of the unit η in the replication $(R, G, \eta, \varepsilon)$ at the τ-homomorphism $j\Omega$, one has the commuting diagram

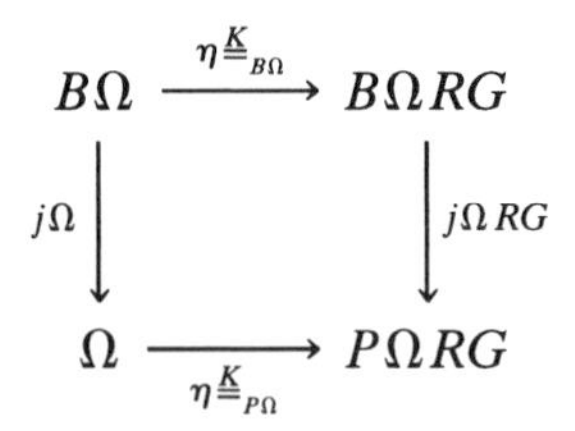

in $\underline{\underline{\tau}}$. Since $(w_1, w_2) \in \ker \eta^{\underline{\underline{K}}}_{B\Omega}$, one has $(u, v) := (w_1 j^{\Omega}, w_2 j^{\Omega}) \in \ker \eta^{\underline{\underline{K}}}_{P\Omega}$. Define $a : \{x_1, \ldots, x_n\} \to A;\ x_i \mapsto a_i$. Since A satisfies $u = v$, one obtains $w_1 \varepsilon_A = x_1^a \ldots x_n^a (u, n)_A = x_1^a \ldots x_n^a (v, n)_A = w_2 \varepsilon_A$. Thus $(w_1, w_2) \in \ker \varepsilon_A$.
□

The proof of Birkhoff's Variety Theorem 2.3.3 shows that each variety $\underline{\underline{K}}$ of τ-algebras, with replication $(R_{\underline{\underline{K}}}, G_{\underline{\underline{K}}}, \eta^{\underline{\underline{K}}}, \varepsilon^{\underline{\underline{K}}})$, is specified uniquely as the class of τ-algebras satisfying the *equational theory* Eq($\underline{\underline{K}}$), the relation

$$\text{(2.3.2)} \qquad \operatorname{Eq}(\underline{\underline{K}}) = \ker\left(\eta^{\underline{\underline{K}}}_{P\Omega} : P\Omega \to P\Omega R_{\underline{\underline{K}}} G_{\underline{\underline{K}}}\right)$$

of synonymity in the free $\underline{\underline{K}}$-algebra over P. For a variety $\underline{\underline{L}}$ contained in the variety $\underline{\underline{K}}$, the Third Isomorphism Theorem 1.2.9 yields a congruence Eq $(\underline{\underline{L}}; \underline{\underline{K}})$ on the free algebra $P\Omega R_{\underline{\underline{K}}}$ in $\underline{\underline{K}}$ with (nat Eq $(\underline{\underline{K}})$)* Eq($\underline{\underline{L}}; \underline{\underline{K}}$) = Eq($\underline{\underline{L}}$), as displayed in the following diagram:

$$\text{(2.3.3)} \qquad \begin{array}{ccc} P\Omega & \xrightarrow{k = \text{nat Eq}(\underline{\underline{K}})} & P\Omega R_{\underline{\underline{K}}} \\ \Big\| & & \Big\downarrow {\scriptstyle r = \text{nat Eq}(\underline{\underline{L}};\, \underline{\underline{K}})} \\ P\Omega & \xrightarrow[l = \text{nat Eq}(\underline{\underline{L}})]{} & P\Omega R_{\underline{\underline{L}}} \end{array} .$$

Note that Eq $(\underline{\underline{L}}; \underline{\underline{\tau}})$ = Eq $(\underline{\underline{L}})$. A natural question arises: Which congruences on $P\Omega R_{\underline{\underline{K}}}$ are of the form Eq $(\underline{\underline{L}}; \underline{\underline{K}})$ for a subvariety $\underline{\underline{L}}$ of a variety $\underline{\underline{K}}$? To answer this question, consider a τ-algebra (A, Ω). The monoid $\underline{\underline{\tau}}(A, A)$ of endomorphisms of (A, Ω) may be construed as a domain of unary operations on A. Then congruences of the algebra $(A, \Omega \cup \underline{\underline{\tau}}(A, A))$ are called *fully invariant congruences* of the algebra (A, Ω).

Theorem 2.3.4. *Let* $\tau : \Omega \to \mathbb{N}$ *be a finitary type. Let* $\underline{\underline{K}}$ *be a variety of* τ*-algebras. Then a congruence on* $P\Omega R_{\underline{\underline{K}}}$ *is fully invariant iff it is of the form* Eq $(\underline{\underline{L}}; \underline{\underline{K}})$ *for a subvariety* $\underline{\underline{L}}$ *of* $\underline{\underline{K}}$.

Proof. Use the notation of (2.3.3). First, consider an identity (u, v) with $(uk, vk) \in$ Eq $(\underline{\underline{K}}; \underline{\underline{L}})$, and an endomorphism t of $P\Omega R_{\underline{\underline{K}}}$. By the initiality of $l : P\Omega \to P\Omega R_{\underline{\underline{L}}} G_{\underline{\underline{L}}}$ in $(P\Omega, G_{\underline{\underline{L}}})$, there is an endomorphism t' of $P\Omega R_{\underline{\underline{L}}}$ with $ktr = lt'$. Since $(ul, vl) \in$ Eq $(\underline{\underline{L}})$, one has $uktr = ult' = vlt' = vktr$. Thus $(ukt, vkt) \in$ Eq $(\underline{\underline{K}}; \underline{\underline{L}})$, so that Eq $(\underline{\underline{K}}; \underline{\underline{L}})$ is fully invariant.

Conversely, let θ be a fully invariant congruence of $P\Omega R_{\underline{\underline{K}}}$. Let $\underline{\underline{L}}$ be the variety of all τ-algebras satisfying $k^*\theta$. It will be shown that $k^*\theta = \text{Eq}\,\underline{\underline{L}}$, whence $\theta = \text{Eq}\,(\underline{\underline{L}}; \underline{\underline{K}})$. Certainly $k^*\theta \leq \text{Eq}\,\underline{\underline{L}}$. Conversely, consider the non-empty $\underline{\underline{K}}$-algebra $A = P\Omega R_{\underline{\underline{K}}}{}^{\theta}$. It will be shown that A lies in $\underline{\underline{L}}$, and that A satisfies no identity in $P\Omega^2 - k^*\theta$. Thus $\text{Eq}\,\underline{\underline{L}} \leq k^*\theta$.

Consider an identity (u, v) with $(uk, vk) \in \theta$, and an element f of $\underline{\underline{\tau}}(P\Omega, A)$ or of $(P\Omega, G_{\underline{\underline{K}}})_0$. By the initiality of k in $(P\Omega, G_{\underline{\underline{K}}})$, there is a $\underline{\underline{K}}$-morphism h such that the diagram

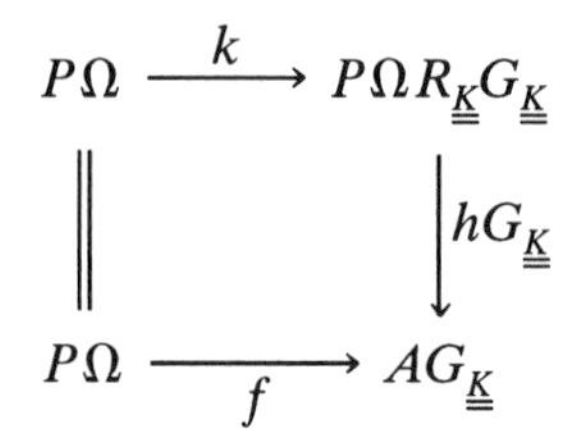

commutes. By the projectivity of $P\Omega R_{\underline{\underline{K}}}$ in $\underline{\underline{K}}$ (Proposition 2.1.6), there is then an endomorphism $\overline{h}$ of $P\Omega R_{\underline{\underline{K}}}$ such that the diagram

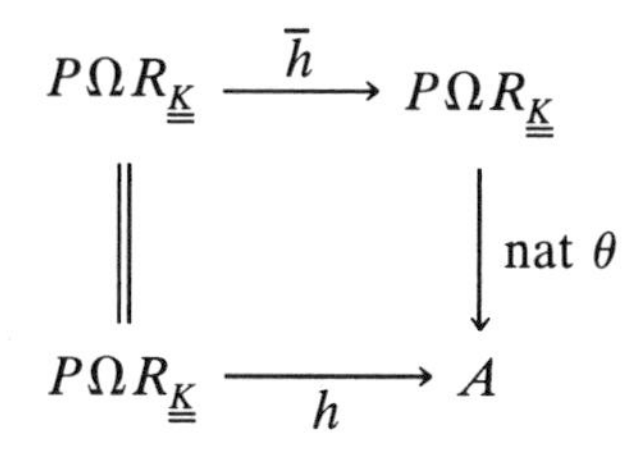

commutes. The full invariance of θ then yields $uf = ukh = uk\overline{h}(\text{nat } \theta) = vf$, so that A satisfies $u = v$. Thus $A \in \underline{\underline{L}}_0$.

Now consider $(u, v) \in P\Omega^2 - k^*\theta$, i.e. $(uk, vk) \notin \theta$. For $k(\text{nat } \theta): P\Omega \to A$, one has $uk(\text{nat } \theta) \neq vk(\text{nat } \theta)$, so that $u = v$ is not satisfied by the $\underline{\underline{L}}$-algebra A. □

Corollary 2.3.5 (Completeness Theorem of Equational Logic). *A relation Σ on $P\Omega$ is an equational theory iff it is closed under the following* deduction rules:

(a) $\forall i > 0, (x_i, x_i) \in \Sigma$;
(b) $(u, v) \in \Sigma \Rightarrow (v, u) \in \Sigma$;
(c) $[(u, v) \in \Sigma \text{ and } (v, w) \in \Sigma] \Rightarrow (u, w) \in \Sigma$;
(d) $\forall \omega \in \Omega, [\forall 1 \leq i \leq \omega\tau, (u_i, v_i) \in \Sigma] \Rightarrow (u_1 \ldots u_{\omega\tau}\, \omega, v_1 \ldots v_{\omega\tau}\, \omega) \in \Sigma$;
(e) $\forall (u, v) \in \Sigma, \forall (w_1, \ldots, w_n) \in P\Omega^{\max((\arg u) \cup (\arg v))}$,
$(w_1 \ldots w_n(u, n)_{P\Omega}, w_1 \ldots w_n(v, n)_{P\Omega}) \in \Sigma$.

Proof. The relation Σ is closed under the deduction rules iff it is a fully invariant congruence of $P\Omega$. For example, closure under (e) corresponds to closure under an endomorphism f of $P\Omega$ with $x_i f = w_i$ for $1 \leq i \leq n$, while (c) corresponds to transitivity. □

EXERCISES

2.3A. Show that $\underline{\underline{K}}_0$ is a variety iff $\underline{\underline{K}}_0 = HSP\underline{\underline{K}}_0$.

2.3B. Let $\underline{\underline{K}}_0$ be a class of τ-algebras.

(a) Show that $HSP\underline{\underline{K}}_0$ is a variety.

(b) If a variety $\underline{\underline{L}}_0$ contains $\underline{\underline{K}}_0$, show that it also contains $HSP\underline{\underline{K}}_0$.

2.3C. Let τ be the type $\{(\varepsilon, 0)\} \cup (\{\beta, \gamma, \delta\} \times \{2\})$ of loops. Consider the derivation of the left inverse property $x_1 = x_2 \varepsilon \gamma x_2 x_1 \beta\beta$ from the First Moufang Identity $x_1 x_2 \beta x_1 \beta x_3 \beta = x_1 x_2 x_1 x_3 \beta\beta\beta$, from (I 2.2.3) IL in the form $x_1 x_1 x_2 \beta\gamma = x_2$, and from (I 2.2.3) SL in the form $x_1 x_1 x_2 \gamma\beta = x_2$, as given in the proof of Proposition I 4.1.5. Write out the derivation as a step-by-step application of the deduction rules of Corollary 2.3.5. How many steps are required?

2.3D. Let I denote the closed real unit interval $[0,1]$ and I° the open unit interval $(0,1)$. Set $\tau = \mathrm{I}^\circ \times \{2\}$, the type of Example 1.3.4(c). Let $\underline{\underline{K}}_0$ be the class of convex sets, as in Example 2.1.5(a).

(a) Show that finite $\underline{\underline{K}}_0$-algebras have at most one element.

(b) Show that $\theta = (\mathrm{I}^\circ \times \mathrm{I}^\circ) \cup \{(0,0),(1,1)\}$ is a fully invariant congruence on $(\mathrm{I}, \mathrm{I}^\circ)$.

(c) Show that I^θ is not a member of the class $\underline{\underline{K}}_0$.

2.3E. **(Axiom Rank and Base Rank.)** Let $\underline{\underline{K}}$ be a variety of algebras of type $\tau : \Omega \to \mathbb{N}$. For each positive integer n, define

$$\underline{\underline{K}}_n = HSP\big\{\{x_1, \ldots, x_n\}\Omega R_{\underline{\underline{K}}}\big\}$$

and

$$\underline{\underline{K}}^n = HSP\big\{P\Omega\,\mathrm{nat}\,\big(\mathrm{Eq}\,(\underline{\underline{K}}) \cap \{x_1, \ldots, x_n\}\Omega^2\big)\big\}.$$

(a) Prove $\underline{\underline{K}}_1 \subseteq \underline{\underline{K}}_2 \subseteq \cdots \subseteq \underline{\underline{K}} \subseteq \cdots \subseteq \underline{\underline{K}}^2 \subseteq \underline{\underline{K}}^1$.

(b) Prove $\underline{\underline{K}} = \bigcup_{n>0} \underline{\underline{K}}_n = \bigcap_{n>0} \underline{\underline{K}}^n$.

The *axiom rank* of $\underline{\underline{K}}$ is

$$r_a(\underline{\underline{K}}) = \min\{n \mid \underline{\underline{K}} = \underline{\underline{K}}_n\}.$$

The *base rank* of $\underline{\underline{K}}$ is

$$r_b(\underline{\underline{K}}) = \min\{n \mid \underline{\underline{K}} = \underline{\underline{K}}^n\}.$$

(c) Assuming the fact that the free group on 2 generators contains a subgroup that is free on a countable set, prove that the base rank of the variety of groups is 2.

(d) Assuming the fact that 2-generated subloops of the loops Q of Examples I 4.1.3 and I 4.1.4 are associative, prove that the axiom rank of the variety of groups is 3.

2.3F. Let $\tau : \Omega \to \mathbb{N}$ be a finitary type. Let A be a τ-algebra generated by a subset X. Show that A is the free algebra over X in a variety $\underline{\underline{K}}$ of τ-algebras iff each function $f : X \to A$ is the restriction to X of an endomorphism $\bar{f}$ of (A, Ω).

2.3G. Determine the two varieties of algebras of empty type.

2.3H. Let $\underline{\underline{K}}$ be a variety containing an algebra of cardinality greater than 1. Show that $\underline{\underline{K}}$ contains a simple algebra of cardinality greater than 1.

2.3I. A *Steiner triple system* (S, B) is a set S together with a set B of 3-element subsets of S, such that each 2-element subset of S is a subset of a unique element of B. The elements of B are called *triples* or *blocks*. (Thus the crews of Example I 2.5 are the blocks of a 7-element Steiner triple system.)

(a) Let $\underline{\underline{\mathrm{STS}}}$ be the variety of magmas satisfying the identities $x_1x_1\,\mu = x_1$, $x_1x_2\,\mu = x_2x_1\,\mu$ and $x_1x_2\,\mu x_2\,\mu = x_2$. For each element S of $\underline{\underline{\mathrm{STS}}}_0$, show that there is a Steiner triple system (S, B) with $B = \big\{\{x, y, xy\mu\} \mid x \neq y \in S\big\}$.

(b) If (S, B) is a Steiner triple system, show that a magma (S, μ) is defined by $xx\mu = x$ and $\{x, y, xy\mu\} \in B$ for $x \neq y$. Furthermore, show that (S, μ) is an object of $\underline{\underline{\mathrm{STS}}}$.

(c) Let M be the submonoid of $(\mathbb{N}, \cdot, 1)$ generated by $\{3, 7\}$. Show that

$$n \in M \Rightarrow \exists S \in \underline{\underline{\mathrm{STS}}}_0 .\ |S| = n.$$

[Hint: Use (I 2.4) together with $H\underline{\underline{\mathrm{STS}}} = S\,\underline{\underline{\mathrm{STS}}} = P\underline{\underline{\mathrm{STS}}} = \underline{\underline{\mathrm{STS}}}$.]

2.4. Entropic Algebras and Tensor Products

Fix a finitary type $\tau : \Omega \to \mathbb{N}$. Consider an algebra (A, Ω) of type τ. For each operator ω from Ω, the corresponding operation

$$\omega : A^{\omega\tau} \to A \tag{2.4.1}$$

is a map to A from the direct power $A^{\omega\tau}$ of the set A. Generalizing from Exercise I 2.2M, the algebra A is said to be *entropic* if each operation ω gives a τ-homomorphism

$$\omega : (A, \Omega)^{\omega\tau} \to (A, \Omega) \tag{2.4.2}$$

to (A, Ω) from the direct power $(A, \Omega)^{\omega\tau}$ of the algebra (A, Ω). Thus by

(1.2.1), the algebra (A, Ω) is entropic iff it satisfies the identity

$$(2.4.3) \qquad (x_1 \dots x_{\omega\tau}\,\omega) \dots (x_{(\psi\tau-1)\,\omega\tau+1} \dots x_{\psi\tau\cdot\omega\tau}\,\omega)\,\psi$$
$$= (x_1 \dots x_{(\psi\tau-1)\,\omega\tau+1}\,\psi) \dots (x_{\omega\tau} \dots x_{\psi\tau\cdot\omega\tau}\,\psi)\,\omega$$

for each subset $\{\psi, \omega\}$ of Ω. In particular, Proposition 2.3.2 yields part (a) of the following:

Proposition 2.4.1. *(a) Let $\underline{\underline{K}}_0$ be a class of entropic τ-algebras. Then each member of HSP$\underline{\underline{K}}_0$ is entropic.*
(b) Reducts of entropic algebras are entropic. □

Example 2.4.2. (a) **(Modules over Commutative, Unital Rings.)** Let S be a unital ring. Consider the type $\{(0,0),(-,2)\} \cup (S \times \{1\})$ of (right) S-modules. For a subset $\{s, t\}$ of S, (2.4.3) reduces to the identity $x_1ts = x_1st$. Applying the unary derived operations on each side of the identity to 1 in S, one sees that the right S-module S_S is entropic only if S is commutative. Conversely, if S is commutative, then each S-module is entropic.
(b) **(Convex Sets.)** By (a), real vector spaces are entropic. By Proposition 2.4.1(b), reducts of real vector spaces are entropic. Then by Proposition 2.4.1(a), convex sets [as in Example 1.3.4(c)] are entropic. □
Abelian groups are entropic [take $S = \mathbb{Z}$ in Example 2.4.2(a)]. In Example III 2.4.3, abelian groups were identified as groups in the category of groups. One may abstract this observation from groups to universal algebra.

Definition 2.4.3. Let C be a category with finite products. Then a *τ-algebra in the category C* is an object A of C, equipped with a C-morphism $\omega : A^{\omega\tau} \to A$ for each ω in Ω. □

Using Definition 2.4.3, entropic τ-algebras are characterized as τ-algebras in the category $\underline{\underline{\tau}}$.
Consider a τ-algebra A in a category C with finite products. The basic operations ω of the type τ act on A as C-morphisms $\omega : A^{\omega\tau} \to A$. Concatenating the actions of basic operations, one then obtains A as a $\bar{\tau}$-algebra in the category C. [For example, diagonals in (III 2.4.3) provide derived unary operations $x_1x_1J\mu$ and $x_1Jx_1\,\mu$.] The τ-algebra A in C is said *to satisfy an identity* $u = v$ if the n-ary clone operations $(u, n) : A^n \to A$ and $(v, n) : A^n \to A$ are equal elements of $C(A^n, A)$, for $n = \max((\arg u) \cup (\arg v))$. Using the Birkhoff Variety Theorem 2.3.3, one may then refine Definition 2.4.3.

Definition 2.4.4. Let C be a category with finite products. Let $\underline{\underline{K}}$ be a variety of τ-algebras.
(a) A *$\underline{\underline{K}}$-algebra in the category C* is a τ-algebra A in C satisfying each member of the equational theory Eq$(\underline{\underline{K}})$ of identities characterizing $\underline{\underline{K}}$.

(b) A *$\underline{\underline{K}}$-algebra morphism in the category C* is a C-morphism $f: A \to B$ between $\underline{\underline{K}}$-algebras in the category C such that, for each operation ω, the diagram

$$\begin{array}{ccc} A^{\omega\tau} & \xrightarrow{\omega} & A \\ {\scriptstyle (f,\ldots,f)}\downarrow & & \downarrow{\scriptstyle f} \\ B^{\omega\tau} & \xrightarrow{\omega} & B \end{array} \tag{2.4.4}$$

in C commutes.

(c) The *category $C \otimes \underline{\underline{K}}$ of $\underline{\underline{K}}$-algebras in the category C* is the subcategory of C consisting of $\underline{\underline{K}}$-algebras and $\underline{\underline{K}}$-algebra morphisms in the category C. □

Example 2.4.5. For a group G and unital ring S, consider an S-representation $R: G \to \underline{\underline{\text{Mod}}}_S$ of G, with corresponding module $1R = M$. Then the split extension $G[_R M$ is an S-module in the slice category $\underline{\underline{\text{Gp}}}/G$. The unary operations corresponding to elements s of S are morphisms $s:(g, m) \mapsto (g, ms)$. □

For a τ-algebra A and a set X, the set $\underline{\underline{\text{Set}}}(X, A)$ of maps from X to A carries the τ-algebra $(A, \Omega)^X$. One may then recognize entropicity of A by the closure of sets of τ-homomorphisms into A under the operations of $(A, \Omega)^X$.

Proposition 2.4.6. *A τ-algebra A is entropic iff, for each τ-algebra X, the set $\underline{\underline{\tau}}(X, A)$ is a subalgebra of A^X.*

Proof. Let ψ be an m-ary operation and ω an n-ary operation. First, suppose that A is entropic and that X is a τ-algebra. Given $f_1, \ldots, f_n$ in $\underline{\underline{\tau}}(X, A)$ and $a_1, \ldots, a_m$ in A, set $x_{(q-1)n+r} = a_q f_r$ in (2.4.3), for $1 \le q \le m$ and $1 \le r \le n$. One obtains

$$\begin{aligned} &(a_1 f_1 \ldots f_n \omega) \ldots (a_m f_1 \ldots f_n \omega) \psi \\ &\quad = (a_1 f_1 \ldots a_1 f_n \omega) \ldots (a_m f_1 \ldots a_m f_n \omega) \psi \\ &\quad = (a_1 f_1 \ldots a_m f_1 \psi) \ldots (a_1 f_n \ldots a_m f_n \psi) \omega \\ &\quad = (a_1 \ldots a_m \psi f_1) \ldots (a_1 \ldots a_m \psi f_n) \omega \\ &\quad = (a_1 \ldots a_m \psi) \ldots (a_1 \ldots a_m \psi)(f_1 \ldots f_n \omega), \end{aligned}$$

so that $f_1 \ldots f_n \omega$ is again in $\underline{\underline{\tau}}(X, A)$, as required.

Conversely, suppose that $\underline{\underline{\tau}}(X, A)$ is a subalgebra of A^X for each τ-algebra X. Consider the projections $\pi_i : (a_1, \ldots, a_n) \mapsto a_i$ in $\underline{\underline{\tau}}(A^n, A)$. Then for

$(a_1, \ldots, a_{mn}) \in A^{mn}$, one has

$$\begin{aligned}(a_1 \ldots a_n \omega) \ldots (a_{(m-1)n+1} \ldots a_{mn} \omega)\psi \\ &= (a_1, \ldots, a_n)(\pi_1 \ldots \pi_n \omega) \ldots (a_{(m-1)n+1}, \ldots, a_{mn})(\pi_1 \ldots \pi_n \omega)\psi \\ &= (a_1, \ldots, a_n) \ldots (a_{(m-1)n+1}, \ldots, a_{mn})\psi(\pi_1 \ldots \pi_n \omega) \\ &= (a_1 \ldots a_{(m-1)n+1}\psi, \ldots, a_n \ldots a_{mn}\psi)(\pi_1 \ldots \pi_n \omega) \\ &= (a_1 \ldots a_{(m-1)n+1}\psi) \ldots (a_n \ldots a_{mn}\psi)\omega,\end{aligned}$$

so that A satisfies the identities (2.4.3). □

For the rest of this section, let $\underline{\underline{K}}$ be a prevariety of entropic τ-algebras. Proposition 2.4.6 shows that for each fixed $\underline{\underline{K}}$-algebra Y, there is a functor $R_Y : \underline{\underline{K}} \to \underline{\underline{K}}$ with morphism part $(f : X \to X') \mapsto (\underline{\underline{K}}(Y, X) \to \underline{\underline{K}}(Y, X'); g \mapsto gf)$. The functor R_Y is a generalization of the functor R_Y of Section III 3.6 defined on the category of modules over a commutative, unital ring K, since this category is itself a prevariety of entropic algebras. The functor R_Y of Section III 3.6 was shown to have a left adjoint S_Y, whose object part assigned the tensor product $Z \otimes_K Y$ to each K-module Z. Now the general functor $R_Y : \underline{\underline{K}} \to \underline{\underline{K}}$ also possesses a left adjoint $S_Y : \underline{\underline{K}} \to \underline{\underline{K}}$. The proof carries over almost verbatim from Section III 3.6, replacing K and $\underline{\underline{\mathrm{Mod}}}_K$ by $\underline{\underline{K}}$ as appropriate. The set dominating the comma category (Z, R_Y) of *τ-bihomomorphisms* $Z \times Y \to X$ is

$$\bigcup_{\alpha \in \mathrm{Cg}_{\underline{\underline{K}}} M} \underline{\underline{K}}(Z, \underline{\underline{K}}(Y, M^\alpha)), \tag{2.4.5}$$

the union being taken over the set of $\underline{\underline{K}}$-congruences α on the free $\underline{\underline{K}}$-algebra M over the set $Z \times Y$. One replaces quotients M/N in Section III 3.6 with quotients M^α by such congruences, for example in (III 3.6.2) and (III 3.6.3). Summarizing:

Theorem 2.4.7. *Let $\underline{\underline{K}}$ be a prevariety of entropic τ-algebras. Let Y be a $\underline{\underline{K}}$-algebra. Then the functor $R_Y : \underline{\underline{K}} \to \underline{\underline{K}}$ with morphism part $(f : X \to X') \mapsto (\underline{\underline{K}}(Y, X) \to \underline{\underline{K}}(Y, X'); g \mapsto gf)$ possesses a left adjoint $S_Y : \underline{\underline{K}} \to \underline{\underline{K}}$.* □

The image ZS_Y of a $\underline{\underline{K}}$-algebra Z under the functor S_Y is called the *tensor product algebra* $Z \otimes_{\underline{\underline{K}}} Y$ or $Z \otimes Y$. Some key properties of the tensor product are listed in the following:

Proposition 2.4.8. *Let $\underline{\underline{K}}$ be a prevariety of entropic τ-algebras.*

(*a*) *For $\underline{\underline{K}}$-algebras X, Y and Z, the commutative law $X \otimes Y \cong Y \otimes X$ and associative law $(X \otimes Y) \otimes Z \cong X \otimes (Y \otimes Z)$ hold.*

(*b*) *For a $\underline{\underline{K}}$-algebra Y and a multiset $\langle Z_i \mid i \in I \rangle$ of $\underline{\underline{K}}$-algebras, the distributive law $Y \otimes \coprod_{i \in I} Z_i \cong \coprod_{i \in I} Y \otimes Z_i$ holds.*

(*c*) *For a* $\underline{\underline{K}}$*-algebra* Y *and the free* $\underline{\underline{K}}$*-algebra* $T\Omega R_{\underline{\underline{K}}}$ *over a singleton set* T *(terminal object of* $\underline{\underline{\mathrm{Set}}}$*), there is a natural isomorphism* $Y \otimes T\Omega R_{\underline{\underline{K}}} \cong Y$.

(*d*) *Tensor products of free* $\underline{\underline{K}}$*-algebras are free. Indeed* $(I \times J)\Omega R_{\underline{\underline{K}}} \cong I\Omega R_{\underline{\underline{K}}} \otimes J\Omega R_{\underline{\underline{K}}}$ *for sets* I *and* J.

Proof. (b) follows by cocontinuity of the left adjoint S_Y [Theorem III 3.4.1(b)]. Note that $\underline{\underline{K}}$ is cocomplete, by Theorem 2.2.3. The other results follow by the uniqueness of adjoints to within isomorphism. Note that (a) is the analogue of (III 3.6.6) and Exercise III 3.6B, that (c) is the analogue of Example III 3.6.2, and that (d) is the analogue of Example III 3.6.3. □

An interesting application of the concept of tensor product of entropic algebras is obtained by taking $\underline{\underline{K}}$ to be the prevariety of convex sets, as in Example 2.1.5(a). If a finite set X represents the set of (pure) states of a (deterministic) system, then the free $\underline{\underline{K}}$-algebra $X\Omega R_{\underline{\underline{K}}}$ over X [the $(|X| - 1)$-simplex in the terminology of Example 2.1.5(a) and Exercise 2.1D] represents the set of probabilistically mixed states of a corresponding non-deterministic system. For instance, given pure states x and y, and for $p \in \mathrm{I}^{\circ} = (0, 1)$, the state $xy\underline{p}$ [notation of Example 1.3.4(c)] is the mixed state yielding y with probability p and yielding x with probability $1 - p$. Now if A and B are convex sets representing probabilistically mixed systems in this sense, the direct product $A \times B$ represents a combined system in which the constituents A and B remain independent. On the other hand, the tensor product $A \otimes B$ represents a combined system in which the constituents A and B are coupled, no longer remaining independent. As a simple example, take $A = \{a_1, a_2\}\Omega R_{\underline{\underline{K}}}$ and $B = \{b_1, b_2\}\Omega R_{\underline{\underline{K}}}$. Compare the mixed state $(0.01)(a_1, b_1) + (0.97)(a_1, b_2) + (0.01)(a_2, b_1) + (0.01)(a_2, b_2)$ of $A \otimes B$ with the mixed state $(0.98a_1 + 0.02a_2, 0.50b_1 + 0.50b_2)$ of $A \times B$. Note that one is most likely to observe a_1 in the constituent A of $A \otimes B$, and then one is most likely to observe b_2 in its second constituent. On the other hand, although one is most likely to observe a_1 in the first constituent of $A \times B$, this does not make any one state of the second constituent more likely than the other. The distributions of the states in the constituents are coupled in $A \otimes B$ but independent in $A \times B$.

EXERCISES

2.4A. Prove part (b) of Proposition 2.4.1.

2.4B. Give a complete verification that unital modules over a commutative ring are entropic.

2.4C. Give a direct verification that convex sets (A, I°) are entropic [cf. Example 1.3.4(c)].

2.4D. Let M be a monoid. Show that each M-set is entropic iff M is commutative.

2.4E. Verify the details of Example 2.4.5.

2.4F. Write out a proof of Theorem 2.4.7.

2.4G. (a) Show that commutative monoids are entropic.

(b) In Section I 1.5, the free commutative monoid $X^{*\kappa}$ on a set X was interpreted as the set of finite multisubsets of X. Given sets X and Y, interpret the coproduct $X^{*\kappa} \oplus Y^{*\kappa}$ and the tensor product $X^{*\kappa} \otimes Y^{*\kappa}$.

2.4H. (a) Show that semilattices are entropic magmas.

(b) Show that the poset of non-empty subsets of a finite set X is the free (join) semilattice over X.

(c) Let S be the join semilattice with Hasse diagram $a \longrightarrow b \longleftarrow c$. Show that $S \otimes S$ has 15 elements.

2.4I. Let X be a finite set, and let $\underline{\underline{K}}$ be the prevariety of convex sets. If $X\Omega R_{\underline{\underline{K}}}$ represents the mixed states of a system with state space X, show that the pure states are represented by the minimal non-empty walls of the algebra $(X\Omega R_{\underline{\underline{K}}}, \mathrm{I}^{\circ})$.

2.4J. For τ-algebras A, B, C in a prevariety $\underline{\underline{K}}$ of entropic algebras, determine whether or not the "mixed associative law" $(A \times B) \otimes C \cong A \times (B \otimes C)$ is satisfied.

2.4L. Let C be a locally small category with finite products. Let $\underline{\underline{K}}$ be a variety of τ-algebras. Show that a C-object A is a $\underline{\underline{K}}$-algebra in C iff, for each C-object Y, the set $C(Y, A)$ lies in $\underline{\underline{K}}$. [Hint: $C(Y, A)^{\omega\tau} \cong C(Y, A^{\omega\tau})$.]

2.4M. Let M be a monoid. Show that an M-set in $\underline{\underline{\mathrm{Set}}}^{\mathrm{op}}$ is an M^{op}-set in $\underline{\underline{\mathrm{Set}}}$.

2.4N. An undirected graph T is a *tree* if each pair of vertices in T is connected by a unique path of edges followed without retracing. The *centroid* of a triple (x, y, z) of vertices in a tree T is the unique vertex common to the paths connecting x with y, y with z, and z with x.

(a) Show that (T, γ) is an algebra of type $\{(\gamma, 3)\}$ under the operation $(x, y, z) \mapsto (x, y, z)\gamma$ mapping a triple of vertices to its centroid.

(b) Show that there is a tree T for which (T, γ) is not entropic.

3. ALGEBRAIC THEORIES

Let S be a unital ring. Let $P = \{x_1 < x_2 < \cdots\}$ be the set of positive integers, written as in (1.3.4). For each natural number n, consider the free (right) S-module nS on the initial subset $\{x_1, \ldots, x_n\}$ of P. Elements of nS have the form

(3.1) $$x_1 s_1 + x_2 s_2 + \cdots + x_n s_n$$

with $s_i \in S$. In other words, nS is the set of n-linear combinations (cf.

Example II 2.3.1). Note that $0S$ is the singleton $\{0\}$. Now let S (same symbol as for the ring) denote the full subcategory of $\underline{\underline{\mathrm{Mod}}}_S$ whose set S_0 of objects is $\{nS \mid n \in \mathbb{N}\}$. This category is called the *algebraic theory* S. It is small. Moreover, each object nS of S is the coproduct of the finite number n of copies of $1S$. For positive integers l and m, the morphism set $S(lS, mS)$ may be identified with the set S_l^m of $(l \times m)$-matrices over the ring S. Thus the algebraic theory S comprises both the set S_1 of matrices over the ring S and the set S_0 of indexed sets of linear combinations over the ring S.

Now consider a right S-module V. For each natural number m, recall the isomorphism $V^m \cong S(S, V)^m \cong S(mS, V)$ of Section II 2.3 and Exercise II 2H. Under this isomorphism, the element $(v_1, \dots, v_m)$ of V^m is identified with the unique S-module homomorphism $v: mS \to V$; $x_1 s_1 + \cdots + x_m s_m \mapsto v_1 s_1 + \cdots + v_m s_m$ specified by the substitutions $x_i \mapsto v_i$ for $1 \le i \le m$. Each morphism $f: lS \to mS$ in the algebraic theory S yields a map

$$S(f, V): S(mS, V) \to S(lS, V);\ v \mapsto fv. \tag{3.2}$$

In particular, the morphism $\lambda: 1S \to mS; x_1 \mapsto x_1 s_1 + \cdots + x_m s_m$ yields the map $S(\lambda, V): S(mS, V) \to S(1S, V)$; $v \mapsto \lambda v$ that takes the form

$$S(\lambda, V): V^m \to V^1; (v_1, \dots, v_m) \mapsto v_1 s_1 + \cdots + v_m s_m \tag{3.3}$$

of the corresponding m-ary linear combination on identifying v with $(v_1, \dots, v_m)$ and $\lambda v: 1S \to V$; $x_1 \mapsto x_1 s_1 + \cdots + x_m s_m \mapsto v_1 s_1 + \cdots + v_m s_m$ with $v_1 s_1 + \cdots + v_m s_m$ [cf. (II 2.3.1) with $I = \{1, \dots, m\}$]. As in (II 2.1.4), one has $S(g, V)S(f, V) = S(fg, V)$ for morphisms $f: lS \to mS$ and $g: mS \to nS$ in the algebraic theory S. Thus the right S-module V determines the functor $S(\underline{\quad}, V): S^{\mathrm{op}} \to \underline{\underline{\mathrm{Set}}}$. This functor sends the finite coproducts mS in S to the products V^m in $\underline{\underline{\mathrm{Set}}}$. In other words, it is a finite-product-preserving functor from S^{op} to $\underline{\underline{\mathrm{Set}}}$.

Conversely, suppose that a functor $F: S^{\mathrm{op}} \to \underline{\underline{\mathrm{Set}}}$ preserves products. Let V be the set $1S^F$. Then mS^F is the power V^m. Consider an S-morphism $\lambda: 1S \to mS; x_1 \mapsto x_1 s_1 + \cdots + x_m s_m$. Its image under the functor F is an m-ary operation $\lambda^F: V^m \to V$ on the set V. Consider the type $\tau = (S \times \{1\}) \cup \{(0,0), (+,2)\}$ of S-modules. For each operation ω of this type, there is a corresponding element of $\omega\tau S$. The unary operations (scalar multiplications) s appear as $x_1 s$ in $1S$, the nullary operation 0 is the unique element of $0S$, and the binary operation $+$ corresponds to $x_1 + x_2$ in $2S$. Thus each operation ω determines the S-morphism $\omega: 1S \to \omega\tau S$; $x_1 \mapsto x_1 \dots x_{\omega\tau}\omega$. The images $\omega^F: V^{\omega\tau} \to V$ of these S-morphisms under the contravariant functor F make V into a τ-algebra. Moreover, V satisfies the identities for an S-module. As an example, consider the identity $x_1 + x_1(-1) = 0$. The composite of the S-morphisms $+: 1S \to 2S; x_1 \mapsto x_1 + x_2$ and $2S \to 1S; x_1 \mapsto x_1, x_2 \mapsto x_1(-1)$ is the S-morphism $0: 1S \to 1S$. The image of the composite under the functor F is the derived operation $x_1 + x_1(-1)$ on V, which thus coincides with the scalar multiplication 0 on V. Finally, one has $S(mS, V) \cong V^m = mS^F$ for natural numbers m, yielding a natural isomorphism between the functors $S(\_\_\_, 1S^F)$ and F. Summarizing, S-modules V correspond exactly to product-preserving functors $S^{\mathrm{op}} \to \underline{\underline{\mathrm{Set}}}$, via the assignments $V \mapsto S(\_\_, V)$ and $F \mapsto 1S^F$.

Now consider an S-module homomorphism $\theta : V \to W$. One obtains a natural transformation $\theta : S(\_\_\_, V) \to S(\_\_\_, W)$ whose component at the S-object mS is $\theta_{mS} : S(mS, V) \to S(mS, W)$; $v \mapsto v\theta$. At an S-morphism [i.e. $(l \times m)$-matrix] $f : lS \to mS$, the commuting of the diagram (III 1.4.2) follows as $vS(f, V)\theta_{lS} = (fv)\theta = f(v\theta) = v\theta_{mS}S(f, W)$. Conversely, consider a natural transformation $\tau : S(\_\_\_, V) \to S(\_\_\_, W)$. For each S-morphism $\lambda : 1S \to mS$, the naturality (III 1.4.2) of τ yields the commuting diagram

$$
\begin{array}{ccc}
S(mS, V) & \xrightarrow{\tau_{mS}} & S(mS, W) \\
{\scriptstyle S(\lambda, V)} \downarrow & & \downarrow {\scriptstyle S(\lambda, W)} \\
S(1S, V) & \xrightarrow[\tau_{1S}]{} & S(1S, W)
\end{array}
\tag{3.4}
$$

of functions. Setting $\lambda : 1S \to mS$ to be the insertion $\iota_i : 1S \to mS$; $x_1 \mapsto x_i$ for each i in $\{1, \dots, m\}$, one obtains $S(\lambda, V)$ and $S(\lambda, W)$ in (3.4) as the respective projections $\pi_i : V^m \to V$ and $\pi_i : W^m \to W$. Thus τ_{mS} is the product $(\tau_{1S}, \dots, \tau_{1S})$ of m copies of $\tau_{1S} : V \to W$. Now considering general $\lambda : 1S \to mS$, yielding $S(\lambda, V)$ and $S(\lambda, W)$ as the corresponding m-ary operations (3.3) on V and W, the diagram (3.4) shows that $\tau_{1S} : V \to W$ is an S-module homomorphism. In summary, one obtains the following:

Theorem 3.1. *Let S be the algebraic theory of a unital ring. Let $\underline{\underline{S}}$ be the full subcategory of the functor category $\underline{\underline{\mathrm{Set}}}^{S^{\mathrm{op}}}$ whose objects preserve finite products. Then $\underline{\underline{S}}$ is isomorphic to the category $\underline{\underline{\mathrm{Mod}}}_S$.*

Proof. For F in $\underline{\underline{S}}_0$, define

$$\Omega_F = \{\lambda^F : mS^F \to 1S^F \mid m \in \mathbb{N}, \lambda \in S(1S, mS)\}. \tag{3.5}$$

The isomorphism between $\underline{\underline{S}}$ and $\underline{\underline{\mathrm{Mod}}}_S$ is then given by the functors with morphism parts

$$(\tau : F \to F') \mapsto [\tau_{1S} : (1S^F, \Omega_F) \to (1S^{F'}, \Omega_{F'})] \tag{3.6}$$

and

(3.7)

$$(\theta : V \to W) \mapsto \left[\theta : S_0 \to \underline{\underline{\mathrm{Set}}}_1; mS \mapsto (S(mS,V) \to S(mS,W); v \mapsto v\theta)\right].$$

Note that the isomorphism commutes with the forgetful functors $\underline{\underline{\mathrm{Mod}}}_S \to \underline{\underline{\mathrm{Set}}}$ and $\underline{\underline{S}} \to \underline{\underline{\mathrm{Set}}}$; $F \mapsto 1S^F$. □

EXERCISES

3A. Verify that the τ-algebra $V = 1S^F$ satisfies the S-module identities $x_1(-1) + (x_1 + x_2) = x_2$ and $(x_1 s)t = x_1(st)$ for $s, t \in S$.

3B. Consider the type $\sigma = (S \times \{1\}) \cup \{(0,0), (-,2)\}$ of S-modules. Show that each product-preserving functor $F : S \to \underline{\underline{\mathrm{Set}}}$ yields a σ-algebra structure on the set $1S^F$. Show that the σ-algebra $1S^F$ satisfies the identity $x_1 - (x_2 - x_3) = (x_1 - x_2) - x_3(-1)$.

3C. For an S-module homomorphism $\theta : V \to W$, verify the naturality (III 1.4.2) of the transformation $\theta : S(\_\_, V) \to S(\_\_, W)$ with components $\theta_{mS} : S(mS, V) \to S(nS, V)$; $v \mapsto v\theta$ at each natural number m.

3D. Let Sl be the small full subcategory of the category $\underline{\underline{\mathrm{Sl}}}$ of semilattices whose objects are of the form $n\mathrm{Sl} = (\mathscr{P}\{x_1, \ldots, x_n\} - \{\varnothing\}, \cup)$ for natural numbers n.

(a) Show that $n\mathrm{Sl}$ is the coproduct in Sl of n copies of the object $1\mathrm{Sl} = (\{\{x_1\}\}, \cup)$, for each natural number n.

(b) Show that each semilattice $(H, +)$ may be identified with $\underline{\underline{\mathrm{Sl}}}(1\mathrm{Sl}, H)$ via $\underline{\underline{\mathrm{Sl}}}(1\mathrm{Sl}, H) \to H$; $\theta \mapsto \{x_1\}\theta$ (cf. Exercise II 2A and Section II 2.3).

(c) For each semilattice $(H, +)$ and natural number n, show that there is a natural isomorphism $H^n \to \underline{\underline{\mathrm{Sl}}}(n\mathrm{Sl}, H)$; $(h_1, \ldots, h_n) \mapsto (\{x_i\} \mapsto h_i)$.

(d) Show that a semilattice H yields a product-preserving functor $\underline{\underline{\mathrm{Sl}}}(\_\_, H)$: $\mathrm{Sl}^{\mathrm{op}} \to \underline{\underline{\mathrm{Set}}}$ with morphism part

$$(f : l\mathrm{Sl} \to m\mathrm{Sl}) \mapsto (\underline{\underline{\mathrm{Sl}}}(f, H) : \underline{\underline{\mathrm{Sl}}}(m\mathrm{Sl}, H) \to \underline{\underline{\mathrm{Sl}}}(l\mathrm{Sl}, H); h \mapsto fh)$$

[cf. (3.2)].

(e) Show that a product-preserving functor $F : \mathrm{Sl}^{\mathrm{op}} \to \underline{\underline{\mathrm{Set}}}$ yields a semilattice $(1\mathrm{Sl}^F, +)$ with join $(1\mathrm{Sl} \to 2\mathrm{Sl}; \{x_1\} \mapsto \{x_1, x_2\})^F$.

(f) Show that the full subcategory of the functor category $\underline{\underline{\mathrm{Set}}}^{\mathrm{Sl}^{\mathrm{op}}}$ whose objects preserve products is isomorphic to the category $\underline{\underline{\mathrm{Sl}}}$ of semilattices.

3.1. Set-Valued Functors

The algebraic theory S of a unital ring is a small category. Theorem 3.1 identifies modules over the ring with certain functors from S^{op} to $\underline{\underline{\mathrm{Set}}}$, namely those that preserve products. In preparation for the study of these and more general algebraic theories, as well as for other purposes, this section and the next consider the relationship between a small category V and the large functor category

$$\hat{V} := \underline{\underline{\mathrm{Set}}}^{V^{\mathrm{op}}}. \tag{3.1.1}$$

The objects of $\hat{V}$ are sometimes described as *presheaves* over the category V. Note that the large category $\hat{V}$ is locally small. For objects $F : V^{\mathrm{op}} \to \underline{\underline{\mathrm{Set}}}$ and $G : V^{\mathrm{op}} \to \underline{\underline{\mathrm{Set}}}$ of $\hat{V}$, the morphism class $\hat{V}(F, G)$ is the class of natural transformations $\tau : F \to G$. Such a natural transformation is given by the function $\tau : V_0 \to \bigcup_{A \in V_0} \underline{\underline{\mathrm{Set}}}(AF, AG)$ of (III 1.4.1). Thus the class $\hat{V}(F, G)$ is a subset of the set $\underline{\underline{\mathrm{Set}}}(V_0, \bigcup_{A \in V_0} \underline{\underline{\mathrm{Set}}}(AF, AG))$.

It is convenient to establish some notation. The arrow in V^{op} corresponding to an arrow $f : A_0 \to A_1$ in V will occasionally be written as $f^{\mathrm{op}} : A_1 \to A_0$. The objects of $\hat{V}$ are covariant functors from V^{op} to $\underline{\underline{\mathrm{Set}}}$, but they will sometimes be considered as contravariant functors from V to $\underline{\underline{\mathrm{Set}}}$. Given such a contravariant functor $F : V \to \underline{\underline{\mathrm{Set}}}$, the image of a V-morphism $f : A \to B$ will be written as

$$f^F : BF \to AF; x \mapsto f^F(x). \tag{3.1.2}$$

In other words, the image function f^F is written on the left of its argument. This has the advantage that, given a second V-morphism $g : B \to C$ and an element y of CF, one may write $fg^F(y) = (fg)^F(y) = f^F \circ g^F(y) = f^F(g^F(y))$.

The connection between V and $\hat{V}$ is given by the (covariant!) *Yoneda functor*

$$\exists : V \to \hat{V}; (g : A \to B) \mapsto (V(\_, g) : V(\_, A) \to V(\_, B)).$$

Thus for an object A of V, the contravariant functor $A\exists$ from V to $\underline{\underline{\mathrm{Set}}}$ has morphism part

$$A\exists : (f : A_0 \to A_1) \mapsto (fA\exists : V(A_1, A) \mapsto V(A_0, A); h \mapsto fh).$$

Using the notation (3.1.2), one has $f^{A\exists}(h) = fh$ for h in $V(A_1, A)$. Given a morphism $g : A \to B$ in V, the natural transformation $G\exists : A\exists \to B\exists$ has component

$$g\exists_{A_0} : V(A_0, A) \to V(A_0, B); h \mapsto hg$$

at an object A_0 of V^{op}. The naturality (III 1.4.2) of $g\exists$ at a V^{op}-morphism

$f^{\mathrm{op}} : A_1 \to A_0$ is the commuting of the diagram

$$
(3.1.3) \qquad \begin{array}{ccc} A_1^{A\exists} & \xrightarrow{g\exists_{A_1}} & A_1^{B\exists} \\ {\scriptstyle f^{A\exists}}\downarrow & & \downarrow{\scriptstyle f^{B\exists}} \\ A_0^{A\exists} & \xrightarrow[g\exists_{A_0}]{} & A_0^{B\exists} \end{array} .
$$

Consider an element h of $A_1^{A\exists} = V(A_1, A)$. Then $f^{A\exists}(h)(g\exists_{A_0}) = (fh)g = f(hg) = f^{B\exists}(h(g\exists_{A_1}))$ as required: The naturality of $g\exists$, i.e. the commutativity of (3.1.3), reduces to the associativity (III 1.4) in the category V.

The crucial property of the Yoneda functor $\exists : V \to \hat{V}$ is described by the following:

Yoneda Lemma 3.1.1. *Given objects B of V and F of $\hat{V}$, there are mutually inverse bijections*

$$\hat{V}(B\exists, F) \to BF; \tau \mapsto 1_B \tau_B$$

and

$$BF \to \hat{V}(B\exists, F); x \mapsto \big(A \mapsto (V(A, B) \to AF; f \mapsto f^F(x))\big).$$

□

As an immediate application of the Yoneda Lemma, one obtains

Theorem 3.1.2. *The Yoneda functor $\exists : V \to \hat{V}$ is a full and faithful embedding of V in $\hat{V}$.*

Proof. For objects B and C of V, the Yoneda Lemma yields the bijection $\hat{V}(B\exists, C\exists) \cong B^{C\exists} = V(B, C)$ on setting $F = C\exists$. □

In the light of Theorem 3.1.2, the Yoneda functor $\exists : V \to \hat{V}$ is usually described as the *Yoneda embedding* of V in $\hat{V}$. Objects of the image $V\exists$ are described as (contravariant) *representable* functors. For an object A of V, the representable functor $A\exists = V(\_\_, A)$ is *represented* by A.

Example 3.1.3 (Monoids). Let M be a monoid, realized as a small category according to Example III 1.2. The object part $\exists : \{1\} \to \hat{M}_0; 1 \mapsto 1\exists$ of the Yoneda embedding yields the left regular representation

$$1\exists : M_1^{\mathrm{op}} \to \underline{\underline{\mathrm{Set}}}_1; m \mapsto \big(m^{1\exists} : M(1,1) \to M(1,1); h \mapsto mh\big)$$

of M, while the morphism part

$$\exists : M(1,1) \to \hat{M}_1;\ m \mapsto (m\exists : 1\exists \to 1\exists)$$

yields the right regular representation

$$m\exists_1 : M(1,1) \to M(1,1);\ h \mapsto hm$$

of M [recall $M(1,1) = 1^{1\exists}$]. As in Example III 1.4.2, the right actions are homomorphisms of the left actions. □

The Yoneda embedding is a functor $\exists : V \to \hat{V}$. Its object part sends objects to functors, and its morphism part sends morphisms to natural transformations. A functor $\int_V : \hat{V} \to \underline{\underline{\text{Cat}}}$ will now be constructed. Its object part sends functors to categories, and its morphism part sends natural transformations to functors:

V	$\xrightarrow{\exists}$	$\hat{V}$	$\xrightarrow{\int_V}$	$\underline{\underline{\text{Cat}}}$
object	$\mapsto$	functor	$\mapsto$	category
morphism	$\mapsto$	natural transformation	$\mapsto$	functor

Together, the functors $\exists$ and $\int_V$ connect all the fundamental concepts of category theory introduced in Section III 1.

For an object F of $\hat{V}$, i.e. a contravariant functor $F : V \to \underline{\underline{\text{Set}}}$, the small category $\int_V F$ has

$$\left(\int_V F\right)_0 = \{(A, x) \mid A \in V_0, x \in AF\} \tag{3.1.4}$$

as its set of objects. For objects (A, x) and (B, y) of $\int_V F$, the morphisms $f : (A, x) \to (B, y)$ are elements of

$$\left(\int_V F\right)((A, x), (B, y)) = \{f \in V(A, B) \mid f^F(y) = x\}. \tag{3.1.5}$$

It is often convenient to write them in the form $f : (A, x = f^F(y)) \to (B, y)$. The composition of $f : (A, x = f^F(y)) \to (B, y)$ with $g : (B, y = g^F(z)) \to (C, z)$ is $fg : (A, x = f^F \circ g^F(z)) \to (C, z)$. Note that this is well defined by the contravariance of F. The identity at (A, x) is $1_A : (A, x) \to (A, x)$. Now consider a morphism $\tau : F \to G$ in $\hat{V}$, a natural transformation between the functors F and G. The functor $\int_V \tau$ has object part

$$\int_V \tau : \left(\int_V F\right)_0 \to \left(\int_V G\right)_0 ;\ (A, x) \mapsto (A, x\tau_A).$$

Note $x \in AF \Rightarrow x\tau_A \in AG$. The morphism part of $\int_V \tau$ is

$$\int_V \tau : (f : (A, x = f^F(y)) \to (B, y)) \mapsto (f : (A, x\tau_A = f^G(y\tau_B)) \to (B, y\tau_B)),$$

well defined since $x\tau_A = f^F(y)\tau_A = f^G(y\tau_B)$ by the naturality of $\tau : F \to G$. The functoriality of $\int_V : \hat{V} \to \underline{\underline{\mathrm{Cat}}}$ is readily verified (Exercise 3.1D). For an object F of $\hat{V}$, the small category $\int_V F$ is called the *category of elements* of F. Note that it comes equipped with the *projection functor*

$$(3.1.6) \quad P_F : \int_V F \to V; (f : (A, x = f^F(y)) \to (B, y)) \mapsto (f : A \to B)$$

to the original category V.

For an object A of V, (3.1.4) specializes to $(\int_V A\exists)_0 = \{(A_0, g) \mid A_0 \in V_0, g \in V(A_0, A)\}$. One may identify $(\int_V A\exists)_0$ with $d_1^{-1}\{A\}$, the set of V-morphisms with codomain A, via the bijection $d_1^{-1}\{A\} \to (\int_V A\exists)_0; g \mapsto (gd_0, g)$. Composing $\exists$ with $\int_V$, one thus obtains a natural underlying set functor $U : V \to \underline{\underline{\mathrm{Set}}}$ with morphism part

$$(3.1.7) \quad U : V \to \underline{\underline{\mathrm{Set}}}; (f : A \to B) \mapsto \left(d_1^{-1}\{A\} \to d_1^{-1}\{B\}; g \mapsto gf\right).$$

The faithfulness $f = 1_A f^U$ of this functor yields

Cayley's Theorem 3.1.4. *Each small category is concrete.* □

Example 3.1.5. If V is a poset category, then (3.1.7) is essentially the down-set functor of Example III 1.2.3. □

EXERCISES

3.1A. Verify the functoriality of the Yoneda functor, and prove that its object part injects.

3.1B. Prove the Yoneda Lemma 3.1.1.

3.1C. Let M be a monoid, considered as a left M-set (M, M^{op}) via the left regular representation. Show that each M^{op}-homomorphism $\theta : (M, M^{\mathrm{op}}) \to (M, M^{\mathrm{op}})$ is of the form $R(m) : M \to M;\ x \mapsto xm$ for an element m of M.

3.1D. Verify that $\int_V : \hat{V} \to \underline{\underline{\mathrm{Cat}}}$ is a functor.

3.1E. Let M be a monoid realized as a small category according to Example III 1.2. Show that Cayley's Theorem 3.1.4 specializes to Theorem O 4.3.

3.1F. Give an example of a small reinforced concrete category $(V, G : V \to \underline{\underline{\mathrm{Set}}})$ for which the underlying set functor is not naturally isomorphic to the functor $U : V \to \underline{\underline{\mathrm{Set}}}$ of (3.1.7).

3.1G. Let S be the algebraic theory of a unital ring, with $\underline{\underline{S}}$ as in Theorem 3.1. Show that the Yoneda embedding $\exists : S \to \hat{S}$ corestricts to $\exists : S \to \underline{\underline{S}}$, i.e. that the image $S\exists$ is a subcategory of $\underline{\underline{S}}$.

3.1H. For each object A of V, show that $(A, 1_A)$ is a terminal object of $\int_V A\exists$.

3.2. Limits of Functors

Let V be a small category. As shown in the previous section, the Yoneda embedding embeds V into the category $\hat{V}$ of contravariant set-valued functors on V. This section shows that the Yoneda embedding is the universal functor from V into a locally small, cocomplete category. First (Corollary 3.2.4 below), it will be shown that $\hat{V}$ is bicomplete. The proof to be given uses Currying for categories and functors that generalizes Currying for sets and functions (Example III 3.1.2 and references from there).

Given small categories C and D, the *product category* $C \times D$ has object set $C_0 \times D_0$, morphism set $C_1 \times D_1$, identity function $e : C_0 \times D_0 \to C_1 \times D_1; (x, y) \mapsto (1_x, 1_y)$, domain d_0 and codomain d_1 functions $d_i :\ C_1 \times D_1 \to C_0 \times D_0;\ (f, g) \mapsto (fd_i, gd_i)$, and composition $(C \times D)((x_0, y_0), (x_1, y_1)) \times (C \times D)((x_1, y_1), (x_2, y_2)) \to (C \times D)((x_0, y_0), (x_2, y_2));\ ((f_0, g_0), (f_1, g_1)) \mapsto (f_0 f_1, g_0 g_1)$. In other words, the category structure required by Definition III 1.1 is imposed componentwise on $C \times D$. (Cf. Exercise 3.2A.) Note that the switching functor $S : C \times D \to D \times C$ with morphism part

$$S : (f, g) \mapsto (g, f) \tag{3.2.1}$$

is an isomorphism.

Theorem 3.2.1 (Currying Functors). *Let E be a category, and let C and D be small categories. Then the functor categories $E^{C \times D}$ and $(E^D)^C$ are isomorphic.*

Proof. In the isomorphism, a functor $F : C \to E^D$ corresponds to a functor $\dot{F} : C \times D \to E$ (the functor F "evaluated pointwise"). The morphism part of $\dot{F}$ sends a $(C \times D)$-morphism $(f, g) : (x, y) \to (x', y')$ to the composite

$$y^{xF} \xrightarrow{(fF)_y} y^{x'F} \xrightarrow{gx'F} y'^{x'F}$$

of the component of the natural transformation $fF : xF \to x'F$ at y with the image of the D-morphism $g : y \to y'$ under the morphism part of the functor $x'F : D \to E$. In the inverse of the isomorphism, an arbitrary functor $\dot{F} : C \times D \to E$ corresponds to the functor $F : C \to E^D$ whose object part sends a C-object x' to the functor $x'F : D \to E$ with morphism part $(g : y \to y') \mapsto ((1_{x'}, g)\dot{F} : (x', y)\dot{F} \to (x', y')\dot{F})$. The morphism part of $F : C \to E^D$ sends a C-morphism $f : x \to x'$ to the natural transformation $fF : xF \to\ x'F$ with component $(fF)_y = (f, 1_y)\dot{F}$ at an object y of D. One may readily check that these object parts of the isomorphism and its putative inverse really are mutually inverse. The verification [Exercise 3.2B(a)] depends on the factorization $(f, g) = (f, 1_y)(1_{x'}, g)$ of a typical $(C \times D)$-morphism $(f, g); (x, y) \to (x', y')$. Now consider $E^{C\times D}$-objects $\dot{F}$ and $\dot{G}$, with corresponding $(E^D)^C$-objects F and G. Under the bijection

$$E^{C\times D}(\dot{F}, \dot{G}) \cong (E^D)^C(F, G), \tag{3.2.2}$$

an $E^{C\times D}$-morphism $\dot{\tau} : \dot{F} \to \dot{G}$ with component $\dot{\tau}_{(x,y)} : (x, y)F \to (x, y)G$ at an object (x, y) of $C \times D$ corresponds to an $(E^D)^C$-morphism $\tau : F \to G$ with component

$$\tau_x : (xF : D \to E) \to (xG : D \to E) \tag{3.2.3}$$

at an object x of C. The component $(\tau_x)_y : y^{xF} \to y^{xG}$ of (3.2.3) at an object y of D is just $\dot{\tau}_{(x,y)} : (x, y)\dot{F} \to (x, y)\dot{G}$. □

Corollary 3.2.2. *Let E be a category, and let C and D be small categories. Then the functor categories $(E^C)^D$ and $(E^D)^C$ are isomorphic.*

Proof. There is an isomorphism of $E^{D\times C}$ with $E^{C\times D}$ whose object part sends a functor $F : D \times C \to E$ to its composite $SF : C \times D \to E$ with the switching functor S of (3.2.1). Theorem 3.2.1 then yields isomorphisms $(E^C)^D \cong E^{D\times C} \cong E^{C\times D} \cong (E^D)^C$. □

The isomorphisms of Theorem 3.2.1 and Corollary 3.2.2 may be displayed by their morphism parts as follows:

$$\begin{gathered} (E^D)^C \longrightarrow E^{C\times D} \longrightarrow (E^C)^D, \\ (\tau : G \to F) \mapsto (\dot{\tau} : \dot{G} \to \dot{F}) \mapsto (\tilde{\tau} : \tilde{G} \to \tilde{F}), \end{gathered} \tag{3.2.4}$$

extending the notation used in the proof of Theorem 3.2.1. For example, one has $y^{xF} = (x, y)\dot{F} = x^{y\tilde{F}}$ in E_0 for x in C_0 and y in D_0. For $g : y \to y'$ in D_1, one has $g^{xF} = (g\tilde{F})_x$. This notation is useful in the statement and proof of the

Pointwise Limit Theorem 3.2.3. *Let E be a category, and let C and D be small categories. Let $F : C \to E^D$ be a functor.*

(a) *If $y\tilde{F} : C \to E$ has a limit $\pi^{y\tilde{F}}$ for each object y of D_0, then F has a limit object*

$$\varprojlim F : D \to E; (g : y \to y') \mapsto \left(\varprojlim \left(\pi^{y\tilde{F}} g^{\tilde{F}}\right) : \varprojlim y\tilde{F} \to \varprojlim y'\tilde{F}\right).$$

(b) *If* $y\tilde{F} : C \to E$ *has a colimit* $\iota^{y\tilde{F}}$ *for each object* y *of* D_0, *then* F *has a colimit object*

$$\varinjlim F : D \to E; (g : y \to y') \mapsto \left(\varinjlim \left(g^{\tilde{F}}\iota^{y'\tilde{F}}\right) : \varinjlim y\tilde{F} \to \varinjlim y'\tilde{F}\right).$$

Proof. (a) Let $\Delta : E^D \to (E^D)^C$ be the diagonal functor. Let $\Gamma : E^D \to E^{C\times D}$ be the functor whose object part is $B \mapsto ((x, y) \mapsto yB)$, and whose morphism part is $(\varphi : B \to B') \mapsto (\varphi\Gamma : B\Gamma \to B'\Gamma)$, where the component $(\varphi\Gamma)_{(x,y)}$ of the natural transformation $\varphi\Gamma$ at an object (x, y) of $C \times D$ is just the component φ_y of φ at y. It must be shown that $\varprojlim F$ comes equipped with a natural transformation $\pi^F : (\varprojlim F)\Delta \to F$ yielding a terminal object of $E^D\Delta/F$. In other words, given an object $\varphi : B\Delta \to F$ of $E^D\Delta/F$, a unique element

$$\begin{array}{ccc} B\Delta & \xrightarrow{(\varprojlim \varphi)\Delta} & (\varprojlim F)\Delta \\ \varphi \downarrow & & \downarrow \pi^F \\ F & = & F \end{array} \qquad (3.2.5)$$

of $(E^D\Delta/F)(\varphi, \pi^F)$ must be exhibited. Now (3.2.5) represents a commuting square in $(E^D)^C$. Under the isomorphism of Theorem 3.2.1, the diagram (3.2.5) corresponds to a commuting diagram

$$\begin{array}{ccc} B\Gamma & \xrightarrow{(\varprojlim \varphi)\Gamma} & (\varprojlim F)\Gamma \\ \dot{\varphi} \downarrow & & \downarrow \dot{\pi}^F \\ \dot{F} & = & \dot{F} \end{array} \qquad (3.2.6)$$

in $E^{C\times D}$. Thus it suffices to exhibit a unique element (3.2.6) of the morphism class $(\Gamma, \dot{F})(\dot{\varphi} : B\Gamma \to \dot{F},\ \dot{\pi}F : (\varprojlim F)\Gamma \to \dot{F})$ of the comma category $(\Gamma, \dot{F})$. Let $\nabla : C \to E^C$ be the diagonal functor. For a fixed object y of D, (3.2.6) Curries to the commuting diagram

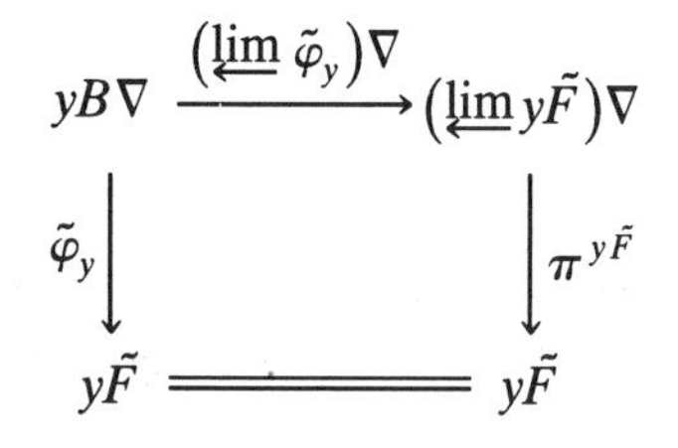

in E^C, the top row being specified uniquely by the terminality of $\pi^{y\tilde{F}}$ in

$C\nabla/y\tilde{F}$. Thus (3.2.6) indeed exhibits a unique element of $(\Gamma, \dot{F})(\dot{\varphi}, \dot{\pi}^F)$, namely with $\dot{\pi}^F_{(x,y)} = \pi_x^{y\tilde{F}}$ for x in C_0, and with $(\varprojlim \varphi)_y = \varprojlim \tilde{\varphi}_y$.

(b) Dual to (a). $\square$

Corollary 3.2.4. *For a small category V, the category $\hat{V} = \underline{\underline{\mathrm{Set}}}^{V^{\mathrm{op}}}$ is bicomplete.*

Proof. Consider a functor $F : J \to \underline{\underline{\mathrm{Set}}}^{V^{\mathrm{op}}}$ with small domain. Now $\underline{\underline{\mathrm{Set}}}$ is bicomplete. Thus for each object A of V, the functor $A\tilde{F} : J \to \underline{\underline{\mathrm{Set}}}$ has a limit and colimit. $\square$

The universality of the Yoneda embedding amongst functors from V to locally small, cocomplete categories follows from the

Extension Theorem 3.2.5. *Each functor $S : V \to C$ with locally small, cocomplete codomain extends via the Yoneda embedding to a cocontinuous functor*

$$\hat{S} : \hat{V} \to C; (\tau : F \to G) \mapsto \Big(\varinjlim \big((A, x) \mapsto \iota^{P_G S}_{(A, x\tau_A)}\big) : (\varinjlim P_F S) \to (\varinjlim P_G S)\Big),$$

unique up to natural isomorphism, that is left adjoint to the functor

$$\check{S} : C \to \hat{V}; X \mapsto \big((f : A \to B) \mapsto (C(BS, X) \to C(AS, X); h \mapsto f^S h)\big).$$

Proof. The natural isomorphism

$$C(F\hat{S}, Y) \cong \hat{V}(F, Y\check{S})$$

required by (III 3.1.1) for the adjointness is given by the mutually inverse bijections

$$C(F\hat{S}, Y) \to \hat{V}(F, Y\check{S}); k \mapsto \iota^{P_F S} \cdot k\Delta$$

(using the diagonal functor $\Delta : C \to C^{\int_V F}$) and

$$\hat{V}(F, Y\check{S}) \to C(F\hat{S}, Y); \tau \mapsto \varinjlim \big((A, x) \mapsto x\tau_A\big).$$

The uniqueness of $\hat{S}$ follows by Exercise III 3.1I, and its cocontinuity follows by Theorem III 3.4.1(b). $\square$

Corollary 3.2.6. *Each functor F in $\hat{V}$ is a colimit*

$$F \cong \varinjlim \big(P_F \exists : (A, x) \mapsto A\exists\big)$$

of representable functors.

Proof. Apply the Extension Theorem 3.2.5 to the Yoneda embedding $\exists : V \to \hat{V}$. By the Yoneda Lemma 3.1.1, the right adjoint

$$\check{\exists} : \hat{V} \to \hat{V}; F \mapsto \big(A \mapsto \hat{V}(A\exists, F)\big)$$

is isomorphic to the identity functor

$$1_{\hat{V}} : \hat{V} \to \hat{V}; F \mapsto (A \mapsto AF).$$

Thus the left adjoint

$$\hat{\exists} : \hat{V} \to \hat{V}; F \mapsto \varinjlim P_F \exists$$

is also isomorphic to the identity functor. □

Corollary 3.2.6 clarifies the way a functor $S : V \to C$ with locally small, cocomplete codomain extends to $\hat{S} : \hat{V} \to C$. For an object A of V, one has

$$A\exists\hat{S} = \varinjlim \left(P_{A\exists} S : \int_V A\exists \to C \right) = AS \tag{3.2.7}$$

by Exercise 3.1H and Exercise III 2.5O(a). For an object F of $\hat{V}$, the cocontinuity of $\hat{S}$ then yields

$$F\hat{S} = (\varinjlim P_F \exists)\hat{S} = \varinjlim \left(P_F \exists \hat{S}\right) = \varinjlim (P_F S). \tag{3.2.8}$$

EXERCISES

3.2A. Consider the forgetful functor $U : \underline{\underline{\text{Cat}}} \to \underline{\underline{\text{Set}}}$ whose morphism part sends a functor $F : C \to D$ to its morphism part $F : C_1 \to D_1$. Show that U creates products.

3.2B. (a) Verify that the respective object parts $F \mapsto \dot{F}$ and $\dot{F} \mapsto F$ of the functors $(E^D)^C \to E^{C \times D}$ and $E^{C \times D} \to (E^D)^C$ of the proof of Theorem 3.2.1 are indeed mutually inverse.
(b) Verify that (3.2.2) is a bijection.

3.2C. Suppose that the limit of the functor $F : C \to E^D$ of Theorem 3.2.3(a) exists. For x in C_0, specify the projection $\pi_x^F : \varprojlim F \to xF$.

3.2D. Write out a proof of Theorem 3.2.3(b). In particular, do the dual of Exercise 3.2C.

3.2E. Verify that $\hat{S}$ and $\check{S}$ in Theorem 3.2.5 really are functors. In particular, for an object Y of C, specify the morphism part of the contravariant functor $Y\check{S} : V \to \underline{\underline{\text{Set}}}$.

3.2F. Let R be a unital ring. Show that each R-module is a colimit of finitely generated free R-modules. (Hint: Use Exercise 3.1G.)

3.3. Finitary Algebraic Theories

Theorem 3.1 described modules over a ring in terms of the algebraic theory of the ring (the small category of linear combinations and matrices over the ring) and in terms of finite-product-preserving presheaves over the algebraic theory of the ring. With the preparation furnished by the previous two sections, the general version of this description may be presented. It offers the approach to universal algebra that extends the treatment of M-sets over a monoid M as functors from the category M to the category of sets.

Definition 3.3.1. A (*finitary*) *algebraic theory* V is a small category whose object set

$$V_0 = \{nV \mid n \in \mathbb{N}\}$$

contains, for each natural number n, the coproduct nV of n copies of the object $1V$. □

Example 3.3.2. Let $\underline{\underline{K}}$ be a prevariety of algebras of given finitary type $\tau : \Omega \to \mathbb{N}$. For each natural number n, let nK be the free $\underline{\underline{K}}$-algebra on the subset $\{x_1, \ldots, x_n\}$ of (1.3.4). Note that the set $\{x_1, \ldots, x_n\}$ is the disjoint union or coproduct of n copies of the set $\{x_1\}$. Since the free algebra functor $\Omega R_{\underline{\underline{K}}} : \underline{\underline{\mathrm{Set}}} \to \underline{\underline{K}}$ is cocontinuous, being left adjoint to the underlying set functor $U : \underline{\underline{K}} \to \underline{\underline{\mathrm{Set}}}$, it follows that nK is the coproduct in $\underline{\underline{K}}$ of n copies of $1K$. Let K be the small full subcategory of $\underline{\underline{K}}$ with object set $\{nK \mid n \in \mathbb{N}\}$. Then K is a finitary algebraic theory, the *algebraic theory of the prevariety* $\underline{\underline{K}}$. □

The general version of Theorem 3.1 is obtained by reversing the construction of Example 3.3.2. Let V be an algebraic theory. Define the *type* τ_V of V to be the disjoint union

$$\tau_V = \dot{\bigcup_{n \in \mathbb{N}}} \, [V(1V, nV) \to \{n\}], \tag{3.3.1}$$

with operator domain $\Omega_V = \dot{\bigcup}_{n \in \mathbb{N}} V(1V, nV)$. Given the insertions $\iota_i : 1V \to lV$ for $1 \le i \le l$, an arbitrary V-morphism $f : lV \to mV$ is described as the coproduct

$$f = \sum_{i=1}^{l} f_i \tag{3.3.2}$$

with $f_i = \iota_i f \in \tau_V^{-1}\{m\}$. Thus the operators from Ω_V yield the full set V_1 of morphisms of the small category V via (3.3.2).

A *model* of the algebraic theory V is a (covariant) functor $A : V^{\mathrm{op}} \to \underline{\underline{\mathrm{Set}}}$ that preserves finite products. It is convenient to denote the image of $1V$ under the functor A by the same symbol A, i.e. using A for the set $1V^A$.

Since A preserves finite products, one then has $mV^A = A^m$ for natural numbers m. An element $\omega : 1V \to mV$ of $\tau_V^{-1}\{m\}$ thus yields the m-ary operation $\omega^A : A^m \to A; (a_1, \ldots, a_m) \mapsto \omega^A(a_1, \ldots, a_m)$ on the set A. [Note the use of the convention (3.1.2).] In this way, the V-model $A : V^{\text{op}} \to \underline{\underline{\text{Set}}}$ gives a τ_V-algebra (A, Ω_V).

Consider a pair $(f : lV \to mV, g : mV \to nV)$ of composable V-morphisms, an element of the pullback $V_1 \times_{V_0} V_1$ (cf. Exercise III 2.3O). By (3.3.2), one has $f = \Sigma_{i=1}^{l} f_i$ and $g = \Sigma_{j=1}^{m} g_j$ with m-ary operations f_i and n-ary operations g_j. Similarly, one has $fg = \Sigma_{i=1}^{l} (fg)_i$ with n-ary operations $(fg)^i$, and

$$\forall 1 \le i \le l, \, (fg)_i = f_i\left(\sum_{j=1}^{m} g_j \right) \tag{3.3.3}$$

in $\tau_V^{-1}\{n\}$ by (III 2.1.3). Applying the covariant functor $A : V^{\text{op}} \to \underline{\underline{\text{Set}}}$ to (3.3.3), one obtains

$$\forall 1 \le i \le l, \, (fg)_i^A = f_i^A \circ \prod_{j=1}^{m} g_j^A. \tag{3.3.4}$$

In other words, the τ_V-algebra (A, Ω_V) satisfies the identities

$$(fg)_i(x_1, \ldots, x_n) = f_i(g_1(x_1, \ldots, x_n), \ldots, g_m(x_1, \ldots, x_n)) \tag{3.3.5}$$

for each composable pair $(f : lV \to mV, \, g : mV \to nV)$ of V-morphisms. The variety of τ_V-algebras satisfying all such identities is called the *variety* $\underline{\underline{V}}$ *of the algebraic theory* V. Let $\underline{\underline{\text{Mod}}}_V$ denote the full subcategory of $\hat{V}$ consisting of models of V. The general version of Theorem 3.1 may then be formulated.

Theorem 3.3.3. *Let* V *be a finitary algebraic theory. Then the categories* $\underline{\underline{V}}$ *and* $\underline{\underline{\text{Mod}}}_V$ *are isomorphic.*

Proof. A model $A : V^{\text{op}} \to \underline{\underline{\text{Set}}}$ has already been observed to yield a $\underline{\underline{V}}$-algebra (A, Ω_V). Conversely, given a $\underline{\underline{V}}$-algebra (A, Ω_V), one obtains a functor $A : V^{\text{op}} \to \underline{\underline{\text{Set}}}$ with morphism part

$$\left(\sum_{i=1}^{l} f_i : lV \to mV \right) \mapsto \left(A^m \to A^l; (a_1, \ldots, a_m) \mapsto (f_1^A(a_1, \ldots, a_m), \ldots, f_l^A(a_1, \ldots, a_m)) \right). \tag{3.3.6}$$

The functoriality of $A : V^{\text{op}} \to \underline{\underline{\text{Set}}}$ follows from the satisfaction of the identities (3.3.5). Homomorphisms $\theta : (A, \Omega_V) \to (B, \Omega_V)$ of $\underline{\underline{V}}$-algebras correspond to the components at $1V$ of natural transformations $\theta : A \to B$. For general natural numbers n, the homomorphism $\theta : (A, \Omega_V) \to (B, \Omega_V)$ yields the component

$$\theta_{nV} : A^n \to B^n; (a_1, \ldots, a_n) \mapsto (a_1\theta, \ldots, a_n\theta).$$

Conversely, for a natural transformation $\theta : A \to B$, the naturality

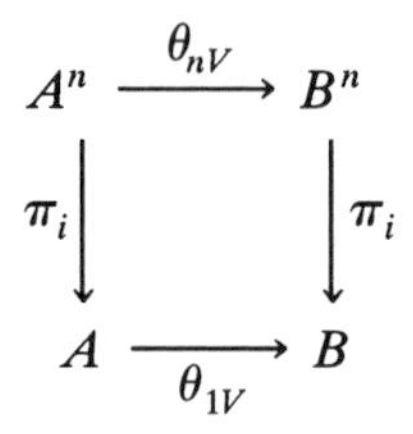

at the insertion $\iota_i : 1V \to nV$ for each i guarantees that the function $\theta_{nV} : A^n \to B^n$ has the form $A^n \to B^n; (a_1, \ldots, a_n) \mapsto (a_1\theta_{1V}, \ldots, a_n\theta_{1V})$. □

By virtue of Theorem 3.3.3, the categories $\underline{\underline{V}}$ and $\underline{\underline{\mathrm{Mod}}}_V$ determined by the algebraic theory V will be identified. One consequence is the following:

Corollary 3.3.4. *The full subcategory $\underline{\underline{V}}$ of $\hat{V}$ consisting of finite-product-preserving functors is (small) bicomplete.*

Proof. By Theorems 2.1.3 and 2.2.3, the variety $\underline{\underline{V}}$ is bicomplete. □

Specializing Example 3.3.2, let $\underline{\underline{K}}$ be a variety of algebras of given finitary type $\tau : \Omega \to \mathbb{N}$, with corresponding algebraic theory K. Let $\underline{\underline{V}}$ be the variety of the algebraic theory K. Thus $\underline{\underline{V}}$ is a variety of algebras of type $\tau_K : \Omega_K \to \mathbb{N}$. For each element (w, n) of the closed operator domain $\overline{\Omega}$, there are corresponding elements $x_1 \ldots x_n(w, n)$ of nK and $(w, n)_K : 1K \to nK$; $x_1 \mapsto x_1 \ldots x_n(w, n)_K$ of $K(1K, nK)$. If $u = v$ with $n = \max((\{\arg u) \cup (\arg v))$ is an identity satisfied by every $\underline{\underline{K}}$-algebra, then $(u, n)_K = (v, n)_K$. Consider a $\underline{\underline{V}}$-algebra (A, Ω_K), identified with the model $A : K^{\mathrm{op}} \to \underline{\underline{\mathrm{Set}}}$ of K. Then the underlying set A (or $1K^A$) becomes an $\overline{\Omega}$-algebra under the action $(w, n)_K^A : A^n \to A$ of the element (w, n) of $\overline{\Omega}$. By reduction, one obtains the algebra (A, Ω) of type $\tau : \Omega \to \mathbb{N}$. Since $(u, n)_K^A = (v, n)_K^A$, the τ-algebra (A, Ω) satisfies each identity $u = v$ of $\underline{\underline{K}}$. Thus the $\underline{\underline{V}}$-algebra (A, Ω_K) yields a $\underline{\underline{K}}$-algebra (A, Ω).

Conversely, consider a $\underline{\underline{K}}$-algebra (A, Ω). For each natural number l, let η_l be the component at $\{x_1, \ldots, x_l\}$ of the unit η of the adjunction $(\Omega R_{\underline{\underline{K}}}, U, \eta, \varepsilon)$ determined by the (pre-)variety $\underline{\underline{K}}$. Then (A, Ω) determines a model $A : K^{\mathrm{op}} \to \underline{\underline{\mathrm{Set}}}$ with morphism part

$$(3.3.7) \qquad (f : lK \to mK) \mapsto \Big(\underline{\underline{\mathrm{Set}}}(\{x_1, \ldots, x_m\}, A) \to \underline{\underline{\mathrm{Set}}}(\{x_1, \ldots, x_l\}, A); a \mapsto \eta_l\Big[f\big(a\Omega R_{\underline{\underline{K}}}\big)\varepsilon_A\Big]^U\Big).$$

This model is identified (according to Theorem 3.3.3) with a $\underline{\underline{V}}$-algebra (A, Ω_K).

The upshot of the correspondence between $\underline{\underline{K}}$-algebras (A, Ω) and ($\underline{\underline{V}}$-algebras or) K-models (A, Ω_K) is that the K-models provide an "invariant" description of the algebras, independent of any arbitrary choice $\tau : \Omega \to \mathbb{N}$ of a type for the algebras. This invariance is normally preferable due to its elegance and naturality.

EXERCISES

3.3A. Determine the algebraic theory of each of the following prevarieties:

(a) the variety $\underline{\underline{\mathrm{Set}}}$ of algebras of empty type;

(b) for a monoid M, the variety $\underline{\underline{M}}$ of M-sets of type $M \times \{1\}$;

(c) the prevariety $\underline{\underline{K}}$ of convex sets [as in Example 2.1.5(a)];

(d) the variety $\underline{\underline{\mathrm{Sl}}}$ of semilattices of type $\{(\cdot, 2)\}$.

3.3B. Consider the set $P = \{x_1, x_2, \ldots\}$ of positive integers as in (1.3.4). Define a small subcategory Set^2 of $\underline{\underline{\mathrm{Set}}}$ by

$$\mathrm{Set}_0^2 = \left\{\{x_1, \ldots, x_n\}^2 \,\middle|\, n \in \mathbb{N}\right\}$$

and

$$\mathrm{Set}^2\left(\{x_1, \ldots, x_m\}^2, \{x_1, \ldots, x_n\}^2\right)$$
$$= \left\{(f, g) \mid f, g \in \underline{\underline{\mathrm{Set}}}\left(\{x_1, \ldots, x_m\}^2, \{x_1, \ldots, x_n\}\right)\right\}$$

for natural numbers m, n.

(a) Show that Set^2 is an algebraic theory.

(b) Determine the variety $\underline{\underline{\mathrm{Set}^2}}$ of Set^2.

(c) A *rectangular band* is a semigroup satisfying the identity $x_1 x_2 x_3 = x_1\, x_3$. Show that $\underline{\underline{\mathrm{Set}^2}}$ is the variety of rectangular bands.

3.3C. Give a detailed verification of the functoriality of (3.3.6) in the proof of Theorem 3.3.3.

3.3D. Give a detailed verification of the functoriality of (3.3.7).

3.3E. Let V be a finitary algebraic theory. Let C be a locally small category, and let $S : V \to C$ be a functor. For a C-object X, show that there is a functor with object part

$$C(\_\, S, X) : V^{\mathrm{op}} \to \underline{\underline{\mathrm{Set}}};\ nV \mapsto C(nVS, X).$$

Show that $C(\_\, S, X)$ is a presheaf over V that yields a $\underline{\underline{V}}$-algebra if S preserves finite coproducts.

3.3F. Let V be the poset category with Hasse diagram $0 \to 1$.
(a) Show that V is an algebraic theory, with $1V = 1$.
(b) Is V the algebraic theory of a prevariety?

3.4. Theory Maps

The full power of the approach to universal algebra via algebraic theories becomes apparent when studying transformations between theories.

Definition 3.4.1. Let V and W be finitary algebraic theories. Then a *theory map* $M : W \to V$ is a functor preserving finite coproducts such that $nWM = nV$ for each natural number n. □

Given a theory map $M : W \to V$, the composite with the Yoneda embedding $\exists_V : V \to \hat{V}$ yields a functor $M\exists_V : W \to \hat{V}$; $nW \mapsto V(\_\_\_, nV)$ with locally small, cocomplete codomain. By the Extension Theorem 3.2.5, there is then a unique extension $\widehat{M\exists_V}$ or

$$(3.4.1) \qquad \hat{M} : \hat{W} \to \hat{V};\ F \mapsto \varinjlim\left(P_F M\exists_V : \int_W F \to \hat{V}\right)$$

that is left adjoint to the functor $\check{M\exists_V}$ or $\check{M} : \hat{V} \to \hat{W}$; $F \mapsto (lW \mapsto \hat{V}(lV\exists_V, F))$ or

(3.4.2)

$$\check{M} : \hat{V} \to \hat{W} : F \mapsto ((f : lW \to mW) \mapsto (f^{MF} : mVF \to lVF)).$$

[Note the use of the Yoneda Lemma 3.1.1 to obtain the simple form (3.4.2) of the object part of $\check{M}$. Also see Exercise 3.4F(a).]

Proposition 3.4.2. *Given a theory map $M : W \to V$, the right adjoint $\check{M} : \hat{V} \to \hat{W}$ restricts to a functor*

$$(3.4.3) \qquad M^r : \underline{\underline{V}} \to \underline{\underline{W}}$$

between the corresponding varieties.

Proof. Let $A : V^{\mathrm{op}} \to \underline{\underline{\mathrm{Set}}}$ be a V-model. Then for each natural number n, one has

$$(3.4.4) \qquad (nW)(A\check{M}) = nVA = (1VA)^n = (1W)(A\check{M})^n. \qquad □$$

Corollary 3.4.3. *The functor M^r of* (3.4.3) *commutes with the underlying set functors $U : \underline{\underline{V}} \to \underline{\underline{\mathrm{Set}}}$; $A \mapsto 1VA$ and $U : \underline{\underline{W}} \to \underline{\underline{\mathrm{Set}}}$; $B \mapsto 1WB$.*

Proof. Set $n = 1$ in (3.4.4). □

Starting from the prevariety $\underline{\underline{\mathrm{Set}}}$ of algebras of empty type, the construction of Example 3.3.2 yields the algebraic theory Set of $\underline{\underline{\mathrm{Set}}}$. For any algebraic theory V, there is a theory map

$$J : \mathrm{Set} \to V; (\iota_i : 1\,\mathrm{Set} \to n\,\mathrm{Set}) \mapsto (\iota_i : 1V \to nV). \tag{3.4.5}$$

Those V-morphisms that lie in the image of J are described as *combinatorial.* By (3.4.2), the functor $J^r : \underline{\underline{V}} \to \underline{\underline{\mathrm{Set}}}$ is the forgetful functor $U : \underline{\underline{V}} \to \underline{\underline{\mathrm{Set}}}$; $(A, \Omega_V) \mapsto A$. Since this functor appears in the adjunction $(\Omega_V R_{\underline{\underline{V}}}, U, \eta, \varepsilon)$, it has a left adjoint $\Omega_V R_{\underline{\underline{V}}} : \underline{\underline{\mathrm{Set}}} \to \underline{\underline{V}}$ or $J^l : \underline{\underline{\mathrm{Set}}} \to \underline{\underline{V}}$, the free algebra functor for the variety $\underline{\underline{V}}$. Theorem 3.4.4 below constructs a left adjoint $M^l : \underline{\underline{W}} \to \underline{\underline{V}}$ to each functor $M^r : \underline{\underline{V}} \to \underline{\underline{W}}$ given via (3.4.3) by a theory map $M : W \to V$. The construction is based on a study of the embedding of the variety $\underline{\underline{V}}$ in $\hat{V}$. It turns out that this embedding is analogous to the embedding of a prevariety $\underline{\underline{K}}$ of τ-algebras in the category $\underline{\underline{\tau}}$ of all τ-algebras. (In the terminology of Exercise III 3.2E, both embeddings are reflections.)

For an algebraic theory V, let $E_V : V \to \underline{\underline{V}}$ denote the embedding of V in $\underline{\underline{V}}$ via the Yoneda embedding. By Corollary 3.3.4, the locally small category $\underline{\underline{V}}$ is cocomplete. By the Extension Theorem 3.2.5, the embedding $E_V : V \to \underline{\underline{V}}$ extends to a functor $\hat{E}_V : \hat{V} \to \underline{\underline{V}}$ that is left adjoint to the embedding $\breve{E}_V : \underline{\underline{V}} \to \hat{V}$. The functor $\hat{E}_V : \hat{V} \to \underline{\underline{V}}$; $F \mapsto \varinjlim (P_F E_V : \int_V F \to \underline{\underline{V}})$ is analogous to the replication functor $R_{\underline{\underline{K}}} : \underline{\underline{\tau}} \to \underline{\underline{K}}$. In particular, note that the colimit of a functor $F : I \to \underline{\underline{V}}$ differs from the colimit of the composite functor $F\breve{E}_V : I \to \hat{V}$, just as colimits of $\underline{\underline{K}}$-algebras in $\underline{\underline{K}}$ differ from their colimits in $\underline{\underline{\tau}}$.

A theory map $M : W \to V$ yields the following system of functors:

$$\begin{array}{ccccc}
 & & \exists_V \ (V \to \hat{V}) & & \\
V & \xrightarrow{E_V} & \underline{\underline{V}} & \underset{\hat{E}_V}{\overset{\breve{E}_V}{\rightleftarrows}} & \hat{V} \\
\uparrow M & & M^l \uparrow\downarrow M^r & & \hat{M} \uparrow\downarrow \breve{M} \\
W & \xrightarrow{E_W} & \underline{\underline{W}} & \underset{\hat{E}_W}{\overset{\breve{E}_W}{\rightleftarrows}} & \hat{W} \\
 & & \exists_W \ (W \to \hat{W}) & &
\end{array}$$

Note the description of $M^r : \underline{\underline{V}} \to \underline{\underline{W}}$ as

$$M^r = \breve{E}_V \breve{M} \hat{E}_W$$

by Proposition 3.4.2 and the definition of

$$M^l = \breve{E}_W \hat{M} \hat{E}_V. \tag{3.4.6}$$

Theorem 3.4.4. *Given a theory map $M : W \to V$, the functor $M^r : \underline{\underline{V}} \to \underline{\underline{W}}$ has the left adjoint $M^l : \underline{\underline{W}} \to \underline{\underline{V}}$; $B \mapsto \varinjlim (P_B M E_V : \int_W B \to \underline{\underline{V}})$*

Proof. By composition of adjunctions (cf. Exercise III 3.1K), the functor $\hat{M}\hat{E}_V : \hat{W} \to \underline{\underline{V}}$ is left adjoint to $\breve{E}_V \breve{M} : \underline{\underline{V}} \to \hat{W}$. Thus for A in $\underline{\underline{V}}_0$ and B in $\underline{\underline{W}}_0$, one has $\underline{\underline{V}}(BM^l, A) = \underline{\underline{V}}(B\breve{E}_W \hat{M}\hat{E}_V, A) \cong \hat{W}(B\breve{E}_W, A\breve{E}_V\breve{M}) = \underline{\underline{W}}(B, AM^r)$, as required. □

Example 3.4.5. (a) **(Replication.)** Let $\underline{\underline{K}}$ be a variety of τ-algebras for a certain type $\tau : \Omega \to \mathbb{N}$. The replica functor $R_{\underline{\underline{K}}} : \underline{\underline{\tau}} \to \underline{\underline{K}}$ restricts to a theory map $R : \tau \to K$ between the corresponding algebraic theories determined as in Example 3.3.2. Then $R^r : \underline{\underline{K}} \to \underline{\underline{\tau}}$ is just the inclusion, while $R^l : \underline{\underline{\tau}} \to \underline{\underline{K}}$ is the replica functor. In other words, the adjunction of Theorem 3.4.4 yields the adjunction of Theorem 2.1.3 in this case.

(b) **(Free Algebras.)** Taking the theory map $J : \text{Set} \to V$ of (3.4.5), Theorem 3.4.4 builds the free $\underline{\underline{V}}$-algebra XJ^l over a set X as a colimit of finitely generated free algebras.

(c) **(Geometric Realizations.)** Let K be the algebraic theory of convex sets [cf. Exercise 3.3A(c)]. Let $\mathbb{R}$ be the algebraic theory of the ring of real numbers, as in Section 3. For each natural number n, the embedding of the $(n-1)$-simplex nK in $n\mathbb{R}$ (cf. Exercise 2.1D) yields a theory map $R : K \to \mathbb{R}$. Then $R^r : \underline{\underline{\mathbb{R}}} \to \underline{\underline{K}}$ construes real vector spaces as convex sets. Its left adjoint $R^l : \underline{\underline{K}} \to \underline{\underline{\mathbb{R}}}$ builds the so-called *geometric realization* of convex sets (and other members of the variety $\underline{\underline{K}}$). (Note that the variety $\underline{\underline{K}}$ properly contains the prevariety of convex sets—cf. Exercise 3.4G.)

(d) **(Knot Groups.)** Let $\underline{\underline{\text{Kw}}}$ be the variety of *quandles*, right quasigroups satisfying the identities $x_1 x_1 = x_1$ of idempotence and $x_1 x_3 \cdot x_2 x_3 = x_1 x_2 \cdot x_3$ of *distributivity*. Let M: Kw $\to$ Gp be the theory map given by $(x_1 \mapsto x_1 \cdot x_2) \mapsto (x_1 \mapsto x_2^{-1} x_1 x_2)$ and $(x_1 \mapsto x_1/x_2) \mapsto (x_1 \mapsto x_2 x_1 x_2^{-1})$. Thus for a group G, the quandle GM^r has multiplication $x_1 \cdot x_2 = x_2^{-1} x_1 x_2$ and right division $x_1/x_2 = x_2 x_1 x_2^{-1}$. For a quandle Q, the group QM^l is the group Adconj Q constructed by D. Joyce ["A Classifying Invariant of Knots, the Knot Quandle," *J. Pure Appl. Algebra* **23** (1982), 37–65]. Joyce showed how to associate a quandle $Q(K)$ to a knot K such that Adconj $Q(K)$ becomes the knot group $\pi_1(S^3 - K)$, the fundamental group of the complement of the knot in S^3. Joyce's construction is a precise version of a method outlined by W. Wirtinger at the 1905 meeting of the German Mathematical Association in Meran ("Über die Verzweigungen bei Funktionen von zwei Veränderlichen," *Jahresbericht der DMV* **14** (1905), 517.) □

The applicability of Theorem 3.4.4 depends on recognizing which functors $R : \underline{\underline{V}} \to \underline{\underline{W}}$ are of the form $R = M^r$ for a theory map $M : V \to W$. To this end, the following characterization of right adjoints with codomain $\underline{\underline{W}}$ proves useful. Consider a category C with finite coproducts. For a natural number n, write nZ for the coproduct of n copies of an object Z of C. A *$\underline{\underline{W}}$-coalgebra* in C is defined (via Definition 2.4.4) as an object Z of $C^{\text{op}} \otimes \underline{\underline{W}}$. Thus an

n-ary operator ω of $\underline{\underline{W}}$ yields a *cooperation* ω_Z or $\omega : Z \to nZ$ on Z. If C is locally small, then (as in Exercise 2.4L) one has that Z is an object of $C^{\text{op}} \otimes \underline{\underline{W}}$ iff, for each X in $C_0 = C_0^{\text{op}}$, the set $C^{\text{op}}(X, Z) = C(Z, X)$ lies in $\underline{\underline{W}}$.

Theorem 3.4.6. *Let W be an algebraic theory. Let C be a locally small, cocomplete category. Then a functor $R : C \to \underline{\underline{W}}$ is a right adjoint iff there is a $\underline{\underline{W}}$-coalgebra Z in C such that $R : C \to \underline{\underline{W}}$ is naturally isomorphic to $C \to \underline{\underline{W}}$; $X \mapsto C(Z, X)$.*

Proof. If R has a left adjoint S, then

$$C(1WS, X) \cong \underline{\underline{W}}(1W, XR) \cong XR \tag{3.4.7}$$

for X in C_0. On the one hand, (3.4.7) shows that one may take $Z = 1WS$ to obtain the required form for R. On the other hand, (3.4.7) shows that $1WS$ is a $\underline{\underline{W}}$-coalgebra in C.

Conversely, consider the functor $R : C \to \underline{\underline{W}}$; $X \mapsto C(Z, X)$ for a $\underline{\underline{W}}$-coalgebra Z in C. Define

$$S : W \to C; (\omega : 1W \to nW) \mapsto (\omega_Z : Z \to nZ). \tag{3.4.8}$$

Note that $S : W \to C$ preserves finite coproducts. By the Extension Theorem 3.2.5, the functor $S : W \to C$ extends to a functor $\hat{S} : \hat{W} \to C$ that is left adjoint to the functor $\breve{S} : C \to \hat{W}$; $X \mapsto (nW \mapsto C(nWS, X))$. Now $\breve{S} : C \to \hat{W}$ corestricts to the functor $R : C \to \underline{\underline{W}}$ (cf. Exercise 3.3E). Thus for X in C_0 and Y in $\underline{\underline{W}}_0$, one has $C(Y\hat{S}, X) \cong \hat{W}(Y, X\breve{S}) = \underline{\underline{W}}(Y, XR)$. In other words, the restriction $S : \underline{\underline{W}} \to C$ to $\underline{\underline{W}}$ of $\hat{S} : \hat{W} \to C$ is the required left adjoint. □

Corollary 3.4.7. *Let $M : W \to V$ be a theory map. Then $1V$ is a $\underline{\underline{W}}$-coalgebra in $\underline{\underline{V}}$.*

Proof. Apply (3.4.7) with $(R : C \to \underline{\underline{W}}) = (M^r : \underline{\underline{V}} \to \underline{\underline{W}})$ and $S = M^l$. Thus $\underline{\underline{V}}(1V, X) = \underline{\underline{V}}(1WS, X) \cong \underline{\underline{W}}(1W, XR) \cong XR \in \underline{\underline{W}}_0$ for X in $\underline{\underline{V}}_0$. □

Theorem 3.4.6 serves to characterize functors $R : \underline{\underline{V}} \to \underline{\underline{W}}$ that appear as $M^r : \underline{\underline{V}} \to \underline{\underline{W}}$ for a theory map $M : W \to V$.

Corollary 3.4.8. *Let V and W be algebraic theories, yielding underlying set functors U_V or $U : \underline{\underline{V}} \to \underline{\underline{\text{Set}}}$; $A \mapsto 1VA$ and U_W or $U : \underline{\underline{W}} \to \underline{\underline{\text{Set}}}$; $B \mapsto 1WB$. Then a functor $R : \underline{\underline{V}} \to \underline{\underline{W}}$ is of the form $M^r : \underline{\underline{V}} \to \underline{\underline{W}}$ for a theory map $M : W \to V$ (and so possesses a left adjoint) iff it commutes with the underlying set functors: $U_V = RU_W$.*

Proof. The "only if" direction is Corollary 3.4.3. Conversely, suppose $U_V = RU_W$. Then for each $\underline{\underline{V}}$-algebra X, the set $\underline{\underline{V}}(1V, X)$, naturally isomorphic with XU and XRU, carries the $\underline{\underline{W}}$-algebra structure XR. Thus $1V$ is a

$\underline{\underline{W}}$-coalgebra such that $R: C \to \underline{\underline{W}}$ takes the form $X \mapsto \underline{\underline{V}}(1V, X)$. The required theory map is then defined by (3.4.8) with $C = V$ and $Z = 1V$. □

Example 3.4.9. Let K be a unital, commutative ring. As in Exercise 2.2H, let $\underline{\underline{\text{CAlg}}}_K$ be the variety of commutative K-algebras. Recall that coproducts in $\underline{\underline{\text{CAlg}}}_K$ are given by the tensor product construction, and that the polynomial algebra $K[x_1]$ is the free commutative K-algebra 1CAlg_K on $\{x_1\}$. The reduction functor $R: \underline{\underline{\text{CAlg}}}_K \to \underline{\underline{\text{Mon}}}$ forgetting the K-module structure preserves underlying sets, and is thus of the form $M^r: \underline{\underline{\text{CAlg}}}_K \to \underline{\underline{\text{Mon}}}$ for a theory map $M: \text{Mon} \to \text{CAlg}_K$ by Corollary 3.4.8. By Corollary 3.4.7, it follows that $K[x_1]$ is a comonoid in $\underline{\underline{\text{CAlg}}}_K$. The comultiplication is given by the K-algebra homomorphism

$$\text{(3.4.9)} \qquad \delta: K[x_1] \to K[x_1] \otimes K[x_1]; x_1 \mapsto x_1 \otimes 1 + 1 \otimes x_1,$$

while the counit is given by $\varepsilon: K[x_1] \to K; x_1 \mapsto 0$. Objects of $\underline{\underline{\text{Mon}}} \otimes \underline{\underline{\text{CAlg}}}_K^{\text{op}}$ such as $K[x_1]$ are called *Hopf algebras*. □

EXERCISES

3.4A. For an algebraic theory V, verify that the functor $\breve{E}_V: \underline{\underline{V}} \to \hat{V}$ is just the inclusion of the category of V-models in the category of presheaves over V.

3.4B. Show that each homomorphism $H: R \to S$ of unital rings yields a theory map between the corresponding algebraic theories. Determine $H^r: \underline{\underline{\text{Mod}}}_S \to \underline{\underline{\text{Mod}}}_R$ and $H^l: \underline{\underline{\text{Mod}}}_R \to \underline{\underline{\text{Mod}}}_S$.

3.4C. Show that the inclusion $H: \mathbb{R} \to \mathbb{C}$ yields $H^l: \underline{\underline{\text{Mod}}}_{\mathbb{R}} \to \underline{\underline{\text{Mod}}}_{\mathbb{C}}$ as the complexification of Exercise III 3.6L.

3.4D. Let B be a subgroup of a group A. Consider the varieties $\underline{\underline{B}}$ of B-sets and $\underline{\underline{A}}$ of A-sets. Denote the corresponding algebraic theories by B and A. Show that the inclusion of B in A yields a theory map $M: B \to A$ such that the adjunction of Theorem 3.4.4 between M^l and M^r becomes the adjunction (III 3.1).

3.4E. Let $\underline{\underline{V}}$ and $\underline{\underline{W}}$ be varieties of algebras of given type $\tau: \Omega \to \mathbb{N}$ such that each $\underline{\underline{V}}$-algebra is a $\underline{\underline{W}}$-algebra. Show that the replica functor $R_{\underline{\underline{V}}}: \underline{\underline{W}} \to \underline{\underline{V}}$ restricts to a theory map $R: W \to V$. Determine the adjoint functors $R^l: \underline{\underline{W}} \to \underline{\underline{V}}$ and $R^r: \underline{\underline{V}} \to \underline{\underline{W}}$.

3.4F. Let $M: W \to V$ be a theory map.

(a) Show that the morphism part of $\breve{M}: \hat{V} \to \hat{W}$ takes the form

$$(\tau: F \to G) \mapsto (nW \mapsto (nVF \to nVG; x \mapsto x\tau_{nV})).$$

(b) Using Theorem 3.3.3, show that the morphism part of $M^r : \underline{\underline{V}} \to \underline{\underline{W}}$ takes the form

$$(\theta : (A, \Omega_V) \to (B, \Omega_V)) \mapsto (\theta : (A, \Omega_W) \to (B, \Omega_W)).$$

Thus note that the non-triviality of $\check{M}$ and M^r is encapsulated in (3.4.2).

3.4G. Let K be the algebraic theory of convex sets, as in Example 3.4.5(c).

(a) Let $(H, \cdot)$ be a semilattice. Show that H yields a model of K with $(p : 1K \to 2K)H = (H^2 \to H; (x, y) \mapsto xy)$ for a K-morphism p whose image is an interior point of the closed bounded real interval 2K.

(b) Show that H is an element of $\underline{\underline{K}}$ that is not a convex set if $|H| > 1$.

(c) Determine the geometric realization of the free semilattice 2Sl over a 2-element set. (Hint: Describe 2Sl as a coequalizer of convex sets, and use the cocontinuity of R^l.)

3.4H. Realize the free semilattice nSl on $\{x_1, \ldots, x_n\}$ as the poset of non-empty subsets of $\{x_1, \ldots, x_n\}$. Realize the free Boolean algebra nBAlg on $\{x_1, \ldots, x_n\}$ as the poset of all subsets of $\{x_1, \ldots, x_n\}$.

(a) Show that inclusion yields a theory map $M : \mathrm{Sl} \to \mathrm{BAlg}$.

(b) Determine $M^r : \underline{\underline{\mathrm{BAlg}}} \to \underline{\underline{\mathrm{Sl}}}$ and $M^l : \underline{\underline{\mathrm{Sl}}} \to \underline{\underline{\mathrm{BAlg}}}$. In particular, identify the Boolean algebra $\{1 < 2 < 3\}M^l$.

3.4I. Prove the following *Extension Theorem* for varieties: Let $\underline{\underline{V}}$ be a variety, with corresponding algebraic theory V. Then each functor $S : V \to C$ with locally small, cocomplete codomain extends to a unique cocontinuous functor $S^l : \underline{\underline{V}} \to C$ with $E_V S^l = S$.

3.4J. Let $M : W \to V$ be a theory map, and let B be a $\underline{\underline{W}}$-algebra. Using Corollary III 2.5.6 and Theorem 2.2.3, give an explicit construction of the $\underline{\underline{V}}$-algebra BM^l as a quotient of a free $\underline{\underline{V}}$-algebra.

3.4K. Construe the functor $\mathrm{Mon}^\tau : \underline{\underline{\mathrm{Mon}}} \to \underline{\underline{\tau}}$ of Exercise 1.2O as the right adjoint $M^r : \underline{\underline{\mathrm{Mon}}} \to \underline{\underline{\tau}}$ determined by a theory map $M : \tau \to$ Mon. What is the left adjoint $M^l : \underline{\underline{\tau}} \to \underline{\underline{\mathrm{Mon}}}$?

3.4L. Let $\tau : \Omega \to \mathbb{N}$ be a finitary type, and let $\sigma : \Psi \to \mathbb{N}$ be the restriction of $\overline{\tau} : \overline{\Omega} \to \mathbb{N}$ to a subset Ψ of $\overline{\Omega}$. Let $R : \underline{\underline{\tau}} \to \underline{\underline{\sigma}}$ be the functor sending a τ-algebra (A, Ω) to its (*non-basic*) *reduct* (A, Ψ) [cf. Section 1.3]. Show that the *reduction functor* $R : \underline{\underline{\tau}} \to \underline{\underline{\sigma}}$ has a uniquely defined left adjoint $S : \underline{\underline{\sigma}} \to \underline{\underline{\tau}}$. The left adjoint is known as the *augmentation functor* adjoint to R. [Hint: Note that the "geometric realization" of Example 3.4.5(c) is an instance of such an augmentation functor.]

3.4M. Let $R : \underline{\underline{\mathrm{Ring}}} \to \underline{\underline{\mathrm{Mon}}}$ be the (restriction of the) reduction functor forgetting the abelian group structure of a ring.

(a) Show that $R : \underline{\underline{\mathrm{Ring}}} \to \underline{\underline{\mathrm{Mon}}}$ has a left adjoint $S : \underline{\underline{\mathrm{Mon}}} \to \underline{\underline{\mathrm{Ring}}}$. The left adjoint is known as the *monoid ring functor*.

(b) Show that the monoid ring functor sends the symmetric group S_3 to the ring $\mathbb{Z}[S_3]$ of Exercise III 2.2D.

3.4N. Verify the details of Example 3.4.9. In particular, show that the image of a power x_1^n under the comultiplication (3.4.9) is $\Sigma_{i=0}^{n}\binom{n}{i}x_1^{n-i} \otimes x_1^i$.

3.4O. Let n be a positive integer. Consider the functor $R_n : \underline{\underline{\text{Mod}}}_{\mathbb{Z}} \to \underline{\underline{\text{Mod}}}_{\mathbb{Z}}$ with object part $A \mapsto \{x \in A \mid nx = 0\}$, and with morphism part given by restriction. Use Theorem 3.4.6 to show that R_n has a left adjoint. Determine the left adjoint.

3.4P. Let G be a group.

(a) Show that the trivial homomorphism $G \to \{1\}$ yields a theory map $M : G \to$ Set whose left adjoint $M^l : \underline{\underline{G}} \to \underline{\underline{\text{Set}}}$ maps a G-set X to the orbit set X/G.

(b) Show that the right adjoint $M^r : \underline{\underline{\text{Set}}} \to \underline{\underline{G}}$ is left adjoint to the functor $\text{Fix} : \underline{\underline{G}} \to \underline{\underline{\text{Set}}}$; $(X, G) \mapsto \text{Fix}(X, G)$.

4. MONADS

Let $(M, \cdot, e)$ be a monoid. As in Example III 1.2.2, the fixed set M yields a functor $T : \underline{\underline{\text{Set}}} \to \underline{\underline{\text{Set}}}$ with morphism part

(4.1)

$$(f : Z_1 \to Z_2) \mapsto ((f, 1_M) : Z_1 \times M \to Z_2 \times M; (z, m) \mapsto (zf, m)).$$

However, the monoid structure on the set M associates extra structure with the functor T. For example, the multiplication in the monoid yields a natural transformation $\mu : T^2 \to T$ whose component at a set X is

$$(4.2) \qquad \mu_X : X \times M \times M \to X \times M; (x, m_1, m_2) \mapsto (x, m_1 m_2).$$

(Cf. Exercise III 3E.) Note that (4.2) is the action of the monoid M on the free M-set $X \times M$ over the set X. The natural transformation $\mu : T^2 \to T$ satisfies the *associative law*

$$(4.3) \qquad (T\mu)\mu = (\mu T)\mu.$$

Indeed, the component at a set X of the left-hand side of (4.3) is $(T\mu)_X \mu_X$ or

$$(x, m_1, m_2, m_3) \overset{\mu_{XT}}{\longmapsto} (x, m_1, m_2 m_3) \overset{\mu_X}{\longmapsto} (x, m_1 \cdot m_2 m_3),$$

while the component of the right-hand side is $(\mu T)_X \mu_X$ or

$$(x, m_1, m_2, m_3) \overset{\mu_X^T}{\longmapsto} (x, m_1 m_2, m_3) \overset{\mu_X}{\longmapsto} (x, m_1 m_2 . m_3).$$

These components agree by the usual associative law (O 4.2) in the monoid M. The identity element e of the monoid yields a natural transformation $\eta : 1_{\underline{\underline{\text{Set}}}} \to T$ whose component at a set X is

$$(4.4) \qquad \eta_X : X \to X \times M;\ x \mapsto (x, e).$$

Using the identity transformation $1_T : T \to T$, the *left unit law*

$$(4.5) \qquad (\eta T)\mu = 1_T$$

and *right unit law*

$$(4.6) \qquad (T\eta)\mu = 1_T$$

are satisfied. For example, the left-hand side of (4.5) has component $(\eta T)_X \mu_X$ or

$$(x, m) \stackrel{\eta_X^T}{\longmapsto} (x, e, m) \stackrel{\mu_X}{\longmapsto} (x, em),$$

while the component of the right-hand side is the identity function on $X \times M$. These components agree by the validity of the left hand side of (O 4.3) in the monoid M. Overall, the monoid structure on the set M is embodied in the *triple*

$$(4.7) \qquad (T, \mu, \eta)$$

or *monad* in the category of sets, i.e. a functor $T : \underline{\underline{\text{Set}}} \to \underline{\underline{\text{Set}}}$ with natural transformations $\mu : T^2 \to T$ and $\eta : 1_{\underline{\underline{\text{Set}}}} \to T$ satisfying the associative law (4.3) and unit laws (4.5), and (4.6).

EXERCISES

4A. Determine the component of the left-hand side of (4.6) at a set X, and verify that it agrees with the component of the right-hand side of (4.6) at X.

4B. Suppose that the monoid M is idempotent. Express its idempotence as a property of the natural transformation $\mu : T^2 \to T$.

4C. For the fixed monoid M, the functor $T : \underline{\underline{\text{Set}}} \to \underline{\underline{\text{Set}}}$ of (4.1) has a right adjoint $T' : \underline{\underline{\text{Set}}} \to \underline{\underline{\text{Set}}}$ with object part $X \mapsto X^M$ (cf. Example III 3.1.2). Does the monoid structure on the set M associate extra structure with the functor T'? [Hint: Dualize (4.7).]

4D. Let K be a commutative, unital ring, and let Y be a K-algebra. For the functor S_Y of Section III 3.6, show that there are natural transformations $\mu : S_Y^2 \to S_Y$ and $\eta : 1_{\underline{\underline{\text{Mod}_K}}} \to S_Y$ satisfying the associative law

$(S_Y\mu)\mu = (\mu S_Y)\mu$ and unit laws $(\eta S_Y)\mu = 1 = (S_Y\eta)\mu$. The component of μ at a K-module Z is

$$\mu_Z : Z \otimes Y \otimes Y \to Z \otimes Y;\ z \otimes y \otimes y' \mapsto z \otimes yy',$$

while the component of η at Z is

$$\eta_Z : Z \to Z \otimes Y;\ z \mapsto z \otimes 1.$$

4E. (a) Give an example of two monads (T, μ, η) and (T, μ', η') in the category of sets for which $\mu \neq \mu'$ and $\eta = \eta'$.

(b) Give an example of two monads (T, μ, η) and (T, μ', η') in the category of sets for which $\eta \neq \eta'$.

4F. Let (T, μ, ε) be a monad in the category of sets. Show that the image under T of the terminal object of $\underline{\underline{\text{Set}}}$ is a set carrying a monoid structure.

4.1. Monads and Their Algebras

The functor $T : \underline{\underline{\text{Set}}} \to \underline{\underline{\text{Set}}};\ Z \mapsto Z \times M$ of (4.1) yielded the monad (T, μ, η) of (4.7) that reflected the monoid structure $(M, \cdot, e)$ on the set M. The definition (4.7) may be extended to more general categories.

Definition 4.1.1. A *monad* or *triple* (T, μ, η) *on a category* B is a functor $T : B \to B$ equipped with natural transformations $\mu : T^2 \to T$ (called *composition*) and $\eta : 1_B \to T$ (called the *unit*), such that the *associative law* $(T\mu)\mu = (\mu T)\mu$, the *left unit law* $(\eta T)\mu = 1_T$, and the *right unit law* $(T\eta)\mu = 1_T$ are satisfied. In this context, the category B is called the *base* category. □

Example 4.1.2. (a) **(Closure Operators.)** A monad T on a poset category $(B, \leq)$ is called a *closure operator*. The existence of the unit yields $\forall x \in B$, $x \leq x^T$. The existence of the composition yields $\forall x \in B, x^{TT} \leq x^T$. On the other hand, the monotonicity of the functor T applied to $x \leq x^T$ yields $x^T \leq x^{TT}$. Thus $x^T = x^{TT}$ for each x in B.

(b) **(K-Algebras.)** For an algebra Y over a commutative, unital ring K, the construction of Exercise 4D yields a monad (S_Y, μ, η) on the category $\underline{\underline{\text{Mod}}}_K$ of K-modules. □

Given a Galois connection $(S, R, \eta, \varepsilon)$ between functors $S : (B, \leq) \to (A, \leq)$ and $R : (A, \leq) \to (B, \leq)$, the composite functor $T = SR$ yields a monad or closure operator on the poset $(B, \leq)$, as corroborated by (III 3.3.2)(a) and (III 3.3.3)(a). This phenomenon is quite general: Each adjunction yields a monad.

Proposition 4.1.3. *Let $(F, G, \eta, \varepsilon)$ be an adjunction between functors $F : B \to A$ and $G : A \to B$. Then $(FG, F\varepsilon G, \eta)$ is a monad on the base category B.*

Proof. For X in B_0, the naturality of $\varepsilon : GF \to 1_A$ at $\varepsilon_{XF} : XFGF \to XF$ yields the commutativity of the diagram

$$(4.1.1) \qquad \begin{array}{ccc} XFGFGF & \xrightarrow{\varepsilon_{XFGF}} & XFGF \\ {\scriptstyle \varepsilon^{GF}_{XF}}\downarrow & & \downarrow{\scriptstyle \varepsilon_{XF}} \\ XFGF & \xrightarrow[\varepsilon_{XF}]{} & XF \end{array}$$

in A. Applying the functor G then yields $\varepsilon^G_{XFGF}\varepsilon^G_{XF} = \varepsilon^{GFG}_{XF}\varepsilon^G_{XF}$, i.e. the equality of the respective components of $(FG \cdot F\varepsilon G)(F\varepsilon G)$ and $(F\varepsilon G \cdot FG)(F\varepsilon G)$ at X. Thus the associative law holds. Now by Proposition III 3.1.7, one has $\eta^F_X \varepsilon_{XF} = 1_{XF}$. Applying the functor G yields $\eta^{FG}_X \varepsilon^G_{XF} = 1_{XFG}$, i.e. the equality of the respective components of $(\eta FG)(F\varepsilon G)$ and 1_{FG} at X. Thus the left unit law holds. The right unit law follows similarly (Exercise 4.1A). □

Example 4.1.4 (Monad of a Prevariety). Let $\underline{\underline{K}}$ be a prevariety of algebras of type $\tau : \Omega \to \mathbb{N}$. According to Proposition 4.1.3, the adjunction $(\Omega R_{\underline{\underline{K}}}, U, \eta, \varepsilon)$ of Section 2.1 then yields a monad $(\Omega R_{\underline{\underline{K}}} U, \Omega R_{\underline{\underline{K}}} \varepsilon U, \eta)$ on the category $\underline{\underline{\mathrm{Set}}}$, the so-called *monad of the prevariety* $\underline{\underline{K}}$. For instance, if $\underline{\underline{K}}$ is the prevariety of M-sets for a monoid M, the monad of $\underline{\underline{K}}$ is the monad (4.7). □

Proposition 4.1.3 shows how to obtain a monad from an adjunction. Conversely, each monad $(T : B \to B, \mu, \eta)$ yields an adjunction, constructed by analogy with the relationship between the adjunction $(\Omega R_{\underline{\underline{K}}}, U, \eta, \varepsilon)$ and the monad $(\Omega R_{\underline{\underline{K}}} U, \Omega R_{\underline{\underline{K}}} \varepsilon U, \eta)$ of a variety $\underline{\underline{K}}$ of algebras.

Definition 4.1.5. Let $T : B \to B$ be a functor on a base category B.

(a) A *T-algebra* (X, a) is an object X of B equipped with a B-morphism a_X or $a : XT \to X$, known as the *action* of T on X.

(b) A *T-homomorphism* $h : (X, a) \to (X', a')$ or *homomorphism of T-algebras* is a B-morphism $h : X \to X'$ between T-algebras, such that the diagram

$$(4.1.2) \qquad \begin{array}{ccc} XT & \xrightarrow{a} & X \\ {\scriptstyle hT}\downarrow & & \downarrow{\scriptstyle h} \\ X'T & \xrightarrow[a']{} & X' \end{array}$$

commutes.

(c) The *category ${}^TB$ of T-algebras* has all T-algebras as its objects and all homomorphisms between them as its morphisms. □

Definition 4.1.6. Let μ or $(T : B \to B, \mu, \eta)$ be a monad on a base category B.

(a) A *μ-algebra* is a T-algebra (X, a) such that the *associative law*, i.e. the commutativity of

$$\begin{array}{ccc} XT^2 & \xrightarrow{\mu_X} & XT \\ {\scriptstyle aT}\downarrow & & \downarrow{\scriptstyle a} \\ XT & \xrightarrow[a]{} & X \end{array} \tag{4.1.3}$$

in B, and the *unit law*, i.e. the commutativity of

(4.1.4)

in B, are both satisfied.

(b) A *μ-homomorphism* $h : (X, a) \to (X', a')$ or *homomorphism of μ-algebras* is a T-homomorphism between μ-algebras.

(c) The *Eilenberg-Moore category* $^{\mu}B$ *of the monad* μ or *category of μ-algebras* has all μ-algebras as its objects and all homomorphisms between them as its morphisms. □

The Eilenberg-Moore category of the monad of a prevariety $\underline{\underline{K}}$ includes objects (XU, ε_X^U) for each $\underline{\underline{K}}$-object X, the action being given by the image under $U : \underline{\underline{K}} \to \underline{\underline{\text{Set}}}$ of the component at X of the counit ε of the adjunction $(\Omega R_{\underline{\underline{K}}}, U, \eta, \varepsilon)$. (Cf. Exercise 4.1F.)

The Eilenberg-Moore category of a monad is the domain for the right adjoint in an adjunction. This adjunction then returns the monad by the construction of Proposition 4.1.3. One thus obtains the following converse of Proposition 4.1.3.

Eilenberg-Moore Theorem 4.1.7. *Let μ or $(T : B \to B, \mu, \eta)$ be a monad, with Eilenberg-Moore category $^{\mu}B$. Then there is an adjunction $(S : B \to {}^{\mu}B, R : {}^{\mu}B \to B, \eta, \bar{\varepsilon})$, called the* Eilenberg-Moore adjunction *of the monad (T, μ, η), such that $(T, \mu, \eta) = (SR, S\bar{\varepsilon}R, \eta)$.*

Proof. The putative right adjoint $R : {}^{\mu}B \to B$ is the forgetful functor with object part $(X, a) \mapsto X$. The putative left adjoint $S : B \to {}^{\mu}B$ is defined to have morphism part

$$(4.1.5) \qquad (f : X' \to X) \mapsto (fT : (X'T, \mu_{X'}) \to (XT, \mu_X)).$$

Note that the component at X of the associativity of the monad (T, μ, η) yields the associativity of the T-algebra (XT, μ_X). Similarly, the component at X of the left unit law of the monad (T, μ, η) yields the unit law in the T-algebra (XT, μ_X). The component at a μ-algebra (X, a) of the counit $\bar{\varepsilon}$ is defined as $\bar{\varepsilon}_{(X,a)} = (a : (XT, \mu_X) \to (X, a))$. Note that this component is indeed a homomorphism of T-algebras, by the associativity of (X, a). Moreover, the naturality of $\bar{\varepsilon}$ at a ${}^{\mu}B$-morphism $h : (X, a) \to (X', a')$ follows since the B-morphism $h : X \to X'$ is a homomorphism of T-algebras.

If X is an object of B, the component at X of the left unit law of the monad (T, μ, η), namely $\eta_X^T \cdot \mu_X = 1_{XT}$, yields the component $\eta_X^S \cdot \bar{\varepsilon}_{XS} = 1_{XS}$ at X of the identity $(\eta S)(S\bar{\varepsilon}) = 1$ of Proposition III 3.1.7. Similarly, if (X, a) is an object of ${}^{\mu}B$, then the unit law $\eta_X \cdot a = 1_X$ in the μ-algebra (X, a) yields the component $\eta_{(X,a)R} \cdot \bar{\varepsilon}^R_{(X,a)} = 1_{(X,a)R}$ at (X, a) of the identity $(R\eta)(\bar{\varepsilon}R) = 1$ of Proposition III 3.1.7. With its hypotheses satisfied, Proposition III 3.1.7 then shows that $(S, R, \eta, \bar{\varepsilon})$ is an adjunction.

Finally, the component at a B-object X of the natural transformation $S\bar{\varepsilon}R$ is $\bar{\varepsilon}^R_{XS} = \bar{\varepsilon}^R_{(XT,\mu_X)} = (\mu_X : (XT^2, \mu_{XT}) \to (XT, \mu_X))^R = (\mu_X : XT^2 \to XT)$, so that $S\bar{\varepsilon}R = \mu$, as required. □

Example 4.1.8. Let K be a commutative unital ring, and let Y be a K-algebra. Let (S_Y, μ, η) be the monad in $\underline{\underline{\mathrm{Mod}}}_K$ given by Exercise 4D and Example 4.1.2(b). Then the Eilenberg-Moore category ${}^{\mu}\underline{\underline{\mathrm{Mod}}}_K$ is the category of Y-modules (Exercise 4.1G). □

EXERCISES

4.1A. Complete the verification of the right unit law in the proof of Proposition 4.1.3.

4.1B. For a monoid M, verify that the monad of the variety $\underline{\underline{M}}$ of M-sets is the monad (4.7).

4.1C. Let A be a set, and let T be a closure operator on the power set $(\mathscr{P}(A), \subseteq)$. The operator T is said to be *topological* if $\varnothing^T = \varnothing$ and $\forall X, Y \subseteq A, (X \cup Y)^T = X^T \cup Y^T$. Show that the set $\{X \subseteq A \mid X = X^T\}$ of fixed points of a topological closure operator T is the set of closed subsets of A in a topology on A.

4.1D. Verify that ${}^{T}B$ in Definition 4.1.5(c) and ${}^{\mu}B$ in Definition 4.1.6(c) really are categories.

4.1E. Let T be a closure operator on a poset $(B, \leq)$. Show that the corresponding Eilenberg-Moore category is the induced poset of *closed* elements, i.e. of fixed points of T.

4.1F. Let $\underline{\underline{K}}$ be a prevariety. Show that each $\underline{\underline{K}}$-morphism $f: X \to X'$ yields a morphism $fU:(XU, \varepsilon_X^U) \to (X'U, \varepsilon_{X'}^U)$ in the Eilenberg-Moore category of the monad $(\Omega R_{\underline{\underline{K}}}U, \Omega R_{\underline{\underline{K}}}\varepsilon U, \eta)$ of $\underline{\underline{K}}$.

4.1G. Verify the claim of Example 4.1.8.

4.2. Monadic Adjunctions

Starting with a monad (T, μ, η) on a category B, the Eilenberg-Moore Theorem 4.1.7 builds the adjunction $(S, R, \eta, \overline{\varepsilon})$ that recovers the monad as $(SR, S\overline{\varepsilon}R, \eta)$. Conversely, one may start with an adjunction $(F: B \to A, G: A \to B, \eta, \varepsilon)$ and consider the monad $(T, \mu, \varepsilon) = (FG, F\varepsilon G, \eta)$ on the base category B. The following question then arises: How is the original category A related to the Eilenberg-Moore category ${}^{\mu}B$? To address the question, one may summarize the data by the outer rectangle in the following diagram:

$$(4.2.1) \qquad \begin{array}{ccccc} B & \xrightarrow{S} & {}^{\mu}B & \xrightarrow{R} & B \\ \Big\| & & \Big\uparrow \overline{G} & & \Big\| \\ B & \xrightarrow[F]{} & A & \xrightarrow[G]{} & B \end{array} .$$

The bottom row records the original adjunction $(F, G, \eta, \varepsilon)$, while the top row records the Eilenberg-Moore adjunction $(S, R, \eta, \overline{\varepsilon})$ of the monad

$$(4.2.2) \qquad (FG, F\varepsilon G, \eta) = (T, \mu, \eta) = (SR, S\overline{\varepsilon}R, \eta).$$

The relationship between A and ${}^{\mu}B$ is then expressed by a uniquely specified functor $\overline{G}: A \to {}^{\mu}B$.

Proposition 4.2.1. *Given an adjunction* $(F: B \to A, G: A \to B, \eta, \varepsilon)$ *yielding the monad* (4.2.2) *and Eilenberg-Moore adjunction* $(S, R, \eta, \overline{\varepsilon})$*, there is a unique functor* $\overline{G}: A \to {}^{\mu}B$*, called the* Eilenberg-Moore comparison *of the adjunction* $(F, G, \eta, \varepsilon)$*, such that* $F\overline{G} = S$ *and* $\overline{G}R = G$.

Proof. The Eilenberg-Moore comparison is defined (by analogy with Exercise 4.1F) to have morphism part

$$(4.2.3) \qquad \overline{G}:(k:Y \to Y') \mapsto \big(kG:\big(YG,\, \varepsilon_Y^G\big) \to \big(Y'G,\, \varepsilon_{Y'}^G\big)\big).$$

For Y in A_0, verification of the associative law $(F\varepsilon G)_{YG}\,\varepsilon_Y^G = \varepsilon_Y^{GFG}\varepsilon_Y^G$ in the FG-algebra $(YG,\, \varepsilon_Y^G)$ proceeds by applying the functor G to the naturality $\varepsilon_{YGF}\varepsilon_Y = \varepsilon_Y^{GF}\varepsilon_Y$ of $\varepsilon: GF \to 1_A$ at the A-morphism $\varepsilon_Y: YGF \to Y$. [Note that (4.1.1) is an instance of this verification, with $Y = XF$.] The unit law $\eta_{YG}\,\varepsilon_Y^G = 1_{YG}$ in the FG-algebra $(YG,\, \varepsilon_Y^G)$ is the component at Y of the identity $(G\eta)(\varepsilon G) = 1$ of Proposition III 3.1.7. Thus $Y\overline{G}$ lies in ${}^{\mu}B_0$. For $k: Y \to Y'$ in A_1, application of the functor G to the naturality $\varepsilon_Y k = k^{GF}\varepsilon_{Y'}$ of $\varepsilon: GF \to 1_A$ at k shows that $k\overline{G}$ lies in ${}^{\mu}B_1$. Verification of the functoriality of $\overline{G}: A \to {}^{\mu}B$ is straightforward (Exercise 4.2A). The equation $\overline{G}R = G$ is immediate from the definitions of $\overline{G}$ and R. On the other hand, for a B-morphism $f: X' \to X$, substituting $fG: X'G \to XG$ for $k: Y \to Y'$ in (4.2.3) yields (4.1.5) by virtue of (4.2.2), so that $F\overline{G} = S$ also holds.

To confirm the uniqueness of $\overline{G}$, consider a functor $K: A \to {}^{\mu}B$ with $FK = S$ and $KR = G$. By the latter equation, the morphism part of K takes the form

$$K:(k:Y \to Y') \mapsto (kG:(YG, a) \to (Y'G, a')),$$

in which a and a' are respective μ-algebra actions on YG and $Y'G$, connected by a T-homomorphism kG. For Y in A_0, it must thus be shown that the action a of YK coincides with ε_Y^G. First note that the naturality of $\overline{\varepsilon}: RS \to 1_{\mu_B}$ at $\varepsilon_Y^K: YKRS = YGS = YGFK \to YK$ yields

$$\overline{\varepsilon}_{YKRS}\,\varepsilon_Y^K = \varepsilon_Y^{KRS}\,\overline{\varepsilon}_{YK}.$$

By Proposition III 3.1.7 and the definition of the counit $\overline{\varepsilon}$ in the Eilenberg-Moore adjunction, one then has $(a:(YGT, \mu_{YG}) \to (YG, a)) = \overline{\varepsilon}_{(YG,a)} = \overline{\varepsilon}_{YK}$ $= (\eta_{YG}\,\varepsilon_Y^G)^S\,\overline{\varepsilon}_{YK} = \eta_{YKR}^S\,\varepsilon_Y^{KRS}\,\overline{\varepsilon}_{YK} = \eta_{YKR}^S\,\overline{\varepsilon}_{YKRS}\,\varepsilon_Y^K = \varepsilon_Y^K = \varepsilon_Y^G$, as required. □

Definition 4.2.2. An adjunction $(F: B \to A,\ G: A \to B,\ \eta,\ \varepsilon)$, and the right adjoint $G: A \to B$ in such an adjunction, are said to be *monadic* if the Eilenberg-Moore comparison $\overline{G}: A \to {}^{F\varepsilon G}B$ is an isomorphism of categories. The category A is said to be *monadic over the base category* B if there is a monadic right adjoint $G: A \to B$. □

Example 4.2.3. As noted in Section III 3.4, the forgetful functor $G: \underline{\underline{\mathrm{Top}}} \to \underline{\underline{\mathrm{Set}}}$ is right adjoint to the discrete topology functor $D: \underline{\underline{\mathrm{Set}}} \to \underline{\underline{\mathrm{Top}}}$.

However, this adjunction yields the trivial monad $(1_{\underline{\underline{\text{Set}}}}, 1:1_{\underline{\underline{\text{Set}}}} \to 1_{\underline{\underline{\text{Set}}}}, 1:1_{\underline{\underline{\text{Set}}}} \to 1_{\underline{\underline{\text{Set}}}})$ on the base category $\underline{\underline{\text{Set}}}$. The Eilenberg-Moore category of this monad is just $\underline{\underline{\text{Set}}}$, and the Eilenberg-Moore comparison is just the forgetful functor $G:\underline{\underline{\text{Top}}} \to \underline{\underline{\text{Set}}}$, which is not an isomorphism. Thus the right adjoint $G:\underline{\underline{\text{Top}}} \to \underline{\underline{\text{Set}}}$ is not monadic. □

Theorem 4.2.7 below shows that, for any theory map $M: W \to V$, the functor $M^r: \underline{\underline{V}} \to \underline{\underline{W}}$ (known by Theorem 3.4.4 to be a right adjoint) is monadic. The proof of Theorem 4.2.7 relies on a certain characterization (Proposition 4.2.6) of those T-algebras that are μ-algebras for a monad (T, μ, η). The characterization uses a special kind of coequalizer.

Prosition 4.2.4. *If the diagram*

(4.2.4)

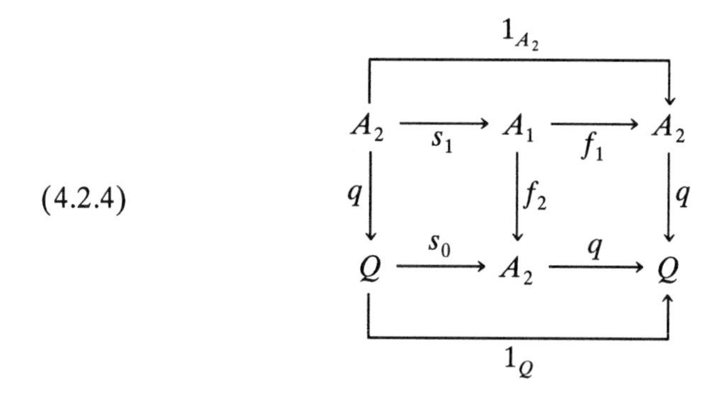

in a category D commutes, then $q: A_2 \to Q$ *is the coequalizer of the parallel pair* (f_1, f_2).

Proof. Given a D-morphism $f: A_2 \to B$ with $f_1 f = f_2 f$, define $\bar{f} = s_0 f$. Then $q\bar{f} = qs_0 f = s_1 f_2 f = s_1 f_1 f = f$. Moreover, if $f': Q \to B$ satisfies $qf' = f$, one has $f' = s_0 qf' = s_0 f = \bar{f}$. □

Corollary 4.2.5. *If* (4.2.4) *commutes in D, then qE is the coequalizer of the parallel pair* $(f_1 E, f_2 E)$ *for each functor* $E: D \to C$ *with domain D.*

Proof. Under the functor E, the commuting diagram (4.2.4) in D translates to a commuting diagram in E. □

A coequalizer $q: A_2 \to Q$ of a parallel pair $(f_1: A_1 \to A_2, f_2: A_1 \to A_2)$ is described as *split* by morphisms $s_0: Q \to A_2$ and $s_1: A_2 \to A_1$ if the diagram (4.2.4) commutes. For example, if α is an equivalence relation on a set A, then the coequalizer nat $\alpha: A \to A^\alpha$ in $\underline{\underline{\text{Set}}}$ of the parallel pair (π_1, π_2) of projections, with $\pi_i : A^2 \to A; (a_1, a_2) \mapsto a_i$ for $1 \le i \le 2$, is split by a section $s : A^\alpha \to A$ of nat α and the diagonal embedding $\Delta : A \to A^2$.

Proposition 4.2.6. *Let (T, μ, η) be a monad on a base category B. Then a T-algebra (X, a) is a μ-algebra iff $a : XT \to X$ is a coequalizer of the parallel pair (μ_X, aT) split by $\eta_X : X \to XT$ and $\eta_{XT} : XT \to XT^2$.*

Proof. Consider the instance

(4.2.5)

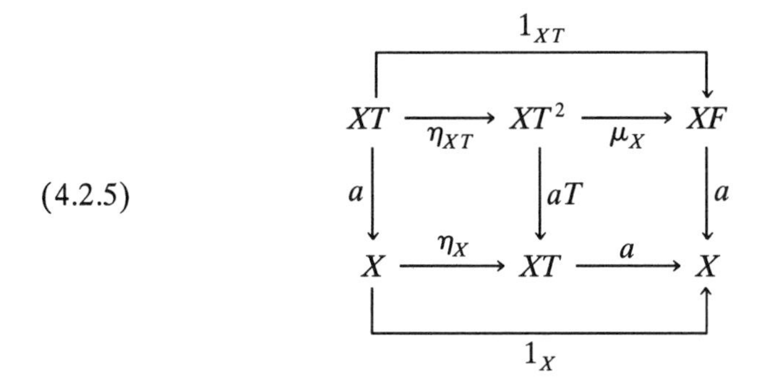

of (4.2.4). Suppose, for the T-algebra (X, a), that (4.2.5) commutes. Then the associative law is satisfied by the commuting of the right-hand square, while the unit law is satisfied by the commuting of the bottom rectangle. Conversely, suppose that (X, a) is a μ-algebra. The right-hand square and bottom rectangle of (4.2.5) commute respectively by the associative law and unit law in the μ-algebra (X, a). The left-hand square commutes by the naturality of $\eta : 1_B \to T$ at the action $a : XT \to X$, while the top rectangle commutes by the component at X of the right unit law of the monad (T, μ, η). $\square$

Theorem 4.2.7. *Let $M : W \to V$ be a theory map. Then the right adjoint $M^r : \underline{\underline{V}} \to \underline{\underline{W}}$ is monadic.*

Proof. By Theorem 3.4.4, the theory map $M : W \to V$ yields an adjunction $(M^l, M^r, \eta, \varepsilon)$. For the monad $(T, \mu, \eta) = (M^l M^r, M^l \varepsilon M^r, \eta)$ of this adjunction, Proposition 4.2.1 yields the Eilenberg-Moore comparison functor $\overline{M^r} : \underline{\underline{V}} \to {}^\mu\underline{\underline{W}}; Y \mapsto (YM^r, \varepsilon_Y M^r)$. The proof of Theorem 4.2.7 thus consists of the provision of a two-sided inverse functor $H : {}^\mu\underline{\underline{W}} \to \underline{\underline{V}}$ to $\overline{M^r}$. To this end, it is most convenient to recognize $\underline{\underline{V}}$ as a variety of algebras of the type $\tau : \Omega \to \mathbb{N}$ of V [cf. (3.3.1)], according to Theorem 3.3.3.

Let (X, a) be a μ-algebra, and let ω be an operator from Ω. The first step in the construction of $(X, a)H$ is to provide an operation ω_X or $\omega : X^{\omega\tau} \to X$ of ω on X. Note that the $\underline{\underline{V}}$-algebra structure on XM^l yields the operation $\omega : (XM^l)^{\omega\tau} \to XM^l$. (Here and elsewhere in this proof, explicit mention of the forgetful functors $U : \underline{\underline{V}} \to \underline{\underline{\underline{\text{Set}}}}$ and $U : \underline{\underline{W}} \to \underline{\underline{\underline{\text{Set}}}}$ is often suppressed.) Since the right adjoint M^r is continuous, and commutes with the forgetful functors by Corollary 3.4.3 [and Exercise 3.4F(b)], one thus obtains $\omega : (XM^lM^r)^{\omega\tau} \to XM^lM^r$ or $\omega_{XT} : (XT)^{\omega\tau} \to XT$, and similarly $\omega_{XTT} : (XT^2)^{\omega\tau} \to XT^2$. These operations provide the two left verticals in the following diagram in $\underline{\underline{\mathrm{Set}}}$:

$$
\begin{array}{ccccc}
(XT^2)^{\omega\tau} & \underset{(aT)^{\omega\tau}}{\overset{\mu_X^{\omega\tau}}{\rightrightarrows}} & (XT)^{\omega\tau} & \xrightarrow{a^{\omega\tau}} & X^{\omega\tau} \\
\downarrow{\scriptstyle \omega_{XTT}} & & \downarrow{\scriptstyle \omega_{XT}} & & \downarrow{\scriptstyle \omega_X} \\
XT^2 & \underset{aT}{\overset{\mu_X}{\rightrightarrows}} & XT & \xrightarrow[a]{} & X
\end{array} \tag{4.2.6}
$$

In each of the horizontal rows, the right-hand arrow is the coequalizer of the parallel pair to its left, according to Corollary 4.2.5. Indeed, starting with the split coequalizer (4.2.5) in $\underline{\underline{W}}$, the bottom row is obtained by applying the functor $U : \underline{\underline{W}} \to \underline{\underline{\mathrm{Set}}}$. The top row is obtained by applying the composite functor

$$
(\omega\tau) = \left(\underline{\underline{W}} \xrightarrow{\Delta} \underline{\underline{W}}^J \xrightarrow{\underleftarrow{\lim}} \underline{\underline{W}} \xrightarrow{U} \underline{\underline{\mathrm{Set}}}\right), \tag{4.2.7}
$$

with J an antichain poset category or set of cardinality $\omega\tau$, with Δ the diagonal functor as in Section III 2.2, and with $\underleftarrow{\lim}$ as in Example III 3.1.6 (b). Now ε_{XM^l} and aM^l are $\underline{\underline{V}}$-morphisms. Recalling the definitions of T and μ, one thus has $\mu_X^{\omega\tau}\omega_{XT}a = \varepsilon_{XM^l}^{\omega\tau}\omega_{XT}a = \omega_{XTT}\varepsilon_{XM^l}a = \omega_{XTT}\mu_{XM^l}a = \omega_{XTT}a^Ta = \omega_{XTT}(aM^l)a = (aM^l)^{\omega\tau}\omega_{XT}a = (aT)^{\omega\tau}\omega_{XT}a$. Since the top row of (4.2.6) is a coequalizer in $\underline{\underline{\mathrm{Set}}}$, it follows that there is a unique function ω_X or $\omega : X^{\omega\tau} \to X$ with $\omega_{XT}a = a^{\omega\tau}\omega$. This function provides the operation of ω on X. Thus one obtains an Ω-algebra $(X, \Omega) = (X, a)H$. In particular, note that $(YM^r, \varepsilon_Y M^r)H = Y$ for a $\underline{\underline{V}}$-algebra Y.

Now let $u = v$ with $n = \max((\arg u) \cup (\arg v))$ be an identity of the variety $\underline{\underline{V}}$. The commuting of the right-hand square of (4.2.6) shows that $a : (XT, \Omega) \to (X, \Omega)$ is an Ω-homomorphism. Thus

$$
a^n(u,n)_X = (u,n)_{XT}a = (v,n)_{XT}a = a^n(v,n)_X, \tag{4.2.8}
$$

since the $\underline{\underline{V}}$-algebra XT or XM^l satisfies $u = v$. Application of the functor $(n) = ((u,n)\overline{\tau})$ of (4.2.7) to the bottom rectangle of (4.2.5) yields $\eta_X^n a^n = 1_{X^n}$, so that $a^n : (XT)^n \to X^n$ surjects. Equation (4.2.8) thus implies $(u,n)_X = (v,n)_X$: The Ω-algebra (X, Ω) satisfies $u = v$. In other words, $(X, a)H$ is an object of $\underline{\underline{V}}$. Moreover, $(X, a)HM^r$ recovers (X, a), since $\varepsilon_{(X,a)H}$ is a coequalizer of μ_X and aT.

Finally, consider a μ-homomorphism $h:(X,a) \to (Y,b)$, so that $ah = h^T b$. For an operator ω from Ω, application of the functor (4.2.7) yields $a^{\omega\tau} h^{\omega\tau} = hT^{\omega\tau} b^{\omega\tau}$. Moreover, $hT:(XT,\Omega) \to (YT,\Omega)$ is a homomorphism. Using the commutativity of the right hand square of (4.2.6) for (X,a) and (Y,b), one has $a^{\omega\tau}\omega_X h = \omega_{XT} ah = \omega_{XT} h^T b = hT^{\omega\tau}\omega_{YT} b = hT^{\omega\tau} b^{\omega\tau}\omega_Y = a^{\omega\tau} h^{\omega\tau}\omega_Y$. Since $a^{\omega\tau}: XT^{\omega\tau} \to X^{\omega\tau}$ surjects, this shows that $\omega_X h = h^{\omega\tau}\omega_Y$. Thus $hH := (h:(X,\Omega) \to (Y,\Omega))$ becomes a $\underline{\underline{V}}$-morphism, thereby yielding the morphism part of $H: {}^{\mu}\underline{\underline{W}} \to \underline{\underline{V}}$. The remaining verifications are straightforward (Exercise 4.2F). □

Corollary 4.2.8. *Let $\underline{\underline{V}}$ be a variety of algebras of type $\tau:\Omega \to \mathbb{N}$. Then $\underline{\underline{V}}$ is the Eilenberg-Moore category of the monad of the (pre-)variety $\underline{\underline{V}}$.*

Proof. Apply Theorem 4.2.7 to the theory map J: Set $\to V$ (3.4.5) for the algebraic theory of the (pre-)variety $\underline{\underline{V}}$. □

As a result of Theorem 4.2.7, one may identify $\underline{\underline{V}}$-algebras as $\underline{\underline{W}}$-algebras equipped with a μ-algebra structure, for the monad $(T,\mu,\eta) = (\overline{M}^l M^r, M^l\varepsilon M^r, \eta)$. In particular, Corollary 4.2.8 and Theorem 3.3.3 combine to yield the equivalence of the following three concepts:

(4.2.9)
- (a) A variety $\underline{\underline{V}}$ of algebras of type $\tau:\Omega \to \mathbb{N}$;
- (b) The category $\underline{\underline{V}}$ of models of the algebraic theory V;
- (c) The Eilenberg-Moore category $\underline{\underline{V}}$ of the monad of the adjunction $\left(\Omega R_{\underline{\underline{V}}}, U, \eta, \varepsilon\right)$.

EXERCISES

4.2A. Complete the verification of the functoriality of the Eilenberg-Moore comparison $\overline{G}: A \to {}^{\mu}B$ in the proof of Proposition 4.2.1.

4.2B. Consider a pair $(F:B \to A, G:A \to B, \eta, \varepsilon)$ and $(S:B \to C, R:C \to B, \eta, \overline{\varepsilon})$ of adjunctions with common unit η. If there is a functor $K:A \to C$ with $FK = S$ and $KR = G$, show that the two monads $(FG, F\varepsilon G, \eta)$ and $(SR, S\overline{\varepsilon}\, R, \eta)$ on B agree.

4.2C. Show that a Galois connection $(S:B \to A, R:A \to B, \eta, \varepsilon)$ is not monadic if A contains elements that are not closed.

4.2D. (a) If $q:A_2 \to Q$ is a coequalizer of the parallel pair $(f_1:A_1 \to A_2, f_2:A_1 \to A_2)$, show that it is also the coequalizer of the parallel pair (f_2, f_1).

(b) Give an example of a morphism $a:A_2 \to Q$ that is a split coequalizer of a parallel pair $(f_1:A_1 \to A_2, f_2:A_1 \to A_2)$, but not of the parallel pair (f_2, f_1).

4.2E. Let 1 and 2 respectively denote the posets with Hasse diagrams 0 and $0 \to 1$, considered as small categories. Let C be a small category with terminal object T. Show that the unique functor $C \to 1$ is the coequalizer in $\underline{\underline{\text{Cat}}}$ of the parallel pair $(C^2 \to C; F \mapsto 0F, C^2 \to C; F \mapsto 1F)$ of functors, split by the functors $1 \to C$ with object part $0 \mapsto T$ and $C \to C^2$ with object part $x \mapsto ((0 \to 1) \mapsto (x \to T))$.

4.2F. Complete the verifications required for the proof of Theorem 4.2.7.

4.2G. Show that the Eilenberg-Moore comparison of the adjunction $(\Omega R_{\underline{\underline{K}}}, U, \eta, \varepsilon)$ determined by a prevariety $\underline{\underline{K}}$ is the inclusion of $\underline{\underline{K}}$ in the variety $H\underline{\underline{K}}$.

4.2H. Let (T, μ, η) be a monad on the base category $\underline{\underline{\text{Set}}}$ such that ${}^{\mu}\underline{\underline{\text{Set}}}$ is a variety of algebras. Characterize those properties of the monad (T, μ, η) that correspond to entropicity of the algebras in the variety ${}^{\mu}\underline{\underline{\text{Set}}}$. (Hint: Cf. Exercise 2.4D.)

4.2I. Let $\tau : \Omega \to \mathbb{N}$ be a finitary type. Let (T, μ, η) be the monad of the adjunction $(\Omega : \underline{\underline{\text{Set}}} \to \underline{\underline{\tau}}, U : \underline{\underline{\tau}} \to \underline{\underline{\text{Set}}}, \eta, \varepsilon)$. Show that the categories $\underline{\underline{\tau}}$, ${}^{T}\underline{\underline{\text{Set}}}$, and ${}^{\mu}\underline{\underline{\text{Set}}}$ coincide.

4.2J. Let (T, μ, η) be a monad on the base category $\underline{\underline{\text{Set}}}$. Show that ${}^{\mu}\underline{\underline{\text{Set}}}$ is a variety of algebras of finitary type iff

$$\forall Y \in \underline{\underline{\text{Set}}}_0, \forall y \in YT, \exists(\varphi : \{x_1, \dots, x_n\} \to Y) \in \underline{\underline{\text{Set}}}_1 .$$
$$\exists x \in \{x_1, \dots, x_n\}T .\ x\varphi^T = y.$$

4.2K. A subcategory A of a category B is called *abstract* (cf. Definition 2.1.1) if each B-object isomorphic to an A-object in B is itself an object of A.

(a) If the embedding of an abstract subcategory A in B is a reflection (cf. Exercise III 3.2.E), show that the embedding is monadic.

(b) For an algebraic theory V, show that $\underline{\underline{V}}$ is monadic over $\hat{V}$.

4.2L. Let $R : A \to B$ be monadic, and let $F : J \to A$ be a functor with small domain. Show that R creates the limit of F (cf. Exercise III 3.4B).

INDEX

A page number repeated within an entry denotes that the keyword on that page has been used in different contexts.

PURE AND APPLIED MATHEMATICS

A Wiley-Interscience Series of Texts, Monographs, and Tracts

ADÁMEK, HERRLICH, and STRECKER—Abstract and Concrete Catetories
ADAMOWICZ and ZBIERSKI—Logic of Mathematics
AKIVIS and GOLDBERG—Conformal Differential Geometry and Its Generalizations
ALLEN and ISAACSON—Numerical Analysis for Applied Science
*ARTIN—Geometric Algebra
AZIZOV and IOKHVIDOV—Linear Operators in Spaces with an Indefinite Metric
BERMAN, NEUMANN, and STERN—Nonnegative Matrices in Dynamic Systems
BOYARINTSEV—Methods of Solving Singular Systems of Ordinary Differential Equations
BURK—Lebesgue Measure and Integration: An Introduction
*CARTER—Finite Groups of Lie Type
CASTILLO, COBO, JUBETE and PRUNEDA—Orthogonal Sets and Polar Methods in Linear Algebra: Applications to Matrix Calculations, Systems of Equations, Inequalities, and Linear Programming
CHATELIN—Eigenvalues of Matrices
CLARK—Mathematical Bioeconomics: The Optimal Management of Renewable Resources, Second Edition
COX—Primes of the Form $x^2 + ny^2$: Fermat, Class Field Theory, and Complex Multiplication
*CURTIS and REINER—Representation Theory of Finite Groups and Associative Algebras
*CURTIS and REINER—Methods of Representation Theory: With Applications to Finite Groups and Orders, Volume I
CURTIS and REINER—Methods of Representation Theory: With Applications to Finite Groups and Orders, Volume II
*DUNFORD and SCHWARTZ—Linear Operators
Part 1—General Theory
Part 2—Spectral Theory, Self Adjoint Operators in Hilbert Space
Part 3—Spectral Operators
FOLLAND—Real Analysis: Modern Techniques and Their Applications
FRÖLICHER and KRIEGL—Linear Spaces and Differentiation Theory
GARDINER—Teichmüller Theory and Quadratic Differentials
GREENE and KRATZ—Function Theory of One Complex Variable
*GRIFFITHS and HARRIS—Principles of Algebraic Geometry
GROVE—Groups and Characters
GUSTAFSSON, KREISS and OLIGER—Time Dependent Problems and Difference Methods
HANNA and ROWLAND—Fourier Series, Transforms, and Boundary Value Problems, Second Edition
*HENRICI—Applied and Computational Complex Analysis
Volume 1, Power Series—Integration—Conformal Mapping—Location of Zeros
Volume 2, Special Functions—Integral Transforms—Asymptotics—Continued Fractions
Volume 3, Discrete Fourier Analysis, Cauchy Integrals, Construction of Conformal Maps, Univalent Functions

*HILTON and WU—A Course in Modern Algebra
*HOCHSTADT—Integral Equations
JOST—Two-Dimensional Geometric Variational Procedures
*KOBAYASHI and NOMIZU—Foundations of Differential Geometry, Volume I
*KOBAYASHI and NOMIZU—Foundations of Differential Geometry, Volume II
LAX—Linear Algegra
LOGAN—An Introduction to Nonlinear Partial Differential Equations
McCONNELL and ROBSON—Noncommutative Noetherian Rings
NAYFEH—Perturbation Methods
NAYFEH and MOOK—Nonlinear Oscillations
PANDEY—The Hilbert Transform of Schwartz Distributions and Applications
PETKOV—Geometry of Reflecting Rays and Inverse Spectral Problems
*PRENTER—Splines and Variational Methods
RAO—Measure Theory and Integration
RASSIAS and SIMSA—Finite Sums Decompositions in Mathematical Analysis
RENELT—Elliptic Systems and Quasiconformal Mappings
RIVLIN—Chebyshev Polynomials: From Approximation Theory to Algebra and Number Theory, Second Edition
ROCKAFELLAR—Network Flows and Monotropic Optimization
ROITMAN—Introduction to Modern Set Theory
*RUDIN—Fourier Analysis on Groups
SENDOV—The Averaged Moduli of Smoothness: Applications in Numerical Methods and Approximations
SENDOV and POPOV—The Averaged Moduli of Smoothness
*SIEGEL—Topics in Complex Function Theory
Volume l—Elliptic Functions and Uniformization Theory
Volume 2—Automorphic Functions and Abelian Integrals
Volume 3—Abelian Functions and Modular Functions of Several Variables
SMITH and ROMANOWSKA—Post-Modern Algebra
STAKGOLD—Green's Functions and Boundary Value Problems, Second Editon
*STOKER—Differential Geometry
*STOKER—Nonlinear Vibrations in Mechanical and Electrical Systems
*STOKER—Water Waves: The Mathematical Theory with Applications
WESSELING—An Introduction to Multigrid Methods
WHITMAN—Linear and Nonlinear Waves
†ZAUDERER—Partial Differential Equations of Applied Mathematics, Second Edition

*Now available in a lower priced paperback edition in the Wiley Classics Library.
†Now available in paperback.